AGROECOLOGY, ECOSYSTEMS, AND SUSTAINABILITY

Advances in Agroecology

Series Editor: Clive A. Edwards

AGROECOLOGY, ECOSYSTEMS, and SUSTAINABILITY

EDITED BY NOUREDDINE BENKEBLIA

CRC Press
Taylor & Francis Group
Boca Raton London New York

CRC Press is an imprint of the
Taylor & Francis Group, an **informa** business

CRC Press
Taylor & Francis Group
6000 Broken Sound Parkway NW, Suite 300
Boca Raton, FL 33487-2742

First issued in paperback 2019

ISBN-13: 978-1-4822-3301-8 (hbk)
ISBN-13: 978-0-367-43598-1 (pbk)

Visit the Taylor & Francis Web site at
http://www.taylorandfrancis.com

and the CRC Press Web site at
http://www.crcpress.com

Contents

Preface...ix
Editor xi
Contributors xiii

Chapter 1
Soil Biogeochemistry: From Molecular to Ecosystem Level Using Terra Preta and Biochar
as Examples 1

Bruno Glaser

Chapter 2
Factors and Mechanisms Regulating Soil Organic Carbon in Agricultural Systems 41

Yadunath Bajgai, Paul Kristiansen, Nilantha Hulugalle, and Melinda McHenry

Chapter 3
Carbon Capture and Use as an Alternative to Carbon Capture and Storage................................. 57

Bruno Glaser and Per Espen Stoknes

Chapter 4
Agroecology of Agromicrobes 81

Manindra Nath Jha, Shankar Jha, and Sanjeet Kumar Chourasia

Chapter 5
Management of Rhizosphere Microorganisms in Relation to Plant Nutrition and Health............ 103

John Larsen, Miguel Nájera Rincón, Carlos González Esquivel, and Mayra E. Gavito

Chapter 6
Mechanized Rain-Fed Farming and Its Ecological Impact on the Drylands: The Case of
Gedarif State, Sudan... 121

Yasin Abdalla Eltayeb Elhadary

Chapter 7
The Paradox of Arable Weeds: Diversity, Conservation, and Ecosystem Services
of the Unwanted 139

Jaime Fagúndez

Chapter 8
Proteomics Potential and Its Contribution toward Sustainable Agriculture.................................. 151

**Abhijit Sarkar, Md. Tofazzal Islam, Sajad Majeed Zargar, Vivek Dogra,
Sun Tae Kim, Ravi Gupta, Renu Deswal, Ganesh Bagler, Yelam Sreenivasulu,
Rungaroon Waditee-Sirisattha, Sophon Sirisattha, Jai Singh Rohila,
Manish Raorane, Ajay Kohli, Dea-Wook Kim, Kyoungwon Cho, Abdiani Attiq
Saidajan, Ganesh Kumar Agrawal, and Randeep Rakwal**

Chapter 9
The Food System Approach in Agroecology Supported by Natural and Social Sciences:
Topics, Concepts, Applications.. 181

Alexander Wezel, Philippe Fleury, Christophe David, and Patrick Mundler

Chapter 10
Agroecology Applications in Tropical Agriculture Systems... 201

Noureddine Benkeblia and Charles A. Francis

Chapter 11
Agroforestry Adaptation and Mitigation Options for Smallholder Farmers Vulnerable
to Climate Change 221

Brenda B. Lin

Chapter 12
Agroecology for Sustainable Coastal Ecosystems: A Case for Mangrove Forest Restoration...... 239

Mona Webber, Dale Webber, and Camilo Trench

Chapter 13
Suggesting an Interdisciplinary Framework for the Management of Integrated Production
and Conservation Landscapes in a Transfrontier Conservation Area of Southern Africa............ 265

Munyaradzi Chitakira, Emmanuel Torquebiau, Willem Ferguson, and Kevin Mearns

Chapter 14
Agroecology in Central Appalachia: Framing Problems and Facilitating Solutions 279

Sean Clark

Chapter 15
Can Agroecological Practices Feed the World?: The Bio- and Ecoeconomic Paradigm in
Agri-Food Production... 309

Lummina G. Horlings

Chapter 16
Vermont Agricultural Resilience in a Changing Climate: A Transdisciplinary and
Participatory Action Research (PAR) Process..325
**Rachel Schattman, Ernesto Méndez, Katherine Westdijk, Martha Caswell, David
Conner, Christopher Koliba, Asim Zia, Stephanie Hurley, Carol Adair, Linda Berlin,
and Heather Darby**

Chapter 17
Experiential Learning Using the Open-Ended Case: Future Agroecology Education..................347
**Charles A. Francis, Lennart Salomonsson, Geir Lieblein,
Tor Arvid Breland, and Suzanne Morse**

Index 359

Preface

The real value of a man should be seen in what he is able to give, and not in what he is able to receive, because only a life lived for others is a life worthwhile.

A. Einstein

Ecology and sustainable agriculture are at the forefront of environmental and food production issues, because conventional agriculture is heavily dependent on fossils energy, while the development of industry is having serious effects on our environment, such as ozone layer depletion, soil erosion, global warming, and climate change. These changes have become our main concerns and the scientific community should solve these issues by developing sustainable agrosystems to ameliorate the worst consequences, and to genuinely feed the world. Consequently, ecological challenges have to be addressed if agriculture is to be truly sustainable, and to attain these "sustainable agricultural and ecological development" goals, urgent action on ecological, environmental, and food production issues is required. Ecological agriculture should be given the chance it deserves, because its success depends on how much importance it is given.

This book describes different aspects of how ecology and agriculture can be allied to ensure food production and security without threatening our environment. It also describes how natural resources can be used in a manner to create a "symbiosis" to preserve ecological systems and develop agriculture.

Noureddine Benkeblia
University of West Indies

Editor

Dr. Noureddine Benkeblia is a professor of crop science involved in food science, focusing on food-plants biochemistry and physiology. His work is mainly devoted to the pre- and postharvest metabolism of crops. A few years ago, he introduced the new concept of systems biology—metabolomics—to investigate the mechanisms of the biosynthesis and accumulation of fructans in liliaceous plants. Professor Benkeblia received his bachelor of science, master of philosophy, and doctor in food sciences from the Institute National Agronomique, Algeria, and a doctor of agriculture (PhD) from Kagoshima University, Japan. After a few years teaching in Algeria, he joined the French National Institute for Agricultural Research (INRA), Avignon, France, as a postdoctoral scientist from 2000. From 2002 to 2007, he worked as a visiting professor at the University of Rakuno Gakuen, Ebetsu, Japan, and as a research associate at Hokaido University. Professor Benkeblia joined the Department of Life Sciences, the University of the West Indies, Jamaica, in 2008, continuing his work on the physiology, biochemistry, and metabolomics of fructan-containing plants in Jamaica. He also works on the postharvest physiology and biochemistry of local fruits. Professor Benkeblia has published over 150 papers and over 37 books and book chapters. He is a recipient of many awards among them the University of the West Indies award for the Most Outstanding Researcher in 2011 and 2013.

Contributors

Carol Adair
Rubenstein School of Environment and Natural
  Resources
University of Vermont
Burlington, Vermont

Ganesh Kumar Agrawal
Research Laboratory for Biotechnology and
  Biochemistry
Kathmandu, Nepal

Ganesh Bagler
Institute of Himalayan Bioresource
  Technology
Council of Scientific and Industrial Research
Palampur, India

Yadunath Bajgai
School of Environmental and Rural Science
University of New England
Armidale, Australia

and

Department of Agriculture
Renewable Natural Resources-Research and
  Development Centre, Bajo
Wangdue Phodrang, Bhutan

Noureddine Benkeblia
Department of Life Sciences
University of the West Indies
Kingston, Jamaica

Linda Berlin
Center for Sustainable Agriculture
University of Vermont Extension
Burlington, Vermont

Tor Arvid Breland
Department of Plant Science
Norwegian University of Life Sciences
Ås, Norway

Martha Caswell
Agroecology and Rural Livelihoods Group
University of Vermont
Burlington, Vermont

Munyaradzi Chitakira
Department of Environmental Sciences
University of South Africa
Pretoria, South Africa

Kyoungwon Cho
Seoul Center
Korea Basic Science Institute
Seoul, South Korea

Sanjeet Kumar Chourasia
Department of Microbiology
Rajendra Agricultural University
Samastipur, India

Sean Clark
Agriculture and Natural Resources Program
Berea College
Berea, Kentucky

David Conner
Department of Community Development and
  Applied Economics
University of Vermont
Burlington, Vermont

Heather Darby
Agronomy and Nutrient Management
University of Vermont Extension
Saint Albans, Vermont

Christophe David
Department of Agroecology and Environment
ISARA Lyon
Lyon, France

Renu Deswal
Department of Botany
University of Delhi
Delhi, India

Vivek Dogra
Institute of Himalayan Bioresource
  Technology
Council of Scientific and Industrial Research
Palampur, India

Yasin Abdalla Eltayeb Elhadary
Faculty of Geographical and Environmental
  Science
University of Khartoum
Khartoum, Sudan

and

Department of Geography
University of Imam Mohmmed Ibn Saoud
Riyadh, Saudi Arabia

Carlos González Esquivel
Center for Ecosystem Research
National Autonomous University of Mexico
Morelia, Mexico

Jaime Fagúndez
Department of Animal, Plant and Ecological
  Biology
University of A Coruña
A Coruña, Spain

Willem Ferguson
Department of Zoology and Entomology
University of Pretoria
Pretoria, South Africa

Philippe Fleury
Department of Social Sciences
ISARA Lyon
Lyon, France

Charles A. Francis
Department of Agronomy and Horticulture
University of Nebraska
Lincoln, Nebraska

and

Department of Plant Science
Norwegian University of Life Sciences
Ås, Norway

Mayra E. Gavito
Center for Ecosystem Research
National Autonomous University of Mexico
Morelia, Mexico

Bruno Glaser
Institute of Agronomy and Nutritional Sciences
Martin-Luther-University of Halle-Wittenberg
Halle, Germany

Ravi Gupta
Department of Botany
University of Delhi
Delhi, India

Lummina G. Horlings
Wageningen University and Research
  Centre
Wageningen, The Netherlands

Nilantha Hulugalle
Australian Cotton Research Institute
New South Wales Department of Primary
  Industries
Narrabri, Australia

Stephanie Hurley
Department of Plant and Soil Science
University of Vermont
Burlington, Vermont

Md. Tofazzal Islam
Department of Biotechnology
Bangabandhu Sheikh Mujibur Rahman
  Agricultural University
Gazipur, Bangladesh

Manindra Nath Jha
Department of Microbiology
Rajendra Agricultural University
Samastipur, India

Shankar Jha
Department of Soil Science
Rajendra Agricultural University
Samastipur, India

Dea-Wook Kim
Department of Crop Sciences
National Institute of Crop Science
Suwon, South Korea

Sun Tae Kim
Department of Plant Bioscience
Pusan National University
Miryang, South Korea

Ajay Kohli
Plant Breeding, Genetics and Biotechnology
   Division
International Rice Research Institute
Manila, The Philippines

Christopher Koliba
Department of Community Development and
   Applied Economics
University of Vermont
Burlington, Vermont

Paul Kristiansen
School of Environmental and Rural Science
University of New England
Armidale, Australia

John Larsen
Center for Ecosystem Research
National Autonomous University of Mexico
Morelia, Mexico

Geir Lieblein
Department of Plant Science
Norwegian University of Life Sciences
Ås, Norway

Brenda B. Lin
Land and Water Flagship
Commonwealth Scientific and Industrial Research
   Organisation
Aspendale, Australia

Melinda McHenry
Centre for Plant and Water Sciences
Central Queensland University
Bundaberg, Australia

Kevin Mearns
Department of Environmental Sciences
University of South Africa
Pretoria, South Africa

Ernesto Méndez
Agroecology and Rural Livelihoods Group
University of Vermont
Burlington, Vermont

Suzanne Morse
Department of Plant Science
Norwegian University of Life Sciences
Ås, Norway

and

College of the Atlantic
Bar Harbor, Maine

Patrick Mundler
Department of Food Economy and
   Consumption Sciences
University of Laval
Laval, Québec, Canada

and

Department of Social Sciences
University of Lyon
Lyon, France

Randeep Rakwal
Research Laboratory for Biotechnology and
   Biochemistry
Kathmandu, Nepal

and

Graduate School of Life and Environmental
   Sciences
University of Tsukuba
Tsukuba, Ibaraki, Japan

and

Department of Anatomy I
Showa University School of Medicine
Tokyo, Japan

Manish Raorane
Plant Breeding, Genetics and Biotechnology
   Division
International Rice Research Institute
Manila, The Philippines

Miguel Nájera Rincón
National Institute of Forestry, Agriculture and
   Livestock Research
Uruapan, Mexico

Jai Singh Rohila
Department of Biology and Microbiology
South Dakota State University
Brookings, South Dakota

Abdiani Attiq Saidajan
Department of Horticulture
Nangarhar University
Nangarhar, Afghanistan

Lennart Salomonsson
Department of Rural and Urban Development
Swedish University of Agricultural Sciences
Uppsala, Sweden

Abhijit Sarkar
Research Laboratory for Biotechnology and
  Biochemistry
Kathmandu, Nepal

Rachel Schattman
Agroecology and Rural Livelihoods Group
University of Vermont
Burlington, Vermont

Sophon Sirisattha
Faculty of Life and Environmental Sciences
Thailand Institute of Scientific and
  Technological Research
Changwat Pathum Thani, Thailand

Yelam Sreenivasulu
Institute of Himalayan Bioresource
  Technology
Council of Scientific and Industrial Research
Palampur, India

Per Espen Stoknes
Norwegian Business School
University of Oslo
Oslo, Norway

Emmanuel Torquebiau
French Agricultural Research Centre for
  International Development, CIRAD
Montpellier, France

Camilo Trench
Department of Life Sciences
The University of the West Indies
Kingston, Jamaica

Rungaroon Waditee-Sirisattha
Department of Microbiology
Chulalongkorn University
Bangkok, Thailand

Dale Webber
Department of Life Sciences
The University of the West Indies
Kingston, Jamaica

Mona Webber
Department of Life Sciences
The University of the West Indies
Kingston, Jamaica

Katherine Westdijk
Agroecology and Rural Livelihoods Group
University of Vermont
Burlington, Vermont

Alexander Wezel
Department of Agroecology and
  Environment
ISARA Lyon
Lyon, France

Sajad Majeed Zargar
Department of Biotechnology
Sher-E-Kashmir University of Agricultural
  Sciences and Technology
Jammu, India

Asim Zia
Department of Community Development and
  Applied Economics
University of Vermont
Burlington, Vermont

CHAPTER 1

Soil Biogeochemistry
From Molecular to Ecosystem Level Using Terra Preta and Biochar as Examples

Bruno Glaser

CONTENTS

1.1 Introduction 2
1.2 The Terra Preta Phenomenon: Facts and Myths..2
    1.2.1 Milestones of Terra Preta Discovery and Research...........................2
        1.2.1.1 Discovery of Terra Preta...2
        1.2.1.2 Geogenic versus Anthropogenic Origin 3
        1.2.1.3 Modern Terra Preta Research 3
    1.2.2 Molecular Markers: The Only Way to Detangle the Terra Preta Secret 7
        1.2.2.1 Aquatic versus Terrestrial Plant Material 7
        1.2.2.2 Composted Garbage...8
        1.2.2.3 Human and Animal Excrements...8
        1.2.2.4 Bones from Mammals and Fish 13
        1.2.2.5 Ash and Charred Organic Material 14
        1.2.2.6 Microbiology of Terra Preta...17
    1.2.3 Future of Terra Preta Research...19
1.3 Biochar Systems: Copying the Terra Preta Concept................................19
    1.3.1 The Problem of Upscaling in Space and Time 19
    1.3.2 The Need for Fast, Reliable, and Cheap Analytical Tools 20
1.4 Biochar Systems: Do They Really Make Sense?......................................24
    1.4.1 Economy versus Ecology..24
        1.4.1.1 General Overview 24
        1.4.1.2 Nutrient Retention...24
        1.4.1.3 Water Retention..25
    1.4.2 Life Cycle Assessment..25
        1.4.2.1 Introduction...25
        1.4.2.2 Economic Assessment 26
        1.4.2.3 C Sequestration Potential..27
        1.4.2.4 Avoided Greenhouse Gas Emissions 28

          1.4.2.5   Improved Fertilizer Use Efficiency...28
          1.4.2.6   Summary and Conclusions 28
1.5   Biochar Systems: Do We Understand Them at All?..29
      1.5.1   Lack of Basic Understanding of Biochar Reactions in the Environment..................29
      1.5.2   Material Properties...30
      1.5.3   Physicochemical Interactions with Soil, Nutrients, and Water 31
      1.5.4   Biological Interaction..31
      1.5.5   Environmental Toxicology 32
      1.5.6   The Urgent Need for Further Inter- and Transdisciplinary Biochar Research..........33
References...33

1.1 INTRODUCTION

Soil biogeochemistry deals with biological and geochemical processes in ecosystems, including soil–plant–water–atmosphere interactions. The focus of this chapter is the identification and quantification of soil processes. Depending on the research question, are used well-established analytical methods such as molecular markers or biomarkers and stable isotope techniques. In addition, modern agroecology also requires the development of new analytical tools such as nondestructive and noninvasive techniques for rapid data acquisition in the field. On the other hand, more sophisticated and resource-intensive technologies such as position-specific isotope analyses are also required for process identification. This method spectrum allows us to cover research questions from the molecule to the ecosystem level or from basic to applied research. In the following, this concept is explained using the famous terra preta/biochar story as a case study.

Terra preta is man-made black soil left behind by pre-Columbian people in Amazonia, occurring in a region dominated by highly weathered infertile soils and is still sustainably fertile today (Glaser et al. 2001a,b). Therefore, knowing how terra preta was made and how it works could help us to solve our problem of soil degradation and increasing food demand.

1.2 THE TERRA PRETA PHENOMENON: FACTS AND MYTHS

1.2.1 Milestones of Terra Preta Discovery and Research

1.2.1.1 Discovery of Terra Preta

The history of terra preta discovery is summarized in Table 1.1. Francisco de Orellana mentioned large cities with millions of people during his journey along the Amazon River in 1542, which gives some indication of the existence of terra preta (de Carvajal 1934). However, it is strange that such cities were never reported by later travelers. Deep black fertile soils containing pottery were first mentioned in Brazil by Hartt (1872) and in Guyana by Brown (1876). The first scientific report mentioning the term *terra preta* was published by Brown and Lidstone (1878). In this early stage of research, the relation between indigenous occupation and soil fertility was not known. However, Smith (1879), who was an assistant of Charles Hartt, suggested that terra preta was a product of Indian kitchen middens that had accumulated "the refuse of a thousand kitchens for maybe a thousand years." Hartt (1885) was the first to report that terra preta contains remains of human occupation (ceramic fragments, lithic artifacts, and charcoal). The next important publication about terra preta was by Friedrich Katzer from Leipzig. Based on his three years of fieldwork in Amazonia, Katzer recognized the fertility of terra preta covering about 50,000 ha in the region south of Santarém between the Tapajós and the Curuá Una rivers (Katzer 1903). Based on his pioneering analytical investigations of terra preta, Katzer (1903) compared terra preta with chernozems, clearly stating that terra preta was of

Table 1.1 Milestones of Terra Preta Discovery

Year	Discoverer	Discovery	References
1542	Francisco Orellana	Cities with millions of people along the Amazon River	de Carvajal (1934)
1868	James Orton	"The soil is black and very fertile"	Orton (1875)
1870	Charles Hartt	Deep black fertile soils with pottery in Brazil	Hartt (1872)
1876	Barrington Brown	Deep black fertile soils with pottery in Guyana	Brown (1876)
1878	Barrington Brown	First scientific report in which the name "terra preta" was used	Brown and Lidstone (1878)
1879	Herbert H. Smith	Terra preta is a product of Indian kitchen middens having accumulated "the refuse of a thousand kitchens for maybe a thousand years"	Smith (1879)
1885	Charles Hartt	Terra preta contains ceramic fragments, lithic artifacts, and charcoal	Hartt (1885)
1895–1898	Friedrich Katzer	50,000 ha of terra preta south of Santarém; first report on anthropogenic origin	Katzer (1903)

anthropogenic origin while chernozems originated naturally. He found that terra preta was a mixture of soil mineral matrix with charred plant material and decomposed organic material (Katzer 1903).

Thus, by the end of the nineteenth century, several scientists reported the presence of terra preta at various sites within Amazonia. They connected the occurrence of Indian artifacts to an anthropogenic origin of terra preta. Additionally, a link was made between prior burning activities and charcoal as a major feature of these soils, and it was established that these soils were highly fertile and productive and probably used for agriculture in the pre-European past. However, very little further progress was made during the first half of the twentieth century.

1.2.1.2 Geogenic versus Anthropogenic Origin

The next era of terra preta research took place from 1940 to 1960, when mostly descriptive work was published rather than analytical results. During that time, the main focus was on the discussion of whether terra preta was of natural (geogenic) or human (anthropogenic) origin rather than on providing detailed analytical results (Glaser et al. 2004a; Woods and Denevan 2009). A summary of these reports is given in Table 1.2. The Brazilian agronomist Felisberto Camargo (1941) believed that terra preta developed on volcanic ash. Archaeologist Barbosa de Faria (1944) and pedologists Cunha Franco (1962) and Ítalo Falesi (1967) argued that terra preta was formed by the sedimentation of organic material in past lakes and ponds, and that such sites attracted Indian settlement, which explained the cultural midden material present. Gourou (1949) and Hilbert (1968) reviewed various origin theories and concluded an "archaeological" origin. Ranzani et al. (1962) suggested that terra preta is similar to the European plaggen soils.

The Dutch soil scientist Wim Sombroek was a pioneer in terra preta research, based on his dissertation containing detailed maps and analytical data of soils of the Belterra Plateau at the lower Rio Tapajós (Sombroek 1966). He differentiated between black terra preta derived from village middens and brown terra mulata, which was used for long-term cultivation. Since then, an anthropogenic origin of terra preta has been favored as the basis for modern terra preta research.

1.2.1.3 Modern Terra Preta Research

Modern terra preta research was initiated by the German soil scientist Wolfgang Zech, focusing more on the investigation of soil processes rather than simply describing observations and basic soil properties (Zech et al. 1979). This, together with the fact that at least three later terra preta and biochar scientists (Bruno Glaser, Johannes Lehmann, and Christoph Steiner) worked in his group, justifies the attribute "godfather of terra preta research," which is often used for Wim Sombroek instead.

Table 1.2   Milestones in Early Terra Preta "Research," Which Was Mainly Based on Field Observations

Year	Scientist	Origin	Natural	Human	References
1941	Felisberto Camargo	Volcanic ash	X		Camargo (1941)
1944	Barbosa de Faria	Sedimented organic matter in dry lakes attracted people to settle	X		Barbosa de Faria (1944)
1949	Pierre Gourou	Archaeological		X	Gourou (1949)
1958	Zimmermann	Fluvial sedimentation	X		Zimmermann (1958)
1962	Cunha Franco	Sedimented organic matter in dry lakes attracted people to settle	X		Franco (1962)
1962	G. Ranzani	Plaggen epipedon		X	Ranzani et al. (1962)
1965	Ítalo Falesi	Sedimented organic matter in dry lakes attracted people to settle	X		Falesi (1967)
1966	Sombroek	Kitchen midden (terra preta) Long-term cultivation (terra mulata)		X	Sombroek (1966)
1968	Hilbert	Archaeological		X	Hilbert (1968)

Note: Please note that most of these studies favored a natural (geogenic) and only a few favored a human (anthropogenic) origin.

Terra preta developed on arenosols, acrisols, ferralsols, plinthosols, and cambisols, among other soil types (Kämpf et al. 2003). Terra preta has a similar texture and mineralogical composition (dominance of low-activity clays and iron and aluminum oxides) to surrounding soils, supporting its anthropogenic origin (Glaser et al. 2004a). Volcanic sedimentation would lead to tephra layers coupled with a different particle size distribution, the occurrence of volcanic glasses and their weathering products (allophanes), and different heavy minerals (Gillespie et al. 1992) in terra preta. A fluvial sedimentation would also result in a different texture compared with surrounding soils, which has never been reported (Zech et al. 1979; Sombroek et al. 1993; Glaser et al. 2004a), also excluding a genesis from anthropogenically applied allochthonous soil material such as plaggen. Archaeological evidence and radiocarbon dating proved the pre-Columbian origin of terra preta. Radiocarbon dates of terra preta sites at the Upper Xingu region ranged between AD 60 and 1640 (Heckenberger et al. 2003) and in Central Amazonia between AD 550 and 1450 BC (Neves et al. 2003).

Terra preta is very famous around the world, but the total extent in Amazonia is still a matter of speculation because remote sensing tools failed to identify terra preta (Thayn et al. 2011). Although a systematic ground check is impossible, there are some indications that terra preta is a widespread phenomenon within Amazonia. For instance, Thayn et al. (2011) reported 40 terra preta sites along a 670 km long gas pipeline from Manaus to Urucu, resulting in a terra preta site every 17 km along this track (Figure 1.1).

The fascinating thing about terra preta is the fact that it still maintains its fertility although it was created some 2000 years ago. This is partly due to the tremendous nutrient levels and soil organic matter (SOM) stocks acting as a long-term slow-release fertilizer. Glaser (1999) reported on average 17 Mg ha^{-1} total N, 13 Mg ha^{-1} total P, and 250 Mg ha^{-1} total organic C in 1 m soil depth (Figure 1.2) for five terra preta profiles representative of the region around Manaus and Santarém (numbers 1–5 in Figure 1.1). These values were two times higher for N, four times higher for P and three times higher for TOC when compared with surrounding soils (Figure 1.2). It is interesting to note that about half of these nutrients and TOC stocks were stored in the agronomically important upper 30 cm soil depth (Figure 1.2).

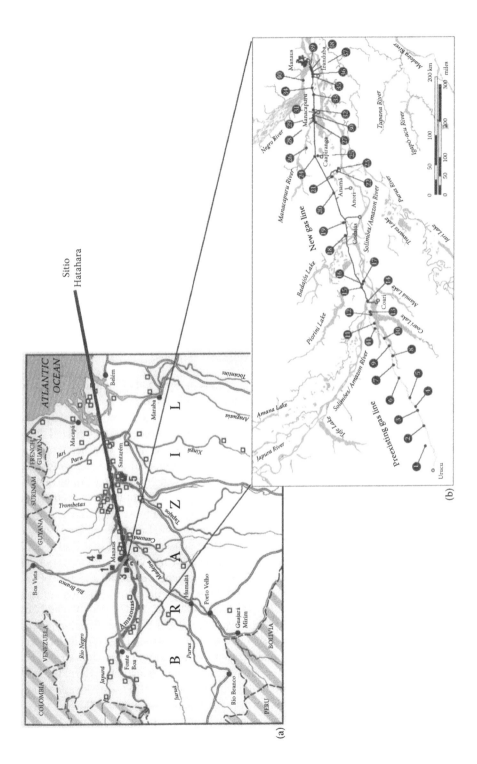

Figure 1.1 (a) Known terra preta sites in Central Amazonia. (From Glaser, B., Haumaier, L., Guggenberger, G., and Zech, W., Naturwissenschaften, 88, 37, 2001b.) (b) Recently discovered terra preta sites along a 670 km Petrobras pipeline from Manaus to Urucu. (From Thayn, J., Price, K.P., and Woods, W.I., *Int J Remote Sens*, 32, 6713–6729, 2011.)

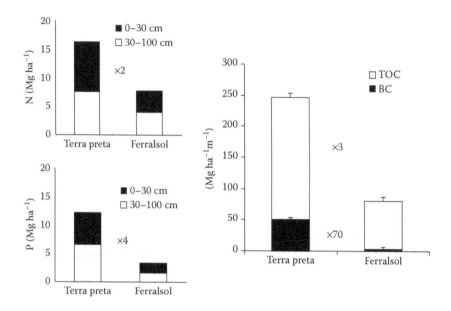

Figure 1.2 Stocks of nutrients, total organic C (TOC), and biochar (BC) compared with surrounding fer-
ralsols. Mean ± standard error of five profile pairs near Manaus and Santarém (numbers 1–5 in
Figure 1.1). Numbers indicate enrichment compared with surrounding soils. (Data from Glaser,
B., *Eigenschaften und Stabilität des Humuskörpers der Indianerschwarzerden Amazoniens*,
Bayreuther Bodenkundliche Berichte 68, Bayreuther, 1999.)

It is also exciting that terra preta still contains higher levels of SOM today, although it is known
that the mean residence time of organic matter in the humid tropics is around four years (Tiessen
et al. 1994). This is mainly due to the chemical composition of terra preta SOM, dominated by stable
condensed aromatic moieties derived from charred organic material or biochar (Glaser et al. 2000,
2001a,b). Terra preta contains on average 50 Mg ha^{-1} biochar, which is about 70 times more than
surrounding soils (Figure 1.2). From these results it is evident that biochar is a key ingredient in
making terra preta so special. However, biochar comprises only about 20% of SOM in terra preta,
which makes clear that organic compounds other than biochar are important for terra preta formation and its unique properties (Glaser 1999). Therefore, it is strange that most experiments copying
the terra preta concept are based on pure biochar amendments to soil (Jeffery et al. 2011), instead
of combining it with labile organic matter and materials rich in nutrients (Schulz and Glaser 2012).

With respect to potential sources for the high nutrient stocks, *in situ* weathering as a source of
P, Mg, K, Zn, and Mn can be excluded, at least for heavily weathered ferralsols and acrisols and
poorly developed infertile arenosols, since these soils do not contain high concentrations of these
elements (Sombroek 1966; Zech et al. 1990; Glaser et al. 2001b; Glaser 2007). Only C and N can
be incorporated in higher amounts via photosynthesis and biological N immobilization, respectively (Glaser 2007). Other elements must be incorporated from the surroundings for nutrient accumulation (Glaser 2007). Mapping of element concentrations of terra preta sites showed different
distribution patterns of elements indicating different nutrient sources (Costa and Kern 1999). As
pre-Columbian populations had no access to mineral fertilizers such as NPK, only local resources
can be responsible for nutrient accumulation in terra preta. Theoretically, the following natural
resources can potentially contribute to the high nutrient stocks in terra preta:

1. Aquatic plant material including algae (being rich in C and N)
2. Composted garbage (being rich in C and N)
3. Human and animal excrements (being rich in P and N)

4. Bones from mammals and fish (being rich in P and Ca)
5. Ash and charred organic material (being rich in Ca, Mg, K, P, and biochar)

As biochar contains only traces of nutrients, it does not significantly contribute to the nutrient status (Glaser 2007). Arroyo-Kalin et al. (2009) and Woods (2003) indicated that ash may have been a significant input into terra preta. It is likely that ash was applied on terra preta by human activities. However, it is unlikely that the application of ash was a key process leading to terra preta formation, as slash-and-burn agriculture does not generate terra preta despite the fact that a lot of ash is applied to such sites. In addition, ash contains significant amounts of Ca, K, Mg, and P, while terra preta is highly enriched in P, but other elements are less enriched or even depleted, especially K (Glaser and Birk 2012). Therefore, the quantities and ratios of nutrients in terra preta indicate that plant materials were not the only nutrient sources. It has been hypothesized that plant biomass from rivers and organic matter from floodplains were transported to the sites (Denevan 1996; Lima et al. 2002; Glaser 2007). However, the question is whether it is possible to unambiguously identify specific nutrient sources of terra preta. The major constraint to identifying specific materials incorporated in the soil 2000 years ago is the fact that it is most likely that these materials are already degraded.

1.2.2 Molecular Markers: The Only Way to Detangle the Terra Preta Secret

Although terra preta is deeply black, there are almost no charcoal particles visible in the soil. The only macroscopic artifacts are potsherds and bones. Therefore, the challenge is to use stable specific molecules (so-called biomarkers) in terra preta that tell us the story of its formation or let us unambiguously identify the materials that give this soil its unique properties, which will be addressed in this section.

The occurrence of archaeological remains such as human and animal bones, fish bones, and turtle backs helps us to identify major nutrient input paths, especially of P. By analysis of lipid biomarkers, which are especially stable in the environment, a differentiation between input of human and animal excrements as well as between terrestrial and aquatic biomass can be obtained.

1.2.2.1 Aquatic versus Terrestrial Plant Material

Terra preta is predominantly located at river bluffs (Figure 1.1). Therefore, it is important to know whether organic matter in terra preta derived from aquatic (e.g., aquatic algae) or from terrestrial (e.g., plant residues) resources. To differentiate between these two sources of organic matter, the pattern of n-alkanes looks promising. Cuticular waxes of terrestrial plants are dominated by long-chain n-alkanes (>C20), while short chain n-alkanes (<C20) are typical for algae (Bourbonniere et al. 1997). Therefore, the ratio of terrestrial (C27 + C29 + C31) to aquatic (C15 + C17 + C19) n-alkanes (TAR) can be used as a measure for the relative contribution of terrestrial or aquatic plant residues to SOM formation in a given soil. A TAR > 4 indicates terrestrial SOM, while a TAR < 1 indicates aquatic SOM (Bourbonniere et al. 1997). Similar indices for terrestrial biomass are the carbon preference index (CPI) being in the range of 4–30 (Collister et al. 1994) and the odd-over-even predominance (OEP) being in the range of 4–8 (Hoefs et al. 2002).

Terra preta shows TAR, CPI and OEP around 1 throughout the whole profile (Figure 1.3), indicating a predominance of aquatic SOM. However, sorption of long-chain n-alkanes to charcoal was observed (data not shown), so that interpretation of these ratios is not valid for biochar-containing soils such as terra preta. On the other hand, the ^{13}C isotope signatures of individual n-alkanes were in the range between −30‰ and −40‰ typical for terrestrial plants (Glaser and Zech 2005). Aquatic plants should have more positive δ^{13}C values (Glaser and Zech 2005). Therefore, it is more likely that SOM in terra preta is of terrestrial origin, despite the fact that the n-alkane ratios indicate the contrary. Further studies are required to specifically trace the importance of aquatic biomass for terra preta formation.

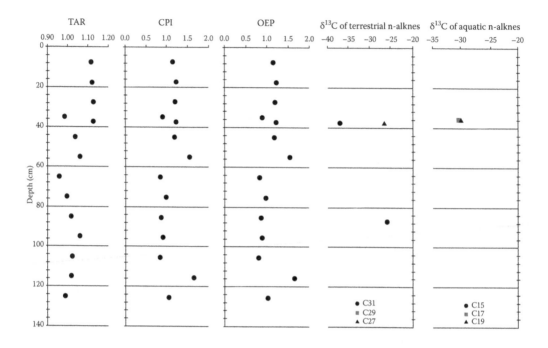

Figure 1.3 Terrestrial-to-aquatic n-alkane ratio (TAR), carbon preference index (CPI), odd-over-even predominance (OEP), and compound-specific $\delta^{13}C$ values of terrestrial (C27, C29, C31) and aquatic (C15, C17, C19) n-alkanes of the terra preta Profile "Hatahara" near Iranduba (Figure 1.1).

1.2.2.2 Composted Garbage

Composting leads to partial C and N loss with an enrichment of stable C and N compounds. While no alteration of C isotope signature was observed, composting leads to an enrichment of ^{15}N (Lynch et al. 2006). Yun and Ro (2009) reported an increase of $\delta^{15}N$ of soil nitrate from 8‰ to 16‰ and of plant $\delta^{15}N$ from 14‰ to 25‰ after excessive compost application (>1000 mg N kg^{-1} soil). Bulk soil $\delta^{13}C$ value of terra preta was around −26‰, which is typical for C3 plants like trees. Bulk soil $\delta^{15}N$ value of terra preta ranged between 13‰ and 15‰, increasing with increasing soil depth (Figure 1.4), which is typical for soil. These high $\delta^{15}N$ values indicate a strong enrichment of ^{15}N in terra preta. However, it should be stressed that an adjacent ferralsol showed bulk $\delta^{15}N$ value in the topsoil of around 11‰, being typical for soils of the humid tropics (Amundson et al. 2003). Compound-specific $\delta^{15}N$ analysis of individual amino acids was shown to be more specific in detecting ancient organic manuring practices (Simpson et al. 1997). These analyses showed that terra preta exhibit more positive $\delta^{15}N$ values of hydrophobic amino acids than the control soil (Figure 1.5), which is characteristic for soils fertilized with organic manure (Simpson et al. 1997). This effect could only be detected in deeper horizons of terra preta, reflecting ancient land use, while the topsoil could be influenced by recent land use, with mineral N fertilization leading to lower ^{15}N values.

Further studies should involve methods that are more specific to detailed cultivation practices, such as phytolith analysis (Piperno and Becker 1996).

1.2.2.3 Human and Animal Excrements

If we look first at the potential to differentiate between human and animal excrements as nutrient sources for terra preta formation, sterols and bile acids have been proven to be useful (Figure 1.6).

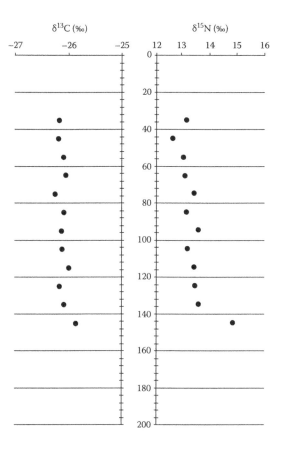

Figure 1.4 Bulk soil $\delta^{13}C$ and $\delta^{15}N$ depth profiles of the terra preta "Hatahara" near Iranduba (Figure 1.1). These values are typical for other terra preta profiles as well (data not shown).

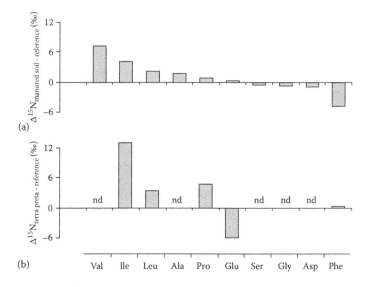

Figure 1.5 Differences of $\delta^{15}N$ values of individual amino acids between (a) manured and control soil in Europe and (b) terra preta "Hatahara" and surrounding soil at 30–40 cm depth. (From Glaser, B. and Birk, J.J., *Geochim Cosmochim Acta*, 82, 39–51, 2012. With permission.)

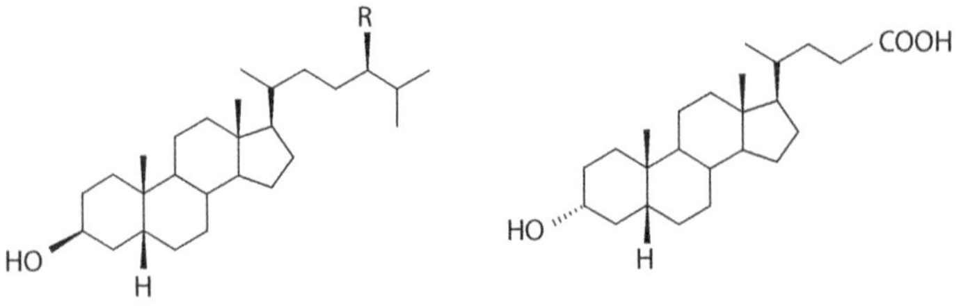

Figure 1.6 Chemical structure of stanols and bile acids.

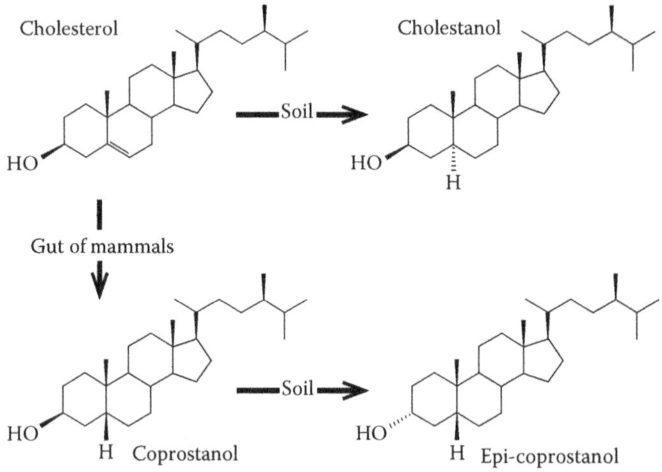

Stanol

R = H, coprostanol => omnivores

R = Me, 5β-campestanol => herbivores

R = Et, 5β-stigmastanol => herbivores

Bile acid

Lithocholic acid => human feces

Deoxycholic acid => human feces

Figure 1.7 Products of cholesterol reduction in the gut of mammals (=fecal indicator) and in soils. (From Birk, J.J., Teixeira, W.G., Neves, E.G., and Glaser, B., *J Archaeol Sci*, 38, 1209–1220, 2011. With permission; Bull, I.D., Lockheart, M.J., Elhmmali, M.M., Roberts, D.J., and Evershed, R.P., *Environ Int*, 27, 647–654, 2002. With permission.) Please note that a minor amount of coprostanol may also be produced in soil by microbial reduction of cholesterol (not shown here).

However, stanols were also found in soils where no anthropogenic deposition of feces was assumed, showing that the presence of stanols alone is not sufficient to prove ancient fecal deposition on soils because its precursors (sterols) can also be reduced microbially to corresponding stanols (Bethell et al. 1994; Bull et al. 2001; Evershed et al. 1997). However, the majority of sterols (e.g., cholesterol) are transformed to 5α-stanols (e.g., cholestanol) by way of reduction in the environment and only a minor amount is reduced to 5β-stanols (e.g., coprostanol, Figure 1.7). In addition, coprostanol may be microbially transformed to epi-coprostanol in soil (Figure 1.7). Due to these transformations, Grimalt et al. (1990) suggested using the ratio of coprostanol to (coprostanol + epi-coprostanol) for detection of feces in water and sediment samples.

Grimalt et al. (1990) reported a coprostanol to (coprostanol + epi-coprostanol) ratio of <0.3 and 0.7 for samples from coastal areas without and with anthropogenic feces input, respectively. These values were used as threshold for feces contamination in further studies (e.g., Bull et al. 2001;

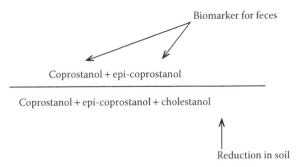

Figure 1.8 Formula for fecal indicator based on fecal stanol (coprostanol).

Elhmmali et al. 2000; Simpson et al. 1998). However, the application of these threshold values in archaeological samples was critically discussed by Bull et al. (2001) and Simpson et al. (1998) and a comparison with local background samples was recommended instead by Bull et al. (1999).

Birk et al. (2011) followed an alternative approach including cholestanol (Figure 1.8). With this approach, the reaction of cholesterol and coprostanol in soil (Figure 1.7) and the sterol background concentrations can be leveled out (Figure 1.8).

The sterol ratios presented in Figure 1.9 indicate that feces were deposited on terra preta and the presence of coprostanol indicates feces of omnivores (Figure 1.6, Birk et al. 2011). Feces of carnivores and birds contain less coprostanol than feces of omnivorous mammals (Leeming et al. 1996; Shah et al. 2007). In addition to coprostanol, feces of herbivores contain 5β-stigmastanol (Figure 1.6, Evershed et al. 1997; Gill et al. 2010; Leeming et al. 1996; Shah et al. 2007). Based on this observation, several criteria were defined to distinguish fecal inputs from herbivores versus feces of omnivores, namely humans and pigs (Bethell et al. 1994; Evershed and Bethell 1996; Leeming et al. 1997). In terra preta, the absence of 5β-stigmastanol at 30–40 cm soil depth in most samples showed that feces of herbivores were not generally deposited on terra preta to the same extent as the feces of omnivores. Pigs were brought to South America by the Europeans, and thus their feces cannot have contributed to the nutrient stocks in terra preta. Therefore, the fecal material in terra preta is most likely of human origin.

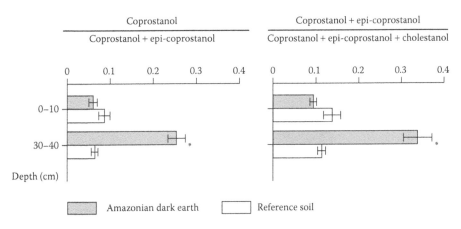

Figure 1.9 Comparison of different fecal indicators in terra preta (Amazonian dark earth) and reference soils. Higher values indicate greater deposition of fecal material (* indicates significant differences ($p < .05$) between terra preta and reference soils; error bars show standard errors; $N = 5$). (Modified from Birk, J.J., Teixeira, W.G., Neves, E.G., and Glaser, B., *J Archaeol Sci*, 38, 1209–1220, 2011. With permission.)

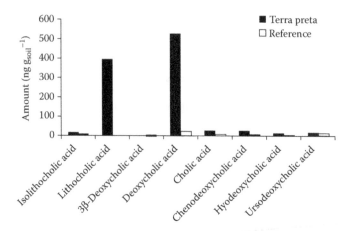

Figure 1.10 Bile acid concentrations in a terra preta sampled near Santarém (Number 5 in Figure 1.1) compared with surrounding soil (0–50 cm). (From Glaser, B. and Birk, J.J., *Geochim Cosmochim Acta*, 82, 39–51, 2012. With permission.)

For further identification of feces-derived nutrient input into terra preta, bile acids were analyzed. Bile acids in a terra preta were dominated by deoxycholic acid and lithocholic acid (Figure 1.10), typical for feces of humans (Bull et al. 1999, 2002). Therefore, all available steroid data so far unambiguously indicate deposition of human feces, which could explain the nutrient pattern dominated by large amounts of P.

Figure 1.11 Bone fragments isolated from the terra preta "Hatahara" near Iranduba (Figure 1.1). Upper bone fragments are from a turtle back and the lower ones are fish bones. (Neves, personal communication.)

1.2.2.4 Bones from Mammals and Fish

Another potential nutrient source is waste from food production including animal remains such as mammal and fish bones, which likely contribute to the nutrient stocks, especially Ca and P. Mammal and fish bones can be even observed macroscopically in terra preta profiles today (Figure 1.11). Using scanning electron microscopy in combination with energy-dispersive x-ray spectroscopy (SEM/EDS), high Ca and P concentrations could be identified both in bone samples and in terra preta samples (Figure 1.12), similarly to the findings of Lima et al. (2002) and Schaefer et al. (2004). It is interesting to note that terra preta soil samples also revealed high Ca and P concentrations, which most probably also derived from weathered bones (Figure 1.12).

Ancient DNA extracted directly from the soil could serve as a biomarker for nutrients, even in the case that archaeological fragments like bones are missing. In the soil environment DNA is not completely digested by soil DNases, but may be stabilized by adsorption to clay minerals and sesquioxides (Demanèche et al. 2001). In several matrices, like papyrus (Marota et al. 2002), coprolite (Hofreiter et al. 2000), caveman painting (Reese et al. 1996), and bones (Meyer et al. 2000), ancient DNA can remain detectable for thousands of years. Recent investigations indicate that ancient DNA may be also preserved in permafrost and temperate sediments for long periods ranging from 400,000 to 10,000 years before present (BP) (Willerslev et al. 2003). However, nothing is known about the stabilization of ancient DNA in highly weathered soils of the humid tropics. From literature, no appropriate methods for extraction, purification, and identification of such DNA from terra preta samples could be found. A further complication was the fact that soil also contains recent DNA from plant

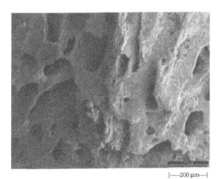
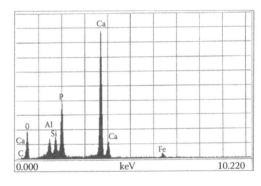

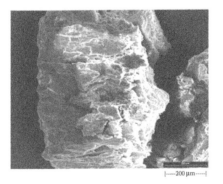
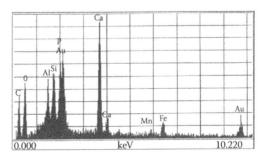

Figure 1.12 Scanning electron microscopic picture of a bone fragment presented in Figure 1.11 (*upper left*) and a terra preta soil aggregate (*lower left*) and corresponding energy-dispersive x-ray spectra (*right*). Bone fragments isolated from the terra preta "Hatahara" near Iranduba (Figure 1.1). Please note the similar composition of both bone and soil particle, especially with respect to Ca and P.

and microbial material. Therefore, this recent DNA material needs to be eliminated prior to ancient DNA extraction. For this purpose, a sequential extraction with sodium pyrophosphate and the commercial DNA kit NucleoSpin Food turned out to be most successful. For identification of ancient DNA, mammal-specific primers were amplified by polymerase chain reaction (PCR). In detail, different 18S rRNA and 12S rRNA genes were used (12 Sa/12 So 161 base pairs, Cytb L14925/Cytb H15052 149 base pairs). Amplified PCR products were separated by agarose and polyacrylamide gel electrophoresis, the latter being more sensitive. The amplified PCR products were cloned in competent *Escherichia coli* colonies, which were subsequently sequenced. DNA sequences were identified by comparison with the GenBank database by means of the computer program BLASTN. Although DNA extraction was generally extremely low, two mammal species could be unambiguously identified in all investigated terra preta samples (100% confidence of clone sequence with target sequence): *Sus scrofa* (wild pig) and *Bos taurus* (cow). Both species could be identified in all investigated terra preta samples. In one sample *Homo sapiens* (human) DNA was also detected. However, as DNA analysis of bones extracted from terra preta also revealed *S. scrofa* and *B. taurus*, DNA contamination might have occurred during sampling, sample processing, or both. In addition, it is known that cows were only brought by Europeans to South America. Therefore, a pre-Columbian origin of cow bones is unlikely. Nevertheless, the method of ancient DNA analysis in terra preta would enable a deeper insight into animal species that may have played a role in pre-Columbian Indian nutrition and as nutrient sources for terra preta formation.

1.2.2.5 Ash and Charred Organic Material

To my knowledge, there is no specific molecular marker indicating ash contribution to terra preta formation. Tracing total element composition revealed an enrichment of C, N, P, S, Ca, and Mg compared with surrounding soil, while K, which is typical for plant ash, was partly depleted in terra preta (Glaser and Birk 2012). Therefore, it is most unlikely that ash significantly contributed to terra preta formation, although it might have been leached from the soil while the other elements were stabilized.

Charred organic matter gives terra preta its black color and enriches the soil with stable aromatic compounds. It is not easy to specifically trace and quantify charred organic matter in soil, as these compounds follow a continuum from slightly torrified to graphitic carbon (Hedges et al. 2000). Many methods exist for detection of different parts of this continuum in diverse environmental matrices, giving a wide range of results for identical samples (Hammes et al. 2007). As a consequence, these discrepancies make a direct comparison of data obtained by different methods difficult. Using benzenepolycarboxylic acids (BPCA; Figure 1.13) as molecular markers, condensed aromatic moieties of the charred organic matter continuum can be specifically identified and quantified (Glaser et al. 1998). In addition, the BPCA pattern allows a characterization of the degree of condensation (Glaser et al. 1998; Brodowski et al. 2005; Schneider et al. 2010). The principle of the method is based on the fact that polycondensed aromatic moieties of charred organic matter are oxidized to BPCA by means of concentrated nitric acid at 170°C and pressure (Figure 1.13, Glaser et al. 1998). Before analysis of the BPCAs by gas chromatography flame ionization detection (GC-FID), the extract needs to be cleaned in several steps and derivatized (Figure 1.13, Brodowski et al. 2005; Glaser et al. 1998). Recently the procedure was simplified, substituting GC-FID by high pressure liquid chromatography, which makes the time-consuming sample cleanup and derivatization redundant (Figure 1.13, Schneider et al. 2011).

The BPCA method has been widely applied to characterize charred organic matter in soils (Andreeva et al. 2011; Czimczik et al. 2003; Dai et al. 2005; Glaser and Amelung 2003; Glaser et al. 2000, 2001b; Guggenberger et al. 2008; Hammes et al. 2008; Rodionov et al. 2006) and charcoals or biochar (Glaser et al. 1998; Kaal et al. 2008; Schneider et al. 2010; Schimmelpfennig and Glaser 2012; Wiedner et al. 2012). Biochar is a certain fraction of the charring continuum similar to charcoal, but it may be produced not only from woody residues but also from biomass wastes (Schimmelpfennig and Glaser 2012).

Terra preta contains on average about 50 Mg biochar per hectare in the upper 100 cm soil depth (Figure 1.14), while adjacent reference soils contain only 4 Mg biochar, which is about ten times less

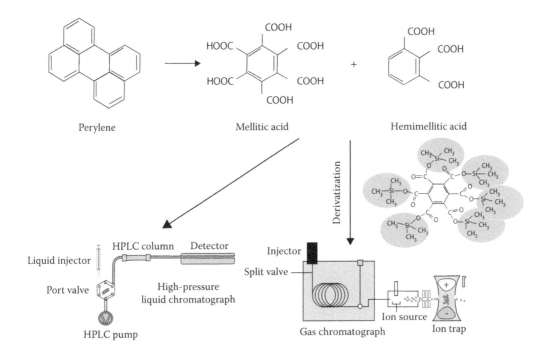

Figure 1.13 Principle of biochar identification and quantification in soil based on benzenepolycarboxylic acids as molecular markers followed by gas chromatographic separation or high-pressure liquid chromatographic (HPLC) separation.

than in terra preta (Figure 1.14). Below 50 cm, biochar amounts were lower and differences between terra preta and adjacent soils more pronounced (Figure 1.14). It is interesting to note that the biochar content is more or less constant in the upper 50 cm soil depth in terra preta, being about 10 Mg ha^{-1} per 10 cm soil depth, while biochar amount in adjacent soils decreases exponentially with increasing soil depth (Figure 1.14).

It is known that biochar is chemically and biologically more stable than other forms of SOM due to its chemical structure (Figure 1.15). For the same reason, biochar is less reactive in soil than other organic molecules. Therefore, it is interesting how and to what extent biochar interacts with soil. Soil density fractionation allows separation of SOM into light, medium, and heavy-density fractions representing particulate, organomineral, and physically stabilized organic matter (Figure 1.14, Glaser et al. 2000). It is interesting to note that about 50% of biochar still occurs in particulate form in terra preta (Figure 1.14), although it was incorporated about 2,000 years ago, corroborating the recalcitrance of biochar in the environment. In the upper 50 cm soil depth, about 40% of biochar is stabilized in organomineral forms, decreasing with increasing soil depth (Figure 1.14). Only a minor part is physically stabilized at the surface of soil minerals (Figure 1.14, Glaser et al. 2000). From these results it is clear that biochar is chemically and biologically recalcitrant and that biochar aging leads only slowly and to a minor extent into organomineral stabilization (Figure 1.14).

Nevertheless, it is known that biochar-rich soils, such as chernozems and terra preta, are sustainably fertile. Therefore, it is most likely that there is a link between physicochemical properties of biochar and ecosystem functions (Figure 1.15). The most striking feature of biochar is the condensed aromatic moieties that give biochar its black color and are responsible for its stability, which makes biochar an interesting compound for C sequestration (Figure 1.15). In addition, aging or biological degradation leads to partial oxidation, resulting in the formation of functional groups on the edges of biochar, causing reactivity in soil such as nutrient retention or organomineral stabilization (Figure 1.15). Last but not least, biochar is a highly porous material, leading to enhanced air and water storage in soil (Figure 1.15).

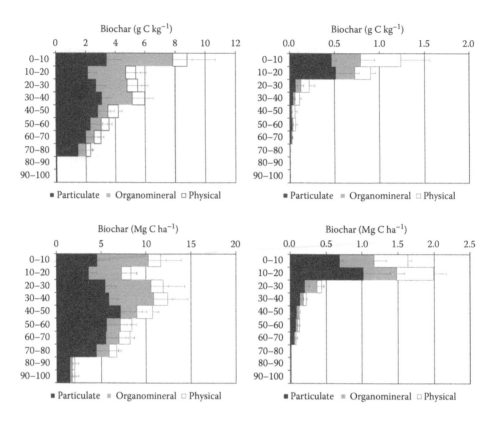

Figure 1.14 Mean biochar concentration (*top*) and stocks (*bottom*) in terra preta (*left*) and reference soils (*right*) separated by stabilization mechanism in soil obtained by density fractionation. Light fraction (<2.0 kg dm^{-3}) contains particulate biochar, medium-density fraction (2.0–2.4 kg dm^{-3}) contains organominerally stabilized biochar, and heavy fraction (>2.4 kg dm^{-3}) contains biochar physically entrapped in soil minerals ($N = 5$). Data from Glaser, B., *Eigenschaften und Stabilität des Humuskörpers der Indianerschwarzerden Amazoniens*, Bayreuther Bodenkundliche Berichte 68, Bayreuther, 1999. With permission.)

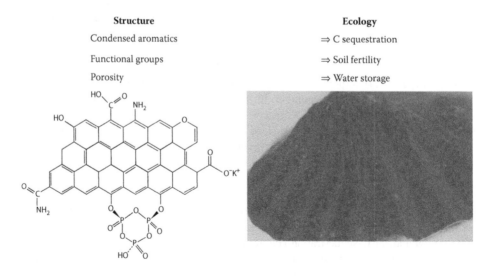

Figure 1.15 Chemical and physical properties of biochar and their translation into ecosystem functions.

1.2.2.6 Microbiology of Terra Preta

It is clear that deposition to soil of the materials discussed in the previous sections without further microbial degradation and/or modification would lead to enormous waste accumulation, and no nutrient and matter recycling would be possible in nature. Therefore, it is important to know whether there is a difference in the microbiology of terra preta compared with adjacent "normal" soils. Microbial cultivation methods and direct DNA extraction of terra preta revealed a higher microbial diversity and species richness in terra preta compared with reference soils (Kim et al. 2007; O'Neill et al. 2009; Ruivo et al. 2009; Tsai et al. 2009). Kim et al. (2007) identified 396 different operational taxonomic units (OTU) in terra preta compared with 291 in adjacent soil. The Shannon index as a measure for microbial diversity was 5.2 in terra preta compared with 4.37 in adjacent soil (Kim et al. 2007). Kim et al. (2007) showed that terra preta and adjacent soil had similar bacterial community composition over a range of phylogenetic distances, among which Acidobacteria were predominant, but terra preta had about 25% higher species richness. Identity of the major bacterial groups by clone sequencing representing different clusters revealed 14 phylogenetic groups in terra preta under forest, as compared with 9 from the adjacent forest soil (Figure 1.15). The dominating microbial group was Acidobacterium, comprising approximately 30% of the bacteria in terra preta and 50% of the forest soil (Figure 1.15). The phylogenetic trees included two possible new clades of Acidobacterium that were distinct from the previously described subgroups for this phylum. Other common bacterial groups for both soils were Proteobacteria, Actinobacteria, Planctomycetes, and Verrucomicrobia (Figure 1.15). Acidobacterium is common in forest soils around the world, comprising from 12% of the noncultured species surveyed in Austrian forest soils under pine to 35% in spruce–fir–beech soils in Europe (Hackl et al. 2004).

Also Tsai et al. (2009) identified bacteria in two terra preta and adjacent sites by cloning and sequencing (Figure 1.16). This study revealed that 42%–85% of bacterial DNA sequences could not

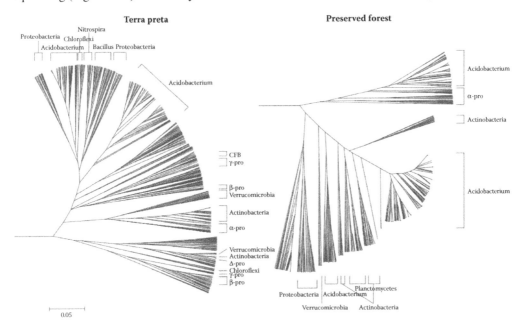

Figure 1.16 Taxonomic cluster analysis of 16S rRNA gene sequences from terra preta and adjacent pristine forest soil based on oligonucleotide fingerprinting of 16S rRNA gene sequences. The 16S rRNA clone libraries were generated for single samples consisting of three composited replicate soil cores (litter layer removed) taken from the top 10 cm of the soil profile at each location. (From Kim, J.S., Sparovek, S., Longo, R.M., De Melo, W.J., and Crowley, D.E., *Soil Biol Biochem*, 39, 648–690, 2007. With permission.)

be identifed either in terra preta or in adjacent soil (Figure 1.16). This result clearly indicates that a high number of soil microorganisms are still unknown, and that this is not restricted to terra preta. The results of Tsai et al. (2009) further revealed that all identified microorganisms in terra preta are "normal" soil microorganisms, of which Firmicutes are the dominant group. Acidobacteria comprised only 2.5% of bacteria in terra preta compared with 13%–27% in adjacent soils (Figure 1.16). It is, further, interesting to note that bacterial community structure was different between the two terra preta sites and between each terra preta and its adjacent soil (Figure 1.17). From these results it is clear that there is no "terra preta code" claimed by "terra preta producers" and that each soil has its own microbial signature, which is dominated by soil properties such as SOM content and pH value, climate and land use (Paul 2007).

Cultivation experiments indicated a predominance of fungi over bacteria in terra preta compared with surrounding soils (Ruivo et al. 2009). However, cultivation experiments and direct extraction of DNA from soils deliver information about the recent microbial community composition, which is not necessarily comparable to the composition during the times of terra preta generation. Therefore, analyses of more stable microbial biomarkers such as amino sugars and muramic acid, which are indicators for fungal and bacterial residues, respectively (Amelung 2001), seems more promising with respect to reconstruction of microbial impact on terra preta formation. Terra preta showed elevated glucosamine to muramic acid ratio compared with reference soils (Glaser et al. 2004b; Glaser and Birk 2012), indicating that the key microorganisms for terra preta formation are fungi but not bacteria.

In conclusion, microbial processes in terra preta are similar to those observed in "normal" soils and consist mainly of aerobic processes similar to those that occur during composting. There is no evidence for anaerobic processes such as fermentation, which has been postulated recently (Factura et al. 2010). Furthermore, it is even more unlikely that fermentation was done in ceramic pots to produce terra preta, because it is most unlikely that thousands of cubic meters of terra preta could have been produced without mechanization under these conditions. To produce a terra preta of 20 ha with a depth of 1 m, it would have been necessary to move 200,000 m^3 of earth, corresponding to 2,000,000 pots containing 100 l each.

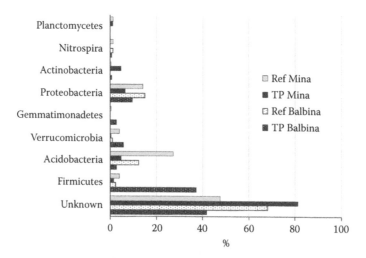

Figure 1.17 Microbial diversity of terra preta (TP) and adjacent soils (Ref) from Balbina Lake (Central Amazonia) and Caxiuana National Forest (Mina, Eastern Amazonia). (Data from Tsai, S.M., O'Neill, B., Cannavan, F.S. et al., *Amazonian Dark Earths: Wim Sombroek's Vision*, Springer, Berlin, 2009.)

1.2.3 Future of Terra Preta Research

As mentioned above, many details are already known with respect to the properties and genesis of terra preta. Future terra preta research should, therefore, focus on still unknown aspects such as:

- More details on land use identification (e.g., phytolith analysis)
- Stability of SOM
- Sustainability of soil fertility
- Similarity to terra preta-like soils in temperate regions
- Role of soil organisms in biochar degradation and formation

1.3 BIOCHAR SYSTEMS: COPYING THE TERRA PRETA CONCEPT

1.3.1 The Problem of Upscaling in Space and Time

The existence of terra preta in Amazonia today proves that it is, in principle, possible to convert infertile soils into sustainably fertile soils even under intensive agriculture. Therefore, terra preta is a general model for sustainably improving soil fertility and ecosystem services while storing large amounts of C in soil for a long time (Glaser et al. 2001a,b; Glaser 2007; Glaser and Birk 2012). A key factor for maintaining sustainable soil fertility is increased levels of SOM and nutrient stocks by using a circular economy with biogenic "wastes" as sources of natural resources, as outlined in Figure 1.18a (Glaser 2007; Glaser and Birk 2012). Biochar is a key factor of the terra preta concept, together with the input of tremendous amounts of nutrients and microbial processes, as discussed in Section 1.2, and this scheme can be translated to modern society (Figure 1.18b). From this concept, it is clear that it makes no sense to work with pure biochar to mimic terra preta effects. This would be like working with pure flour while studying the principle of making a cake. Therefore, for eco-logical studies it is important to include nutrients and microorganisms to study biochar (terra preta) effects. In order to understand biochar-related processes, studies on pure biochar should only be undertaken when it is extracted from nutrient- and microorganism-containing environments, from either laboratory or field incubations.

As terra preta was created about 2000 years ago over a period of about 500 years, both formation and properties require long-term studies. The problem in ecological research and modern society is, however, that there is not enough time. It is very difficult to devote more than three years to a research project. Investigating long-term effects within such a short period of time is impossible. Therefore,

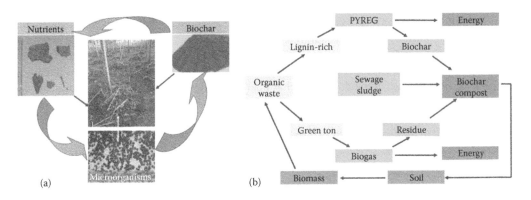

Figure 1.18 Simplified terra preta principle (a) and translation into modern circular economy recycling waste as source of natural resources (b).

the concept of "false time series" has been used to overcome this constraint. For instance, the stability of biochar in soil was investigated by comparison of black carbon content of archived Russian steppe soils with "the same soil" sampled recently (Hammes et al. 2008). This study revealed a biochar degradation of 25% within 100 years, translated into a biochar turnover of around 300 years, which is much lower than real measurements. This might be due to the fact that biochar contains a labile C pool, which is degraded much more rapidly than the stable pool. The study of Hammes et al. (2008) comprised ecosystem processes such as decomposition (the target variable), leaching and erosion. On the other hand, it is very difficult to sample exactly the same place 100 years later. Nguyen et al. (2008) investigated biochar decomposition in a false time series covering 100 years in Kenya. The authors found a rapid decrease of biochar content from 12.7 to 3.8 g C kg^{-1} soil during the first 30 years following deposition, after which it slowly decreased to a steady state at 3.5 g C kg^{-1} soil. From these results the authors calculated a biochar loss of 6 Mg ha^{-1} in the top 10 cm of soil within one century. The initial rapid biochar degradation resulted in a mean residence time of only around 8.3 years, which was likely a function of decomposition as well as transport processes. The protection by physical and chemical stabilization was apparently sufficient to minimize not only decomposition, which was below detection between 30 and 100 years after deposition, but also physical export by erosion and vertical transport below 0.1 m (Nguyen et al. 2008).

Therefore, there is an urgent need for real long-term biochar/terra preta studies under controlled conditions lasting longer than 30 years. However, such sites should be carefully selected, as black carbon is a ubiquitous compound prone to atmospheric deposition, especially in urban regions (Marschner et al. 2008).

Biochar addition to agricultural soils can potentially enhance their fertility and is compatible with sustainable agriculture, in particular when the tremendously porous biochar matrix is soaked in, or co-applied with, nutrient-rich wastes such as slurries or digestate from biogas production. Furthermore, it is claimed that biochar reduces erosion, nutrient leaching and greenhouse gas (GHG) emissions, and binds toxic agents such as heavy metals or organic pollutants. Several ideas for C sequestration and GHG mitigation options are in their infancy but developing rapidly. For instance, it is suggested that peat substrates could be substituted by biochar-composts for horticulture, which would take the pressure off bogs, leaving this giant C pool untouched instead of being mineralized and emitted as CO_2. Further ideas are to develop biochar into a nutrient-loaded carbon-based slow-release fertilizer, or to use biochar as an animal food supplement for detoxification of food-chain pollutants in animal feeding, which automatically creates C-based fertilizer slurry.

1.3.2 The Need for Fast, Reliable, and Cheap Analytical Tools

In Section 1.2, the significant contribution of sophisticated analytical tools such as molecular marker analysis to knowledge expansion in terra preta research was discussed in depth. However, when applying these tedious methods to modern biochar research, only a limited number of samples can be processed. A forecast of biochar effects under real conditions requires large-scale field experiments. A sampling campaign of a biochar field experiment with ten different treatments and five field replicates sampled at two different soil depths results in 100 samples at each sampling time. Within a three-year project, a minimum of four sampling campaigns are necessary (prior to application, right thereafter, one and two years later). Therefore, a total of 400 samples need to be processed just for one field experiment. For typical molecular marker analysis, about 20 samples can be processed per week, meaning that for just one molecular marker (e.g., black carbon) about half a year is necessary without further complications, which often arise during laboratory work. Therefore, frequent sampling and real field replicates produce many samples, which cannot be analyzed with the time-consuming biomarker analysis. For this reason, fast, reliable, and cheap analytical tools are required, such as spectroscopic methods.

In this respect, near-infrared (NIR) and mid-infrared (MIR) spectroscopy is very promising, being a routine method for quality control of food (Williams and Norris 1987). Soil is a very diffusive and absorptive medium. Therefore, soil spectra must be acquired by diffuse reflectance, not by transmission, except of samples diluted in nonabsorbing materials like KBr. Near-infrared reflectance spectroscopy (NIRS) and mid-infrared reflectance spectroscopy (MIRS) measure the reflectance of samples in the spectral range 800–2,500 nm (400–4,000 cm^{-1}) and 2,500–25,000 nm (4,000–12,500 cm^{-1}), respectively. The main difference between the ranges is that MIR absorption corresponds to molecular vibrations (Table 1.3), whereas NIR absorption is based on overtones and band combinations (Williams and Norris 1987). Therefore, MIR bands are more specific than NIR bands. Another disadvantage of NIR is the fact that light diffusion is higher in NIR than in MIR. Therefore, NIR spectra will be more affected by factors which affect the diffusion of light, such as the physical structure (size of aggregates, porosity), but also the presence of water, which changes the refractive index and therefore the diffusion of light (Williams and Norris 1987). The practical consequence is that NIR requires less sample preparation than MIR and is best fitted for in-field analysis, with lower specificity requirements, whereas MIR generally shows a better specificity and reproducibility, but requires more sample preparation in order to optimize the sample/light interaction.

Table 1.3 Characteristic Chemical Bindings Identifiable by Infrared Spectroscopy

Wavenumber (cm^{-1})	Vibration	Functional Group/ Component
3700–3200	SiO-H stretching	Silica
3700–3400	O-H stretching	Hydroxyl groups, water
2920	C-H stretching	Aliphatic methylene group
2850	C-H stretching	Aliphatic methylene group
2590–2520	S-H stretching	Thiols
1740–1700	C=O stretching	Aldehyde, ketone, carboxylic acid, ester
1685–1630	C=O, COO- stretching C=C stretching	Amide, carboxylate Aromatic ring, alkene
1635	O-H bending	Water
1600–1590	C=C	Aromatic skeleton
1570–1540	N-H in plane bending	Amide, secondary amine
1515–1505	Aromatic skeletal	Lignin from lignocellulose material
1450–1410	C-O stretching	Carbonate
1430–1420	COO- stretching	Carboxylic acid
1400–1340 1384	N-O stretching	Nitrate (leachate)
1350–1250	C-N stretching	Aromatic amine
1265–1240	C-O-C stretching C-N stretching	Ester Amide
1250–900	C-O-C, C-O, C-O-P	Polysaccharide Phosphodiester
1140–1080	S-O stretching	Sulfate
1080	Si-O stretching	Quartz
1030	Si-O stretching Si-O-Si stretching	Clay minerals Silica
875	C-O out of the plane bending	Carbonate
713	C-O in the plane bending	Carbonate
680–610	S-O bending	Sulfate

The two classical modes of MIR analysis of soils are diffuse reflectance on dried/ground soils, as shown in several articles cited by Viscarra Rossel et al. (2006), or attenuated total reflectance (ATR) applied to soil pastes, especially when ions, like nitrate, ammonium and so on, are assayed (Jahn et al. 2006; Du and Zhou 2009). Therefore, up to now, MIR has been restricted to laboratory analysis, although its use in the field has been reported (Reeves and Smith 2009). Modern IR spectrometers transform their spectra by a mathematical procedure called Fourier transformation (FT), which allows a better separation of small absorptions, similar to derivations of spectra used for compound quantification.

Infrared spectroscopy can be used for qualitative and quantitative analysis. Qualitative analysis comprises the identification of chemical bindings and compounds by comparison of observed absorption bands with databases (Table 1.3).

Another option is to use multivariate statistical tools such as principal component analysis to group sample sets or to identify unknown sample sources. For instance, compost derived from different sources can be separated by this method (Figure 1.19).

The quantification of material properties based on NIRS or MIRS requires multivariate statistical methods such as partial least squared calibration (Figure 1.20). For this approach, the calibration sample set needs to be investigated both with IR spectroscopy and with reference methods (e.g., CN analysis, molecular marker analysis). Once such a calibration has been developed, NIRS is a rapid, inexpensive and nondestructive method for sample characterization, which does not require any consumables. It has already been used to determine the total C and N contents of soils (Barthès et al. 2006), to assess the elemental and biochemical composition of plant materials (Stenberg et al. 2004; Stuth et al. 2003; Thuriès et al. 2005), composted materials (Albrecht et al. 2009; Michel et al. 2006; Vergnoux et al. 2009) and animal manures (Malley et al. 2002; Reeves and Van Kessel 2002; Sørensen et al. 2007), and to predict the C and N mineralization of plant materials (Borgen et al. 2010; Bruun et al. 2005; Shepherd et al. 2005) or the mineralization of soil organic matter (Thomsen et al. 2009).

Near-infrared spectroscopy also made it possible to differentiate between different types of char in soil–char–organic matter mixtures (Figure 1.21). Therefore, IR spectroscopic techniques even have advantages over biomarker analysis, as differentiation between different chars is difficult with classical methods (Hammes et al. 2007).

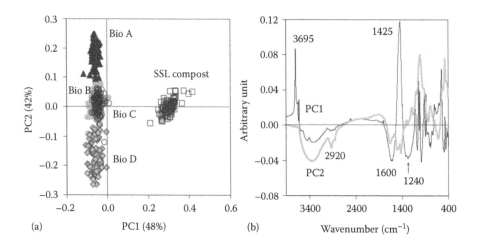

Figure 1.19   (a) Scores plot and (b) loadings spectra (PC1 and PC2) of a PCA based on spectral characteristics of five different composting processes. Bio, biowaste; SSL, sewage sludge. (From Smidt, E., Böhm, K., and Schwanninger, M., Ch 19 in *Fourier Transforms: New Analytical Approaches and FTIR Strategies*, G. S. Nikolić, ed., InTech, Rijeka, 2011. With permission.)

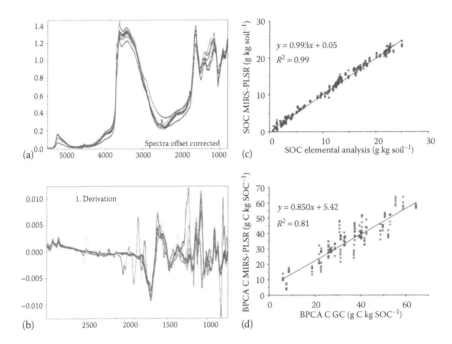

Figure 1.20   Principle of soil organic carbon (SOC) and biochar (BPCA) quantification based on mid-infrared reflectance spectra (a) followed by first derivation (b) and partial least-squares correlation with reference methods such as elemental analysis (c) or benzenepolycarboxylic acid (BPCA) analysis (d). (From Bornemann, L., Welp, G., Brodowski, S., Rodionov, A., and Amelung, W., *Org Geochem*, 39, 1537–1544, 2008. With permission.)

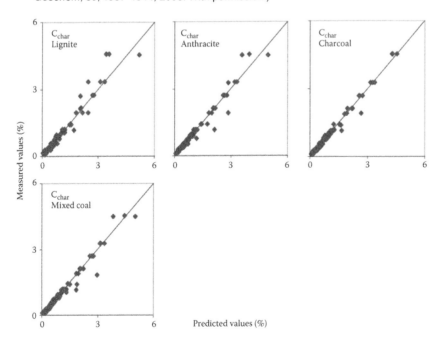

Figure 1.21   Partial least-squares calibration of near-infrared (NIR) spectra for different chars in soil–char–organic matter mixtures (*N* = 108). The 1:1 line is indicated on each figure. (From Michel, K., Terhoeven-Urselmans, T., Nietschke, R., Stefan, P., and Ludwig, B., *J Plant Nutr Soil Sci*, 172, 63–70, 2009. With permission.)

1.4 BIOCHAR SYSTEMS: DO THEY REALLY MAKE SENSE?

1.4.1 Economy versus Ecology

1.4.1.1 General Overview

Already Sombroek (1966) questioned whether it was economically feasible to create and cultivate terra preta today. In the meantime, the idea of creating terra preta-like soils (Terra Preta Nova, Biochar Technologies) as carbon stores and sinks for intensive cultivation is hotly debated (Glaser et al. 2002; Glaser 2007; Lehmann and Joseph 2009). Despite the current biochar hype, the question is whether this makes sense from both ecological and economic points of view.

The fascinating thing about terra preta has to do with the fact that it occurs in an environment where notorious land degradation and infertile soils occur. Therefore, in the humid tropics, the ecological and economic advantages of terra preta are obvious when compared with the surrounding landscape dominated by ferralsols (Glaser et al. 2001b). However, if we now try to create terra preta all over the world, the situation may be different, especially in regions where fertile soils occur, such as the chernozem regions in the Russian steppe, in Central Europe, the American prairie and the Argentinian pampa. There, ecological and economic effects of implementing terra preta technologies might be limited due to the naturally high soil fertility, which would be hard to improve. In other regions, resources of biomass and nutrients for terra preta formation might be limited. Therefore, terra preta technologies should be tested in regions with degraded soils or in reclamation areas with highest priority. From an ecological point of view, terra preta has the potential to improve or maintain ecosystem services such as crop production, nutrient and water retention and to be a habitat for soil organisms (Glaser et al. 2002).

Biochar application to soil can increase crop yields (Glaser et al. 2002). Tremendous yield increases were especially observed in degraded or low-fertility soils rather than already fertile soils (Glaser et al. 2002). All over the world, a mean crop production increase of about 10% was observed when using 10–100 Mg ha⁻¹ biochar in agricultural systems (Jeffery et al. 2011). Crop yield increases were higher when additional nutrients were added or when biochar was made from nutrient-rich material such as poultry litter (Jeffery et al. 2011). However, nutrient supply, pH and other soil properties alone were not always sufficient to fully explain the observed positive or negative effects of biochar on yields. It is interesting to note that no single biochar application rate exhibited a statistically significant negative effect on the crops from the range of soils, feedstock and application rates compared (Jeffery et al. 2011). No comprehensive studies of biochar effects on temperate soils are available (Atkinson et al. 2010).

1.4.1.2 Nutrient Retention

As biochar generally has a low nutrient content, its nutrient retention capacity is of greater interest. The principal nutrient retention mechanisms, such as pores, surface adsorption, cationic and anionic interaction, are determined by the physical and chemical structure of biochar.

Although fresh biochar has only a low number of functional groups, such as carboxylic acid (Figure 1.15), higher cation retention was observed when mixing soil with biochar (Glaser et al. 2002). The underlying mechanism for this observation is still unclear. Nevertheless, cation exchange capacity (CEC) of biochars can be increased by chemical (e.g., spraying with oxidizing acids during biochar production) or biological aging (e.g., composting of biochar). Biochar in terra preta was exposed to, on average, 2000 years of biological aging, significantly increasing its surface reactivity (Figure 1.14). The higher cation exchange capacity of terra preta is partly a "simple" pH effect, as it is known that variable (pH-dependent) cation exchange sites increase with increasing pH, and terra preta has a higher pH compared with surrounding soils (Figure 1.22a). However, the potential CEC is also increased in terra preta (Figure 1.22b), corroborating the fact that CEC of SOM can be increased

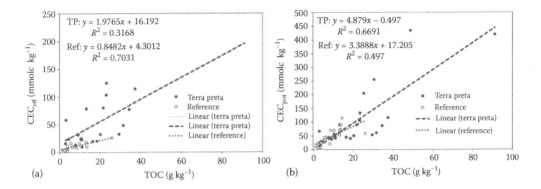

Figure 1.22   Effective (at soil pH, a) and potential (at pH 7, b) cation exchange capacity (CEC) of terra preta and adjacent sites in central Amazonia. (Data from Glaser, B., *Eigenschaften und Stabilität des Humuskörpers der Indianerschwarzerden Amazoniens*, Bayreuther Bodenkundliche Berichte 68, Bayreuther, 1999. With permission.)

when biochar is present. However, this effect is only of minor importance, despite the fact that about 50% of biochar is organominerally complexed in the upper 50 cm of terra preta (Figure 1.14).

The nutrient retention of biochar systems can be further increased by higher crop production. Steiner et al. (2008) found a 60%–80% higher total N retention in the ecosystem (plant and soil) when organic amendments (biochar and compost) were used compared with pure mineral fertilizer.

One important process in this retention was found to be the recycling of N taken up by the crop. In another study, biochar addition did not reduce ammonium, nitrate, and phosphate leaching compared with mineral and organic fertilizers, but it reduced nitrification (Schulz and Glaser 2012).

1.4.1.3 Water Retention

Biochar has a porous physical structure, which can absorb and retain water, although its chemical structure, being dominated by condensed aromatic moieties, suggests hydrophobicity (Figure 1.15). The water retention of terra preta was 18% higher compared with adjacent soils (Glaser et al. 2002): addition of 20 Mg ha^{-1} biochar to a sandy soil in northeast Germany increased water-holding capacity by 100% (Liu et al. 2012). Major et al. (2010) suggested that, due to the physical characteristics of biochar, there will be changes in soil pore size distribution, and this could alter percolation patterns, residence time, and flow paths of the soil solution. In a field trial in northeast Germany, application of 5–20 Mg ha^{-1} biochar together with 30 Mg ha^{-1} compost significantly increased plant-available water, during both wet and dry conditions, when compared with the pure compost treatment or the control site, which did not receive any organic amendment (Figure 1.23). This result was quite surprising, as it was anticipated that the fine pores of biochar would retain water that was not plant available, which obviously was not the case (Figure 1.23).

1.4.2 Life Cycle Assessment

1.4.2.1 Introduction

Life cycle assessment (LCA) is a methodology to evaluate the environmental impact associated with a product, process, or activity throughout its full life cycle (materials, manufacture, use, end of life) by quantifying energy, resources, and emissions and assessing their impact on the global environment. LCA has been standardized by the International Organization for Standardization (ISO 14040). Because of its "cradle-to-grave" approach and transparent methodology, LCA is an appropriate tool for estimating the energy and climate change impacts of biochar systems. For instance,

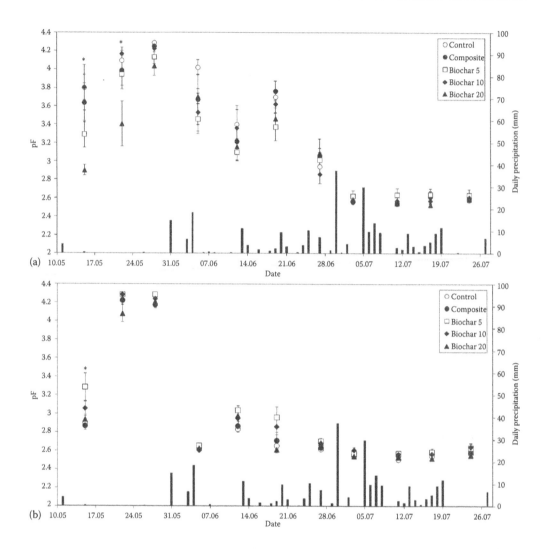

Figure 1.23 Plant-available water (pF) in 10 cm (a) and 30 cm soil depth (b) in a sandy soil in northeast Germany to which 30 Mg ha^{-1} compost and 5–20 Mg ha^{-1} biochar were added. Please note that the permanent wilting point is at pF 4.2. (From Bromm, T., Auswirkungen von Biochar-Applikationen auf den Bodenwasserhaushalt eines Sandbodens unter Freilandbedingungen. BSc Thesis, Martin-Luther-University Halle-Wittenberg, Halle (Saale), 2012. With permission.)

LCA can be used to estimate the full life cycle energy, GHG emissions balance, and economic feasibility of biochar made from different feedstock.

Attributional LCA (ALCA) focuses on describing the environmentally relevant physical flows to and from a product or process, while consequential LCA (CLCA) describes how relevant environmental flows will change in response to possible decisions (Table 1.4). These two approaches aim to answer different questions, and failure to distinguish them can result in the wrong method being applied, a mixture of the two approaches within a single assessment, or misinterpretation of the results (Brander et al. 2009).

1.4.2.2 Economic Assessment

Currently, the economics of biochar production relative to composting is uncertain, but clearly advantageous when pyrolysis feedstock comprises waste material (Roberts et al. 2010). It is possible

Table 1.4 Characteristics of Attributional (ALCA) versus Consequential Life Cycle Assessment (CLCA)

Attributional LCA	Consequential LCA
Include full life cycle	Include affected processes
Use average data	Use data reflecting expected effects
Allocate in proportion to (e.g., economic value)	Avoid allocation through system expansion

that biochar created from low-value feedstock could add a high value to low-quality compost, or reduce the costs of compost production. Alternatively, biochar from waste feedstock may be used in the same manner as the waste, or applied independently to soil or within a crop rotation. There are a number of anticipated policy developments that would be relevant, including some mechanism for incentivizing better carbon balance in food production. The adoption of carbon labeling by such companies based on detailed life cycle analysis of the production chain has been an important demonstration of this.

The primary costs of biochar production are buying and maintaining the pyrolysis facility and collection of feedstock. Transport of feedstock and biochar and application of biochar contribute little to the total economic costs (Meyer et al. 2011). Biochar production costs per ton of biochar varied between $51 for yard waste and $386 for retort charcoal (Meyer et al. 2011).

1.4.2.3 C Sequestration Potential

Biochar is assumed to be stable in the environment. The stability of biochar carbon in soils makes it a highly promising tool for climate change mitigation. However, mean residence times varying from centennial to millennial timescales have been reported (Figure 1.24). This discrepancy might be due to the facts that (i) different technology produces biochar with different stability and (ii) individual biochars are not homogeneous with respect to degradation but contain both labile and stable carbon. C sequestration potential could be calculated as the amount of biochar carbon that is expected to remain stable after 100 years (BC + 100). As this is very difficult to determine experimentally for individual biochars, more simple methods to estimate biochar stability (BC + 100) are necessary. As shown in Figure 1.24, the molar ratio of H/Corg varies significantly with the relative stability of biochar. Therefore, by means of the molar H/Corg ratio of a given biochar, the amount of stable biochar C can be determined, which can be used for C offset payments.

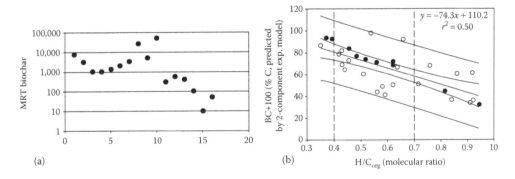

Figure 1.24 (a) Mean residence time (MRT) of biochar. (b) correlation between molecular ratio of H/Corg and the fraction of biochar which is more stable than 100 years measured in two incubation studies. (Data from Budai, A., Zimmerman, A.R., Cowie, A.L. et al., *Int Biochar Initiat*, 20, 1–20, 2013. With permission.)

1.4.2.4 Avoided Greenhouse Gas Emissions

In addition to the assessment of the true carbon sequestration potential, several indirect effects of biochar application have to be taken into account, such as fertilizer use, N_2O and CH_4 emissions, change in soil organic carbon (SOC), and increased productivity. Libra et al. (2011) reviewed the effects of biochar additions to different agroecosystems on greenhouse gas emissions. They found a reduction in N_2O release after biochar addition in seven out of nine reported studies. In addition, they reported an exponential decrease of N_2O emission from soil with increasing biochar addition (ng N_2O-N kg^{-1} h^{-1} = $206.19e^{-0.122\,g}$ biochar/100 g soil, $R^2 = 0.9705$).

Little is known about the effect of biochar on CH_4 emission. Zhang et al. (2010) reported 34%–41% increased CH_4-C emissions when paddy soils were amended with biochar at 40 Mg ha^{-1}, while N_2O-N emissions were reduced by 40%–51% and by 21%–28% in biochar-amended soils with and without N fertilization, respectively. Using biochar, the N_2O-N emission could be reduced from 4.2 to 1.3 g kg^{-1} N.

1.4.2.5 Improved Fertilizer Use Efficiency

Besides improved crop production, it is anticipated that biochar reduces nutrient leaching and, thus, improves fertilizer use efficiency (Glaser et al. 2002). However, a meta-analysis of biochar systems across the tropics and subtropics showed no additional fertilizer effect on crop productivity independent of the type of fertilizer used (Jeffery et al. 2011). On the other hand, Schulz and Glaser (2012) showed that crop production could be significantly increased when biochar was combined with organic fertilizer (compost) compared with pure biochar, pure mineral fertilizer, and biochar combined with mineral fertilizer. Also, Steiner et al. (2008) showed improved N retention in the soil–plant system when biochar was used with organic fertilizer compared with mineral fertilizer. Economic benefits strongly depend on mineral fertilizer and crop prices.

1.4.2.6 Summary and Conclusions

In summary, several biomass pyrolysis systems with biochar returned to soil have potential for C sequestration, GHG emission reductions, renewable energy generation, and economic viability. Careful feedstock selection is required to avoid unintended consequences such as net GHG emissions or consuming more energy than is generated, and also to ensure economic and environmental sustainability throughout the process life cycle. Waste biomass streams such as yard waste have the greatest potential to be economically viable while still being net energy positive and reducing GHG emissions. Agricultural residues such as corn stover have high yields of energy generation and GHG reductions, and have moderate potential to be profitable, depending on the value of C offsets and feedstock collection costs. If energy crops such as switch grass are grown on land diverted from annual crops, indirect land-use-change impacts could mean that more GHG are actually emitted than sequestered. Even if switch grass is grown on marginal lands, the economics for switch grass biochar are unfavorable. The primary barriers to the economic viability of pyrolysis–biochar systems are the pyrolysis process and the feedstock production costs. A comprehensive assessment of the technical, economic, and environmental strengths and weaknesses of biochar production technologies is, unfortunately, still not possible on the basis of the available scientific peer-reviewed literature. This is at least partly explainable by the fact that the production of biomass-based chars for the improvement of agricultural soils is still a relatively new topic of scientific interest. It will be necessary for both the public and the private sector to carry out and publish research on the indicated knowledge gaps in order to support project developers, technology developers, and policy makers with a comprehensive and detailed picture of the different options to produce biochar for soil improvement and climate change mitigation (Meyer et al. 2011).

The value arising from the agronomic impact of biochar after its application to soil is one value stream on which considerable attention is focused. However, biochar amendments offer the potential for carbon-neutral or carbon-negative food production and concomitant development of brand value and an income stream from acquisition of carbon certificates traded on the voluntary carbon market. The ecoregion Kaindorf in Austria has successfully implemented such a system. In addition, when combined with plant nutrients from waste materials, such as green wastes or slurries, additional benefit arises from saving money on buying mineral fertilizers such as NPK and by reducing resource allocation for water purification, for example, when nitrate is leached into groundwater after improper slurry application to agricultural fields. This is predicted to increase in the near future due to rapid growth in biogas production followed by the disposal of huge amounts of biogas slurry.

1.5 BIOCHAR SYSTEMS: DO WE UNDERSTAND THEM AT ALL?

1.5.1 Lack of Basic Understanding of Biochar Reactions in the Environment

Biochar addition to agricultural soils can potentially enhance their fertility and is compatible with sustainable agriculture, in particular when the tremendously porous biochar matrix is soaked in, or co-applied with, nutrient-rich wastes such as slurries or digestate from biogas production. Furthermore, it is claimed that biochar reduces erosion, nutrient leaching, and GHG emissions, and binds toxic agents such as heavy metals or organic pollutants. Several ideas for C sequestration and GHG mitigation options are in their infancy but developing rapidly. For instance, it is suggested that composted biochar could substitute for peat substrates in horticulture, which would take the pressure off bogs, leaving this giant C pool untouched instead of being mineralized and emitted as CO_2. Further ideas are to develop biochar into a nutrient-loaded carbon-based slow-release fertilizer, or to use biochar as an animal feed supplement for detoxification of food-chain pollutants in animal feeding, which automatically creates C-based fertilizer slurry.

However, all these ideas lack proper detailed, mechanistic understanding to evaluate and bring forward their full potential. Moreover, biochar effects in different ecosystems or in the food chain, particularly under different climatic conditions, soil types, and land management systems, have been assessed poorly. The major limitations of current biochar research are: (i) published research is almost exclusively focused on (sub)tropical regions; (ii) available data often only relate to initial (short-term) biochar effects; (iii) published results (e.g., life cycle analyses) are based on theoretical concepts and are not proven experimentally; and, most importantly, (iv) the mechanisms of physical, chemical, and biological effects of biochar are poorly understood.

Theoretical concepts of biochar mechanisms are summarized in several recent review papers (Atkinson et al. 2010; Joseph et al. 2010; Knicker 2011; Kookana 2010; Libra et al. 2011; McHenry 2009), but proof of these concepts is still lacking. The overall conclusion drawn in these reviews is that biochar research is an emerging field, still in its infancy, and that substantially more data, in particular basic mechanistic understanding, is required before robust predictions can be made about how biochar acts mechanistically, especially in temperate regions. Despite these basic knowledge gaps, public interest is enormous due to the economic and political potentials of biochar technologies, as demonstrated by the rapidly emerging companies and pending patents.

A proper process understanding of the underlying principles will enable production of biochars with desired properties and reliable prediction of the effects of biochar once applied. Therefore, there is an urgent need for inter- and transdisciplinary biochar research in order to: (i) develop and advance appropriate analytical methods for biochar characterization (material properties); (ii) identify and quantify processes of biochar physicochemical interaction with soils, plants, and water (physicochemical interaction); (iii) identify and quantify processes of biochar biological interaction (biological processes); and (iv) identify potential threats of biochar application to the environment, to animals, and to humans (environmental toxicology; Figure 1.25). Such studies should aim at revealing *mechanisms and processes* behind biochar effects and interactions with the environment, and thus enable generalization and prediction of biochar benefits and risks in integrative approaches.

1.5.2 Material Properties

As biochar is a porous and hydrophobic material of low density, sophisticated analytical procedures (both field and laboratory based) are required for achieving mechanistic understanding. Routine methods, for example, in soil science, are often not appropriate for biochar analysis due to the peculiar material properties of biochar. This is true both for standard methods such as the determination of nutrient retention (e.g., CEC, microbial biomass, polycyclic aromatic hydrocarbon (PAH) extraction) and for sophisticated approaches such as the detection of biochar decomposition in the field due to the expected high stability of biochar. From a material properties point of view, appropriate variables and thresholds need to be defined for a proper differentiation of biochar from non-biochar materials.

Since recent discussions propose the mixture of biochars with organic fertilizers such as compost, biochar (organic) fertilizer interactions need to be identified, while pure biochar experiments should be exceptions for separate process/mechanism identification. This will make biochar experiments methodically more complicated with respect to interpretation of obtained results, requiring more sophisticated methodical approaches.

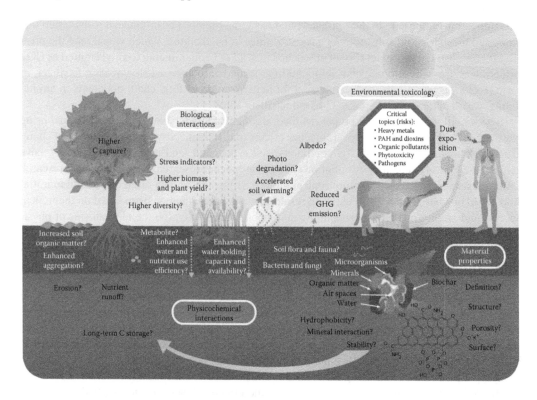

Figure 1.25   (See color insert) Most important unknowns on the interaction of biochar within ecosystem components. An integrative approach covers the whole range from the molecular to the ecosystem level and could be divided into four thematic modules: material properties, physicochemical interaction with soil and water, biological interaction, and environmental toxicology (Layout by MiSoo Kim).

Adaptation of existing and development of new analytical methods, in particular rapid nondestructive and noninvasive methods, require collaboration of interdisciplinary teams to achieve rapid progress in this challenging topic. In addition, a standard set of biochars from a ^{13}C- and ^{15}N-labeled stockpile should be produced, jointly characterized, and used among cooperating biochar groups for laboratory and field studies. Such an approach does not only increase each group's efficiency (synergism) but induces scientific communication between different disciplines.

Long-term studies using isotope labeling provide a unique tool for tracing biochar's biotic and abiotic transformations and the fate of its metabolites in the environment, including soil, water, plants, air, and animals.

1.5.3 Physicochemical Interactions with Soil, Nutrients, and Water

There is a discrepancy or convergence between material properties expected from the chemical structure of biochar (recalcitrance, hydrophobicity) and the mechanistic understanding of claimed positive effects for ecosystem services such as enhanced water storage capacity and soil stabilization. The same is true for the reported high surface area of biochars, related especially to the dominance of nanopores, which cannot store plant-available water. Therefore, soil hydraulic properties of various biochars under laboratory and field conditions need to be investigated in combination with 3-D imaging of the porous architecture of biochars. In addition, increased storage time of biochar in soil may result in steric changes of aromatic regions, release of char-specific metabolites, and oxidation in hot spots of biochar particles. Another important issue is the theoretical recalcitrance due to the polyaromatic backbone, and thus nonreactivity of biochar in relation to a potential organomineral stabilization and reactivity. This aspect is also related to water infiltration, surface runoff, and wind erosion. A further important physical aspect is interaction with solar radiation (albedo effect), which could contribute to warming the soil and the atmosphere near the soil, potentially resulting in accelerated SOM degradation due to enhanced soil biological activity.

1.5.4 Biological Interaction

Key drivers and products of biochar transformations should be identified and quantified ("metabolic tracing"). Position-specific labeling is a unique tool to gain insights into submolecular transformation pathways of low-molecular-weight organic substances in soil (Dijkstra et al. 2011). As all high-molecular weight organic substances pass the stage of low-molecular-weight organic substances (LMWOS) during their decomposition, the kinetics, metabolization pathway, and formation of new SOM compounds are determining steps in the SOM cycle. LMWOS availability, microbial activity, and microbial community structure are expected to change after biochar amendment to soils, and thus a change in the microbial metabolization pathways can be expected. Stabilization of LMWOS on the sorbents of the biochar, changing metabolization pathways as well as new microbial transformation products, can be detected. Thus, position-specific labeling is a very sensitive tool to detect changes in C transformation pathways and C fluxes induced by biochar amendment. For this purpose, multiply and highly labeled biochars will be needed (e.g., ^{14}C, ^{13}C, and ^{15}N) due to the expected complexity of the processes involved, the necessary measurement resolution and sensitivity (large background C pool), and the expected low turnover rate, respectively.

Again, complications are expected due to the fact that decomposition rates of biochar cannot be simply calculated according to the concept of biologically active time (BAT), which was elaborated for decomposition of plant litter, being microbially much more easily available compared with biochar. Additionally, other factors, such as high variation of environmental conditions and the presence of soil animals, may contribute to faster mineralization rates of biochar-C under field conditions as compared with controlled laboratory incubation conditions. Furthermore, biochar may consist of various "pools" with different stability. Therefore, for more realistic determination of biochar stability, long-term field studies are required including modern sophisticated analytical tools such as (stable) isotope labeling combined with compound-specific stable isotope analysis, which, to date, is a challenge under field conditions, again requiring an interdisciplinary team of scientists.

Within both methanotrophy (CH_4 oxidation) and methanogenesis (CH_4 production), there is considerable lack of experimental evidence concerning the effects of biochar, as well as lack of mechanistic understanding. A key milestone of an improved process understanding will be achieved if a method can be applied that allows the simultaneous quantification of gross CH_4 consumption and production in soils. This may be important, for example, when the char itself gives off CH_4, but the soil consumes it. Such a method has been proposed by von Fischer and Hedin (2002), but has never been broadly applied. In addition, it would need further adaptation and optimization for biochar research, involving an interdisciplinary approach, including the contribution of soil fauna (Kammann et al. 2009). The cavity ring down spectroscopy (CRDS) methodology has only recently reached sufficient maturity to be taken from the lab into the field, probably enabling for the first time determination of the simultaneous gross CH_4 exchange between (biochar-amended) soils and the atmosphere.

Biochar C–N interactions and modification of soil gross N turnover processes are important for understanding and predicting long-term effects of biochar on soil fertility, GHG fluxes, and plant performance. However, soil N transformation processes in particular appear to be a "black box" to date (Clough and Condron 2010). Even the proportion of physicochemical and biological contributions to observed changes is not understood. Thus, the application of state-of-the-art ^{15}N labeling–techniques based on the pool dilution approach (Kirkham and Bartholomew 1954) and developed into a complex, sophisticated mathematical tool (Müller et al. 2007) could shed light on the unknown but essential fate of N in the presence of biochar. The interdisciplinary use of ^{15}N-labeled field plots as well as ^{15}N-chars could enable going one step further and allow the identification of organisms involved in the soil–biochar N turnover from the molecular/cellular level to soil food webs and plant N export.

The control of biochar degradation by fungi deserves further research attention in order to predict the potential of long-term C storage with biochar in different environments. A large knowledge gap exists with respect to the responses of communities of soil and root-inhabiting fungi to biochar additions: so far effects have only been documented for overall abundances. Here, stable isotope probing of soils amended with highly labeled biochars may offer a window into soil microbial and faunal food webs and abundance shifts.

Another important aspect concerns the influence of biochar on soil microbial diversity and on the interaction between soil microorganisms and plants, as well as plant resilience, stress, or both. Hardly any information is available on the ecological effects that biochar may have when incorporated into soils at the community level, that is, ranging from plant ecophysiological performance (e.g., plant strategies to cope with stressors such as water stress, salinity, heavy metal toxicity or herbivores), to the plant/faunal community composition and diversity, up to ecosystem functioning. Besides ecological investigations in biochar field experiments, another possibility to peer into this "black box" of long-term biochar effects may be the investigation of old charcoal-making sites and old, overgrown kiln plates that can be found in German forests (e.g., "Haubergswirtschaft"). Charcoal making ("Köhlerwirtschaft") often has several hundreds of years of tradition in Germany.

1.5.5 Environmental Toxicology

On the one hand, biochar production might produce toxic compounds such as PAHs or dioxins. On the other hand, biochar can also reduce the bioavailability and/or mobility of these and other compounds (e.g., pathogens) due to its physical (nanopores) or chemical structure (condensed aromatic backbone, functional groups). Knowledge of the behavior and fate of pathogens, particularly of microbial transport and retention in substrates and soils, is required for the safe use of new approaches in agriculture in addition to soil application only, such as the use of biochars in animal housing filters, as a slurry-fermenting matrix in staples, or as animal food additives. A primary goal is to assess whether potentially harmful microorganisms from biochar additions (compost, manure) are able to pass the layer of biochar-amended soils and reach deeper soil layers or groundwater resources.

Another important issue is to test whether biochar itself might be a toxic compound to plants (e.g., by uptake of dissolved compounds), animals (e.g., by ingestion), and humans (e.g., by inhalation after wind drift).

According to the European Community Regulation on chemicals and their safe use (REACH), safety information on biochar must deal with toxic compounds (i.e., heavy metals, PAHs, dioxins) and their effects on plant, animal, and human health (e.g., genotoxicity) and effects on GHG emissions (i.e., N_2O, CH_4, CO_2). One problem in this regard is, however, that the available standard methods do not allow a quantitative extraction of available PAHs due to the strong sorption characteristics of most biochars, demanding method adaptation. Moreover, it must be evaluated how much of the total toxic compound load that may be present on a biochar will effectively be bioavailable for plants or (soil) animals.

1.5.6 The Urgent Need for Further Inter- and Transdisciplinary Biochar Research

From a scientific point of view, it is imperative to address these mentioned biochar unknowns as soon as possible, and in coordinated and interdisciplinary projects, before biochar technologies can be disseminated to a great extent, which is already ongoing (biochar producers and pending terra preta patent). This coordinated biochar research should cover the following criteria:

1. Upscaling from the molecular (detail) to the ecosystem level (complexity) using standardized, common state-of-the-art as well as newly developed experimental tools with the same set of different, well-characterized biochars.
2. Process identification and quantification (e.g., by isotope labeling such as ^{14}C, ^{13}C, and ^{15}N) combined with compound-specific and/or position-specific isotope analysis with low detection limits, as biochar turnover might be low. For this purpose, new innovative methods need to be integrated, such as liquid chromatography (LC) linked via an oxidation device (O) to an isotope ratio mass spectrometer (LC-O-IRMS), and integration of nondestructive and noninvasive methods directly in the field, such as FTIR and CRDS, x-ray (μ-XRT) and nuclear magnetic resonance tomography (MRT), atomic force microscopy (AFM), and (nano) secondary ion mass spectrometry (SIMS).
3. Exploring long-term mechanistic effects under real (field) conditions. For this purpose, a range of long-term field trials are necessary and the advantage of already existing biochar field experiments covering a range of biochar application amounts, different biochar/fertilizer combinations, and climatic gradients should be used.

REFERENCES

Albrecht, R., R. Joffre, J. Petit, G. Terrom, and C. Perissol. 2009. Calibration of chemical and biological changes in cocomposting of biowastes using near-infrared spectroscopy. *Environ Sci Technol* 43:804–811.

Amelung, W. 2001. Methods using amino sugars as markers for microbial residues in soil. In R. Lal, J. M. Kimble, R. F. Follett, and B. A. Stewart (eds), *Assessment Methods for Soil Carbon*, pp. 233–272. Boca Raton (FL): Lewis.

Amundson, R., A. T. Austin, E. A. G. Schuur et al. 2003. Global patterns of the isotopic composition of soil and plant nitrogen. *Global Biogeochem Cycles* 17:1031.

Andreeva, D. B., K. Leiber, B. Glaser et al. 2011. Genesis and properties of black soils in Buryatia, southeastern Siberia, Russia. *Quatern Int* 243:313–326.

Arroyo-Kalin, M., E. G. Neves, and W. I. Woods. 2009. Anthropogenic dark earths of the Central Amazon region: Remarks on their evolution and polygenetic composition. In W. I. Woods, W. G. Teixeira, J. Lehmann, C. Steiner, A. M. G. A. Winkler-Prins, and L. Rebellato (eds), *Amazonian Dark Earths: Wim Sombroek's Vision*, pp. 99–125. Berlin: Springer.

Atkinson, C. J., J. D. Fitzgerald, and N. A. Hipps. 2010. Potential mechanisms for achieving agricultural benefits from biochar application to temperate soils: A review. *Plant Soil* 337:1–18.

Barbosa de Faria, J. 1944. A cerâmica da tribo Uaboí dos rios Trombetas e Jamundá: Contribuição para o estudo da arqueologia pré-histórica do Baixo Amazonas. Anais III, *9° Congresso Brasileiro de Geografía*, vol. 3, pp. 141–165. Rio de Janiero: Conselho Nacional de Geografia.

Barthès, B. G., D. Brunet, H. Ferrer, J. L. Chotte, and C. Feller. 2006. Determination of total carbon and nitrogen content in a range of tropical soils using near infrared spectroscopy: Influence of replication and sample grinding and drying. *J Near Infrared Spec* 14:341–348.

Bethell, P. H., J. Ottaway, L. J. Goad, and R. P. Evershed. 1994. The study of molecular markers of human activity: The use of coprostanol in the soil as an indicator of human faecal material. *J Archaeol Sci* 21:619–632.

Birk, J. J., W. G. Teixeira, E. G. Neves, and B. Glaser. 2011. Faeces deposition on Amazonian Anthrosols as assessed from 5[β]-stanols. *J Archaeol Sci* 38:1209–1220.

Borgen, S. K., L. Molstad, S. Bruun, T. A. Breland, L. R. Bakken, and M. A. Bleken. 2010. Estimation of plant litter pools and decomposition-related parameters in a mechanistic model. *Plant Soil* 338:205–222.

Bornemann, L., G. Welp, S. Brodowski, A. Rodionov, and W. Amelung. 2008. Rapid assessment of black carbon in soil organic matter using mid-infrared spectroscopy. *Org Geochem* 39:1537–1544.

Bourbonniere, R. A., S. L. Telford, L. A. Ziolkowski, J. Lee, M. S. Evans, and P. A. Meyers. 1997. Biogeochemical marker profiles in cores of dated sediments from large North American lakes. In R. P. Eganhouse (ed.), *Molecular Markers in Environmental Geochemistry*, ACS Symposium Series, vol. 671, pp. 133–150. Washington, DC: American Chemical Society.

Brander, M., R. Tipper, C. Hutchison, and G. Davis. 2009. Consequential and attributional approaches to LCA: A guide to policy makers with specific reference to greenhouse gas LCA of biofuels. Technical Paper TP-090403-A. Ecometrica (http://ecometrica.com).

Brodowski, S., A. Rodionov, L. Haumaier, B. Glaser, and W. Amelung. 2005. Black carbon assessment using benzenepolycarboxylic acids: Revised method. *Org Geochem* 36:1299–1310.

Bromm, T. 2012. Auswirkungen von Biochar-Applikationen auf den Bodenwasserhaushalt eines Sandbodens unter Freilandbedingungen. BSc Thesis, Martin-Luther-University Halle-Wittenberg, Halle (Saale).

Brown, C. B. 1876. *Canoe and Camp Life in British Guiana*. London: Edward Stanford.

Brown, C. B. and W. Lidstone. 1878. *Fifteen Thousand Miles on the Amazon and Its Tributaries*. London: Edward Stanford.

Bruun, S., B. Stenberg, T. A. Breland et al. 2005. Empirical predictions of plant material C and N mineralization patterns from near infrared spectroscopy, stepwise chemical digestion and C/N ratios. *Soil Biol Biochem* 37:2283–2296.

Budai, A., A. R. Zimmerman, A. L. Cowie et al. 2013. Biochar carbon stability test method: An assessment of methods to determine biochar carbon stability. *Int Biochar Initiat* 20:1–20.

Bull, I. D., P. P. Betancourt, and R. P. Evershed. 2001. An organic geochemical investigation of the practice of manuring at a Minoan site on Pseira Island, Crete. *Geoarchaeology* 16:223–242.

Bull, I. D., M. J. Lockheart, M. M. Elhmmali, D. J. Roberts, and R. P. Evershed. 2002. The origin of faeces by means of biomarker detection. *Environ Int* 27:647–654.

Bull, I. D., I. A. Simpson, S. J. Dockrill, and R. P. Evershed. 1999. Organic geochemical evidence for the origin of ancient anthropogenic soil deposits at Tofts Ness, Sanday, Orkney. *Org Geochem* 30:535–556.

Camargo, F. 1941. *Estudo de alguns perfis do solos coletados em diversas regiões da Hiléia*. Belém: Instituto Agronômico do Norte.

Clough, T. J. and L. M. Condron. 2010. Biochar and the nitrogen cycle: Introduction. *J Environ Qual* 39:1218–1223.

Collister, J. W., E. Lichtfouse, G. Hieshima, and J. M. Hayes. 1994. Partial resolution of sources of n-alkanes in the saline portion of the Parachute Creek Member, Green River Formation (Piceance Creek Basin, Colorado). *Org Geochem* 21:645–659.

Costa, M. L. and D. C. Kern. 1999. Geochemical signatures of tropical soils with archaeological black earth in the Amazon, Brazil. *J Geochem Explor* 66:369–385.

Czimczik, C. I., C. M. Preston, M. W. I. Schmidt, and E. D. Schulze. 2003. How surface fire in Siberian Scots pine forests affects soil organic carbon in the forest floor: Stocks, molecular structure, and conversion to black carbon (charcoal). *Global Biogeochem Cycles* 17, doi: 10.1029/2002GB001956.

Dai, X., T. W. Boutton, B. Glaser, R. J. Ansley, and W. Zech. 2005. Black carbon in a temperate mixed-grass savanna. *Soil Biol Biochem* 37:1879–1881.

de Carvajal, F. G. 1934. Discovery of the Orellana river. In H. D. Heaton (ed.), *The Discovery of the Amazon According to the Account of Friar Gaspar de Carvajal and Other Documents with an Introduction by José Toribio Medina*, vol. 17, pp. 167–242. New York: American Geographical Society Publication.

Demanèche, S., L. Jocteur-Monrozier, H. Quiquampoix, and P. Simonet. 2001. Evaluation of biological and physical protection against nuclease degradation of clay-bound plasmid DNA. *Appl Environ Microbiol* 67:293–299.

Denevan, W. M. 1996. A bluff model of riverine settlement in prehistoric Amazonia. *Ann Assoc Am Geogr* 86:654–681.

Dijkstra, P., J. C. Blankinship, P. C. Selmants et al. 2011. Probing carbon flux patterns through soil microbial metabolic networks using parallel position-specific tracer labeling. *Soil Biol Biochem* 43:126–132.

Du, C. and J. Zhou. 2009. Evaluation of soil fertility using infrared spectroscopy: A review. *Environ Chem Lett* 7:97–113.

Elhmmali, M. M., D. J. Roberts, and R. P. Evershed. 2000. Combined analysis of bile acids and sterols/stanols from riverine particulates to assess sewage discharges and other fecal sources. *Environ Sci Technol* 34:39–46.

Evershed, R. P. and P. H. Bethell. 1996. Application of multimolecular biomarker techniques to the identification of fecal material in archaeological soils and sediments. In M. V. Orna (ed.), *Archaeological Chemistry: Organic, Inorganic and Biochemical Analysis*, ACS Symposium Series, vol. 625, pp. 157–172. Washington, DC: American Chemical Society.

Evershed, R. P., P. H. Bethell, P. J. Reynolds, and N. J. Walsh. 1997. 5β-Stigmastanol and related 5β-stanols as biomarkers of manuring: Analysis of modern experimental material and assessment of the archaeological potential. *J Archaeol Sci* 24:485–495.

Factura, H., T. Bettendorf, C. Buzie, H. Pieplow, J. Reckin, and R. Otterpohl. 2010. Terra Preta sanitation: Re-discovered from an ancient Amazonian civilisation—Integrating sanitation, bio-waste management and agriculture. *Water Sci Technol* 61:2673–2679.

Falesi, I. C. 1967. O Estado Atual dos Conhecimentos Sôbre os Solos da Amazônia Brasileira. In H. Lent (ed.), *Atlas do simpósio sôbre a biota Amazônica*, vol. 1, pp. 151–168. Río de Janeiro: Conselho Nacional de Pesquisas.

Franco, E. C. 1962. As "Terras Pretas" do Planalto de Santarém. *Revista da Sociedade dos Agrônomos e Veterinários do Pará* 8:17–21.

Gill, F. L., R. J. Dewhurst, J. A. J. Dungait et al. 2010. Archaeol—A biomarker for foregut fermentation in modern and ancient herbivorous mammals? *Org Geochem* 41:467–472.

Gillespie, R., A. P. Hammond, K. M. Goh et al. 1992. AMS dating of a late quaternary tephra at Graham's Terrace, New Zealand. *Radiocarbon* 34:21–27.

Glaser, B. 1999. *Eigenschaften und Stabilität des Humuskörpers der Indianerschwarzerden Amazoniens*, p. 196. Bayreuther: Bayreuther Bodenkundliche Berichte 68.

Glaser, B. 2007. Prehistorically modified soils of central Amazonia: A model for sustainable agriculture in the twenty-first century. *Philos Trans R Soc B* 362:187–196.

Glaser, B. and W. Amelung. 2003. Pyrogenic carbon in native grassland soils along a climosequence in North America. *Global Biogeochem Cycles* 17:1064.

Glaser, B., E. Balashov, L. Haumaier, G. Guggenberger, and W. Zech. 2000. Black carbon in density fractions of anthropogenic soils of the Brazilian Amazon region. *Org Geochem* 31:669–678.

Glaser, B. and J. J. Birk. 2012. State of the scientific knowledge on properties and genesis of anthropogenic dark earths in Central Amazonia (terra preta de Índio). *Geochim Cosmochim Acta* 82:39–51.

Glaser, B., G. Guggenberger, L. Haumaier, and W. Zech. 2001a. Persistence of soil organic matter in archaeological soils (terra preta) of the Brazilian Amazon region. In R. M. Rees, B. C. Ball, C. D. Campbell, and C. A. Watson (eds), *Sustainable Management of Soil Organic Matter*, pp. 190–194. Wallingford: CAB International.

Glaser, B., G. Guggenberger, and W. Zech. 2004a. Identifying the pre-Columbian anthropogenic input on present soil properties of Amazonian dark earther (terra preta). In B. Glaser and W. I. Woods (eds), *Amazonian Dark Earths: Explorations in Space and Time*, pp. 145–158. Heidelberg: Springer.

Glaser, B., L. Haumaier, G. Guggenberger, and W. Zech. 1998. Black carbon in soils: The use of benzenecar-boxylic acids as specific markers. *Org Geochem* 29:811–819.

Glaser, B., L. Haumaier, G. Guggenberger, and W. Zech. 2001b. The "terra preta" phenomenon: A model for sustainable agriculture in the humid tropics. *Naturwissenschaften* 88:37–41.

Glaser, B., J. Lehmann, and W. Zech. 2002. Ameliorating physical and chemical properties of highly weathered soils in the tropics with charcoal—A review. *Biol Fert Soils* 35:219–230.

Glaser, B., M. B. Turrión, and K. Alef. 2004b. Amino sugars and muramic acid—Biomarkers for soil microbial community structure analysis. *Soil Biol Biochem* 36:399–407.

Glaser, B. and W. Zech. 2005. Reconstruction of climate and landscape changes in a high mountain lake catch-ment in the Gorkha Himal, Nepal during the Late Glacial and Holocene as deduced from radiocarbon and compound-specific stable isotope analysis of terrestrial, aquatic and microbial biomarkers. *Org Geochem* 36:1086–1098.

Glaser, B., W. Zech, and W. I. Woods. 2004c. History, current knowledge and future perspectives of geoecologi-cal research concerning the origin of Amazonian anthropogenic dark earths (terra preta). In B. Glaser and W. I. Woods (eds), *Amazonian Dark Earths: Explorations in Space and Time*, pp. 9–17. Heidelberg: Springer.

Gourou, P. 1949. Observações geograficas na Amazônia. *Rev Bras Geogr* 11:354–408.

Grimalt, J. O., P. Fernandez, J. M. Bayona, and J. Albaiges. 1990. Assessment of faecal sterols and ketones as indicators of urban sewage inputs to coastal waters. *Environ Sci Technol* 24:357–363.

Guggenberger, G., A. Rodionov, O. Shibistova et al. 2008. Storage and mobility of black carbon in permafrost soils of the forest tundra ecotone in northern Siberia. *Glob Change Biol* 14:1367–1381.

Hackl, E., S. Zechmeister-Boltenstern, L. Bodrossy, and A. Sessitsch. 2004. Comparison of diversities and compositions of bacterial populations inhabiting natural forest soils. *Appl Environ Microbiol* 70:5057–5065.

Hammes, K., M. W. I. Schmidt, R. J. Smernik et al. 2007. Comparison of quantification methods to measure fire-derived (black/elemental) carbon in soils and sediments using reference materials from soil, water, sediment and the atmosphere. *Global Biogeochem Cycles* 21, doi: 10.1029/2006GB002914.

Hammes, K., M. S. Torn, A. G. Lapenas, and M. W. I. Schmidt. 2008. Centennial black carbon turnover observed in a Russian steppe soil. *Biogeosciences* 5:1339–1350.

Hartt, C. F. 1872. Recent explorations in the valley of the Amazonas, with map. *Ann Rep Am Geogr Soc NY* 1870–71 3:231–252.

Hartt, C. F. 1885. Contribuições para a ethnologiado Valledo Amazonas. *Ach Museum Nacl Rio de Janeiro* 6:1–174.

Heckenberger, M. J., A. Kuikuro, U. T. Kuikuro et al. 2003. Amazonia 1492: Pristine forest or cultural park-land. *Science* 301:1710–1713.

Hedges, J. I., G. Eglinton, P. G. Hatcher et al. 2000. The molecularly uncharacterized component of nonliving organic matter in natural environments. *Org Geochem* 31:945–958.

Hilbert, P. P. 1968. *Archäologische Untersuchungen am mittleren Amazonas: Beiträge zur Vorgeschichte des südamerikanischen Tieflandes*. Marburger Studien zur Völkerkunde 1. Berlin: Dietrich Reimer Verlag.

Hoefs, M. J. L., W. I. C. Rijpstra, and J. S. S. Damste. 2002. The influence of oxic degradation on the sedimen-tary biomarker record I: Evidence from Madeira Abyssal Plain turbidities. *Geochim Cosmochim Acta* 66:2719–2735.

Hofreiter, M., H. N. Poinar, W. G. Spaulding et al. 2000. A molecular analysis of ground sloth diet through the last glaciation. *Mol Ecol* 9:1975–1984.

Jahn, B. R., R. Linker, S. K. Upadhyaya, A. Shaviv, D. C. Slaughter, and I. Shmulevich. 2006. Mid-infrared spectroscopic determination of soil nitrate content. *Biosys Eng* 94:505–515.

Jeffery, S., F. G. A. Verheijen, M. Van der Velde, and A. C. Bastos. 2011. A quantitative review of the effects of biochar application to soils on crop productivity using meta-analysis. *Agric Ecosys Environ* 144:174–187.

Joseph, S. D., M. Camps-Arbestain, A. Lin et al. 2010. An investigation into the reactions of biochar in soil. *Soil Res* 48:501–515.

Kaal, J., A. Martínez-Cortizas, P. Buurman, and F. Criado Boado. 2008. 8000 yr of black carbon accumulation in a colluvial soil from NW Spain. *Quaternary Res* 69:56–61.

Kammann, C., S. Hepp, K. Lenhart, and C. Müller. 2009. Stimulation of methane consumption by endogenous CH_4 production in aerobic grassland soil. *Soil Biol Biochem* 41:622–629.

Kämpf, N., W. I. Woods, W. Sombroek, D. C. Kern, and T. J. F. Cunha. 2003. Classification of Amazonian dark earths and ancient anthropic soils. In J. Lehmann, C. C. Kern, B. Glaser, and W. I. Woods (eds), *Amazonian Dark Earth: Origin, Properties, Management*, pp. 77–104. Dordrecht: Kluwer Academic.

Katzer, F. 1903. *Grundzüge der Geologie des Unteren Amazonasgebietes (des Staates Pará in Brasilien)*. Leipzig: Verlag von Max Weg.

Kim, J. S., S. Sparovek, R. M. Longo, W. J. De Melo, and D. E. Crowley. 2007. Bacterial diversity of terra preta and pristine forest soil from the western Amazon. *Soil Biol Biochem* 39:648–690.

Kirkham, D. and W. V. Bartholomew. 1954. Equations for following nutrient transformations in soil, utilizing tracer data. *Soil Sci Soc Am Proc* 18:33–34.

Knicker, H. 2011. Pyrogenic organic matter in soil: Its origin and occurrence, its chemistry and survival in soil environments. *Quaternary Int* 243:251–263.

Kookana, R. S. 2010. The role of biochar in modifying the environmental fate, bioavailability, and efficacy of pesticides in soils: A review. *Soil Res* 48:627–637.

Leeming, R., A. Ball, N. Ashbolt, and P. Nichols. 1996. Using faecal sterols from humans and animals to distinguish faecal pollution in receiving waters. *Water Res* 30:2893–2900.

Leeming, R., V. Latham, M. Rayner, and P. Nichols. 1997. Detecting and distinguishing sources of sewage pollution in Australian inland and coastal waters and sediments. In R. P. Eganhouse (ed.), *Molecular Markers in Environmental Geochemistry*, pp. 306–319. Washington, DC: American Chemical Society.

Lehmann, J. and S. Joseph. 2009. *Biochar for Environmental Management: Science and Technology*. London: Earthscan.

Libra, J. A., K. S. Ro, C. Kammann et al. 2011. Hydrothermal carbonization of biomass residuals: A comparative review of the chemistry, processes and applications of wet and dry pyrolysis. *Biofuels* 2:89–124.

Lima, H. N., C. E. R. Schaefer, J. W. V. Mello, R. J. Gilkes, and J. C. Ker. 2002. Pedogenesis and pre-Columbian land use of "terra preta anthrosols" ("Indian black earth") of western Amazonia. *Geoderma* 110:1–17.

Liu, J., H. Schulz, S. Brandl, H. Miethke, B. Huwe, and B. Glaser. 2012. Short-term effect of biochar and compost on soil fertility and water status of a dystric cambisol in NE Germany under field conditions. *J Plant Nutr Soil Sci* 175:698–707.

Lynch, D. H., R. P. Voroney, and P. R. Warman. 2006. Use of ^{13}C and ^{15}N natural abundance techniques to characterize carbon and nitrogen dynamics in composting and in compost-amended soils. *Soil Biol Biochem* 38:103–114.

Major, J., J. Lehmann, M. Rondon, and C. Goodale. 2010. Fate of soil-applied black carbon: Downward migration, leaching and soil respiration. *Glob Change Biol* 16:1366–1379.

Malley, D. F., L. Yesmin, and R. G. Eilers. 2002. Rapid analysis of hog manure and manure-amended soils using near-infrared spectroscopy. *Soil Sci Soc Am J* 66:1677–1686.

Marota, I., C. Basile, M. Ubaldi, and F. Rollo. 2002. DNA decay rate in papyri and human remains from Egyptian archaeological sites. *Am J Phys Anthropol* 117:310–318.

Marschner, B., S. Brodowski, A. Dreves et al. 2008. How relevant is recalcitrance for the stabilization of organic matter in soils? *J Plant Nutr Soil Sci* 171:91–110.

McHenry, M. P. 2009. Agricultural bio-char production, renewable energy generation and farm carbon sequestration in Western Australia: Certainty, uncertainty and risk. *Agric Ecosyst Environ* 129:1–7.

Meyer, E., M. Wiese, H. Bruchhaus, M. Claussen, and A. Klein. 2000. Extraction and amplification of authentic DNA from ancient human remains. *Forensic Sci Int* 113:87–90.

Meyer, S., B. Glaser, and P. Quicker. 2011. Technical, economical, and climate-related aspects of biochar production technologies: A literature review. *Environ Sci Technol* 45:9473–9483.

Michel, K., C. Bruns, T. Terhoeven-Urselmans, B. Kleikamp, and B. Ludwig. 2006. Determination of chemical and biological properties of composts using near infrared spectroscopy. *J Near Infrared Spec* 14:251–259.

Michel, K., T. Terhoeven-Urselmans, R. Nietschke, P. Stefan, and B. Ludwig. 2009. Use of near- and mid-infrared spectroscopy to distinguish carbon and nitrogen originating from char and forest-floor materials in soils. *J Plant Nutr Soil Sci* 172:63–70.

Müller, C., T. Rütting, J. Kattge, R. J. Laughlin, and R. J. Stevens. 2007. Estimation of parameters in complex ^{15}N tracing models by Monte Carlo sampling. *Soil Biol Biochem* 39:715–726.

Neves, E. G., J. B. Petersen, R. N. Bartone, and C. A. Da Silva. 2003. Historical and socio-cultural origins of Amazonian dark earth. In J. Lehmann, C. C. Kern, B. Glaser, and W. I. Woods (eds), *Amazonian Dark Earths: Origin, Properties, Management*, pp. 29–50. Dordrecht: Kluwer Academic.

Nguyen, B. T., J. Lehmann, J. Kinyangi, R. Smernik, S. J. Riha, and M. H. Engelhard. 2008. Long-term black carbon dynamics in cultivated soil. *Biogeochemistry* 92:163–176.

O'Neill, B., J. Grossman, M. T. Tsai et al. 2009. Bacterial community composition in Brazilian anthrosols and adjacent soils characterized using culturing and molecular identification. *Microb Ecol* 58:23–35.

Orton, J. 1875. *The Andes and the Amazon.* New York: Harper and Brothers.

Paul, E. A. 2007. *Soil Microbiology, Ecology, and Biochemistry.* Burlington: Academic.

Piperno, D. R. and P. Becker. 1996. Vegetational history of a site in the central Amazon basin derived from phytolith and charcoal records from natural soils. *Quaternary Res* 45:202–209.

Ranzani, G. 1962. Ocorrencias de "Plaggen Epipedon" no Brasil. *Bo Tecn Cient da Esc Sup da Agric "Luiz de Queirzo" USP Piracicaba,* 5:1–11.

Reese, R. L., E. J. Mawk, J. N. Derr, M. Hyman, M. W. Rowe, and S. K. Davis. 1996. Ancient DNA in Texas rock paintings. In M. V. Orna (ed.), *Archaeological Chemistry, Organic, Inorganic, and Biochemical Analysis,* ACS Symposium Series, vol. 625, pp. 379–390. Washington, DC: American Chemical Society.

Reeves III, J. B. and D. B. Smith. 2009. The potential of mid- and near-infrared diffuse reflectance spectroscopy for determining major- and trace-element concentrations in soils from a geochemical survey of North America. *Appl Geochem* 24:1472–1481.

Reeves, J. B. and J. A. S. Van Kessel. 2002. Spectroscopic analysis of dried manures. Near-versus mid-infrared diffuse reflectance spectroscopy for the analysis of dried dairy manures. *J Near Infrared Spec* 10:93–101.

Roberts, K. G., B. A. Gloy, S. Joseph, N. R. Scott, and J. Lehmann. 2010. Life cycle assessment of biochar systems: Estimating the energetic, economic, and climate change potential. *Environ Sci Technol* 44:827–833.

Rodionov, A., W. Amelung, L. Haumaier, I. Urusevskaja, and W. Zech. 2006. Black carbon in the zonal steppe soils of Russia. *J Plant Nutr Soil Sci* 169:363–369.

Ruivo, M. L. P., C. B. Amarante, M. L. S. Oliveira, I. C. M. Muniz, and D. A. M. Santos. 2009. Microbial population and biodiversity in Amazonian dark earth soils. In W. I. Woods (ed.), *Amazonian Dark Earths: Wim Sombroek's Vision,* pp. 351–362. Berlin: Springer.

Schaefer, C., H. N. Lima, R. J. Gilkes, and J. W. V. Mello. 2004. Micromorphology and electron microprobe analysis of phosphorus and potassium forms of an Indian black earth (IBE) anthrosol from western Amazonia. *Aust J Soil Res* 42:401–409.

Schimmelpfennig, S. and B. Glaser. 2012. One step forward toward characterization: Some important material properties to distinguish biochars. *J Environ Qual* 41:1001–1013.

Schneider, M. P. W., M. Hilf, U. F. Vogt, and M. W. I. Schmidt. 2010. The benzene polycarboxylic acid (BPCA) pattern of wood pyrolyzed between 200°C and 1000°C. *Org Geochem* 41:1082–1088.

Schneider, M. P. W., R. H. Smittenberg, T. Dittmar, and M. W. I. Schmidt. 2011. Comparison of gas with liquid chromatography for the determination of benzenepolycarboxylic acids as molecular tracers of black carbon. *Org Geochem* 42:275–282.

Schulz, H. and B. Glaser. 2012. Effects of biochar compared to organic and inorganic fertilizers on soil quality and plant growth in a greenhouse experiment. *J Plant Nutr Soil Sci* 175:410–422.

Shah, V. G., R. H. Dunstan, P. M. Geary, P. Coombes, T. K. Roberts, and E. Von Nagy-Felsobuki. 2007. Evaluating potential applications of faecal sterols in distinguishing sources of faecal contamination from mixed faecal samples. *Water Res* 41:3691–3700.

Shepherd, K. D., B. Vanlauwe, C. N. Gachengo, and C. A. Palm. 2005. Decomposition and mineralization of organic residues predicted using near infrared spectroscopy. *Plant Soil* 277:315–333.

Simpson, I. A., S. J. Dockrill, I. D. Bull, and R. P. Evershed. 1998. Early anthropogenic soil formation at Tofts Ness, Sanday, Orkney. *J Archaeol Sci* 25:729–746.

Smidt, E., K. Böhm, and M. Schwanninger. 2011. The application of FT-IR spectroscopy in waste management. In G. S. Nikolić (ed.), *Fourier Transforms: New Analytical Approaches and FTIR Strategies,* pp. 431–458. Rijeka: InTech.

Smith, H. H. 1879. *Brazil: The Amazons and the Coast.* New York: Charles Scribner's Sons.

Sombroek, W. G. 1966. *Amazon Soils: A Reconnaissance of the Soils of the Brazilian Amazon Region.* Wageningen: Center for Agricultural Publications and Documentation.

Sombroek, W. G., F. O. Nachtergaele, and A. Hebel. 1993. Amounts, dynamics and sequestering of carbon in tropical and subtropical soils. *Ambio* 22:417–426.

Sørensen, L. K., P. Sørensen, and T. S. Birkmose. 2007. Application of reflectance near infrared spectroscopy for animal slurry analyses. *Soil Sci Soc Am J* 71:1398–1405.

Steiner, C., B. Glaser, W. G. Teixeira, J. Lehmann, W. E. H. Blum, and W. Zech. 2008. Nitrogen retention and plant uptake on a highly weathered central Amazonian ferralsol amended with compost and charcoal. *J Plant Nutr Soil Sci* 171:893–899.

Stenberg, B., L. S. Jensen, E. Nordkvist et al. 2004. Near infrared reflectance spectroscopy for quantification of crop residue, green manure and catch crop C and N fractions governing decomposition dynamics in soil. *J Near Infrared Spec* 12:331–346.

Stuth, J., A. Jama, and D. Tolleson. 2003. Direct and indirect means of predicting forage quality through near infrared reflectance spectroscopy. In *International Workshop on Approaches to Improve the Utilization of Food-feed Crops, Addis Ababa*, pp. 45–56. Amsterdam: Elsevier Science.

Thayn, J., K. P. Price, and W. I. Woods. 2011. Locating Amazonian dark earths (ADE) using vegetation vigour as a surrogate for soil type. *Int J Remote Sens* 32:6713–6729.

Thomsen, I. K., S. Bruun, L. S. Jensen, and B. T. Christensen. 2009. Assessing soil carbon lability by near infrared spectroscopy and NaOCl oxidation. *Soil Biol Biochem* 41:2170–2177.

Thuriès, L., D. Bastianelli, F. Davrieux et al. 2005. Prediction by near infrared spectroscopy of the composition of plant raw materials from the organic fertiliser industry and of crop residues from tropical agrosystems. *J Near Infrared Spec* 13:187–199.

Tiessen, H., E. Cuevas, and P. Chacon. 1994. The role of soil organic matter in sustaining soil fertility. *Nature* 371:783–785.

Tsai, S. M., B. O'Neill, F. S. Cannavan et al. 2009. The microbial world of *terra preta*. In W. I. Woods (ed.), *Amazonian Dark Earths: Wim Sombroek's Vision*, pp. 299–308. Berlin: Springer.

Vergnoux, A., M. Guiliano, Y. Le Dréau, J. Kister, N. Dupuy, and P. Doumenq. 2009. Monitoring of the evolution of an industrial compost and prediction of some compost properties by NIR spectroscopy. *Sci Total Environ* 407:2390–2403.

Viscarra Rossel, R. A., R. N. McGlynn, and A. B. McBratney. 2006. Determining the composition of mineral-organic mixes using UV-vis-NIR diffuse reflectance spectroscopy. *Geoderma* 137:70–82.

von Fischer, J. C. and L. O. Hedin. 2002. Separating methane production and consumption with a field-based isotope pool dilution technique. *Global Biogeochem Cycles* 16:8–13.

Wiedner, K., C. Naisse, C. Rumpel, A. Pozzi, P. Wieczorek, and B. Glaser. 2012. Chemical modification of biomass residues during hydrothermal carbonization—What makes the difference, temperature or feedstock? *Org Geochem* 54:91–100.

Willerslev, E., A. J. Hansen, J. Binladen et al. 2003. Diverse plant and animal genetic records from Holocene and Pleistocene sediments. *Science* 300:791–795.

Williams, P. and K. Norris. 1987. *Near-Infrared Technology in the Agricultural and Food Industries*. St Paul: American Association of Cereal Chemistry.

Woods, W. I. 2003. Soils and sustainability in the prehistoric new world. In B. Benzing and B. Hermann (eds), *Exploitation and Overexploitation in Societies Past and Present*, pp. 143–157. Münster: Lit.

Woods, W. I. and W. M. Denevan. 2009. Amazonian Dark Earths: The first century of reports. In W. I. Woods, W. G. Teixeira, J. Lehmann, C. Steiner, A. M. G. A. WinklerPrins, and L. Rebellato (eds), *Amazonian Dark Earths: Wim Sombroek's Vision*, pp. 309–324. Heidelberg: Springer.

Yun, S. I. and H. M. Ro. 2009. Natural ^{15}N abundance of plant and soil inorganic-N as evidence for over-fertilization with compost. *Soil Biol Biochem* 41:1541–1547.

Zech, W., L. Haumaier, and R. Hempfling. 1990. Ecological aspects of soil organic matter in tropical land use. In P. McCarthy, C. E. Clapp, R. L. Malcolm, and P. R. Bloom (eds), *Humic Substances in Soil and Crop Sciences, Selected Readings*, pp. 187–202. Madison: American Society of Agronomy and Soil Science Society of America.

Zech, W., E. Pabst, and G. Bechtold. 1979. Analytische Kennzeichnung der Terra Preta do Indio. *Mitteilungen der Deutschen Bodenkundlichen Gesellschaft* 29:709–716.

Zhang, A., L. Cui, G. Pan et al. 2010. Effect of biochar amendment on yield and methane and nitrous oxide emissions from a rice paddy from Tai Lake plain, China. *Agr Ecosyst Environ* 139:469–475.

Zimmermann, J. 1958. *Studien zur Anthropogeographie Amazoniens*, vol. 21. Bonn: Bonner Geographische Abhandlungen.

CHAPTER 2

Factors and Mechanisms Regulating Soil Organic Carbon in Agricultural Systems

Yadunath Bajgai, Paul Kristiansen, Nilantha Hulugalle, and Melinda McHenry

CONTENTS

2.1 Introduction 41
2.2 Factors Affecting SOC 42
    2.2.1 Climate 42
    2.2.2 Soil Type 43
    2.2.3 Land Use 43
2.3 Effect of Agricultural Management on SOC 44
    2.3.1 Conservation Agricultural Practices 44
    2.3.2 Crop Residue and Its Quality for SOC Stabilization 44
    2.3.3 Crop Residue Management 45
    2.3.4 Crop Residue and Tillage 46
    2.3.5 Fertilizer Management 47
    2.3.6 Organic and Conventional Farming Systems 47
2.4 Pools of C and C Stabilization Mechanisms 48
    2.4.1 Labile and Particulate Organic C 48
    2.4.2 Soil C Stabilization through Aggregation 48
    2.4.3 Physicochemical Stabilization 49
2.5 Conclusion and Summary 49
Acknowledgments 50
References 50

2.1 INTRODUCTION

Soil organic carbon (SOC) is the part of carbon (C) in the soil that is derived from living organisms and plays an important role in the C cycle (Paustian et al. 1997). Soil is a major reservoir of soil C, at 3.3 times the size of the atmospheric pool of 760 pentagrams (Pg) and 4.5 times the size of the biotic pool of 560 Pg (Lal 2004). Soils act as a reservoir of SOC and the level of storage within an ecosystem is mainly dependent on the soil type, climate, land use history, and current management practices. The quantity of SOC stored in a particular soil is dependent on the quantity and quality of organic matter returned to the soil matrix, the soil's ability to retain SOC (a function of texture and cation exchange capacity), and abiotic influences of both temperature and precipitation (Grace et al. 2005).

SOC is essential for maintaining fertility, water retention, and plant production in terrestrial ecosystems with different land uses (Grace et al. 2006). Soil organic matter (SOM) maintains soil structure and productivity in agroecosystems (Lal 2010). Maintaining high levels of SOM is beneficial for all agriculture and crucial in improving soil quality. SOM has been widely used as an effective indicator of the functional response of soils to land use intensification (Dalal et al. 2003).

2.2 FACTORS AFFECTING SOC

2.2.1 Climate

The decomposition of SOM depends on its nature and abundance and on climatic factors, in particular temperature and humidity, which influence the decomposition processes through their effects on microbial activity in the soil (Leirós et al. 1999). For example, a semiarid moderately grazed rangeland that was a net sink of up to 1.6 Mg C/ha/year during the wetter than average year was a net source of 0.5 Mg C/ha/year during a year that experienced a growing season drought (Sims and Bradford 2001). Potter et al. (2007) conducted a multiyear study at six locations in central Mexico with a wide range of climatic conditions to determine the effect of varying rates of residue removal with no-till on SOC. They found that retaining 100% of the crop residues with no-till always increased or maintained the SOC content. SOC increased in cooler climates, but as mean annual temperature increased, more retained crop residues were needed to increase the SOC. In tropical conditions (mean annual temperature $>20°C$), 100% corn residue retention with no-till only maintained SOC levels but did not increase them. Mean annual temperature had a greater impact on SOC than did annual rainfall. At the higher temperatures, most of the residue would decompose if left on the soil surface without improving soil C content (Potter et al. 2007).

Soil C represents a significant pool of C within the biosphere (Lal 2004) and climatic shifts in temperature and precipitation have a major influence on the decomposition and amount of SOC stored within an ecosystem and that released into the atmosphere (Grace et al. 2006). The temperature and precipitation affects decomposition of plant materials, as both factors increase SOM decomposition to a certain level following sigmoid functions (Parton et al. 1987).

Mean annual precipitation and mean annual temperature have been used to explain spatial variation in global SOM levels (Post et al. 1982). SOC content increases with increasing precipitation; at a fixed precipitation level, it also increases with decreasing temperature (McLauchlan 2006; Post et al. 1982). Owing to generally semiarid climatic conditions, high temperatures and prolonged dry periods, plant primary productivity in Australia is frequently suboptimal and, thus, return of crop residues and accumulation of SOC in soil is lower than that reported for wetter and colder regions of the world (Kern and Johnson 1993; Sanderman et al. 2010; Webb 2002). Dryland farmers of northwestern New South Wales (NSW), Australia, may not even expect to sequester atmospheric C into soil in the short-to-medium term (Wilson et al. 2011; Young et al. 2009).

In the presence of substrate, the decomposition rate of less decomposable materials appears to respond more to temperature than the decomposition rate of readily decomposable substrates (Conant et al. 2011). However, processes controlling substrate availability and the response of those processes to temperature as well as the intrinsic decomposition rate of the less decomposable compounds are also critically important (Kleber et al. 2011). Relevant field experiments are difficult to implement and the incubation studies have a limited capacity to shed additional light on the affects of temperature changes in soil carbon stocks, except to the extent that they help us understand temperature controls on the components of decomposition of available SOM (Conant et al. 2011). Climate change could act as a driver for land use change, thus further altering terrestrial C fluxes. Due to the large size of terrestrial C pools, they have considerable potential to drive large positive climate feedbacks because increased CO_2 concentrations in the atmosphere will accelerate climate change (Cox et al. 2000; Friedlingstein et al. 2006).

2.2.2 Soil Type

Parent material partially determines the composition and content of clay mineral factors, which are considered important in determining soil C longevity of agricultural soils (McLauchlan 2006). Clay minerals stabilize SOM by generally resulting in a positive relationship between SOC and clay content (Chan et al. 2003; Six et al. 1999). The influence of soil type on SOC is mainly on plant growth (Chan et al. 2003) and the protection of organic matter (particulate organic matter aggregates and organochemical complexes) (Hassink 1997; Six et al. 2000b) due to soil texture and cation exchange capacity (CEC). Soil type determines the ability of soils to form soil aggregates, which is a result of the rearrangement of particles, flocculation, and cementation (Duiker et al. 2003) that stabilize SOC and prevent it from being acted on by microbes.

The rate and stability of aggregation generally increase with SOC or clay surface area and CEC. In soils with low SOC or clay concentration, aggregation would be dominated by cations, while the role of cations in aggregation may be minimal in soils with high SOC or clay concentration (Bronick and Lal 2005; Six et al. 2002a). Since texture of the soil has a significant influence on aggregation, increased clay concentration is associated with increased SOC stabilization (Bajgai et al. 2014; Francesca Cotrufo et al. 2013; Hassink 1997).

The CEC is related to stability of soil aggregates (Dimoyiannis et al. 1998; McKenzie 1998), and the polycationic charges balance the repulsive forces between negatively charged clay or SOC complexes (Bronick and Lal 2005). Further, soil texture is an important factor affecting the mineralization due to its effects on pore size distribution, which also affects the moisture release characteristics of soils (Thomsen et al. 1999). Soil pH influences SOC sequestration both through its influence on plant productivity and through the interactive effect on repulsive behavior of negatively charged clay particles (Haynes and Naidu 1998).

2.2.3 Land Use

The global decline in SOC as a result of deforestation, shifting cultivation, and intensive arable cropping has made significant contributions to increased levels of atmospheric CO_2 (Post et al. 1990). Whether it is land use (e.g., woodland, pasture, and crop land), cropping systems (conventional broad acre, intensive vegetable, and organic horticulture), or the intensity of farming (number of crops grown in a year), key drivers of SOC dynamics are the quantity and quality (C:N ratio) of organic material entering soil (Lemke et al. 2010; Valzano et al. 2005; Wilhelm et al. 2004; Yadvinder et al. 2005) and the quantity of CO_2 emitted from the soil by microbial activity, which is partly controlled by the level of soil disturbance (Angers et al. 1993; Six et al. 1999; von Lützow et al. 2006). The type of land use directly affects microclimate, the quantity and quality of C inputs and the pathways of C cycling.

Erosion is controlled by land use and management and may decrease SOC stocks in agricultural systems compared with forests because it can be a major cause of SOC loss, particularly on poorly aggregated soils (Berhe et al. 2007). This is especially true for tropical regions with less-fertile, highly weathered soils managed with few external inputs of nutrients and C, which may be more susceptible to perturbations from land use changes, with SOC turnover twice as high as in temperate regions (Six et al. 2002b). Land use effects across different soil types on soil C for parts of semiarid regions of NSW in Australia were reported to be in the following order: cultivated land < pasture < woodland

(Wilson et al. 2008, 2011). A similar trend is reported for the three land uses in England, a cold temperate region (Saby et al. 2008). Woodland or forest produces the largest plant biomass of all land uses. Pasture does not undergo frequent cultivation, unlike cropland, and thus the above order is generally true for SOC storage capacity of the three broad land use categories.

The effect of land use change may be another way of looking at the changes due to land use. Meta-analyses have found that, if one of the land use changes decreased soil C, the reverse process increased soil C, and vice versa (Don et al. 2011; Guo and Gifford 2002). For example, conversion of cropland to plantation increased SOC, but conversion of native forest to cropland decreased it, because perturbations like tillage disrupt soil aggregates in cropland.

Within a particular land use system, specific agricultural management practice also affects SOC including tillage types, fertilization, bare fallows, crop rotation, residue retention, and so on (Johnson et al. 2007; Luo et al. 2010; Sanderman et al. 2010; Smith et al. 2008). These are presented in the following section.

2.3 EFFECT OF AGRICULTURAL MANAGEMENT ON SOC

2.3.1 Conservation Agricultural Practices

The recent review by Luo et al. (2010) has synthesized published research that compares conservation agriculture practices (CAPs) with conventional practices in Australian agroecosystems. The review has shown that there is more SOC sequestration by CAPs than by conventional methods. Further, it has shown that the introduction of perennial plants into a rotation had the greatest potential to increase soil C: by 18% compared with other CAPs. In a vegetable farming system, after 3.5 years of conservation management including no-till and high inputs of compost with a high content of organic C, the SOC for the CAP was 75% higher than that in conventional management systems (Wells et al. 2000).

Despite adopting CAPs, some studies have found no change or even a decrease in SOC over time (Baker et al. 2007; Heenan et al. 1995; Hoyle and Murphy 2006; Hulugalle et al. 2002) because the soil could have had a delayed response to the implementation of CAP (Sanderman and Baldock 2010). Some studies with time series data suggest that the relative gain is often due to a reduction or cessation of soil carbon losses rather than an actual increase in stocks (Sanderman and Baldock 2010; Sanderman et al. 2010). This is because the soil aggregation process that decreases SOC mineralization could be slower than its destruction during tillage (Jastrow 1996; Six et al. 2000a) and thus SOC may not build up as quickly as it appears to be lost (Balesdent et al. 2000). Pankhurst et al. (2002) demonstrated this concept of hysteresis by applying tillage to previously tilled and no-till plots after a cessation of 3 years; no-tilled plots showed large SOC losses, while the tilled ones did not respond to the tillage.

2.3.2 Crop Residue and Its Quality for SOC Stabilization

The addition of new residues in no-till management promotes organic matter stabilization through the binding of primary soil particles and old microaggregates (<250 µm) into new macroaggregates (>250 µm) (Puget et al. 1995; Six et al. 1999). Fragmented crop residues, that is, particulate organic matter (POM), can form the nuclei for new microaggregates that can be bound together by transient, labile organic matter to form new macroaggregates, or new microaggregates may form within the larger macroaggregates around POM (Golchin et al. 1994; Six et al. 1998). Recently decomposed crop residues and SOM are central to the aggregate hierarchy model of Tisdall and Oades (1982). They proposed that transient forms of SOM could act as binding agents, causing microaggregates to form stable macroaggregates.

Several studies have shown that incorporation of residue-derived C into aggregates increased with increasing aggregate size (Jastrow 1996; Puget and Drinkwater 2001; Six et al. 2000b). Consequently, macroaggregate formation could be enhanced by residue additions throughout the entire plow layer with tilled management. In contrast, under no-till management, residues are concentrated near the soil surface so that a smaller soil volume is exposed to fresh residues, which may limit new aggregate formation. Tillage has a strong influence on soil aggregation and SOM dynamics by increasing macroaggregate turnover and reducing microaggregate formation compared with no-till management (Six et al. 1999). The stabilization of SOM may be enhanced in no-till management because macroaggregates that are formed are less susceptible to disruption from tillage-induced physical disturbances. However, when residues are distributed to a 0.3 m depth, the negative impact of aggregate disruption through tillage appears to be counterbalanced, with similar efficiencies of C stabilization between the no-till and tilled management practices, possibly due to slower decomposition of residues in the deeper soil profile (Baker et al. 2007; Olchin et al. 2008).

When crop residue is applied to soil, the composition of residue strongly affects the rate of crop residue decomposition and nutrient dynamics. As microorganisms have a low C:N ratio, residues with high C:N ratios (e.g., corn) are decomposed more slowly than residues with low C:N ratios (e.g., cabbage), and so corn residues may lead to N immobilization in the microbial biomass (Moritsuka et al. 2004; Trinsoutrot et al. 2000a). Moreover, the decomposition rate is positively correlated with the concentration of soluble sugars, proteins, and, in the later stages of decomposition, the concentration of cellulose (Gunnarsson et al. 2008), whereas high lignin concentration or high lignin:N ratio negatively affects the decomposition rate (Trinsoutrot et al. 2000b; Wang et al. 2004).

2.3.3 Crop Residue Management

There are several options for management of crop residues based on the requirements of farmers. Options include removing from the field, incorporating into the soil, burning *in situ*, composting, or leaving on the soil surface as mulch for succeeding crops (Lemke et al. 2010; Wilhelm et al. 2004; Yadvinder et al. 2005). However, stubble retention, incorporation, and burning are the main three stubble management practices in Australian conditions (Valzano et al. 2005). The off-farm uses of crop residues are increasingly being practiced. Crop residues are removed from the field for feed/ bedding for livestock, a substrate for composting, biogas generation, or mushroom culture or as a raw material for industry (Lemke et al. 2010; Wilhelm et al. 2004; Yadvinder et al. 2005). In the United States, corn has considerable potential for fuel production because of the large amount of residue it produces. Other high-residue crops, such as rice and sugarcane, contribute to biofuel production as a solution to their residue disposal issues (Blanco-Canqui and Lal 2007; Wilhelm et al. 2004). Due to the high demand for crop residues for off-farm uses, maintaining SOC is a critical issue among researchers (Blanco-Canqui and Lal 2007; Lemke et al. 2010).

Crop residue burning converts biomass C into CO_2 and black C, which is similar to charcoal. Burning of crop residue or stubble generally leads to a reduction of soil C (Luo et al. 2010). The high temperatures generated by fire affect microbial activity in the surface soil and alter soil structure and soil hydraulic properties (Kumar and Goh 1999; Valzano et al. 2005). The remaining black C is resistant to microbial decomposition and can persist in the soil for centuries (Harden et al. 2000). For this reason, black C has been proposed as a method to store C and offset the anthropogenic emission of CO_2 (Lehmann 2007; Marris 2006). Stubble retention has been shown to increase relative change in soil C over burning (Luo et al. 2010). The synergy of combining stubble retention and conservation tillage increased SOC content by 16% as compared with stubble burning and conventional tillage. This benefit is significantly higher than the separate application of these two practices, with 6% for stubble retention and only 3% for conservation tillage (Luo et al. 2010). Retention of crop residue can help increase yields, improve soil nutrients, and conserve soil water in semiarid conditions (Johnson et al. 2006; Wilhelm et al. 2004). Crop residue management systems that maintain organic materials *in situ* can benefit SOM (Hulugalle and Scott 2008; Liu et al. 2009; van Groenigen et al. 2011) and could, in the long term, offset the increasing concentration of atmospheric CO_2 by stabilizing more carbon in soil (Mondini et al. 2007).

2.3.4 Crop Residue and Tillage

The effects of tillage and crop residue management can have opposing influences on SOC (Dalal et al. 2011; Dong et al. 2009; Liu et al. 2009) and may be difficult to isolate. The type of tillage determines the rate of SOC breakdown, while crop residue management determines the rate of organic C input to a system (Liu et al. 2009; van Groenigen et al. 2011). For practical assessment, quantification of the effect of each of the two practices individually is desirable to enable evaluation of their separate contributions (Liu et al. 2009; van Groenigen et al. 2011).

Tillage breaks soil clods, redistributes and incorporates crop residue and soil amendments in soil profile, aerates the soil, and recycles nutrients in the rooting zone (Conant et al. 2007; Dao 1998). Soil tillage management can affect factors controlling soil respiration, including substrate availability, soil temperature, water content, pH, oxidation–reduction potential, kind and number of microorganisms, and soil ecology (Kladivko 2001). Tillage also exposes organic C in both the inter- and intra-aggregate zones, where C is vulnerable to rapid oxidation (Roscoe and Buurman 2003). This is because of the improved availability of O_2, water, and decomposition surfaces, thereby stimulating increased microbial activity (Beare et al. 1994; Jastrow 1996).

Whether to incorporate stubble into soil or not is an important farm management decision because terrestrial C sequestration has been suggested as a potential strategy for greenhouse gas mitigation (Valzano et al. 2005). The manner by which crop residues are introduced to the soil matrix differs dramatically between no-till and tilled management, and thus aggregate turnover of the two tillage treatment differs vastly. In no-till management, crop residues are left on the soil surface after harvest, whereas residues are mechanically incorporated into the soil during tillage (Baker et al. 2007; Olchin et al. 2008). Studies tracking stable isotope shifts in SOC fractions have shown that tillage increases the SOM decomposition rates (Kisselle et al. 2001; Paustian et al. 2000). To improve soil C sequestration in rotations, the input of residue and the CO_2 emission should be balanced by adopting appropriate tillage and residue management practices (Bàrberi 2006; Dong et al. 2009).

Clapp et al. (2000) evaluated the roles of residue, tillage, and N fertilization in SOM accrual by determining the source of C in the organic matter over a period of 13 years. All three management factors were reported to affect SOC storage. Corn-derived SOC was shown to be greatest under their no-till system without residue removal but lowest in the no-till when stover was removed. Meanwhile, conventional and reduced tillage treatments had an intermediate effect regardless of whether residues were harvested or returned. In one of the few studies that compared residue removal effects under moldboard plowing, Reicosky et al. (2002) found that total C and N remained virtually unchanged over the 30-year study. This is in agreement with the study by Clapp et al. (2000), which found that when the soil was moldboard plowed, residue with additional N fertilization did not increase SOC. Similarly, Dick et al. (1998) concluded that tillage and rotation had greater effects on C accrual than residue removal did.

Baker et al. (2007) argued that in no-tilled soils the SOC is concentrated near the surface, while in tilled soils it is distributed deeper in the profile. Because of this, apparent SOC gains from no-till based on near-surface samples disappear when deeper samples are included. In another study, where residues are distributed to a 0.3 m depth, the negative impact of aggregate disruption through tillage appears counterbalanced, with similar efficiencies of C stabilization between the no-till and tilled management practices, possibly due to slower decomposition of residues deeper in the profile (Balesdent et al. 2000; Olchin et al. 2008).

2.3.5 Fertilizer Management

Fertilizers and irrigation water are applied to soils in order to maintain or improve crop yields. In the long term, increased crop yields and organic matter returns with regular fertilizer applications result in a higher SOM content and biological activity being attained compared with practices where no fertilizers are applied (Haynes and Naidu 1998). Since N is an important limiting soil nutrient that determines net primary productivity, its effect is of major significance to C sequestration in soil. In a recent comprehensive review of the effect of N management on C sequestration in North American cropland soils, Christopher and Lal (2007) concluded that, in most cases, N-fertilizer addition either had limited effect or, more likely, resulted in increased SOC. This was credited to the fact that SOC was linearly correlated to the amount of crop residues returned to the soil and crop residue production was directly related to N addition. Further, a major review found a positive relationship between N-fertilizer use and SOC storage, but SOC was found to increase only when crop residue was returned to the soil (Alvarez 2005), confirming the important role of crop residue for C sequestration in soil. Similarly, Dalal et al. (2011), in a 40-year study, found that crop residue retention increased SOC only when N fertilizer was applied, and fertilization increased SOC only when crop residue was retained, highlighting the synergy of fertilization and residue retention. Although N fertilizer application increases crop yields, it may also accelerate mineralization of SOM (Khan et al. 2007; Russell et al. 2009) or reduce aggregate stability (Fonte et al. 2009), potentially negating any incremental yield benefits associated with its use.

2.3.6 Organic and Conventional Farming Systems

One of the alternatives to conventional farming systems is organic farming, which is claimed to prevent or to mitigate negative environmental impacts of intensive agriculture (Leifeld et al. 2009; Mäder et al. 2002; Mazzoncini et al. 2010). In organically managed farms, increased levels of SOM and hence increased reservoirs of nutrients are widely reported (Mäder et al. 2002; Stockdale et al. 2001). The regular use of manure is an important means in such cropping systems to maintain these high levels of SOM (Marinari et al. 2010; Watson et al. 2002). However, SOM is a relatively stable parameter that reflects the influence of management over decades (Kirkby et al. 2011; Pulleman et al. 2000) rather than tracking the changes in biologically based soil fertility at a time scale of several years after transitioning from conventional to organic farming (Wander et al. 1994). Biologically active or labile C fractions of SOM are likely to be controlled by management to a much greater extent than total SOM (Marriott and Wander 2006).

While no-till farming is suitable for broadacre crops, it is not commonly used in vegetable cropping systems. The reliance of the vegetable systems on tillage to perform basic management operations are proven to break soil structure and aggregates exposing the physically protected SOM for microbial attack, which stimulates the release of CO_2 to the atmosphere (Angers et al. 1993; Six et al. 1999; von Lützow et al. 2006). Despite requiring multiple tillages, the vegetable systems are characterized by little or no crop residue input (Chan et al. 2007; Jackson et al. 2004) to offset the loss of C induced by the tillage. Soil C in organic systems is generally reported to be higher than in conventional systems (Clark et al. 1998; Mancinelli et al. 2010; Wells et al. 2000), although in some studies no such differences were detected (Hathaway-Jenkins et al. 2011; Leifeld et al. 2009). Hence, the distinction between organic and conventional systems is still not clear, even though the former rely heavily on organic materials for crop nutrition and, thus, are claimed to store more SOC.

Organic vegetable systems rely on tillage for weed control, whereas conventional systems use herbicides to manage weeds (Bond and Grundy 2001; Chirinda et al. 2010). In addition, organic systems use organic sources, such as crop residue and compost, for fertilization but conventional systems use mineral fertilizers as the main source of crop nutrition (Chirinda et al. 2010; Mondelaers et al. 2009). These management differences have a direct impact on the C balance of each system.

The slower nutrient-releasing properties of organic fertilizers may help in retaining more nutrients in soil than the soluble mineral fertilizers (Berry et al. 2002; Marinari et al. 2010). The former may reduce the amount of nutrients lost to the environment via leaching (Poudel et al. 2002; van Diepeningen et al. 2006) or in gaseous forms (Bouwman et al. 2002). In a corn–cabbage rotation, nutrient lost in the conventionally managed plots may be speculated to be equivalent to nutrient retained in an organically grown crop (Bajgai et al. 2013). This was because yields obtained were similar when equivalent quantities of nutrients were applied in both the systems in that study. Soil nutrient reserves and underlying nutrient cycling processes in organically cropped soils are similar to those in conventionally managed soils, however, the former holds nutrients in less-available forms (Berry et al. 2002; Stockdale et al. 2002), which is of a greater significance for soil nutrient retention for sustainable agricultural systems (Bajgai et al. 2013).

2.4 POOLS OF C AND C STABILIZATION MECHANISMS

Crop residue or organic fertilizer undergoes a series of processes in soil leading to different pools: labile or active, intermediate or slow, and stable or inert or recalcitrant pools (Smith et al. 1997). These are conceptual pools with characteristic turnover rates ranging from less than one year to thousands of years. The pools exist in a continuum between labile and stable pools, rather than fixed pools (Davidson and Janssens 2006). These conceptual SOM pools are defined by their different turnover times and their pool sizes (Smith et al. 1997; von Lützow et al. 2007). The SOC functional pools concept has been summarized using the Century and Roth-C models, which capture the main characteristics of the three pools (Davidson and Janssens 2006; Kleber and Johnson 2010).

The labile C fractions are increasingly used in agroecosystems research because they respond more sensitively to changes in land management (Cambardella and Elliott 1992; Chan 1997). Labile fractions of C are closely linked to soil microbial biomass and organic matter turnover and can serve as indicators of the key chemical and physical properties of soils such as infiltration (Bell et al. 1999) and the availability of labile nutrients such as N, P, and S (O'Donnell et al. 2001).

2.4.1 Labile and Particulate Organic C

Labile pools, often referred to as unprotected C, are characterized by a rapid turnover, mainly consist of young SOM, and are sensitive to land management and environmental conditions (Parton et al. 1987). This labile C has a short residence time and is subject to fast mineralization, whereas stable C has long residence time and is recalcitrant in nature (Davidson and Janssens 2006; Kleber and Johnson 2010). Owing to these features, labile SOM pools play an important role in short-term C and N cycling in terrestrial ecosystems (O'Donnell et al. 2001; Schlesinger 1990). The most commonly isolated labile C pools are the light fraction (LF) and POM. The LF consists mostly of mineral-free, partly decomposed plant debris (Spycher et al. 1983) and appears in soils as free POM (Golchin et al. 1994; Six et al. 1998). When free POM is further decomposed, it can be incorporated into aggregates, where its decomposition is restricted (Golchin et al. 1994; Puget et al. 1995). In this case, POM becomes part of a less-labile pool, but is easily decomposable when it is exposed again; hence, the degree of protection of labile C is dependent on aggregate turnover. Stabilization of occluded organic matter is greatest when aggregate stability is high and aggregate turnover is slow (Six et al. 1998).

2.4.2 Soil C Stabilization through Aggregation

For soils to act as a C sink, organic C needs to be stabilized in stable C pools (Paustian et al. 1997). Of the stable C pools, SOC protected within aggregates is sensitive to land management practices such as manure application, tillage, and crop rotation (Six et al. 1999). Therefore, the amount of C stabilized in protected SOC pools is critical for the determination of the extent to which soils can operate as a C sink under specific management conditions. Due to different protection mechanisms, the degree and the duration of stabilization of SOC within macroaggregates and microaggregates vary (Tisdall and Oades 1982). Macroaggregates contain soil C that (a) functions as a transient binding agent holding microaggregates together and (b) is occluded within microaggregates, which results in greater absolute C contents of macroaggregates than microaggregates (Elliott 1986; Tisdall and Oades 1982).

The aggregation model of C stabilization is based on the size of soil aggregates where macroaggregates provide minimal physical protection (Beare et al. 1994; Elliott 1986) and microaggregates, including those within macroaggregates, provide more physical protection against microbial decomposition (Balesdent et al. 2000; Blanco-Canqui and Lal 2004; Six et al. 1999). Stabilization is determined by the silt plus clay content (<53 µm) of the soil and the availability of organic matter to form the matrix of an aggregate. The aggregates physically protect SOM from decomposition by forming spatial barriers that reduce accessibility of the soil microorganisms to their enzymes, substrates, water, and oxygen (Blanco-Canqui and Lal 2004; von Lützow et al. 2006). Thus, confinement of plant debris in the core of microaggregates is the SOC sequestration mechanism through soil aggregation (Blanco-Canqui and Lal 2004; Six et al. 2002a). Microaggregates have a smaller C storage potential, but as they sequester C in the long term (Six et al. 2000a; Skjemstad et al. 1990), the degree of stabilization is greater. Increased organic matter input to soil through management practices such as animal and green manure application and straw incorporation increases the SOM content (Mikha and Rice 2004) and enhances the soil aggregation process (Sommerfeldt and Chang 1985).

2.4.3 Physicochemical Stabilization

The physicochemical stabilization of SOC is due to the sorption of SOC compounds on silt plus clay surfaces or mineral colloids (Christensen 2001; Hassink 1997; von Lützow et al. 2006). The SOC particles bound on the surfaces of organomineral colloids are believed to be stabilized because the SOM is older (Eusterhues et al. 2003) or has a longer turnover time than other SOM fractions (Balesdent 1996; Ludwig et al. 2003). Hence, clay soils are more efficient in stabilizing SOM than sandy soils (Bajgai 2013). The chemical and physical properties of a mineral matrix, as well as the morphology and the chemical structure of SOM, determine the extent to which SOC is stabilized on mineral colloids (Baldock and Skjemstad 2000). However, it is not clearly understood why there is a reduction of microbial decomposition of the SOC particles bound on mineral surfaces (von Lützow et al. 2006). Isotopic studies have shown that early-stage leaf and root litter mineralization products contribute a relatively large amount of C to mineral soil (Bird et al. 2008; Rubino et al. 2010), and that this contribution is mainly from microbial compounds produced during the mineralization process (Mambelli et al. 2011). However, recalcitrant plant parts (e.g., lignin and phenols) do not preferably accumulate in SOM (Marschner et al. 2008). Labile plant components, not the recalcitrant ones, are dominant sources of microbial breakdown products because they are utilized more efficiently by microbes, and thus become the primary precursors of stable SOM by promoting aggregation and through strong physicochemical bonding to the soil matrix (Francesca Cotrufo et al. 2013). In vegetable systems where tillage is intensively and regularly used, protection through physicochemical binding predominates over protection through aggregation (Bajgai et al. 2014) as tillage breaks soil aggregates.

2.5 CONCLUSION AND SUMMARY

The key factors that affect SOC are climate, soil type, land use, and agricultural management of land. Decreasing SOC in land and increasing greenhouse gases due to agricultural activities like intensive tillage and bare fallows have encouraged the adoption of conservation agricultural techniques such as no-tillage, crop rotations, and residue retention (Johnson et al. 2007; Luo et al. 2010; Sanderman et al. 2010; Smith et al. 2008). However, some conservation practices such as no-till farming suitable for broadacre crops are unsuitable for most vegetable crops. Since no herbicides are permitted in organic agriculture, tillage is used to perform basic management operations like preparation of beds and control of weeds (Bond and Grundy 2001; Chirinda et al. 2010). These tillage operations disrupt soil structure and break soil aggregates, exposing the protected SOM to microbial attack, which stimulates the release of CO_2 to the atmosphere (Angers et al. 1993; Six et al. 1999; von Lützow et al. 2006). Also aggravated by the routine use of tillage operations, vegetable systems are characterized by little or no crop residue input (Chan et al. 2007; Jackson et al. 2004). In this light, a relatively high-value vegetable crop like sweet corn may be used as a rotation crop for a vegetable system so that there would be an abundance of residue available as input to soils to compensate for the expected loss associated with tillage and to increase SOC (Bajgai et al. 2014). The soil aggregation model and physicochemical stabilization are the main soil stabilization mechanisms for accumulating SOC in soil.

ACKNOWLEDGMENTS

The first author was funded by the Endeavour Postgraduate Award of Australia Awards at the University of New England, Armidale, Australia, for his PhD study.

REFERENCES

Alvarez, R. 2005. A review of nitrogen fertilizer and conservation tillage effects on soil organic carbon storage. *Soil Use Manage* 21:38–52.

Angers, D. A., A. N'dayegamiye, and D. Côté. 1993. Tillage-induced differences in organic matter of particle-size fractions and microbial biomass. *Soil Sci Soc Am J* 57:512–516.

Bajgai, Y. 2013. Effect of alterative cropping management on soil organic carbon. PhD Thesis, School of Environmental and Rural Sciences, University of New England, Armidale.

Bajgai, Y., P. Kristiansen, N. Hulugalle, and M. McHenry. 2013. Comparison of organic and conventional managements on yields, nutrients and weeds in a corn–cabbage rotation. *Renew Agric Food Syst* FirstView: 1–11. Available at: journals.cambridge.org/action/displayJournal?jid=RAF (accessed: September 7, 2013).

Bajgai, Y., P. Kristiansen, N. Hulugalle, and M. McHenry. 2014. Changes in soil carbon fractions due to incorporating corn residues in organic and conventional vegetable farming systems. *Soil Res* 52:244–252.

Baker, J. M., T. E. Ochsner, R. T. Venterea, and T. J. Griffis. 2007. Tillage and soil carbon sequestration—What do we really know? *Agr Ecosyst Environ* 118:1–5.

Baldock, J. A. and J. O. Skjemstad. 2000. Role of the soil matrix and minerals in protecting natural organic materials against biological attack. *Org Geochem* 31:697–710.

Balesdent, J. 1996. The significance of organic separates to carbon dynamics and its modelling in some cultivated soils. *Eur J Soil Sci* 47:485–493.

Balesdent, J., C. Chenu, and M. Balabane. 2000. Relationship of soil organic matter dynamics to physical protection and tillage. *Soil Till Res* 53:215–230.

Bàrberi, P. 2006. Tillage: How bad is it in organic agriculture? In P. Kristiansen, A. Taji, and J. Reganold (eds), *Organic Agriculture: A Global Perspective*, pp. 295–303. Collingwood, Australia: CSIRO.

Beare, M. H., P. F. Hendrix, and D. C. Coleman. 1994. Water-stable aggregates and organic matter fractions in conventional- and no-tillage soils. *Soil Sci Soc Am J* 58:777–786.

Bell, M. J., P. W. Moody, S. A. Yo, and R. D. Connoly. 1999. Using active fractions of soils organic matter as indicators of the sustainability of ferrosol farming systems. *Aust J Soil Res* 37:279–288.

Berhe, A. A., J. Harte, J. W. Harden, and M. S. Torn. 2007. The significance of the erosion-induced terrestrial carbon sink. *BioScience* 57:337–346.

Berry, P. M., R. Sylvester-Bradley, L. Philipps et al. 2002. Is the productivity of organic farms restricted by the supply of available nitrogen? *Soil Use Manage* 18:248–255.

Bird, J. A., M. Kleber, and M. S. Torn. 2008. [13]C and ^{15}N stabilization dynamics in soil organic matter fractions during needle and fine root decomposition. *Org Geochem* 39:465–477.

Blanco-Canqui, H. and R. Lal. 2004. Mechanisms of carbon sequestration in soil aggregates. *Crit Rev Plant Sci* 23:481–504.

Blanco-Canqui, H. and R. Lal. 2007. Soil and crop response to harvesting corn residues for biofuel production. *Geoderma* 141:355–362.

Bond, W. and A. C. Grundy. 2001. Non-chemical weed management in organic farming systems. *Weed Res* 41:383–405.

Bouwman, A. F., L. J. M. Boumans, and N. H. Batjes. 2002. Emissions of N_2O and NO from fertilized fields: Summary of available measurement data. *Global Biogeochem Cy* 16:1058.

Bronick, C. J. and R. Lal. 2005. Soil structure and management: A review. *Geoderma* 124:3–22.

Cambardella, C. A. and E. T. Elliott. 1992. Particulate soil organic-matter changes across a grassland cultivation sequence. *Soil Sci Soc Am J* 56:777–783.

Chan, K. Y. 1997. Consequences of changes in particulate organic carbon in Vertisols under pasture and cropping. *Soil Sci Soc Am J* 61:1376–1382.

Chan, K. Y., C. G. Dorahy, S. Tyler et al. 2007. Phosphorus accumulation and other changes in soil properties as a consequence of vegetable production, Sydney region, Australia. *Soil Res* 45:139–146.

Chan, K. Y., D. P. Heenan, and H. B. So. 2003. Sequestration of carbon and changes in soil quality under conservation tillage on light-textured soils in Australia: A review. *Aus J Exp Agr* 43:325–334.

Chirinda, N., J. E. Olesen, J. R. Porter, and P. Schjønning. 2010. Soil properties, crop production and greenhouse gas emissions from organic and inorganic fertilizer-based arable cropping systems. *Agr Ecosyst Environ* 139:584–594.

Christensen, B. T. 2001. Physical fractionation of soil and structural and functional complexity in organic matter turnover. *Eur J Soil Sci* 52:345–353.

Christopher, S. F. and R. Lal. 2007. Nitrogen management affects carbon sequestration in North American cropland soils. *Crit Rev Plant Sci* 26:45–64.

Clapp, C. E., R. R. Allmaras, M. F. Layese, D. R. Linden, and R. H. Dowdy. 2000. Soil organic carbon and ^{13}C abundance as related to tillage, crop residue, and nitrogen fertilization under continuous corn management in Minnesota. *Soil Till Res* 55:127–142.

Clark, M. S., W. R. Horwath, C. Shennan, and K. M. Scow. 1998. Changes in soil chemical properties resulting from organic and low-input farming practices. *Agron J* 90:662–671.

Conant, R. T., M. Easter, K. Paustian, A. Swan, and S. Williams. 2007. Impacts of periodic tillage on soil C stocks: A synthesis. *Soil Till Res* 95:1–10.

Conant, R. T., M. G. Ryan, G. I. Ågren et al. 2011. Temperature and soil organic matter decomposition rates—Synthesis of current knowledge and a way forward. *Glob Change Biol* 17:3392–3404.

Cox, P. M., R. A. Betts, C. D. Jones, S. A. Spall, and I. J. Totterdell. 2000. Acceleration of global warming due to carbon-cycle feedbacks in a coupled climate model. *Nature* 408:184–187.

Dalal, R. C., D. E. Allen, W. J. Wang, S. Reeves, and I. Gibson. 2011. Organic carbon and total nitrogen stocks in a Vertisol following 40 years of no-tillage, crop residue retention and nitrogen fertilisation. *Soil Till Res* 112:133–139.

Dalal, R. C., R. Eberhard, T. Grantham, and D. G. Mayer. 2003. Application of sustainability indicators, soil organic matter and electrical conductivity, to resource management in the northern grains region. *Aus J Exp Agr* 43:253–259.

Dao, T. H. 1998. Tillage and crop residue effects on carbon dioxide evolution and carbon storage in a Paleustoll. *Soil Sci Soc Am J* 62:250–256.

Davidson, E. A. and I. A. Janssens. 2006. Temperature sensitivity of soil carbon decomposition and feedbacks to climate change. *Nature* 440:165–173.

Dick, W. A., R. L. Blevins, W. W. Frye et al. 1998. Impacts of agricultural management practices on C sequestration in forest-derived soils of the eastern corn belt. *Soil Till Res* 47:235–244.

Dimoyiannis, D. G., C. D. Tsadilas, and S. Valmis. 1998. Factors affecting aggregate instability of Greek agricultural soils. *Commun Soil Sci Plan* 29:1239–1251.

Don, A., J. Schumacher, and A. Freibauer. 2011. Impact of tropical land-use change on soil organic carbon stocks—A meta-analysis. *Glob Change Biol* 17:1658–1670.

Dong, W., C. Hu, S. Chen, and Y. Zhang. 2009. Tillage and residue management effects on soil carbon and CO_2 emission in a wheat-corn double-cropping system. *Nutr Cycl Agroecosyst* 83:27–37.

Duiker, S. W., F. E. Rhoton, J. Torrent, N. E. Smeck, and R. Lal. 2003. Iron (hydr)oxide crystallinity effects on soil aggregation. *Soil Sci Soc Am J* 67:606–611.

Elliott, E. T. 1986. Aggregate structure and carbon, nitrogen, and phosphorus in native and cultivated soils. *Soil Sci Soc Am J* 50:627–633.

Eusterhues, K., C. Rumpel, M. Kleber, and I. Kögel-Knabner. 2003. Stabilisation of soil organic matter by interactions with minerals as revealed by mineral dissolution and oxidative degradation. *Org Geochem* 34:1591–1600.

Fonte, S. J., E. Yeboah, P. Ofori et al. 2009. Fertilizer and residue quality effects on organic matter stabilization in soil aggregates. *Soil Sci Soc Am J* 73:961–966.

Francesca Cotrufo, M., M. D. Wallenstein, C. M. Boot, K. Denef, and E. Paul. 2013. The Microbial Efficiency-Matrix Stabilization (MEMS) framework integrates plant litter decomposition with soil organic matter stabilization: Do labile plant inputs form stable soil organic matter? *Glob Change Biol* 19:988–995.

Friedlingstein, P., P. Cox, R. Betts et al. 2006. Climate-carbon cycle feedback analysis, results from the C4MIP model inter comparison. *J Climate* 19:3337–3353.

Golchin, A., J. M. Oades, J. O. Skjemstad, and P. Clarke. 1994. Soil structure and carbon cycling. *Aust J Soil Res* 32:1043–1068.

Grace, P. R., J. N. Ladd, G. P. Robertson, and S. H. Gage. 2005. SOCRATES—A simple model for predicting long-term changes in soil organic carbon in terrestrial ecosystems. *Soil Biol Biochem* 38:1172–1176.

Grace, P. R., W. M. Post, and K. Hennessy. 2006. The potential impact of climate change on Australia's soil organic carbon resources. *Carbon Balance Manage* 1:14.

Gunnarsson, S., H. Marstorp, A. Dahlin, and E. Witter. 2008. Influence of non-cellulose structural carbohydrate composition on plant material decomposition in soil. *Biol Fertil Soils* 45:27–36.

Guo, L. B. and R. M. Gifford. 2002. Soil carbon stocks and land use change: A meta analysis. *Glob Change Biol* 8:345–360.

Harden, J. W., S. E. Trumbore, B. J. Stocks et al. 2000. The role of fire in the boreal carbon budget. *Glob Change Biol* 6:174–184.

Hassink, J. 1997. The capacity of soils to preserve organic C and N by their association with clay and silt particles. *Plant Soil* 191:77–87.

Hathaway-Jenkins, L. J., R. Sakrabani, B. Pearce, A. P. Whitmore, and R. J. Godwin. 2011. A comparison of soil and water properties in organic and conventional farming systems in England. *Soil Use Manage* 27:133–142.

Haynes, R. J. and R. Naidu. 1998. Influence of lime, fertilizer and manure applications on soil organic matter content and soil physical conditions: A review. *Nutr Cycl Agroecosyst* 51:123–137.

Heenan, D., W. McGhie, F. Thomson, and K. Chan. 1995. Decline in soil organic carbon and total nitrogen in relation to tillage, stubble management, and rotation. *Aus J Exp Agr* 35:877–884.

Hoyle, F. C. and D. V. Murphy. 2006. Seasonal changes in microbial function and diversity associated with stubble retention versus burning. *Aust J Soil Res* 44:407–423.

Hulugalle, N. R., P. C. Entwistle, T. B. Weaver, F. Scott, and L. A. Finlay. 2002. Cotton-based rotation systems on a sodic Vertosol under irrigation: Effects on soil quality and profitability. *Aus J Exp Agr* 42:341–349.

Hulugalle, N. R. and F. Scott. 2008. Review of the changes in soil quality and profitability accomplished by sowing rotation crops after cotton in Australian vertosols from 1970 to 2006. *Aust J Soil Res* 46:173–190.

Jackson, L. E., I. Ramirez, R. Yokota et al. 2004. On-farm assessment of organic matter and tillage management on vegetable yield, soil, weeds, pests, and economics in California. *Agr Ecosyst Environ* 103:443–463.

Jastrow, J. D. 1996. Soil aggregate formation and the accrual of particulate and mineral-associated organic matter. *Soil Biol Biochem* 28:665–676.

Johnson, J. M. F., R. R. Allmaras, and D. C. Reicosky. 2006. Estimating source carbon from crop residues, roots and rhizodeposits using the national grain-yield database. *Agron J* 98:622–636.

Johnson, J. M. F., A. J. Franzluebbers, S. L. Weyers, and D. C. Reicosky. 2007. Agricultural opportunities to mitigate greenhouse gas emissions. *Environ Pollut* 150:107–124.

Kern, J. S. and M. G. Johnson. 1993. Conservation tillage impacts on national soil and atmospheric arbon levels. *Soil Sci Soc Am J* 57:200–210.

Khan, S. A., R. L. Mulvaney, T. R. Ellsworth, and C. W. Boast. 2007. The myth of nitrogen fertilization for soil carbon sequestration. *J Environ Qual* 36:1821–1832.

Kirkby, C. A., J. A. Kirkegaard, A. E. Richardson et al. 2011. Stable soil organic matter: A comparison of C:N:P:S ratios in Australian and other world soils. *Geoderma* 163:197–208.

Kisselle, K. W., C. J. Garrett, S. Fu et al. 2001. Budgets for root-derived C and litter-derived C: Comparison between conventional tillage and no tillage soils. *Soil Biol Biochem* 33:1067–1075.

Kladivko, E. J. 2001. Tillage systems and soil ecology. *Soil Till Res* 61:61–76.

Kleber, M. and M. G. Johnson. 2010. Advances in understanding the molecular structure of soil organic matter: Implications for interactions in the environment. *Adv Agron* 106:77–142.

Kleber, M., P. S. Nico, A. Plante et al. 2011. Old and stable soil organic matter is not necessarily chemically recalcitrant: Implications for modeling concepts and temperature sensitivity. *Glob Change Biol* 17:1097–1107.

Kumar, K. and K. M. Goh. 1999. Crop residues and management practices: Effects on soil quality, soil nitrogen dynamics, crop yield, and nitrogen recovery. *Adv Agron* 68:197–319.

Lal, R. 2004. Soil carbon sequestration impacts on global climate change and food security. *Science* 304:1623–1627.

Lal, R. 2010. Beyond Copenhagen: Mitigating climate change and achieving food security through soil carbon sequestration. *Food Sec* 2:169–177.

Lehmann, J. 2007. A handful of carbon. *Nature* 447:143–144.

Leifeld, J., R. Reiser, and H. R. Oberholzer. 2009. Consequences of conventional versus organic farming on soil carbon: Results from a 27-year field experiment. *Agron J* 101:1204–1218.

Leirós, M. C., C. Trasar-Cepeda, S. Seoane, and F. Gil-Sotres. 1999. Dependence of mineralization of soil organic matter on temperature and moisture. *Soil Biol Biochem* 31:327–335.

Lemke, R. L., A. J. VandenBygaart, C. A. Campbell, G. P. Lafond, and B. Grant. 2010. Crop residue removal and fertilizer N: Effects on soil organic carbon in a long-term crop rotation experiment on a Udic Boroll. *Agr Ecosyst Environ* 135:42–51.

Liu, D. L., K. Y. Chan, and M. K. Conyers. 2009. Simulation of soil organic carbon under different tillage and stubble management practices using the Rothamsted carbon model. *Soil Till Res* 104:65–73.

Ludwig, B., B. John, R. Ellerbrock, M. Kaiser, and H. Flessa. 2003. Stabilization of carbon from maize in a sandy soil in a long-term experiment. *Eur J Soil Sci* 54:117–126.

Luo, Z., E. Wang, and O. J. Sun. 2010. Soil carbon change and its responses to agricultural practices in Australian agro-ecosystems: A review and synthesis. *Geoderma* 155:211–223.

Mäder, P., A. Fließbach, D. Dubois et al. 2002. Soil fertility and biodiversity in organic farming. *Science* 296:1694–1697.

Mambelli, S., J. A. Bird, G. Gleixner, T. E. Dawson, and M. S. Torn. 2011. Relative contribution of foliar and fine root pine litter to the molecular composition of soil organic matter after in situ degradation. *Org Geochem* 42:1099–1108.

Mancinelli, R., E. Campiglia, A. Di Tizio, and S. Marinari. 2010. Soil carbon dioxide emission and carbon content as affected by conventional and organic cropping systems in Mediterranean environment. *Appl Soil Ecol* 46:64–72.

Marinari, S., A. Lagomarsino, M. C. Moscatelli, A. Di Tizio, and E. Campiglia. 2010. Soil carbon and nitrogen mineralization kinetics in organic and conventional three-year cropping systems. *Soil Till Res* 109:161–168.

Marriott, E. E. and M. M. Wander. 2006. Total and labile soil organic matter in organic and conventional farming systems. *Soil Sci Soc Am J* 70:950–959.

Marris, E. 2006. Putting the carbon back: Black is the new green. *Nature* 442:624–626.

Marschner, B., S. Brodowski, A. Dreves et al. 2008. How relevant is recalcitrance for the stabilization of organic matter in soils? *J Plant Nutr Soil Sci* 171:91–110.

Mazzoncini, M., S. Canali, M. Giovannetti et al. 2010. Comparison of organic and conventional stockless arable systems: A multidisciplinary approach to soil quality evaluation. *Appl Soil Ecol* 44:124–132.

McKenzie, D. 1998. *SOILpak for Cotton Growers*, 3rd edn. Orange, Australia: NSW Agriculture.

McLauchlan, K. 2006. The nature and longevity of agricultural impacts on soil carbon and nutrients: A review. *Ecosystems* 9:1364–1382.

Mikha, M. M. and C. W. Rice. 2004. Tillage and manure effects on soil and aggregate-associated carbon and nitrogen. *Soil Sci Soc Am J* 68:809–816.

Mondelaers, K., J. Aertsens, and G. Van Huylenbroeck. 2009. A meta-analysis of the differences in environmental impacts between organic and conventional farming. *Brit Food J* 111:1098–1119.

Mondini, C., M. L. Cayuela, T. Sinicco et al. 2007. Greenhouse gas emissions and carbon sink capacity of amended soils evaluated under laboratory conditions. *Soil Biol Biochem* 39:1366–1374.

Moritsuka, N., J. Yanai, K. Mori, and T. Kosaki. 2004. Biotic and abiotic processes of nitrogen immobilization in the soil-residue interface. *Soil Biol Biochem* 36:1141–1148.

O'Donnell, A., M. Seasman, A. Macrae, I. Waite, and J. Davies. 2001. Plants and fertilisers as drivers of change in microbial community structure and function in soils. *Plant Soil* 232:135–145.

Olchin, G. P., S. Ogle, S. D. Frey et al. 2008. Residue carbon stabilization in soil aggregates of no-till and tillage management of dryland cropping systems. *Soil Sci Soc Am J* 72:507–513.

Pankhurst, C. E., C. A. Kirkby, B. G. Hawke, and B. D. Harch. 2002. Impact of a change in tillage and crop residue management practice on soil chemical and microbiological properties in a cereal-producing red duplex soil in NSW, Australia. *Biol Fertil Soils* 35:189–196.

Parton, W. J., D. S. Schimel, C. V. Cole, and D. S. Ojima. 1987. Analysis of factors controlling soil organic matter levels in Great Plains grasslands. *Soil Sci Soc Am J* 51:1173–1179.

Paustian, K., O. Andrén, H. H. Janzen et al. 1997. Agricultural soils as a sink to mitigate CO_2 emissions. *Soil Use Manage* 13:230–244.

Paustian, K., J. Six, E. T. Elliott, and H. W. Hunt. 2000. Management options for reducing CO_2 emissions from agricultural soils. *Biogeochemistry* 48:147–163.

Post, W. M., W. R. Emanuel, P. J. Zinke, and A. G. Stangenberger. 1982. Soil carbon pools and world life zones. *Nature* 298:156–159.

Post, W. M., T. H. Peng, W. R. Emanuel et al. 1990. The global carbon cycle. *Am Sci* 78:310–326.

Potter, K. N., J. Velazquez-Garcia, E. Scopel, and H. A. Torbert. 2007. Residue removal and climatic effects on soil carbon content of no-till soils. *J Soil Water Conserv* 62:110–114.

Poudel, D. D., W. R. Horwath, W. T. Lanini, S. R. Temple, and A. H. C. van Bruggen. 2002. Comparison of soil N availability and leaching potential, crop yields and weeds in organic, low-input and conventional farming systems in northern California. *Agr Ecosyst Environ* 90:125–137.

Puget, P., C. Chenu, and J. Balesdent. 1995. Total and young organic matter distributions in aggregates of silty cultivated soils. *Eur J Soil Sci* 46:449–459.

Puget, P. and L. E. Drinkwater. 2001. Short-term dynamics of root- and shoot-derived carbon from a leguminous green manure. *Soil Sci Soc Am J* 65:771–779.

Pulleman, M. M., J. Bouma, E. A. van Essen, and E. W. Meijles. 2000. Soil organic matter content as a function of different land use history. *Soil Sci Soc Am J* 64:689–693.

Reicosky, D. C., S. D. Evans, C. A. Cambardella et al. 2002. Continuous corn with moldboard tillage: Residue and fertility effects on soil carbon. *J Soil Water Conserv* 57:277–284.

Roscoe, R. and P. Buurman. 2003. Tillage effects on soil organic matter in density fractions of a Cerrado oxisol. *Soil Till Res* 70:107–119.

Rubino, M., J. A. J. Dungait, R. P. Evershed et al. 2010. Carbon input belowground is the major C flux contributing to leaf litter mass loss: Evidences from a ^{13}C labelled-leaf litter experiment. *Soil Biol Biochem* 42:1009–1016.

Russell, A. E., C. A. Cambardella, D. A. Laird, D. B. Jaynes, and D. W. Meek. 2009. Nitrogen fertilizer effects on soil carbon balances in midwestern U.S. agricultural systems. *Ecol Appl* 19:1102–1113.

Saby, N. P. A., P. H. Bellamy, X. Morvan et al. 2008. Will European soil-monitoring networks be able to detect changes in topsoil organic carbon content? *Glob Change Biol* 14:2432–442.

Sanderman, J. and J. A. Baldock. 2010. Accounting for soil carbon sequestration in national inventories: A soil scientist's perspective. *Environ Res Lett* 5:1–6.

Sanderman, J., R. Farquharson, and J. Baldock. 2010. Soil carbon sequestration potential: A review for Australian agriculture. Report to the Australian Government Department of Climate Change and Energy Efficiency. Canberra, Australia: CSIRO Publication.

Schlesinger, W. H. 1990. Evidence from chronosequence studies for a low carbon-storage potential of soils. *Nature* 348:232–234.

Sims, P. L. and J. A. Bradford. 2001. Carbon dioxide fluxes in a southern plains prairie. *Agr Forest Meteorol* 109:117–134.

Six, J., R. T. Conant, E. A. Paul, and K. Paustian. 2002a. Stabilization mechanisms of soil organic matter: Implications for C-saturation of soils. *Plant Soil* 241:155–176.

Six, J., E. T. Elliott, and K. Paustian. 1999. Aggregate and soil organic matter dynamics under conventional and no-tillage systems. *Soil Sci Soc Am J* 63:1350–1358.

Six, J., E. T. Elliott, and K. Paustian. 2000a. Soil macroaggregate turnover and microaggregate formation: A mechanism for C sequestration under no-tillage agriculture. *Soil Biol Biochem* 32:2099–2103.

Six, J., E. T. Elliott, K. Paustian, and J. W. Doran. 1998. Aggregation and soil organic matter accumulation in cultivated and native grassland soils. *Soil Sci Soc Am J* 62:1367–1377.

Six, J., C. Feller, K. Denef et al. 2002b. Soil organic matter, biota and aggregation in temperate and tropical soils—Effects of no-tillage. *Agronomie* 22:755–775.

Six, J., K. Paustian, E. T. Elliott, and C. Combrink. 2000b. Soil structure and organic matter I. Distribution of aggregate-size classes and aggregate-associated carbon. *Soil Sci Soc Am J* 64:681–689.

Skjemstad, J. O., R. P. Lefeuvre, and R. E. Prebble. 1990. Turnover of soil organic matter under pasture as determined by ^{13}C natural abundance. *Aust J Soil Res* 28:267–276.

Smith, P., D. Martino, Z. Cai et al. 2008. Greenhouse gas mitigation in agriculture. *Philos T Roy Soc B* 363:789–813.

Smith, P., J. U. Smith, D. S. Powlson et al. 1997. A comparison of the performance of nine soil organic matter models using datasets from seven long-term experiments. *Geoderma* 81:153–225.

Sommerfeldt, T. G. and C. Chang. 1985. Changes in soil properties under annual applications of feedlot manure and different tillage practices. *Soil Sci Soc Am J* 49:983–987.

Spycher, G., P. Sollins, and S. Rose. 1983. Carbon and nitrogen in the light fraction of a forest soil: Vertical distribution and seasonal patterns. *Soil Sci* 135:79–87.

Stockdale, E. A., N. H. Lampkin, M. Hovi et al. 2001. Agronomic and environmental implications of organic farming systems. *Adv Agron* 70:261–327.

Stockdale, E. A., M. A. Shepherd, S. Fortune, and S. P. Cuttle. 2002. Soil fertility in organic farming systems— Fundamentally different? *Soil Use Manage* 18:301–308.

Thomsen, I. K., P. Schjønning, B. Jensen, K. Kristensen, and B. T. Christensen. 1999. Turnover of organic matter in differently textured soils: II. Microbial activity as influenced by soil water regimes. *Geoderma* 89:199–218.

Tisdall, J. M. and J. M. Oades. 1982. Organic matter and water-stable aggregates in soils. *J Soil Sci* 33:141–163.

Trinsoutrot, I., S. Recous, B. Bentz et al. 2000a. Biochemical quality of crop residues and carbon and nitrogen mineralization kinetics under nonlimiting nitrogen conditions. *Soil Sci Soc Am J* 64:918–926.

Trinsoutrot, I., S. Recous, B. Mary, and B. Nicolardot. 2000b. C and N fluxes of decomposing ^{13}C and ^{15}N *Brassica napus* L.: Effects of residue composition and N content. *Soil Biol Biochem* 32:1717–1730.

Valzano, F., B. W. Murphy, and T. Koen. 2005. The impact of tillage on changes in soil carbon density with special emphasis on Australian conditions. National Carbon Accounting System technical report. Canberra: Australian Greenhouse Office.

van Diepeningen, A. D., O. J. de Vos, G. W. Korthals, and A. H. C. van Bruggen. 2006. Effects of organic versus conventional management on chemical and biological parameters in agricultural soils. *Appl Soil Ecol* 31:120–135.

van Groenigen, K. J., A. Hastings, D. Forristal et al. 2011. Soil C storage as affected by tillage and straw management: An assessment using field measurements and model predictions. *Agr Ecosyst Environ* 140:218–225.

von Lützow, M., I. Kögel-Knabner, K. Ekschmitt et al. 2006. Stabilization of organic matter in temperate soils: Mechanisms and their relevance under different soil conditions—A review. *Eur J Soil Sci* 57:426–445.

von Lützow, M., I. Kögel-Knabner, K. Ekschmitt et al. 2007. SOM fractionation methods: Relevance to functional pools and to stabilization mechanisms. *Soil Biol Biochem* 39:2183–2207.

Wander, M. M., S. J. Traina, B. R. Stinner, and S. E. Peters. 1994. Organic and conventional management effects on biologically active soil organic matter pools. *Soil Sci Soc Am J* 58:1130–1139.

Wang, W. J., J. A. Baldock, R. C. Dalal, and P. W. Moody. 2004. Decomposition dynamics of plant materials in relation to nitrogen availability and biochemistry determined by NMR and wet-chemical analysis. *Soil Biol Biochem* 36:2045–2058.

Watson, C. A., D. Atkinson, P. Gosling, L. R. Jackson, and F. W. Rayns. 2002. Managing soil fertility in organic farming systems. *Soil Use Manage* 18:239–247.

Webb, A. A. 2002. Pre-clearing soil carbon levels in Australia. National Carbon Accounting System technical report. Canberra: Australian Greenhouse Office.

Wells, A. T., K. Y. Chan, and P. S. Cornish. 2000. Comparison of conventional and alternative vegetable farming systems on the properties of a yellow earth in New South Wales. *Agr Ecosyst Environ* 80:47–60.

Wilhelm, W. W., J. M. F. Johnson, J. L. Hatfield, W. B. Voorhees, and D. R. Linden. 2004. Crop and soil productivity response to corn residue removal: A literature review. *Agron J* 96:1–17.

Wilson, B. R., I. Growns, and J. Lemon. 2008. Land-use effects on soil properties on the north-western slopes of New South Wales: Implications for soil condition assessment. *Soil Res* 46:359–367.

Wilson, B. R., T. B. Koen, P. Barnes, S. Ghosh, and D. King. 2011. Soil carbon and related soil properties along a soil type and land-use intensity gradient, New South Wales, Australia. *Soil Use Manage* 27:437–447.

Yadvinder, S., S. Bijay, and J. Timsina. 2005. Crop residue management for nutrient cycling and improving soil productivity in rice-based cropping systems in the tropics. *Adv Agron* 85:269–407.

Young, R. R., B. Wilson, S. Harden, and A. Bernardi. 2009. Accumulation of soil carbon under zero tillage cropping and perennial vegetation on the Liverpool Plains, eastern Australia. *Soil Res* 47:273–285.

CHAPTER 3

Carbon Capture and Use as an Alternative to Carbon Capture and Storage

Bruno Glaser and Per Espen Stoknes

CONTENTS

3.1 Introduction 58
3.2 Carbon Capture and Storage...59
3.3 Carbon Capture and Use: Soil Improvement..60
    3.3.1 The Terra Preta Phenomenon 60
    3.3.2 Biochar Systems: Copying the Terra Preta Concept...............................62
    3.3.3 Plasma Carbon...64
    3.3.4 Economic Value of Carbons to Soil 66
    3.3.5 Ecological Value (Ecosystem Services)..67
        3.3.5.1 Carbon Sequestration..67
        3.3.5.2 Non-CO_2 Greenhouse Gas Emissions 68
        3.3.5.3 Crop Yields and Soil Fertility 68
    3.3.6 Scalability 69
    3.3.7 Ongoing and Further Research on Plasma Carbon 70
        3.3.7.1 Available Data on Plasma Carbon Characterization 70
        3.3.7.2 Desired and Necessary Further Research on Biochar and Plasma Carbon 70
3.4 Carbon Capture and Use in Buildings and Civil Infrastructure.....................70
    3.4.1 Construction Materials 70
    3.4.2 Economic Value (Value Chains)...73
    3.4.3 Life Cycle Assessment..75
    3.4.4 Scalability 76
3.5 Other Possible Long-Term Carbon Sequestration Concepts..........................76
    3.5.1 Sequestration of Solid Carbon in (Polluted) Ocean Sediments..............76
    3.5.2 Use of Carbon as Adsorbent (Activated Carbons)..................................77
3.6 Carbon Markets with Long-Term Sequestration Potential 77
3.7 Summary and Conclusions 78
References...78

3.1 INTRODUCTION

A principal global problem caused by humans is atmospheric pollution with greenhouse gases (GHGs) such as carbon dioxide (CO_2), methane (CH_4), and nitrous oxides (especially N_2O). The major contribution to the global anthropogenic greenhouse effect is caused by the emission of high amounts of CO_2 into the atmosphere (corresponding to 11 Pg of carbon [C] per year), of which about 10 Pg of C per year is caused by emissions from fossil fuel burning (about 80% for electricity production and transport and 20% for cement and steel production) and an additional 1 Pg of C per year by land use due to the mineralization of organic matter resulting in soil degradation and desertification (Figure 3.1). Carbon dioxide emissions due to improper land use include those from deforestation, biomass burning, conversion of grassland to agricultural ecosystems, drainage of wetlands, and soil cultivation in general (Lal 2004). GHG emissions will further increase due to the increasing population growth, estimated to be 9 billion in 2050 compared with 7 billion in 2012. In addition, climate change itself will influence terrestrial and marine ecosystems' capacity to act as repositories for anthropogenic CO_2 and hence provide a mostly self-reinforcing feedback to climate change (Fung et al. 2005). For instance, soil organic matter (SOM) mineralization is expected to increase due to increasing temperatures and the potential of CO_2 storage in the oceans is expected to decrease with increasing temperatures, both of which will further increase the anthropogenic greenhouse effect (Fung et al. 2005). Therefore, sustainable solutions for the removal (sequestration) of atmospheric CO_2 and thus for global climate change mitigation are needed.

According to Lal (2004), C sequestration is the long-term removal of CO_2 from the atmosphere by biotic or abiotic processes. Although there is no clear definition of how long is long term, it is generally accepted that it should be in the range of at least hundreds of years, even better thousands of years, thus being in the centennial to millennial timescales of secure storage. Please note that for conversion from CO_2 to C and from C to CO_2, division by and multiplication with 3.67 is necessary, respectively. In this chapter, wherever possible, we refer to C and not to CO_2 for better comparability with the global C sinks presented in Figure 3.1.

The current available agricultural techniques such as conservation tillage, no-tillage, or desertification control will contribute only little to C sequestration into soil (Figure 3.1). Only the conversion of agricultural land into grassland (pasture) could cope at least with agriculturally derived CO_2 emissions (Figure 3.1). However, this is not possible at the large scale due to the tremendous loss

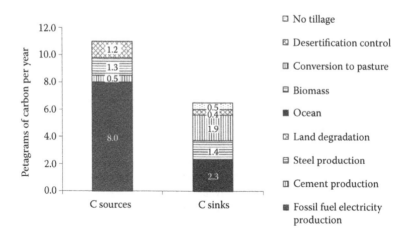

Figure 3.1   Sources and potential agricultural management sinks for atmospheric carbon dioxide (CO_2) in petagrams (10^{15} g) equal to a billion metric tons of C. Note that conversion into CO_2 data requires multiplication by 3.67. (Data from www.iwr.de and Lal, R., *Environ Int*, 30, 981–990, 2004.)

of energy during the conversion of plant to animal food, which would even deteriorate the current problem. Therefore, alternative ways for additional C sequestration are needed, which is the objective of this chapter.

The following potential options for C sequestration are discussed in the subsequent sections:

1. Carbon capture and storage (CCS)
2. Carbon capture and use (CCU) including C sequestration in soil, construction materials, and infrastructure
3. Sequestration of solid carbon in (polluted) ocean sediments
4. Use of carbon as an adsorbent (activated carbons) and then the disposal of carbon in wastelands (landfills, closed mines, deserts, etc.)

3.2 CARBON CAPTURE AND STORAGE

CCS is the removal of CO_2 from fossil fuel combustion in power plants and its subsequent dehydration, compression, transportation, and storage in depleted oil and gas reservoirs or in the deep ocean (Figure 3.2). For this purpose, three different technologies are currently under development: (i) postcombustion capture, (ii) precombustion capture, and (iii) oxyfuel combustion capture (Figure 3.2). In postcombustion capture, CO_2 will be removed after its combustion from the flue gas by CO_2 separation techniques that are still under development (Figure 3.2). The principle of precombustion capture is that fossil fuels are gasified or partially oxidized to carbon monoxide and hydrogen, the former being shifted to CO_2 with pure oxygen and separated while hydrogen is used for heat and power generation (Figure 3.2). The oxyfuel concept works with partial CO_2 recycling (Figure 3.2).

Independent from the used technology, CCS will reduce the energy efficiency of power and heat generation because CO_2 separation is an energy-intensive process. It is estimated that this reduction can be up to one-third, which will increase the demand for primary resources and the price for

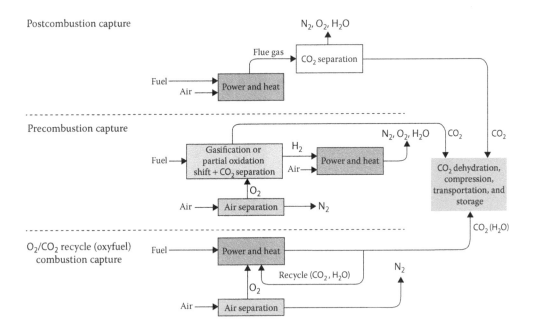

Figure 3.2 Principle of carbon capture and storage (CCS) technologies.

electricity. Although CCS can reduce atmospheric emissions, especially if biomass is co-fired in thermal plants, this technique cannot reduce atmospheric CO_2 levels. There are several additional problems that will not be addressed in this chapter. Therefore, it will be necessary to complement CCS with other approaches, which will be discussed in the following sections.

3.3 CARBON CAPTURE AND USE: SOIL IMPROVEMENT

3.3.1 The Terra Preta Phenomenon

Soil quality is a function of soil organic matter (SOM) content, plant-available water-holding capacity, root depth and density, clay content, and time. As SOM contains nutrients, stores soil water, and aerates the soil providing a habitat for soil organisms (plants and animals), it is a crucial factor for soil fertility (Ross 1993; Zech et al. 1997). However, intensive and improper land use will deteriorate SOM levels and release CO_2 into the atmosphere. This leads to soil degradation (desertification), which sets in motion a vicious circle with increased pressure on agricultural land (Lal 2009). The continuous cultivation of the most fertile natural soils (chernozems) in temperate regions reduced SOM levels (and thus soil fertility) by 50% over 65 years. The same effect took only 6 years in tropical regions (Tiessen et al. 1994).

In the humid tropics, a combination of high temperatures and a favorable soil moisture regime induces high rates of organic matter decomposition and the weathering of primary minerals (Sombroek et al. 1993; Tiessen et al. 1994). For these reasons, naturally occurring Amazonian soils have low SOM levels and low soil fertility (e.g., ferralsols, Figure 3.3). Surprisingly, patches of sustainably fertile, dark-colored soils that are rich in SOM have been found in the same region, especially in Amazonia. They are known as anthropogenic dark earths (ADE) or *terra preta* (*de Índio*) (Figure 3.3). These soils are characterized by large stocks of stable SOM (averaging 250 Mg/ha) and higher levels of nutrients such as nitrogen (averaging 17 Mg/ha) and phosphorus (averaging 12 Mg/ha) compared with the surrounding soils by a factor of 3, 4–10, and 4–50, respectively (Glaser and Birk 2012).

Figure 3.3 **(See color insert)** Typical soil profile of humid tropical soil with no biochar exhibiting low SOM levels and low soil fertility (*left*) due to intensive weathering and a terra preta with biochar (*right*) showing exactly the opposite; namely, high SOM and nutrient contents and thus high soil fertility.

Terra preta contains higher amounts of rapidly mineralizable (labile) and passive/recalcitrant (stable) SOM compared with the surrounding soils (Glaser and Birk 2012). The enhanced input of unaltered plant material is responsible for the higher amounts of labile SOM in terra preta. Mineralization of labile SOM continuously supplies plants with nutrients in terra preta, similar to a continuous natural fertilization (Glaser and Birk 2012). Besides this labile SOM pool, terra preta contains as much as 70 times more stable SOM than the surrounding soils, which can be partly explained by physical and organomineral stabilization and also by higher recalcitrance against microbial degradation (Glaser et al. 2003). However, the main reason for the high stability is the chemical recalcitrance of SOM in terra preta. ^{13}C nuclear magnetic resonance (NMR) spectroscopy and molecular marker analysis revealed that SOM in terra preta is rich in condensed aromatic structures (Glaser et al. 2003; Novotny et al. 2009; Zech et al. 1990). These structures are called black carbon (Glaser et al. 1998) as they are a form of charcoal or biochar (Keiluweit et al. 2010). Terra preta sites in central Amazonia contain on average 50 Mg of biochar per hectare and 1 m soil depth, being enriched by a factor of 70 compared with adjacent soils (Glaser et al. 2001). From these results it is clear that biochar is the key factor for the stability of SOM in terra preta and thus, for the terra preta genesis (Glaser and Birk 2012).

^{13}C NMR spectroscopy further revealed a signal corresponding to aromatic acids such as mellitic acid (Möller et al. 2000), which is considered a metabolite of biochar oxidation (Glaser et al. 1998, 2001). Thus, slow biochar oxidation over time produces carboxylic groups on the edges of the aromatic backbone or aromatic acids as metabolites, which increases the capacity of ionic binding, thereby preventing nutrients from leaching (Glaser et al. 2000). From these results it can be concluded that the biochar found in terra preta is not only responsible for long-term C sequestration, but it is also an indirect key factor for the high fertility of terra preta (Glaser 2007; Glaser et al. 2003, 2001).

The existence of terra preta even several thousand years after its creation shows that an improvement of highly weathered tropical soils by human actions is possible. More importantly, terra preta represents a model for sustainable soil fertility and long-term C sequestration that is still possible today if it is created by the sustainable management of natural resources (Figure 3.4). Most probably, terra preta was not intentionally created to improve soil fertility at large scales, at least not at

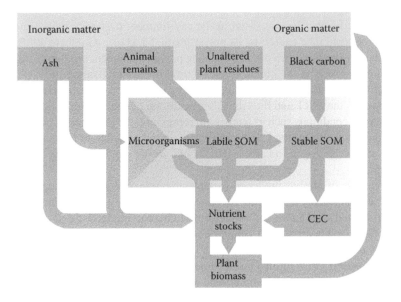

Figure 3.4 Simplified model of terra preta genesis. (From Glaser, B. and Birk, J.J., *Geochim Cosmochim Acta*, 82, 39–51, 2012. With permission.)

the beginning. Instead, more likely it is a genesis from the deposition of organic waste and charred residues combined with home garden agriculture. As a consequence, the creation of a fertile soil (terra preta) attracted more and more people and favored population growth and food production, which set in motion self-reinforcing and self-organizing processes as outlined in Figure 3.4.

The existence of terra preta has been known for more than 100 years, but only in the last decade has it attracted broad scientific interest (Woods and Denevan 2009). The potential of terra preta not only motivated research, but also stimulated a huge number of hypotheses and ideas concerning its genesis, partly in advance of proof based on scientific knowledge. For modern society, terra preta could act as a model for sustainable agriculture in the humid tropics (Glaser 2007; Glaser et al. 2001; Sombroek et al. 2002) and for other soils that exhibit low soil fertility and low SOM levels (degraded or desertified soils). Even strategies to prevent fertile soils from being degraded are possible. Agricultural techniques leading to the formation of terra preta have the potential to reverse land degradation due to intensive agricultural practices and to reclaim degraded areas in many regions of the world (Glaser and Birk 2012; Glaser 2007). Additionally, the high stability of SOM in terra preta and experimental data (e.g., Kuzyakov et al. 2009) show that with biochar, carbon can be sequestered for millennia in soils. Therefore, the terra preta concept has the potential to combine sustainable agriculture with long-term CO_2 sequestration, which will be addressed in the next section.

3.3.2 Biochar Systems: Copying the Terra Preta Concept

Biochar is created by the thermal decomposition of organic materials for use as a soil amendment. Pyrolysis converts diverse and predominantly labile organic molecules into three fractions— one that comprises stable aromatic rings (char) that can be stored long term in the soil (biochar), and two that can be used for energy generation: a liquid bio-oil and a gas that can be used for the synthesis of organic molecules (syngas). The carbon that remains in biochar is considered to reside in stable compounds that are depleted in hydrogen and oxygen, precluding its subsequent degradation to CO_2 in the environment. The nutrient elements that are not volatilized at high temperature are conserved in biochar, and free of carbon they may become soluble and readily available to plants.

It is believed that biochar has persisted in the environment over millennia due to its biological and chemical recalcitrance caused by its polyaromatic backbone (Goldberg 1985; Schmidt and Noack 2000). The existence of terra preta even today proves that biochar is stable over millennia in extreme environments such as the humid tropics. Using the mineralization of ^{14}C-labeled biochar over a 4 year period, Kuzyakov et al. (2009) calculated a mean residence time (MRT) for biochar of 2000 years. However, the MRT and thus the C sequestration potential of different biochars depend on biochar formation conditions such as temperature, pressure, the presence of oxygen, and the process itself. A rough estimate for the C sequestration potential of different biochars can be obtained by plotting the atomic hydrogen to carbon (H/C) ratio versus the atomic oxygen to carbon (O/C) ratio (Figure 3.5).

The traditional use of fire has produced variable amounts of charcoal. It appears likely that all human society adopting stable agriculture for the first time, in the absence of artificial fertilizer or livestock, experimented with the use of some form of charcoal. Indeed, the use of biochar in crop production was described in nineteenth-century handbooks on agricultural management in both the United Kingdom and the United States. Its role in traditional soil management practices is ongoing in several African countries (Whitman and Lehmann 2009) as well as Japan (Ogawa and Okimori 2010). Archaeological evidence for the traditional use of biochar remains perfectly preserved in the Amazon Basin, where it is widely accepted that its original inhabitants created fertile "black" soil (terra preta) by combining their own municipal waste with charcoal in a way that provided sustainable settled agriculture (Glaser and Birk 2012; Glaser 2007; Glaser et al. 2001).

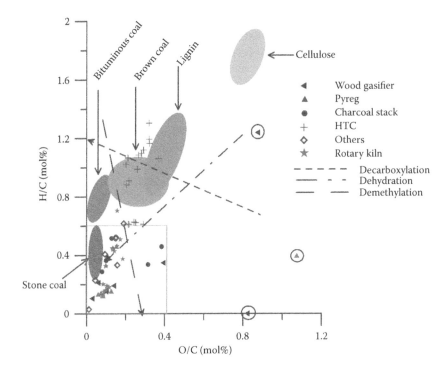

Figure 3.5 Atomic H/C versus atomic O/C plot (From van Krevelen, D.W., *Coal: Typology, Chemistry, Physics, Constitution*, Elsevier, Amsterdam, 1961. With permission.) of different natural materials and materials proposed as biochar. (From Schimmelpfennig, S. and Glaser, B., *J Environ Qual*, 41, 1001–1013, 2012.) Please note that according to the suggested "biochar window" of O/C < 0.4 and H/C < 0.6 drawn as a frame in the lower left of the graph, hydrothermal carbonization material (HTC material) cannot be accepted as biochar under the currently applied conditions of about 180°C and 20 bar.

Biochar in the form of charcoal has long been used in the production of potting media for horticulture. Although the origins of this practice and its objectives are not clear, it is generally perceived that charcoal mitigates the odor that can emanate from the decomposition of other organic materials in horticultural compost. In the context of mainstream agriculture, documented experimentation with composted charcoal extends over 30 years in Japan, with tests of palm-shell charcoal mixed into poultry manure (Ogawa and Okimori 2010). The perception that these mixtures can function as a "biological fungicide" has led to the commercialization of a product targeting crop health rather than soil conditioning or soil fertility. More recent experiments examining the combination of compost and biochar have emerged from the understanding that the function of biochar in the absence of other inputs or soil fractions is limited, and it functions best as a catalyst for other soil processes. Such mixtures were an element of early experiments that sought to rapidly recreate the function of ancient terra preta in new areas of the same soil type (Steiner et al. 2007, 2008).

Because biochar production for agricultural use in the European Union (EU) is still under development, no protocol or standard has been established. Efforts led by the International Biochar Initiative are currently underway to develop a "biochar standard." This will be important to satisfy consumer and policy concerns on the safety of biochar on human and environmental health. A simple definition of biochar can be obtained by elemental composition and ratios (Figure 3.5). Recently, as a threshold, O/C and H/C ratios of ≤0.4 and ≤0.6 have been suggested (Schimmelpfennig and Glaser 2012). Further criteria comprise the absence of toxic compounds such as heavy metals mainly depending on the feedstock material, and polycyclic aromatic hydrocarbons (PAHs) and dioxins mainly depending on the process parameters (Schimmelpfennig and Glaser 2012). Also the degree of aromatization as assessed by O/C and H/C ratios or the black carbon content and the specific surface area are helpful material properties (Schimmelpfennig and Glaser 2012).

Like charcoal, biochar is expected to be made up of highly stable molecular forms of carbon. Due to its resistance to biochemical degradation, the potential for biochar creation to sequester carbon into the soil in the long term (centuries to millennia) is well appreciated, offering clear potential for its use as a tool in climate change mitigation (Woolf et al. 2010). The impacts of other characteristics of biochar on soil productivity have been measured in isolation, or in soil under controlled conditions. These impacts remain incompletely explained or proven in the predictive sense but include effects such as modification of soil pH and cation-exchange capacity (and thus the storage of available nitrogen in the form of ammonium), crop nutrient availability, changes in soil water dynamics and soil moisture, soil microbial community structure and abundance, soil structure and architecture, SOM content and its turnover, crop disease mitigation, and the impacts on the rate of non-CO_2 GHG emissions from soil. Plant productivity reflected by crop yield is the most common aggregate means by which the net result of the intervention has been assessed, although only measurements of more detailed variables allow an insight into how the effect has occurred.

Other means of biochar loss from ecosystems include surface erosion by wind or water or transport to the subsoil, either as small particles with rainwater or through dissolution as (highly aromatic) dissolved organic carbon (DOC). Further mechanisms of biochar movement from the surface to deeper layers include bioturbation, kryoturbation, or anthropogenic management (Major et al. 2010). The losses due to erosion or relocation to deeper soil can be considerable and quick: Major et al. (2010) reported a migration rate of 379 kg of C/ha/year corresponding to 0.3% after biochar application of 116 Mg/ha to the top 10 cm of a grassland soil downward to a depth of 15–30 cm during a 2 year study in Columbia. In the same experiment, respiratory biochar losses or losses via DOC leaching were 2.2% and 1%, respectively. In a temperate agroecosystem, no significant biochar losses were observed after biochar applications of 5, 10, and 20 Mg/ha from the top 10 cm of soil during a 2 year study in Brandenburg, Germany (Schulz and Glaser 2012).

To date, standardized best practice guidelines for the application of biochar have not been established. Challenges include the handling of dry biochar. In particular, losses due to wind erosion and transport are critical as black particles in the atmosphere decrease the albedo effect with a considerable greenhouse potential (Woodward et al. 2009). However, dust development can be easily avoided by biochar wetting or by mixing it with organic wastes such as slurry or compost. Strategies for mixing or composting biochar with a nutrient carrier substance such as green waste (composting), slurry, or manure will have the positive side effect of "loading" biochar with nutrients.

3.3.3 Plasma Carbon

An effective way to prevent CO_2 emission is simply not to burn the carbon. Instead, fossil fuels can be decarbonized producing hydrogen fuel and solid carbon as a "residue." This concept is not new and was proposed by Steinberg (1998) as "fossil fuels as hydrogen ores." We will extend this concept to hydrogen and carbon capture and use (H + CCU). Fossil fuels are approximately CH_x, where $x \approx 4$ for natural gas, $x \approx 2$ for petroleum, and $x \approx 0.7$ for coal. Hydrogen is commonly produced from hydrocarbons such as methane by reforming with steam, according to Equation 3.1

$$CH_4 + 2\,H_2O \rightarrow 4\,H_2 + CO_2 \qquad (3.1)$$

In this conventional reaction, methane/natural gas is used to reduce water, producing more hydrogen, but with the coproduction of CO_2. This is advantageous if CO_2 can be vented into the atmosphere at no cost. However, in the context of climate change and GHG emissions, CO_2 venting at no cost is no longer accepted. Therefore, the water–fuel reaction (Equation 3.1) imposes an additional cost due to CO_2 taxation.

During thermolysis, less hydrogen is produced, but no carbon is oxidized so that no CO_2 is formed and no venting or capturing is required (Equation 3.2).

$$CH_4 \rightarrow 2\,H_2 + \text{solid C} \tag{3.2}$$

As carbon contains a significant amount of energy, less electricity can be produced with the same amount of fuel. However, added value can be obtained from the use of solid carbon as a material.

Plasma carbon platform technology has the potential to apply microwave plasma as a disruptive technology in a wide range of energy and industrial applications. Examples of plasma, the fourth state of matter (solid, fluid, gas, plasma), are the sun, the northern lights, lightning, and neon tubes. A thermal plasma is an ionized gas having a temperature of at least 2000°C.

The innovation of a nonequilibrium or cold plasma is an uneven energy distribution of ions and electrons. The basic principle is that microwaves excite covalent C–H bonds yielding cold plasma (nonequilibrium plasma phase, electron temperature of >10,000°C, ion temperature of 200°C–1000°C). Technically, a cold plasma can be generated by the GasPlas reactor (Figure 3.6). Natural gas or biogas (methane, CH_4) is thus transformed by low-energy plasma cracking into hydrogen, heat, and pure carbon according to Equation 3.3:

$$CH_4 + \text{energy} \rightarrow \text{solid C} + 2\,H_2 + \text{heat} \tag{3.3}$$

The carbon forms carbon nodes of about 30–40 nm mean size, which can further build aggregates of 100–500 nm size. The microwave plasma principle is not new and has already been

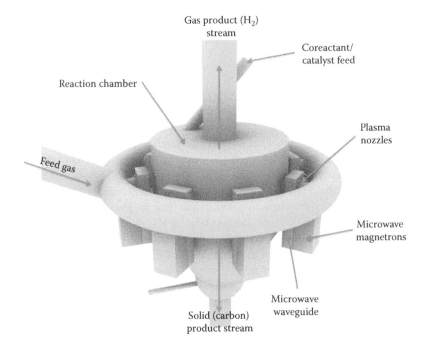

Figure 3.6 Principle of the GasPlas reactor.

investigated for the manufacture of diamond coatings and carbon fibers at small scale (Ting et al. 2007; Yu et al. 2002).

The limitations of the microwave plasma technology for industrial uses have so far been low pressure, high vacuum, and thus low gas throughput. Three industrial challenges have stopped the scaling up of this technology: (i) production of a sufficient and stable plasma volume, (ii) application at atmospheric pressure or above, and (iii) efficient separation of outputs in a continuous process.

GasPlas managed to cope with these constraints by developing a separation cyclone like a centrifuge in which the solid carbon particles circle in the periphery due to higher density, while H_2 can be extracted from the center (GasPlas patent GB2010/050290). To date, only 1.1 L plasma volume reactors exist, but GasPlas will improve this by a factor of 10 or more in the near future. The higher the pressure is, the more difficult it is to get stable, homogeneous plasma. However, for industrial productivity, only this makes sense. GasPlas has thus reconfigured the existing technology into a design that has the potential for (i) scaling large plasma volumes, (ii) controlling carbon formation and collection, (iii) on-demand operation, (iv) continuous process, (v) low capex and opex, and (vi) wide range of applications based on the same principles.

From a material properties point of view, carbon derived from the GasPlas process (plasma carbon) is almost pure carbon. Therefore, it has a higher carbon content compared with biochar and thus a lower content of H, O, N (O/C and H/C near 0, Figure 3.5), and other elements such as potential nutrients. This has two implications for the potential environmental use of this material. First, it can be expected that its MRT in the environment is much longer compared with biochar. This is due to the more inert, graphitic structure of plasma carbon, while biochar is more aromatic. Second, due to the lack of accompanying minerals, it should not be used alone without value-added processing such as composting with organic waste or slurry materials, as is the case for biochar as well. In addition, as plasma carbon size is in the nanoparticle range (10–100 nm), it can be harmful if inhaled. Again, adding value through composting or mixing with slurry could prevent this. In addition, it is similar to carbon black, a chemical that has been successfully used for 100 years in industry, for example, as a pigment in car tires, plastics, and so on. Furthermore, plasma carbon has a similar specific surface area to biochar materials, making it suitable for the adsorption of especially hydrophobic xenobiotics such as PAHs, polychlorinated biphenyls (PCBs), dioxins, pesticides, and so on.

3.3.4 Economic Value of Carbons to Soil

In Europe, there is no specific legislation for biochar or standards to underpin the licensing of its use and it is likely that it will be treated as a waste when created from a waste. This presents a major challenge in the marketing of biochar for soil amelioration (C sequestration and/or soil fertility improvement). The most obvious approach to overcome this challenge is to develop a quality assurance process analogous to the compost protocols. Making the case for the creation of a product is an alternative, but more challenging option. To date, commercial biochar-based soil amendments have been limited to the domestic compost market, rather than the lower-margin agricultural sector.

Currently, the economics of biochar production relative to composting are uncertain, but clearly advantageous when pyrolysis feedstock comprises waste material (Roberts et al. 2010). It is possible that biochar created from low-value feedstock could add high value to low-quality compost, or reduce the costs of compost production. Alternatively, biochar from waste feedstock may be used in the same manner as the waste, or applied independently to soil. There are a number of anticipated policy developments that would be relevant, including some mechanism for incentivizing a better carbon balance in food production. Such developments may be linked to the sustainability objectives of food producers, processors, and retailers in turn driven by consumer preferences. The adoption of carbon labeling by such companies based on a detailed life cycle analysis of the production chain has been an important demonstration of this.

The biochar and CO_2 markets are likely to be driven by the price of energy and carbon trading schemes. The increasing price of waste disposal is likely to make the production and application of biochar for electricity and waste management economically viable. However, for some wastes (such as waste wood from construction) the price of waste disposal is rapidly falling because of competition for the waste. However, even in this case the price for the waste itself will increase. Carbon offsets will play a greater role once biochar is certified under the clean development mechanism (CDM) of the Kyoto Protocol, an issue that was discussed at the World Climate Summit in Durham, South Africa, in December 2011. Uncertainty over market interventions may risk the investment in energy facilities that are able to produce biochar.

The value arising from the agronomic impact of biochar after its application to soil is one value stream to which considerable attention is focused. However, biochar amendments offer the potential for carbon-neutral or carbon-negative food production and the concomitant development of brand value and an income stream from the acquisition of carbon certificates traded on the voluntary carbon market. The ecoregion Kaindorf (Austria) has successfully implemented such a system. In addition, when combined with plant nutrients from waste materials such as from green wastes or slurries, additional benefits arise from saving money to buy mineral fertilizers such as NPK and by reducing resources allocation for water purification, for example, when nitrate is leached into groundwater after the improper application of slurry to agricultural fields. The latter is supposed to increase in the near future due to the rapid growth in biogas production followed by the disposal of huge amounts of biogas slurry.

Plasma carbon principally has the same target applications as biochar (C sequestration and soil fertility improvement), especially when produced from biogas. In addition, the following target markets are even more promising from an economic point of view: (i) transportation (hydrogen on demand, biofuels), (ii) chemicals (ammonia, methanol, ethylene), (iii) materials (carbon black, carbon nanostructures, carbon fibers), (iv) metals, and (v) power applications such as hydrogen power stations being CO_2 free if H_2 is delivered from the GasPlas reactor.

3.3.5 Ecological Value (Ecosystem Services)

3.3.5.1 Carbon Sequestration

Much research has been undertaken in recent years to show that returning carbon to the soil in the form of char can sequester carbon and increase soil fertility. The determination of the carbon mitigation potential requires a defined period of time over which carbon is expected to be sequestered and it crucially depends on suitable reference systems (Libra et al. 2011). In addition, biochar application may possess additional carbon mitigation potential owing to indirect effects. For example, increases in soil organic carbon (SOC) stocks and decreases in GHG emissions and fertilizer use should be considered in addition to the direct benefits of carbon sequestration.

The true C sequestration potential of biochar depends on several factors. The char from natural fires is usually the oldest carbon pool present in ecosystems that are prone to fire. The MRT of naturally generated char in Australian woodlands ranges between 718 and 9259 years (Lehmann et al. 2008). The long-term stability of biochar has been shown in terra preta, which is 500–7000 years old (Neves et al. 2003). Considerable biochar stocks of 50 Mg C/ha have been found down to a depth of 1 m in such soils, despite the climatic conditions strongly favoring decomposition and the lack of additional biochar during the last 500 years (Glaser et al. 2001). However, MRT assessments from natural ecosystems or terra preta can only deliver orders of magnitude in accuracy, since there is no way to quantify the initial (repeated) biochar input to obtain a straightforward mass balance for terra preta. The longest laboratory biochar decomposition study also revealed an MRT of about 2000 years (Kuzyakov et al. 2009). However, MRT depends on material properties, especially the degree of aromatization. Hydrochar (char from hydrothermal carbonization) has a less aromatic structure and a higher percentage of labile carbon species (Schimmelpfennig and Glaser 2012). Therefore, it decomposes similar to SOM within decades (Steinbeiss et al. 2009).

3.3.5.2 Non-CO$_2$ Greenhouse Gas Emissions

In addition to the assessment of the true carbon sequestration potential, several indirect effects of char application have to be taken into account, such as fertilizer use, N$_2$O and CH$_4$ emissions, changes in SOC, and increased productivity. Libra et al. (2011) reviewed the effects of biochar additions to different agroecosystems on GHG emissions (Table 3.1).

3.3.5.3 Crop Yields and Soil Fertility

The application of biochar to soil can increase crop yields (Glaser et al. 2002). A tremendous increase in yield was observed especially in degraded or low-fertility soils compared with already fertile soils (Glaser et al. 2002). The crop yield increase was higher when additional nutrients were added, especially in an organic form such as compost (Figure 3.7). In a literature survey of a large number of biochar studies, mostly conducted in tropical or subtropical regions, Jeffery et al. (2011) calculated a mean crop yield increase of 10% after the addition of biochar to soil with a trend of increasing crop productivity with increasing biochar addition. However, crop productivity was not linearly correlated with the addition of biochar. Instead, the biochar response to crop yield ranged

Table 3.1   Reported Biochar Effects on Greenhouse Gas Fluxes

Greenhouse Gas Flux	Increase	Decrease	Dominating Effect
CO$_2$ emission from soil	4	3	Biochar increased CO$_2$ emission from soil
N$_2$O emission from soil	2	7	Biochar reduced N$_2$O emission from soil
CH$_4$ emission from soil	0	1	Biochar decreased CH$_4$ emission from soil
CH$_4$ sequestration into soil	1	2	Biochar increased CH$_4$ sequestration into soil

Source: Libra, J.A., Ro, K.S., Kammann, C. et al., *Biofuels*, 2, 89–124, 2011.
Note: Numbers indicate number of published studies.

Figure 3.7   Plant (*Avena sativa*) growth in sandy soil. Ko=control, NPK=mineral fertilizer, HK=charcoal, Kompost=compost. Please note a synergism between biochar and compost. (From Schulz, H. and Glaser, B., *J Plant Nutr Soil Sci*, 175, 410–422, 2012.)

from −40% to +100% (Jeffery et al. 2011). Furthermore, a large variation in crop response to similar amounts of biochar additions was observed, especially at low biochar additions (5.5–11 Mg/ha), but also at large biochar additions (above 100 Mg/ha). From these data it can be concluded that medium biochar application rates (10–100 Mg/ha) might be most appropriate for increased crop production. The reason for the large variation observed is likely due to the different biochar feedstock used, the different crops assessed, and the differences in the soil type to which the biochar was added. It is interesting to note that no single biochar application rate exhibited a statistically significant negative effect on the crops from the range of soils, feedstock, and application rates compared. However, the data used for meta-analysis did not cover a wide range of latitudes and they were mainly from tropical and subtropical regions (Jeffery et al. 2011). This means that care should be taken when extrapolating these results to European latitudes, crops, and soil types. There is only one review on biochar effects in temperate agroecosystems (Atkinson et al. 2010); however, it contains mostly theoretical considerations as up to now experimental data have been limited.

Nutrient supply, pH, and other soil properties alone were not always sufficient to fully explain the observed positive or negative biochar effects on yields. Especially at low application rates, negative biochar yield effects were sometimes reported, which were explained by nitrogen immobilization. However, it remains unclear why this effect should occur only at low biochar additions. The opposite would be more logical.

Hydrochars often exhibit higher labile carbon fractions such as carbohydrates and carboxylates compared with biochars (pyrochars), but have higher nitrogen contents. Therefore, hydrochars alone might be better for plant growth from a nutrients point of view. However, growth reduction was often reported when hydrochar was applied to soil. This could be explained by the toxicity of low-molecular-weight substances such as phenolic compounds on plant germination or growth or both (Rillig et al. 2010).

Soil quality may not necessarily be improved by adding biochar. Soil quality can be considered to be relatively high for supporting plant production and the provision of ecosystem services if it already contains a sufficient amount of SOM (at least 4% corresponding to about 2% organic carbon). If biochar is added to soil, a relative portion of the easily mineralizable (active) SOM pool will be reduced. Therefore, simply adding pure biochar (or plasma carbon) to a soil does not increase the SOM quality or soil fertility. Instead, biochar/plasma carbon should be mixed with nutrients and easily available SOM (Figure 3.7). This can be best achieved by composting biochar together with green waste or slurry or both from biogas where the amount of biochar is 10%–50% of the organic carbon present in the final biochar value-added product.

Biochar also has the potential to introduce a wide range of hazardous compounds such as heavy metals, PAHs, polychlorinated dibenzodioxins (PCDDs), and polychlorinated dibenzofurans (PCDFs) into the environment, which can be present as contaminants in biochar that are produced either from contaminated feedstocks or under processing conditions that favor their production (Schimmelpfennig and Glaser 2012). Therefore, tight control of these contaminants can reduce the potential risk for soil contamination through biochar addition. However, little is known about the interaction of biochar with, for example, an already contaminated soil. The few studies available indicate positive biochar effects on both inorganic and organic contaminant adsorption and thus immobilization.

3.3.6 Scalability

For biochar production, photosynthetically fixed CO_2 from the atmosphere is used in the pyrolysis of plant biomass or its residues (Figure 3.8). Therefore, biochar production is not only a "carbon-negative" technology, but it is also limited by the amount of organic matter that is dedicated to biochar production (Figure 3.8).

A conservative estimate for the scalability of biochar production is the conversion of organic wastes. Given the availability of appropriate biochar production technologies, for example, PYREG

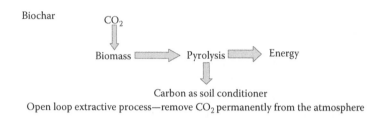

Figure 3.8   Principle of biochar production with respect to C sequestration. Biochar production is an open loop process that can theoretically extract CO_2 long term from the atmosphere. The only limitation is the amount of biomass or organic waste that is available for biochar production. (From Glaser, B., Parr, M., Braun, C., and Kopolo, G., *Nat Geosci*, 2, 2–2, 2009. With permission.)

or Carbon Terra, from about 500 Mt of organic waste across Europe, about 140 Mt of biochar could be produced with a conservative conversion rate of about 30%. This means that an additional C offset of about 10% could be achieved immediately. If additional biomass were to be "sacrificed" (e.g., via a "biomass tax") for biochar production, these numbers would be even higher. With respect to the cold plasma technology, the only limitation is the amount of natural gas or biogas consumed annually (Figure 3.1).

3.3.7 Ongoing and Further Research on Plasma Carbon

3.3.7.1 *Available Data on Plasma Carbon Characterization*

The elemental composition of plasma carbon is >99% C, <0.1% N, and <0.1% H (GasPlas, unpublished). Therefore, plasma carbon can be considered as elemental carbon. Impurities of liquid phase were low molecular benzene compounds. In the gas phase, nitriles and butadiene has been detected in small quantities (GasPlas, unpublished). As little is known about the material properties and the environmental behavior of plasma carbon, further studies are required, which will be addressed in the next section.

3.3.7.2 *Desired and Necessary Further Research on Biochar and Plasma Carbon*

Similar to biochar research, additional information on the material properties, toxicology, and environmental behavior of plasma carbon is necessary (Figure 3.9). From a material properties point of view, similar properties and thresholds as for carbon black and biochar could be used, such as O/C and H/C ratios, PAH, and dioxin contents (Schimmelpfennig and Glaser 2012). With respect to its physical properties, specific surface area, porosity, hydrophobicity, and interaction with soil minerals would have highest priority. From a biological point of view, stability (C sequestration), its influence on soil fauna and flora, and its effect on other GHG emissions would be of greatest interest. Last but not least, the biochar and plasma carbon effects on ecosystem services such as plant growth, water retention in soil, filter capacities for nutrients, and water and toxic compounds would be important to study.

3.4 CARBON CAPTURE AND USE IN BUILDINGS AND CIVIL INFRASTRUCTURE

3.4.1 Construction Materials

Hydrocarbon fossil fuels can be considered as hydrogen ores for CO_2-free energy and carbon ores for carbonaceous construction materials (Halloran 2008). Hydrogen fuel can be extracted from

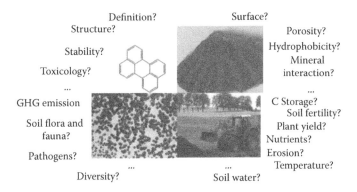

Definition?   Surface?
Structure?   Porosity?
                    Hydrophobicity?
Stability?   Mineral
Toxicology?   interaction?
...   ...
GHG emission   C Storage?
                        Soil fertility?
Soil flora and   Plant yield?
fauna?   Nutrients?
Pathogens?   Erosion?
                      Temperature?
...   ...
Diversity?   Soil water?

Figure 3.9 Unknowns in the physical, chemical, and biological properties of biochar and plasma carbon and their environmental behavior (ecosystem services).

fossil fuels by decarbonization and used as an energy resource. The carbon by-product can be used as a versatile construction material (H + CCU). Carbon materials could sequester carbon while replacing CO_2-generating steel and concrete. Thus, there is a double benefit, and wasting carbon resources by burning fossil fuels is like burning fine ebony wood in a fireplace. Natural gas and the thermolysis gases from petroleum and coal can be converted into hydrogen fuel gas while constructing vapor-deposited carbon (VDC) materials from elemental carbon, in the form of nanoparticles, fibrous materials, or pyrolytic carbon solids.

The H + CCU concept is only practical if the value of the carbon materials and hydrogen is greater than the value of the energy that would be produced if the fuel were simply burned. The concept has a double benefit with respect to carbon dioxide in a life cycle analysis, avoiding CO_2 production by combustion on the energy side, as well as avoiding CO_2 production in the manufacture of steel and cement.

Before the industrial revolution and the steel era, the only available construction materials were stone, brick, or wood. During the last century, these materials were largely replaced by concrete and steel-reinforced concrete. This is why we are currently in the steel-reinforced concrete era, at least from a construction material point of view. Modern construction materials can be principally divided into two categories: materials resisting compressive forces with acceptable brittleness and materials resisting tensile forces with no acceptable brittleness. The major compressive force resistor is concrete, while the major tensile force resistor is steel.

Carbon-based materials already exist for resisting compressive and tensile forces due to the unique material properties of elemental carbon. Thus, carbon fibers are already used in tensile force-resisting construction material to reinforce concrete for special applications. In addition, carbon-based concretes have been used for decades in compressive force-resistant construction material. For instance, the lower part of blast furnaces for iron production consists of carbon brick (Halloran 2007).

Carbon composite materials can substitute steel and other metal although they exhibit different properties. To become a useful construction material, the mechanical strength of carbon materials must be comparable with existing steel and concrete materials. Whereas metals are formable (ductile), carbon fiber composites do not have this property. However, the ductility of metals is not a property that is necessary in the construction material itself but it is necessary for metal forming. Therefore, alternative forming methods are available in the carbon industry. A second important property of steel is fracture resistance (tensile strength) by energy absorption during ductile deformation, which is not available in carbon materials. Therefore, alternative absorption mechanisms are exploited to provide fracture resistance of carbon composite materials. Construction-grade solid shapes (bricks, blocks, tiles, roofing tiles, drainpipes) made of clay or concrete typically have a tensile strength of 1–4 MPa. Ordinary portland cement concrete, a sand/gravel aggregate bonded with portland cement, has a characteristic 28 day tensile strength in the range of 3–4 MPa (Mindess and Young 1981). Thus, the tensile strengths of carbon substitutes should be in the same order of magnitude.

The tensile strength of metallurgical coke depends on the coke microstructure with a typical tensile strength of 4–6 MPa, thus it is stronger than portland concrete (Mindess and Young 1981). Better coke properties can be obtained from fabricated materials, such as compacted pulverized coal, where compressive strengths of 160 MPa for a strongly coking coal and 130 MPa for a weakly coking coal and a tensile strength of about 13–16 MPa were reported (Honda et al. 1966). Extruded forms of activated carbon fibers have been produced by the thermolysis of a coal powder and a coal tar pitch blend at 900°C with a tensile strength of 5.5 MPa (Henning et al. 1987). This product, although produced to be a high surface area activated carbon, is still stronger than concrete.

The composites used for construction materials are already an emerging field. To substitute for steel in the tensile parts used in construction (beams, reinforcement bars, cables, etc.), carbon fiber–reinforced plastics seem to be appropriate steel substitutes, particularly low-cost composites using new "construction-grade" carbon fibers. Carbon fibers have been used in paving materials such as concrete and asphalt (Halloran 2007). Buildings could be made from concrete reinforced with carbon fibers instead of steel reinforcement bars for significant size and weight reduction. Composite fiber wrapping is widespread for seismic retrofit in California. Fiber-reinforced plastics are increasingly used to repair aging bridges that are structurally deficient due to corrosion of the steel and crumbling of the concrete. One of the world's first operational cast-iron bridges, the Tickford Bridge in England built in 1810, has been rehabilitated and strengthened with carbon fiber (Jacob 1999). Complete composite bridges have also been built (Black 2000), including a bridge in Quebec using carbon fibers (Redston 1999).

Carbon fibers are state-of-the-art materials with properties that include lightweight, high strength, and chemical stability, and are applied in various fields including the aeronautical and space sciences. Investigation of the applications of carbon fibers to biomaterials started 30 years ago, and various products have been developed. Because the latest technological progress has realized nanolevel control of carbon fibers, applications to biomaterials have also progressed to the age of nanosize. Carbon fibers with diameters in the nano scale (carbon nanofibers) dramatically improve the functions of conventional biomaterials and make the development of new composite materials possible. Carbon nanofibers also open up possibilities for new applications in regenerative medicine and cancer treatment (Saito et al. 2010).

Carbon has unique electrical and structural properties that make it an ideal material for use in fuel cell construction. In alkaline, phosphoric acid, and proton-exchange membrane fuel cells (PEMFCs), carbon is used in fabricating the bipolar plate and the gas diffusion layer. It can also act as a support for the active metal in the catalyst layer. Various forms of carbon—from graphite and carbon black to composite materials—have been chosen for fuel cell components. The development of carbon nanotubes and the emergence of nanotechnology in recent years have opened new avenues of materials development for low-temperature fuel cells, particularly the hydrogen PEMFC and the direct methanol PEMFC. Carbon nanotubes and aerogels are also being investigated for use as catalyst support, which could lead to the production of more stable, high-activity catalysts, with low platinum loadings (<0.1 Mg/cm) and therefore low cost. Carbon can also be used as a fuel in high-temperature fuel cells based on solid oxide, alkaline, or molten carbonate technology. If used in the direct carbon fuel cell (DCFC), the carbon combustion energy is converted into electrical power with a thermodynamic efficiency close to 100%. The DCFC, producing a near 100% CO_2 stream that could be stored geologically as in CCS, could therefore help to extend the use of fossil fuels for power generation as society moves toward a more sustainable energy future (Dicks 2006).

3.4.2 Economic Value (Value Chains)

The economic benefits of integrating carbon materials into construction systems include decreased labor and energy costs and life cycle improvements related to the elimination of steel corrosion.

Oxidizing methane to H_2O and CO_2 provides 800 kJ/mol of methane or 34 GJ/Mg. Separating carbon as solid carbon and burning just the hydrogen provides 404 kJ/mol of methane corresponding to 51% of the available energy, while producing zero CO_2 emissions (Halloran 2007). The energy efficiency of noncarbon products of petroleum and coal is 33% and 15%, respectively (Halloran 2007). According to Halmann and Steinberg (1999), hydrogen from the thermal decomposition of methane can provide 59% of the combustion heat of natural gas (Table 3.2), after accounting for the endothermic heat of decomposition and the atomic fraction of hydrogen in methane. Hydrogen from the thermal decomposition of petroleum can provide 37% of the heat of combustion of the alkane fraction and 25% of the aromatic fraction, after accounting for the endothermic heat of petroleum decomposition and the atomic fraction of hydrogen in petroleum (Halmann and Steinberg 1999). Assuming typical alkane and aromatic fractions of 70% and 30% for petroleum, respectively, the overall energy efficiency of petroleum is 33% (Table 3.2). Hydrogen from the thermal decomposition of a typical coal can provide 19% of the combustion heat of coal (Halmann and Steinberg 1999).

Based on the abovementioned numbers, calculations were made to produce 1 GJ of energy for the business-as-usual scenario in which fossil fuels are burned completely, compared with an alternative scenario in which only hydrogen is burned for energy production while carbon products are used, for example, as construction material (Table 3.3). As expected, higher amounts of fossil fuels must be consumed if only hydrogen is burned, particularly when using coal, which is poor in hydrogen (Table 3.3). At this point, it seems that the dramatic increase in fossil fuels consumption is economically not feasible. However, the data in Table 3.2 do not include the value of the solid carbon product, and assume that CO_2 can be vented to the atmosphere at zero cost, which should not be the case in the future. Therefore, in the following, the market value of carbon products and a carbon tax for CO_2 emission will be considered as well.

The economic feasibility of CCU has to involve the costs of the additional fossil fuels required, the cost of the decarbonization process (only energy cost used for calculation), and the value of

Table 3.2   Energy Efficiency of Fossil Fuels Assuming That Only Hydrogen Is Used for Energy Production While Solid Carbon Is Separated for Carbon Capture and Use

Coal (%)	Petroleum (%)	Methane (%)	References
19	31	59	Halmann and Steinberg (1999)
15	33	51	Halloran (2007)
17	32	55	Mean

Table 3.3   Fossil Fuel Consumption and Carbon Balance

	Energy (GJ/Mg)	C (%)	H (%)	C + H Burning (kg)	CO_2–C Emission (kg)	H Burning (kg)	Solid C (CCU) (kg)
Coal	29	83	5	34.5	28.4	203	167
Petroleum	44	86	13	22.7	19.5	68	59
Methane	34	75	25	29.4	22.1	53	40

Note: Balance is calculated for 1 GJ energy production.
C+H Burning, fossil fuels are conventionally burned; H Burning, only hydrogen is used for energy production; Solid C (CCU), solid carbon is separated for carbon capture and use purposes.

the carbon product. The following calculations were made using the mean commodity prices for 2010–2011: natural gas 4.2€/GJ or 185€/ton, crude oil 78€/bbl or 348€/ton, coal 66€/ton, steel 572€/ton, reinforcement bar 417€/ton, portland cement 71€/ton, and carbon fiber 28–764€/ton (www.msn.com).

Considering the mean 2010–2011 commodity fuel prices, the production costs of 1 GJ of Gibbs free energy are shown in Table 3.4. From these data, it is obvious that energy production costs generally increase in the order coal < methane < petroleum (oil). In addition, the costs for conventional energy production are two to six times lower compared with energy production from hydrogen alone (Table 3.4). However, in this calculation, the potential benefits of CCU are not yet considered, which will be done in the following.

For conventional burning, 20–28 kg of CO_2–C must be captured and sequestered for 1 GJ of produced energy (Table 3.2), which will have additional cost (currently about 20€ and 92€ per megagram (Mg) of CO_2 and CO_2–C, respectively) and environmental impacts. For the CCU approach, 40–167 kg of solid carbon is produced, which might serve as an added-value material. These impacts are shown in Table 3.5. As the market price for carbon fibers varies widely (28–764€/ ton), a conservative estimate is using the material value of the replaced product, that is, carbon-reinforced concrete versus steel-reinforced concrete. For this purpose, it is assumed that solid carbon from decarbonization can be used as carbon concrete, with both the concrete aggregate and the carbon–cement binder made from decarbonization by-products. For these calculations, the following 2010–2011 mean prices were used: reinforcement bar 417€/ton, portland cement 71€/ton, and carbon fiber 28–764€/ton. For simplicity, the same price for carbon fiber (417€/ton) was used as for the reinforcement bar and 20€/ton of CO_2 as CO_2 tax.

The results in Table 3.5 clearly demonstrate the economic advantage of the H + CCU concept. The biggest economic advantage can be achieved if coal is used as energy source (net benefit of 60€/ GJ of produced energy). The H + CCU concept would also work for methane (plasma carbon) and petroleum (net benefit of 14 and 10€/GJ, respectively, Table 3.5). Therefore, the cost of foregoing the carbon energy content as a fuel can be offset by the value of carbon-based construction materials. However, it should be mentioned here that the economic revenue of H + CCU strongly depends on

Table 3.4   Fossil Fuel Costs for 1 GH of Produced Energy

	Energy (GJ/Mg)	2010–2011 price (€/Mg)	C + H Price (€/GJ)	H-Only Price (€/GJ)	H-Only/C + H
Coal	29	66	2	13	6
Petroleum	44	348	8	24	3
Methane	34	185	5	10	2

Note: C+H, conventional fossil fuel burning; H-Only, only hydrogen is used for energy production

Table 3.5   Comparison of Fossil Fuel Costs for 1 GJ of Produced Energy

	C + H Price (€/GJ)	CO_2-C Emission (kg/kg/GJ)	CO_2 Tax (€/GJ)	Total Conventional (€/GJ)	H-Only Price (€/GJ)	Solid C (kg/GJ)	Solid C Price (€/GJ)	Total H + CCU (€/GJ)
Coal	2	28	2	4	14	167	70	−56
Petroleum	8	20	1	9	24	59	25	−1
Methane	5	22	2	7	10	40	17	−7

Note: C+H, conventional fossil fuel burning; H-Only, only hydrogen is used for energy production; CCU, replacing reinforcement bar by solid carbon; Total Conventional, C+H price + CO_2 tax; Total H+CCU, H-only price − sold C price the market prices for the used energy sources and for the solid carbon. For instance, assuming the lowest available price for carbon fibers (28€/ton), a net loss of 2€/GJ of methane (plasma carbon) would result from the H + CCU concept compared with the conventional energy production for natural gas burning but not taking into account the money saved for replacing reinforcement bars (data not shown). The opposite is true if the CO_2 tax increases, when CO_2 certificates (hopefully) need to be bought in European countries. In addition, the market for large quantities of carbon materials is big enough to consume the carbon-rich residue. Finally, the market value of carbon materials can significantly exceed their fuel value, offsetting both the cost of decarbonization and the foregone fuel value of the carbon.

Another issue should be mentioned here, which will favor carbon-low technologies in the future. The current costs of steel and concrete are artificially low because producers are allowed to vent the CO_2 directly into the atmosphere. If CO_2 mitigation technologies or CO_2 taxes were implemented, these materials would be much more expensive. In these cases, the H + CCU technology would be economically even more feasible.

Of further relevance is the fact that solid carbon is durable and resistant to corrosion. Solid carbon or carbon-rich building materials will avoid CO_2 generation, sequester carbon, and put our carbon resources to excellent use in long-term applications.

3.4.3 Life Cycle Assessment

Cement and steel production generate about 20% of global CO_2 emissions (Figure 3.1). About 0.9 Mg of CO_2 is emitted per megagram of produced cement, so that the total global cement production of 1.6 Pg/year causes the emission of 1.4 Pg of CO_2 corresponding to about 0.5 Pg of CO_2–C covering about 13% of annual anthropogenic CO_2 emissions (Figure 3.1). Some of the emissions are due to cement kiln heating with fossil fuels, which could possibly be avoided. However, 60% is unavoidable due to the reduction of limestone ($CaCO_3$) to cement (CaO) under the release of CO_2. Concrete is made by bonding a low-cost aggregate (sand and gravel) with cement, consisting of about two-thirds of aggregate and one-third of cement.

Halloran (2007) argued that solid carbons could be used in large-scale infrastructure, in place of steel and concrete. Construction-grade solid carbon is already made from fossil fuel–derived materials. Carbon in construction materials offsets the negative aspects of steel production such as iron ore mining, energy consumption during the melting process, and environmental problems associated with slag deposition. Steel is produced by the carbothermic reduction of iron oxide with coke to remove the oxygen as CO_2 ($2 \, Fe_2O_3 + 6 \, C + Energy \rightarrow 2 \, Fe + 6 \, CO_2$), producing 1.5 Mg of CO_2 per megagram of iron for the reduction alone. Even more CO_2 is generated by refining the iron to steel. Global steel production of about 0.85 Pg/year causes a CO_2 emission of about 1.3 Pg of C per year, contributing about 12% to the annual anthropogenic CO_2 emission of about 11 Pg/year (Figure 3.1).

Carbon materials are very resistant to corrosion and degradation at ambient temperature, so it is reasonable to expect that the carbon materials used in construction will sequester the carbon for at least the lifetime of the building (decades to centuries). Post service sequestration will require recycling or burial of the carbon materials. In addition, depending on the material properties, carbon fibers are more or less resistant to burning. Though carbon fiber is constituted of carbon, which is flammable, the fiber itself does not flare up even if it is ignited by a match or gas burners. If heated to higher than 400°C together with some fuel, the fiber slowly burns (oxidized) but stops burning right after the burning fuel is removed. Therefore, in Japan, carbon fiber is categorized as "incombustible" for building purposes.

An additional environmental benefit is the low density of carbon materials compared with their metal counterparts. This saves energy during transportation or when used in transportation (e.g., for vehicle construction), as the lower weight needs less energy for movement.

3.4.4 Scalability

Low-cost carbon materials could be substituted for steel-reinforced concrete as a normal part of a market-driven process. As outlined in Tables 3.2 through 3.5, this process does not need the incentive of CO_2 reduction policies such as a CO_2 tax, as it is already shown to be profitable (Table 3.5). However, a CO_2 tax would accelerate replacing CO_2-intensive technologies such as energy, cement, and steel production with CO_2-low technologies such as H + CCU technology.

A comparison of the global consumption of energy and construction materials suggests a rough mass balance of energy and materials markets. If a substantial fraction of fossil fuels for energy production were decarbonized, it would produce about 8 Pg/year of solid carbon (Figure 3.1). Global cement and steel production is 1.6 and 0.85 Pg/year, respectively (Halloran 2007). Assuming a 100% replacement of steel with carbon, at least 7 Pg of solid carbon per year remains, which needs to be converted to other carbon materials. C sequestration into soils according to the terra preta concept might be a promising option.

At the beginning of the twenty-first century, carbon fibers are expensive, single-walled carbon nanotubes are more expensive, and single-layer carbon (graphene) is most expensive. It is most likely that a combination of increased demand and process innovation will bring these materials into large-scale use within a short period of time. The carbon materials market has already started by introducing reduced-cost versions of existing carbon products. Major automotive companies such as Volkswagen, BMW, and Daimler are already heavily involved in the carbon materials market, for example, by their significant ownership of the global leading carbon materials company (SGL Carbon). Therefore, it is most likely that cars in the future will consist of carbon fiber materials instead of steel and aluminum, which is already the case in Formula 1.

Carbon is fundamentally cheap and abundant, and can be the basis for a post-steel-reinforced concrete materials economy. Carbon ores occur as vast deposits of fossil resources, where they are combined with a clean, high-value fuel, hydrogen. A complete and effective global industry already exists to exploit this resource. Although the petroleum companies are already major material suppliers through their petrochemical divisions, most fossil fuels are misused by simple burning as discussed in the previous sections. Instead, much of what is needed is already in place, or can be easily developed.

3.5 OTHER POSSIBLE LONG-TERM CARBON SEQUESTRATION CONCEPTS

3.5.1 Sequestration of Solid Carbon in (Polluted) Ocean Sediments

Solid carbon can principally also be sequestered into ocean sediments, although there is no additional economic value to such an option. However, ecosystem services might be improved through the sorption of (organic) pollutants to solid carbon, thereby lowering their bioavailability to marine organisms such as plankton and later fish (Ghosh et al. 2011). In field tests in the United States and Norway, 5% of solid carbon (activated carbon) was mixed with the top 10 cm of sediment accounting for 35 Mg/ha. Assuming a solid carbon price of 400€/ton, investments of 14 k€/ha would be necessary. Such costs would only be economically feasible when site remediation is necessary. For instance, the dredging and disposal cost for the Hudson River cleanup has been projected at 1.8 M€/ha (Ghosh et al. 2011). Thus, the material cost of solid carbon would be at least two orders of magnitude lower compared with dredging and disposal. In addition, the technology is especially attractive at locations where dredging is not feasible or appropriate, such as under piers and around pilings, in sediment full of debris, in areas where overdredging is not possible, and at ecologically sensitive sites such as wetlands. *In situ* amendments can also be used in combination with other remediation techniques such as the use of carbon as an adsorbent, which will be discussed in the next section.

3.5.2 Use of Carbon as Adsorbent (Activated Carbons)

A wide range of materials that are known for their adsorption properties have the potential to be used for the removal of trace substances from drinking water and wastewater. These include physically and chemically activated carbons, surface-modified carbons, nonporous resins with ion-exchange capacities, inorganic microporous solids such as zeolites and clays, and mixed organic–inorganic materials such as bone chars. The adsorption capacity that is exhibited by each material relates primarily to its textural and chemical properties. Other factors, however, such as apparent density, regeneration potential, and cost, need to be taken into consideration when selecting one adsorbent over another. Solid carbon normally needs to be activated to increase its sorption capacity, thus becoming "activated carbon." The sorption properties of activated carbons are extremely versatile and can be used to remove a variety of inorganic and organic contaminants from water, such as heavy metals, PAHs, PCBs, PCDDs, PCDFs, pesticides, and many other toxic substances. Activation generally increases the specific surface area and pore space, which can be achieved through physical and chemical methods. Physical activation is achieved with CO_2 or steam while chemical activation is achieved through impregnation with potassium salts, sodium hydroxide, magnesium, calcium chloride, or phosphoric acid.

Activated carbon has an approximately 50 times higher internal surface area compared with the precursor material and is highly microporous. Fourier-transformation infrared spectroscopy analysis showed the development of additional aromatization in the structure of activated carbon. X-ray diffraction data indicated the formation of a small, two-dimensional, graphite-like structure at high temperatures (Azargohar and Dalai 2006).

A comparative investigation of 18 solids and their capacity to remove organics and metals from natural waters showed that zeolites and ion-exchange resins exhibited limited capacities to remove organic matter from a solution but were highly effective with ionic species such as manganese and aluminum. Activated carbons adsorbed organic matter very efficiently, with a correlation between adsorption capacity and specific surface area. The removal of metal was highly variable and was enhanced in activated carbons subjected to acid washing. Owing to its mixed organic/inorganic nature, and despite its poorly developed porous structure, bone char exhibited a strong adsorption capacity for both organic and metal species (San Miguel et al. 2006). Therefore, if both inorganic and organic contaminants have to be removed, either oxidized activated carbon or mixtures of complementary adsorbents for the removal of organic and inorganic species from a solution should be used.

3.6 CARBON MARKETS WITH LONG-TERM SEQUESTRATION POTENTIAL

As outlined in the previous sections, mitigating the anthropogenic greenhouse effect requires the sequestration of about 8 Gt of C from fossil fuel combustion per year (Figure 3.1). Instead of "simply" storing fossil fuel–derived CO_2 in the ground (CCS), it is more intelligent to use carbon in a solid form (CCU). There are principally two different ways to achieve this: (i) using solid C for environmental protection and soil improvement (biochar/terra preta concept) and (ii) using solid C as part of construction materials (carbon fibers).

The carbon market using the terra preta concept is theoretically unlimited and depends only on the availability of raw material (biochar and plasma carbon) and the price of added-value products. The availability of fossil fuel C of about 8 Gt/year could increase the total amount of global soil C by only 0.32% or 3.2% in 10 years. Especially when fuel and fertilizer prices increase, terra preta has strong economic benefits in addition to its ecological ones.

The demand for carbon in construction materials is more limited. Even assuming that all the steel that is currently used for construction is replaced by carbon fibers, the annual demand would be limited to about 1 Gt of C. However, as C fibers could also be used for new construction applications, the potential market for such materials could go well above this 1 Gt/year. Prices for niche materials are likely to stay high.

3.7 SUMMARY AND CONCLUSIONS

Climate change will have serious effects on the planet and on its ecosystems. Current mitigation efforts such as CCS and land use change (especially no-tillage agriculture, desertification control, and agriculture to pasture conversion) are insufficient in reducing anthropogenic CO_2 emissions. Long-term CO_2 removal techniques worth further research and implementation need to address rising atmospheric CO_2 concentrations, the principal cause of climate change. It is suggested that the most promising techniques not only remove atmospheric CO_2 but also use it (CCU) for economic profit. Such techniques comprise the use of solid carbon for soil improvement (biochar), in construction materials, and in infrastructure (plasma carbon). However, future research is needed to unambiguously demonstrate that CCU techniques are economically feasible and have low uncertainties and risks.

REFERENCES

Atkinson, C. J., J. D. Fitzgerald, and N. A. Hipps. 2010. Potential mechanisms for achieving agricultural benefits from biochar application to temperate soils: A review. *Plant Soil* 337:1–18.
Azargohar, R. and A. K. Dalai. 2006. Biochar as a precursor of activated carbon. *Appl Biochem Biotechn* 131:762–773.
Black, S. 2000. A survey of composite bridges. *Composites Technol* 6:14–18.
Dicks, A. L. 2006. The role of carbon in fuel cells. *J Power Sources* 156:128–141.
Fung, I. Y., S. C. Doney, K. Lindsay, and J. John. 2005. Evolution of carbon sinks in a changing climate. *Proc Natl Acad Sci USA* 102:11201–11206.
Ghosh, U., R. G. Luthy, G. Cornelisson, D. Werner, and C. A. Menzie. 2011. In-situ sorbent amendments: A new direction in contaminated sediment management. *Environ Sci Technol* 45:1163–1168.
Glaser, B. 2007. Prehistorically modified soils of central Amazonia: A model for sustainable agriculture in the twenty-first century. *Phil Trans Royal Soc B* 362:187–196.
Glaser, B., E. Balashov, L. Haumaier, G. Guggenberger, and W. Zech. 2000. Black carbon in density fractions of anthropogenic soils of the Brazilian Amazon region. *Org Geochem* 31:669–678.
Glaser, B. and J. J. Birk. 2012. State of the scientific knowledge on properties and genesis of anthropogenic dark earths in Central Amazonia (terra preta de Índio). *Geochim Cosmochim Acta* 82:39–51.
Glaser, B., G. Guggenberger, W. Zech, and M. L. Ruivo. 2003. Soil organic matter stability in Amazonian dark earths. In J. Lehmann, D. Kern, B. Glaser, and W. Woods (eds), *Amazonian Dark Earths: Origin, Properties, and Management*, pp. 141–158. Dordrecht: Kluwer.
Glaser, B., L. Haumaier, G. Guggenberger, and W. Zech. 1998. Black carbon in soils: The use of benzenecarboxylic acids as specific markers. *Org Geochem* 29:811–819.
Glaser, B., L. Haumaier, G. Guggenberger, and W. Zech. 2001. The terra preta phenomenon: A model for sustainable agriculture in the humid tropics. *Naturwissenschaften* 88:37–41.
Glaser, B., J. Lehmann, and W. Zech. 2002. Ameliorating physical and chemical properties of highly weathered soils in the tropics with charcoal: A review. *Biol Fertil Soils* 35:219–230.
Glaser, B., M. Parr, C. Braun, and G. Kopolo. 2009. Biochar is carbon negative. *Nat Geosci* 2:2–2.
Goldberg, E. D. 1985. *Black Carbon in the Environment*. New York: John Wiley.
Halmann, M. M. and M. Steinberg. 1999. Greenhouse gas carbon dioxide mitigation science and technology. Boca Raton, FL: Lewis.
Halloran, J. W. 2007. Carbon-neutral economy with fossil fuel-based hydrogen energy and carbon materials. *Energ Policy* 35:4839–4846.
Halloran, J. W. 2008. Extraction of hydrogen from fossil fuels with production of solid carbon materials. *Int J Hydrogen Energ* 33:2218–2224.

Henning, K. D., W. Bongartz, and K. Knoblauch. 1987. Mechanical properties of pitch-coal extrudates during the carbonization process. *Fuel* 66:1516–1518.

Honda, H., Y. Sanada, and T. Furuta. 1966. Mechanical and thermal properties of heat treated coals. *Carbon* 3:421–428.

Jacob, A. 1999. Composites gives new life to bridges. *Reinf Plast* 43:10.

Jeffery, S., G. G. A. Verheijen, M. van der Velde, and A. C. Bastos. 2011. A quantitative review of the effects of biochar application to soils on crop productivity using meta-analysis. *Agr Ecosyst Environ* 144:175–187.

Keiluweit, M., P. S. Nico, M. G. Johnson, and M. Kleber. 2010. Dynamic molecular structure of plant biomass-derived black carbon (biochar). *Environ Sci Technol* 44:1247–1253.

Kuzyakov, Y., I. Subbotina, H. Q. Chen, I. Bogomolova, and X. L. Xu. 2009. Black carbon decomposition and incorporation into soil microbial biomass estimated by C-14 labeling. *Soil Biol Biochem* 41:210–219.

Lal, R. 2004. Carbon emission from farm operations. *Environ Int* 30:981–990.

Lal, R. 2009. Soils and food sufficiency. A review. *Agron Sust Develop* 29:113–133.

Lehmann, J., J. Skjemstad, and S. Sohi. 2008. Australian climate-carbon cycle feedback reduced by soil black carbon. *Nat Geosci* 1:832–835.

Libra, J. A., K. S. Ro, C. Kammann et al. 2011. Hydrothermal carbonization of biomass residuals: A comparative review of the chemistry, processes and applications of wet and dry pyrolysis. *Biofuels* 2:89–124.

Major, J., J. Lehmann, M. Rondon, and C. Goodale. 2010. Fate of soil-applied black carbon: Downward migration, leaching and soil respiration. *Glob Change Biol* 16:1366–1379.

Mindess, F. and J. F. Young. 1981. *Concrete*. Englewood Cliffs, NJ: Prentice-Hall.

Möller, A., K. Kaiser, W. Amelung et al. 2000. Relationships between C and P forms in tropical soils (Thailand) as assessed by liquid-state ^{13}C- and ^{31}P-NMR spectroscopy. *Aust J Soil Res* 38:1017–1035.

Neves, E. G., J. B. Petersen, R. N. Bartone, and C. A. D. Silva. 2003. Historical and socio-cultural origins of Amazonian dark earths. In J. Lehmann, D. Kern, B. Glaser, and W. Woods (eds), *Amazonian Dark Earths: Origin, Properties, Management*, pp. 29–50. Dordrecht: Kluwer Academic.

Novotny, E. H., T. J. Bonagamba, E. R. Azevedo, and M. H. B. Hayes. 2009. Solid-state nuclear magnetic resonance characterisation of humic acids extracted from Amazonian dark earths (Terra Preta de Índio). In W. Woods, W. G. Teixeira, J. Lehmann, C. Steiner, A. M. G. A. WinklerPrins, and L. Rebellato (eds), *Amazonian Dark Earths: Wim Sombroek's Vision*, pp. 373–391. Berlin: Springer.

Ogawa, M. and Y. Okimori. 2010. Pioneering works in biochar research. *Japan Soil Res* 48:489–500.

Redston, J. 1999. Canada's infrastructure benefits from FRP. *Reinf Plast* 43:34–38.

Rillig, M. C., M. Wagner, M. Salem et al. 2010. Material derived from hydrothermal carbonization: Effects on plant growth and arbuscular mycorrhiza. *Appl Soil Ecol* 45:238–242.

Roberts, K. G., B. A. Gloy, S. Joseph, N. R. Scott, and J. Lehmann. 2010. Life cycle assessment of biochar systems: Estimating the energetic, economic, and climate change potential. *Environ Sci Technol* 44:827–833.

Ross, S. M. 1993. Organic matter in tropical soils: Current conditions, concerns and prospects for conservation. *Progr Phys Geogr* 17:265–305.

Saito, N., K. Aoki, Y. Usui et al. 2010. Application of carbon fibers to biomaterials: A new era of nano-level control of carbon fibers after 30-years of development. *Chem Soc Rev* 40:3824–3834.

San Miguel, G., S. D. Lambert, and N. J. D. Graham. 2006. A practical review of the performance of organic and inorganic adsorbents for the treatment of contaminated waters. *J Chem Technol Biotechnol* 81:1685–1696.

Schimmelpfennig, S. and B. Glaser. 2012. Material properties of biochars from different feedstock material and different processes. *J Environ Qual* 41:1001–1013.

Schmidt, M. W. I. and A. G. Noack. 2000. Black carbon in soils and sediments: Analysis, distribution, implications, and current challenges. *Glob Biogeochem Cycl* 14:777–793.

Schulz, H. and B. Glaser. 2012. Biochar—Benefits for soil and plants as compared with conventional soil amendments. *J Plant Nutr Soil Sci* 175:410–422.

Sombroek, W. G., D. C. Kern, T. Rodrigues et al. 2002. Preta and terra mulata: Pre-Colombian kitchen middens and agricultural fields, their sustainability and replication. In R. Dudal (ed.), *17th World Congress of Soil Science, Symposium 18, Anthropogenic Factors of Soil Formation*, August 2002, Bangkok.

Sombroek, W. G., F. O. Nachtergaele, and A. Hebel. 1993. Amounts, dynamics and sequestering of carbon in tropical and subtropical soils. *Ambio* 22:417–426.

Steinbeiss, S., G. Gleixner, and M. Antonietti. 2009. Effect of biochar amendment on soil carbon balance and soil microbial activity. *Soil Biol Biochem* 41:1301–1310.

Steinberg, M. 1998. Natural gas decarbonization for mitigating global warming. BNL Report 65452. Upton, NY: Brookhaven National Laboratory.

Steiner, C., B. Glaser, W. G. Teixeira, J. Lehmann, W. E. H. Blum, and W. Zech. 2008. Nitrogen retention and plant uptake on a highly weathered central Amazonian ferralsol amended with compost and charcoal. *J Plant Nutr Soil Sci* 171:893–899.

Steiner, C., W. G. Teixeira, J. Lehmann et al. 2007. Long term effects of manure, charcoal and mineral fertilization on crop production and fertility on a highly weathered Central Amazonian upland soil. *Plant Soil* 291:275–290.

Tiessen, H., E. Cuevas, and P. Chacon. 1994. The role of soil organic matter in sustaining soil fertility. *Nature* 371:783–785.

Ting, C. C., T. F. Young, and C. S. Jwo. 2007. Fabrication of diamond nanopowder using microwave plasma torch technique. *Int J Adv Manuf Technol* 34:316–322.

van Krevelen, D. W. 1961. *Coal: Typology, Chemistry, Physics, Constitution.* Amsterdam: Elsevier.

Whitman, T. and J. Lehmann. 2009. Biochar—One way forward for soil carbon in offset mechanisms in Africa? *Environ Sci Policy* 12:1024–1027.

Woods, W. I. and W. M. Denevan. 2009. Amazonian dark earths: The first century of reports. In W. I. Woods, W. G. Teixeira, J. Lehmann, C. Steiner, A. M. G. A. WinklerPrins, and L. Rebellato (eds), *Amazonian Dark Earths: Wim Sombroek's Vision*, pp. 1–14. Berlin: Springer.

Woodward, F. I., R. D. Bardgett, J. A. Raven, and A. M. Hetherington. 2009. Biological approaches to global environment change mitigation and remediation. *Curr Biol* 19:615–623.

Woolf, D., J. E. Amonette, F. A. Street-Perrott, J. Lehmann, and J. Joseph. 2010. Sustainable biochar to mitigate global climate change. *Nat Commun* 1:56.

Yu, J., Q. Zhang, J. Ahn et al. 2002. Synthesis of carbon nanostructures by microwave plasma chemical vapor deposition and their characterization. *Mater Sci Eng B* 90:16–19.

Zech, W., L. Haumaier, and R. Hempfling. 1990. Ecological aspects of soil organic matter in tropical land use. In P. McCarthy, C. E. Clapp, R. L. Malcolm, and P. R. Bloom (eds), *Humic Substances in Soil and Crop Sciences. Selected Readings*, pp. 187–202. Madison, WI: American Society of Agronomy and Soil Science Society of America.

Zech, W., N. Senesi, G. Guggenberger et al. 1997. Factors controlling humification and mineralization of soil organic matter in the tropics. *Geoderma* 79:117–161.

CHAPTER 4

Agroecology of Agromicrobes

Manindra Nath Jha, Shankar Jha, and Sanjeet Kumar Chourasia

CONTENTS

4.1 Introduction 81
    4.1.1 Agromicrobial Diversity 83
        4.1.1.1 Culture-Based Methods 84
        4.1.1.2 Culture-Independent Methods 85
        4.1.1.3 Communication in Agroecosystems 88
4.2 Utilization of Beneficial Agromicrobes for Sustainable Agriculture 88
    4.2.1 Engineering Agricultural Practices 89
        4.2.1.1 Soil Organic Matter 89
        4.2.1.2 Plant Genotype 89
        4.2.1.3 Tillage 90
        4.2.1.4 Soil Type and Land Use History 91
    4.2.2 Artificial Inoculation 92
        4.2.2.1 Nitrogen Acquisition of Microbial Origin 92
        4.2.2.2 Phosphorus Acquisition through Microbe-Mediated Processes 94
4.3 Epilogue 95
References 96

4.1 INTRODUCTION

Agroecological approaches to agricultural production may facilitate the conversion of high-input agricultural systems to more sustainable systems for addressing world food needs while maintaining environmental sustainability. The earliest farmers practiced sustainable agriculture/land management as resource limitations of one form or another affected their ability to produce food during that period. Shifting cultivation was the main agricultural practice, which restored the productivity of soil. The bush fallow system and organic farming later replaced this system, which continued till the arrival of the concept of the green revolution. Thus, agricultural heritage based on the mining of bionutrients for crop feeding was replaced by chemical fertilization to achieve high yield and profitability in the modern agriculture or post–green revolution era. However, due to the unintended introduction of degrading processes, such as the extensive use of agrochemicals and irrigation-mediated salinization, the intrinsic capacity of natural resources (soil) has become suboptimal in developing as well as developed countries. Additionally, the indiscriminate addition of restorer inputs, particularly agrochemicals, and their inefficient use by crops has led to unsustainable farming enterprises. Along with unsustainable yields, the leftover agrochemicals are the principal cause of soil pollution. Soil acidification is a widespread problem in areas that frequently experience high rates of fertilizer application (Guo et al. 2010; Tarkalson et al. 2006). Similarly, the application of synthetic nitrogen (N) in agricultural systems can increase soil carbon dioxide (CO_2) emissions and soil carbon (C) loss (Kwon and Hudson 2010; Russell et al. 2009). A major problem is synthetic N (urea). The nitrogen in urea has an immediate greening effect on crops and many farmers tend to blindly apply urea as soon as the symptoms of chlorosis appear. Fields becomes green overnight and the farmer feels that he or she has effectively managed the problem at hand. However, the green fades in a matter of days. This "boot polish" effect of urea has led to its unscientific use and has upset the nutritional balance of the soil as excess application of the nutrient tends to reduce the efficiency of uptake of other nutrients. The problem is more aggravated in developing and underdeveloped countries where cultivation of fragmented land is quite common. Another major concern is that since 1960, there has been a roughly 30% increase in per capita food production worldwide; however, the real price of food has declined by almost a half and this may result in diminishing returns, reducing the value of mineral fertilizers to farmers (Halweil 2002; Uphoff 2006). These developments and the consequent concerns over the long-term benefits of current modern agricultural strategies have led to the revisiting of the concept of *sustainable agriculture*. Sustainable agriculture means production as well as economic and environmental sustainability. It envisages a system approach and places the primary emphasis on maximizing not only yield but also agrosystem stability. Thus, it represents an integration of traditional techniques with modern advances that are recognized as appropriate. Further, the high fertilizer prices brought about by the energy crisis, the cost of inputs and transportation, import restrictions and controls, and the need to reduce the fiscal deficit and/or introduce new policy directions justify the close attention being given to a technology with both chemical and biological components for the proper management of soil nutrients, pests, and diseases. Thus, some components of modern agriculture, identified as the cause of fatigue of the green revolution, need another alternative or substitute. We believe that the alternation or substitution of agrochemicals with agromicrobes is a healthy and promising approach and will facilitate the achievement of the mandate of the "ever-green revolution."

Agromicrobes are a vital component of agricultural sustainability and the ever green revolution. They are the basis of sustaining life on this planet. They affect all life forms and the physical–chemical makeup of our planet and have done so since the origin of life. No other group of organisms can make such a claim. Life without all other creatures is possible but life without microbes is not. Microbes are the masters of the biosphere and ours indeed is a planet of microbes. Agromicrobes are also known to be the most crucial element in the functioning of the agroecosystem. They mediate many processes that are essential to crop productivity, including nutrient recycling of carbon, nitrogen, phosphorus, sulfur, and so on; organic matter decomposition; plant nutrient availability; plant growth augmentation; plant feeding through a multichannel bionutrient delivery system; enhanced protection against diseases and pests; stress tolerance induction; and soil structure maintenance. Recent estimates of the contribution of biological nitrogen fixation (BNF) to agricultural systems are 2.95 million ton (Tg) for pulses, 18.2 Tg for oil seed legume, 12–25 Tg for pasture and fodder legume, 5 Tg for rice, 0.5 Tg for sugarcane, <4 Tg for nonlegume croplands, and <4 Tg for extensive savannas. Aggregating these individual estimates provides an overall estimate of 50–70 Tg N fixed biologically on agricultural systems (Herridge et al. 2008). Such estimates include symbionts (*Rhizobia*), associative symbionts (*Azospirillum*), and asymbionts (*Azotobacter*, Cyanobacteria, etc.). Other important processes are the colonization of plant roots by the symbiont arbuscular mycorrhizal fungi (AMF). The benefit of this association to the host plants is an enhanced uptake of phosphorus and other poorly mobile nutrients via the extensive hyphal network in soil. Further, a diverse array of free-living microbes, including *Azotobacter*,

Azospirillum, Acetobacter, Bacillus, Burkholderia, Pseudomonas, Serratia, Trichoderma, nuclear polyhedrosis virus (NPV), *Beauveria*, and so on, are known to enhance plant growth by increasing seed emergence, plant biomass, and crop yield, and suppressing bacteria, fungi, and nematode pathogens. These potentialities of agromicrobes are translated into microbial technologies to derive maximum benefit; nevertheless, while microbial technologies have been applied to various agricultural problems with considerable success in recent years, they have not been widely accepted by the farming community because of the difficulty of consistently reproducing their beneficial effects. Further, one of the paradoxes of research in sustainable agriculture has been the inability of low-input practices, especially agromicrobe-based practices, to outperform conventional practices in side-by-side experimental comparisons. Poor agroecological knowledge of agromicrobes is one of the major limiting factors for constant reproducibility and performance in a modern or conventional agricultural production system. The present chapter aims to provide an overview of the current status of agromicrobial ecology with respect to the methods used to detect the diversity of agromicrobes in crop ecosystems and future strategy to ensure a more effective and consistent performance of agromicrobes in a sustainable agricultural production strategy. This is particularly needed when planning for the effective utilization of agromicrobes for the long-term sustainability of our agricultural production system.

4.1.1 Agromicrobial Diversity

In recent years, there has been greater understanding of how to take advantage of biological processes, linking agromicrobes and plants that have been "eclipsed" by the emphasis on chemical inputs over the past century and on genes in this era. The baseline for deriving advantages in generating knowledge about agromicrobes is mediated in soil ecosystems. Agromicrobial diversity in soil ecosystems far exceeds that of other organisms. Soil is a reservoir of microbial activities and occupies about 5% of the total space (Ingham et al. 1985). Ten billion bacteria live in a gram of soil, a mere pinch held between the thumb and forefinger. What is most astounding is that each different gram of soil will contain a different collection of thousands of species.

An analysis of the DNA of a bacterial community in 1990, and a later study by Torsvik et al. (1990, 1996) estimated that there were 4,000–10,000 bacterial species per gram of soil. However, this was superseded by a reanalysis of the data using new analytical methods by Gans et al. (2005), who showed that soil could contain more than 1 million distinct genomes per gram of soil. It should be pointed out that these estimates are for prokaryotes and do not account for eukaryotic species that also must make a significant contribution to soil genomes. In contrast, the genomic complexity recovered by culturing methods was <40 genomes. Further, the microbial biomass is large—in a temperate grassland soil, the bacterial and fungal biomasses amount to 1–2 and 2–5 t ha^{-1}, respectively (Killham 1994). A bacteria may weigh approximately 10^{-11} g, yet, collectively, microbes constitute about 60% of the earth's biomass. These soil microbes always work in consortia mode. It is believed that more than 99% of agromicrobes are unknown and uncultivable. All such information is flooding in due to the advancements in the analytical tools for studying microbial diversity. Phenotypic and genotypic analytical tools are now available to characterize the preponderance and community structure of agromicrobes.

Phenotypic methods that are used to culture microbes include standard plating methods on selective media (who is there?); community-level physiological profiling (CLPP) using a biology system (what is the activity of the soil microbial community?) (Garland 1996a); and phospholipid fatty acid (PLFA) profiling (Tunlid and White 1992) and fatty acid methyl ester (FAME) profiling (Germida et al. 1998). Genotypic approaches or culture-independent molecular techniques are based on the direct extraction of DNA from soil and a 16S rRNA sequence analysis, a bacterial artificial chromosome (BAC) system, or an expression cloning system (Rondon et al. 1999). Such molecular approaches help us to determine community structure (who is there?) by phylogenetic arrays, 16S

RNA tag sequencing, and community image analysis (CMEIAS); genetic potential (what can they do?) through metagenomics and functional gene arrays; gene expression (what are they doing?) by metatranscriptomics, metaproteomics; and functional characterization through stable isotope probing, NanoSIMS, and CARD-FISH. The analytical tool that is chosen to investigate the diversity of agromicrobes may have different perceptions about the composition and preponderance of microbial species.

4.1.1.1 Culture-Based Methods

4.1.1.1.1 Plate and Direct Count

Traditionally, the diversity of agromicrobes is assessed using most probable number (MPN) counting, selective plating, and direct viable counting. In these approaches, a set of culture media is commonly employed and the identification of isolates is based on a variety of morphological typing methods, including FAME and 16S rDNA sequence analysis. The culture-based methods are fast and inexpensive and can provide information on the active heterotrophic/autotrophic component of the population. By applying these methods, specific groups will be selected, depending on the type of media used. The main advantages are that the microbial isolates obtained can be used for studying biotic interactions, for example, with other microbial groups or with the plant host, and for selecting efficient strains for application in agriculture. They also help in making inferences on the physiological and metabolic properties of the microbes. However, microorganisms in the soil do not escape the big plate anomaly by which a huge gap exists between the number of viable cells present in any sample and that of colonies retrieved on culture media. This leaves many species unknown—about 90% of the microbial cells present in plant roots and observed by microscopy are not considered by cultivation *in vitro* (Goodman et al. 1998). In fact, culturing of microorganisms is hampered by difficulties in reproducing natural ecological niches in the laboratory. Furthermore, symbiotic relationships (AMF) might be of crucial importance for community functioning and are often impossible to maintain in the culture media. Other limitations include the difficulty in dislodging bacteria or spores from soil particles or biofilms, growth medium selection, growth conditions (temperature, pH, light, etc.), the inability to culture a large number of bacterial and fungal species with current techniques, and the potential for colony–colony inhibition or colony spreading. In addition, plant growth favors those microorganisms with fast growth rates and those fungi that produce large numbers of spores. Sometimes, collected soil samples along with visible colonies of known cyanobacteria did not appear in the plate count and rather other cyanobacterial species were encountered.

Direct counting using fluorescence microscopy can give 100–1000 times more microbial numbers than the numbers obtained by plate counting (Johnsen et al. 2001). Several strains specific to proteins or nucleic acids viz. fluorescein isothiocyanate (FITC), acridine orange (AO), ethidium bromide (EB), and europium chelate, with a bright fluorescent ever differential stain (DFS) for bacteria and phenol aniline blue (PAB) and fluorescent diacetate (FDA) for fungal hyphae have been used. Bloem et al. (1995) improved the direct counting method using a video camera mounted on an epifluorescence microscope. However, the fluorescent microscopy procedure does not allow the counting of specific microbial species, and some of the strains used do not discriminate between living and dead microbial cells.

4.1.1.1.2 Sole Carbon Source Utilization Pattern/Community-Level Physiological Profiling

A Biolog™ based method for directly analyzing the potential activity of soil microbial communities displaying CLPP has been used to study microbial diversity (Garland 1996b; Garland and Mills 1991). CLPP is the characterization of heterotrophic microbial communities based on an assessment of sole-source carbon use (SSCU) patterns. It represents sensitive and rapid methods for accessing the potential metabolic diversity of microbial communities. It involves the inoculation of a microbial sample onto a plate, which contains a 95-carbon source in addition to a tetrazolium dye. The tetrazolium salt changes color as the substrate is metabolized. Since many fungal species are not capable of reducing tetrazolium salt, Biolog developed the fungal-specific plates SFN_2 and SFP_2, which have the same substrate as bacterial plates but without the tetrazolium salt (Carlsson and Huss-Danell 2003). Inoculated populations are monitored over time for their ability to utilize the substrate and the speed at which the substrate is utilized. A multivariate analysis is applied to the data and the relative differences between soil functional diversity are assessed. CLPPs can differentiate between microbial communities, are relatively easy to use, reproducible, and produce large amounts of data reflecting the metabolic characteristics of the communities. However, the robustness of the culture-based methods and the SSCU procedures used in the various community studies is limited by their ability to detect only 0.1%–10% of the total bacterial population in soil (Amman et al. 1995). Other limitations are that it only represents a culturable fraction of the community, favors fast-growing organisms, represents only those microorganisms that are capable of utilizing the available carbon source, is sensitive to the inoculum density, and represents potential metabolic diversity, not *in situ* diversity.

4.1.1.2 Culture-Independent Methods

Despite improvements in cultivation techniques to study the community composition of agromicrobes and their preponderance in the soil ecosystem, the microorganisms in culture represent only a small fraction of microorganisms that occur *in situ*. This is a common problem in science; our image of the world around us is defined and limited by the methods we use to measure the world. However, such a problem also creates new opportunities, such as the application of molecular approaches to study microbial diversity or microbial molecular ecology.

4.1.1.2.1 Molecular-Based Techniques to Study Agromicrobial Diversity

Molecular techniques based on culture-independent methods open new possibilities for exploring the diversity of agromicrobes and thus represent a step forward in the exploration of agromicrobes diversity for human welfare. They generally involve the extraction of nucleic acid, directly or indirectly from soil, thereby mitigating the biases in isolation and *in vitro* cultivation. The approaches to extracting nucleic acids from soil is either by direct extraction of the nucleic acids after *in situ* cell lysis followed by DNA purification (Ogram et al. 1987) or separation of the cell fraction from soil particles before the lysis of cells and nucleic acid purification (Courtois et al. 2001; Holben et al. 1988). Both approaches have advantages related to DNA yields, DNA purity for molecular purposes, and the ever-questioned representation of the entire agromicrobes diversity.

Low-resolution methods include the analysis of DNA reassociation profiles and the base distribution (mole percentage of G+C) of DNA. In DNA reassociation profiles or hybridization, the reassociation rate of thermally denatured DNA is determined at a fixed temperature (approximately 25°C) below its melting point (Torsvik et al. 1990). Such a reassociation rate will depend on the similarity of the sequence present. As the complexity or diversity of the DNA sequence increases, the rate at which the DNA reassociates will decrease. DNA reassociation only gives an estimate of the total species number in a sample. A percentage of GC profiles gives information about the base composition of the DNA, but they cannot, on their own, be used to determine the presence or abundance of a particular species. This is based on the knowledge that microorganisms differ in their G+C content and that taxonomically related groups only differ between 3% and 5%. These techniques provide an overall indication of agromicrobes diversity.

The basis of molecular microbial ecology is molecular markers. Molecular markers can be genes or gene transcripts that can be identified in a compiled pool of nucleic acids providing information on the group of organisms harboring these genes; the small subunit (SSU) rRNA (16S of prokaryote or 18S of eukaryote) is the principle marker. Over the last decade, the sequencing of SSU rRNA from uncultured organisms has led to the development of databases and it has been verified that only a small fraction of the diversity of microorganisms is known. Polymerase chain reaction (PCR) products generated with primers based on 16S or 18S rDNA using the total DNA of the specific soil microbial community yield a mixture of DNA fragments representing all PCR-accessible species present in the soil sample. The mixed PCR product can be used either for preparing clone libraries (McCaig et al. 1999) or for a range of microbial community fingerprinting. Such clone libraries are useful for identifying and characterizing the dominant bacterial or fungal types in soil, thereby providing a picture of the diversity of agromicrobes (Garbeva et al. 2004), fingerprinting techniques such as denaturing or temperature gradient gel electrophoresis (DGGE/TGGE), terminal restriction fragment length polymorphism (T-RFLP), restriction fragment length polymorphism/amplified ribosomal DNA restriction analysis (RFLP/ARDRA), single-stranded conformational polymorphism (SSCP), ribosomal intragenic spacer analysis/automated ribosomal intragenic spacer analysis (RISA/ARISA), and highly repeated sequence characterization of microsatellite regions (rep-PCR).

DGGE is based on the different melting behavior of double-stranded DNA due to sequential differences in a denaturing gradient during electrophoresis (Muyzer et al. 1993). TGGE uses the same principle as DGGE except that the gradient is temperature rather than chemical denaturing. These techniques allow us to differentiate two molecules at the single base level and they are commonly used for characterizing bacterial communities of the soil. They have the advantage of being reliable, reproducible, rapid, and somewhat inexpensive. Another advantage of DGGE or TGGE is that the bands can be investigated by hybridization with specific probes or by extraction and sequencing (Muyzer and Smalla 1998). However, biases associated with PCR could cause relative under- or overrepresentation of a given taxon. The limit of resolution of these methods seems to be about 1% of the community population. In the case of complex samples producing hundreds of DNA bands, multiple amplicons comigrating to the same position can occur and cannot be distinguished. Further, only the dominant population are revealed and bands from more than one species may be hidden behind a single band, resulting in an underestimation of bacterial diversity (Heuer et al. 2001). Finally, the same isolate can have different bands because multiple copies of an operon in a single species are present (Heuer et al. 2001). In addition, we must keep in mind that fingerprinting depends on the primer used.

T-RFLP separates fragments based on the length of the terminal fragments obtained due to differences in restriction endonuclease sites (as in RFLP and ARDRA) (Liu et al. 1997) and SSCP separates fragments based on the different mobility of single-stranded DNA in nondenaturing gels (Schwieger and Tebbe 1998). An advantage of SSCP approaches is the possibility of isolating specific genetic elements for subsequent sequencing, which is not possible when T-RFLP is used, thus omitting the opportunity to fully identify elements of interest. On the other hand, T-RFLP is suitable for high-throughput analysis, despite the need for a restriction digest step. The banding pattern in T-RFLP can be used to measure the richness and evenness of species as well as the similarities between samples. Care must be taken to select restriction enzymes that produce a broad community pattern. T-RFLP can provide a good view of the most abundant species and is typically limited to the 50 most abundant organisms. A sensitivity limit is estimated for the member of a community that accounts for about 0.5% of total amplified rDNA (Liu et al. 1997). The major drawback of SSCP is the high rate of reannealing of single-stranded DNA during electrophoresis (Nocker et al. 2007). Experimental evidence suggests that SSCP will produce a visible band for members of a community that comprise no less than 1.5% of the total extracted DNA. Since no database exists for SSCP bands, the only way to obtain taxonomic information is to use other methods such as cloning and sequencing of the resulting DNA bands on the pattern.

RFLP, also known as ARDRA, is another tool that is used to study microbial diversity that relies on DNA polymorphism. PCR-amplified rDNA (SSU insulinoma tumor suppressor [ITS] and large subunit [LSU] rRNA genes) is digested by various restriction enzymes or by a combination of restriction enzymes. Different fragment lengths are detected using agarose or nondenaturing poly-acrylamide gel electrophoresis in the case of community analysis. The pattern of the fragment sizes is characteristic of species with microbial genes.

The following techniques allow the detection of microbial strains at the species and subspecies level: rep-PCR (microsatellite region), based on the amplification of a sequence between repetitive elements; ribosomal interspace analysis (RISA), which is based on the length polymorphism of the spacer region between 16S and 23S rRNA genes; and random amplified polymorphic DNA (RAPD), which does not require a primitive knowledge of the genome (Harry et al. 2001). In RISA, PCR products are separated by gel electrophoresis and the separated bands are sequenced. A limit of this technique is the number of spacer sequences in the database. Further, RISA requires large quantities of DNA, is more time consuming, silver staining is somewhat insensitive, and its resolution tends to be low.

In recent years, the variable number of tandem repeats (VNTRs) has proven to be a suitable target for assessing genetic polymorphism within a bacterial species. Other developments include the creation of BACs and the application of DNA microarrays (Rondon et al. 2000; Tiedje et al. 2001). The former technique clones high-molecular-weight soil DNA (metagenome), which can be analyzed at a phylogenetic and a functional level. DNA microarrays could be valuable in the study of agromicrobial diversity since a single array can contain thousands of DNA sequences (Cho and Tiedje 2001) with the possibility of very broad hybridization with a broad identification capacity. The microarray can contain either specific target genes such as nitrate reductase, nitrogenase, or naphthalene dioxygenase to provide functional diversity information or a sample of an environmental "standard" (DNA fragments with <70% hybridization) representing different species found in the environmental sample.

It is challenging for microbiologists to develop techniques to study agromicrobial diversity when it is currently impossible to know the accuracy of these techniques. We do not know what is present in a gram of soil and therefore it is difficult to conclude whether one technique to study diversity is better than another. Given the current state of knowledge, we feel that the best way to study agromicrobial diversity would be to use a variety of tools with different end points and degrees of resolution to obtain the broadest picture possible and the most information regarding the agromicrobial community. Further, we have to be aware of the fact that agromicrobial diversity should not degenerate to gene diversity or base sequence diversity.

4.1.1.2.2 Fatty Acid Methyl Ester (FAME) Analysis

Biochemical approaches that do not rely on microorganism culturing for assessing the agromicrobial community composition are FAME and PLFA analyses (Tunlid and White 1992; Pankhurst et al. 2001; Zelles 1999). These techniques are based on the extraction, fractional methylation, and chromatography of the phospholipid component of soil lipids.

Fatty acids make up a relatively constant proportion of the cell biomass and a signature fatty acid exists that can differentiate major taxonomic groups within a community. Therefore, a change in the fatty acid profile would represent a change in the microbial population and can be interpreted by reference to a database of pure cultures and known biosynthetic pathways (Zelles 1999). The direct extraction of PLFAs or whole fatty acids from soil does not permit detection at the species level and can only be used to estimate gross changes in the community structure. However, the identification of species by fatty acid analysis is possible with standard culture-based media and databases (MIDI, Newark, DE). FAME profiles of different soils can be compared using multivariate analysis. Lechevalier (1989) and Zelles (1999) have listed many fatty acids isolated from specific microbial groups. However, this approach has several limitations. Other organisms can confound FAME analysis (Graham et al. 1995), individual species may have numerous fatty acids (Bossio and Scow 1998), analysis can be influenced by external factors, and the interpretation of individual markers depends on a database derived mostly from information on isolated pure cultures of bacteria (Vestal and White 1989). A direct comparison of PLFA/FAME and SSCU procedures has produced conflicting results (Waldrop et al. 2000; Ibekwe and Kennedy 1998). Additionally, a large quantity of material is required if the method is used for fungal spores.

4.1.1.3 Communication in Agroecosystems

Besides the diversity of agromicrobial species, a fundamental understanding of cell–cell communication by chemical signaling among microorganisms and between microorganisms and plants is required for the effective use of beneficial microorganisms in our sustainable agricultural production system. A large number of chemicals allow microorganisms to coordinate gene expressions on a population-wide scale. One such system is quorum sensing (QS), which involves the production, secretion, and subsequent detection of small hormone-like signaling molecules. The impacts of signaling are multifaceted, with microorganisms communicating with each other to carry out certain functions and two-way communication between plants and microbes. The establishment of AM symbiosis and the root nodule (RN) symbiosis requires an ongoing molecular dialogue that underpins the reprogramming of root cells for compatibility. It starts with a chemical signal exchange between root and microsymbiont, priming both partners for the subsequent association. Plant root exudates contain biological molecules, including humanoids and strigolactones, which are perceived by rhizobial bacteria and AMF. This is followed by distinct phases of other interactions, including a presymbiotic anticipation phase with subsequent interradical accommodation of the microsymbiont. Another example is fungal elicitors detected by plants such as chitin fragments that cause the plant to have a defense reaction to fungal pathogens and to produce phytoalexins (Ebel and Mithofer 1998). At the same time, there is evidence that plants send signals to microorganisms, which in turn results in the stimulation of microorganisms that can have synergistic and positive effects on plants. The neutral, beneficial, or pathogenic influences of microbes depend on the balance of plant–microbes interaction (Raaijmakers et al. 2008). Ample evidence exists that clearly demonstrates the selection of microbes by plants (Hartmann et al. 2009). The study of the nature and function of the molecules mediating these interactions—the area of research defined as microbial chemical ecology—is of great basic scientific interest and is likely to exhibit new ways to create artificial symbiosis and new means to understand the pathogens plant–microbes interaction. It will also offer the possibility of reprogramming plant roots to release components to activate the colonization of beneficial microorganisms for their bionutrient supply system, environmental stress mitigation, and disease suppression in achieving the goal of sustainable agriculture.

4.2 UTILIZATION OF BENEFICIAL AGROMICROBES FOR SUSTAINABLE AGRICULTURE

Chemical-based agricultural production systems have created many sources of pollution that, either directly or indirectly, can contribute to the degradation of the soil environment and the destruction of our natural resource base. Harnessing the potential of soil microbes to manage crop nutrition can be considered as an alternative or a supplementary means for reducing the use of chemicals in agriculture. The extent of the benefits depends on the preponderance of the soil microbes and their composition. One approach should be to generate information on agricultural practices that stimulate microbial preponderance in the soil. We should plan a strategy for encouraging the proliferation of a beneficial autochthonous microbial population that facilitates nutrient acquisition (*Rhizobia*, *Mycorrhiza*, Cyanobacteria, *Azotobacter*), promotes plant growth directly (*Pseudomonas, Azospirillum*), or suppresses plant pathogens. Farmers influence the soil environment and energy each time they irrigate their fields, apply fertilizer, and introduce other agronomical practices, and such interventions have a distinct effect on the soil microbial preponderance and diversity (Lupwayi et al. 1998; Verma et al. 2011). Another approach adopted is to apply efficient microbial strains in high numbers in soil or seed or root, a process called biofertilization (Herridge et al. 2008; Jha and Prasad 2006; Vessey et al. 2004). However, the selection of a region-specific promising strain is the prerequisite for deriving maximum benefit (Bashan et al. 2004). The success of both approaches will be dictated by the knowledge gained in the agroecology of agromicrobes.

4.2.1 Engineering Agricultural Practices

4.2.1.1 Soil Organic Matter

The natural engineering of beneficial microbial populations through agroecological approaches (agronomic practices) will facilitate a reduced dependency on chemical nutrient sources. Such engineering of agricultural practices includes amending the soil (organic fertilization, reduced tillage), breeding plants under low-input agricultural systems, and introducing suitable crop rotation. Ammonium-based fertilizers applied to plants tend to acidify the rhizosphere whereas nitrate-based fertilizers result in a more alkaline rhizosphere. Shifts in pH can alter the soil chemistry and can influence the preponderance of microbial communities. Degens (1998a) showed that the addition of simple organic substances to soil changed the metabolic profiles and metabolic diversity of soil microbial communities, with the characteristics of the responses varying according to the nature of the substrate added. Fungi are considered the most important degraders of high C/N ratio straw in soil (Cheshire et al. 1999) and bacteria are key contributors to the degradation of low C/N ratio substrates (Lundquist et al. 1999). The activities of most enzymes increase as native soil organic matter increases, reflecting large microbial communities and the increased stabilization of enzymes by humic material (Burns 1982). Soil organic matter can be enhanced through organic fertilization. An increased trend was observed in the preponderance of *Azospirillum* with an increase in the organic carbon level from 0.2% to 1.0% (Verma et al. 2011) and the colonization of maize root by *Azospirillum* was more augmented in an organic fertilizer system than a mineral fertilizer system. Distinct differences in the colonization potential of AMF between organically and chemically managed fields were observed with less inoculants of AMF in chemically fertilized fields than in organic systems (Mader et al. 2002). Kahiluoto et al. (2001) observed a decrease in AMF densities under different soil and climate conditions on application of more than 50 kg Pha^{-1}. In another study, several years of sodium (NA) application in the form of NH_4NO_3 at levels between 100 and 170 kg Nha^{-1} yr^{-1} decreased AMF spore densities in four North American grassland soils (Johnson et al. 2003). Elevated bacterial and fungal numbers including *Azotobacter* and PSM were reported by Surekha et al. (2010). Further, a microbial biomass is considered a sensitive indicator of changes in soil quality (Powlson et al. 1987). A higher microbial biomass in organic-fertilized soil was reported by earlier researchers (Liu et al. 2007; Mader et al. 2002). Thus, it can be proposed that the declining trend of soil organic matter in intensive agriculture, especially in tropical countries, has a highly damaging effect on the nutrient carbon of microbial origin for crops. Currently, it seems that organic fertilization is an ideal approach for deriving benefits from agromicrobes in our agricultural production system.

4.2.1.2 Plant Genotype

Plant species and genotype have a considerable influence on both the quality and composition of autochthonous microorganisms. Soil microorganisms, which are mostly heterotrophic, depend on an exogenous supply of carbon substrates for energy as well as nutrient sources for growth and development. Roots provide almost all types of carbon substrate, which are lacking in soil. Root exudates support large microbial populations that afford a basal level of feeding to the crop. To keep control of microbes, the plant restricts or directs their development by excreting quite selective mixtures of substances, which provide selective conditions for agromicrobes. Different plant species select different microbial communities and plant-specific enrichment can be increased by repeated cultivation of the same plant species in the same field (Smalla et al. 2001). Interestingly, the interactions between cultivar-dependent variations and beneficial microorganisms seem to have resulted from evolution over generations. In most cases, the capacity of a plant genotype to positively interact with beneficial microorganisms appears to be an inherited trait. Picard et al. (2008) reported that the maize genotype influences the size of microbial communities involved in N_2 fixation, as well as the diversity of the mycorrhizal fungi colonizing population. A traditional or local cultivar of maize had the highest *Azospirillum* population followed by a composite maize cultivar. The population decreased drastically in hybrid maize or a modern cultivar (Verma et al. 2011). The mycorrhizal competence of several wheat cultivars revealed that those developed prior to 1950 were more reliant on mycorrhizal symbiosis than the modern wheat cultivar (Hetrick et al. 1993). Further, landraces of mycorrhizal wheat grown in low-P soils produced a higher yield than modern varieties grown under the same conditions (Egle et al. 1999). Similarly, root entophytes such as *Azoarcus* spp. preferentially colonized wild species and older varieties than modern cultivars of rice (Engelhard et al. 2000) and wheat (Marshell et al. 1999), respectively. The differential capacity to support symbiotic/associative N_2 fixation/AM colonization has been clearly established among crop species as well as genotypes within crop species (Baldani et al. 2002; Egle et al. 1999; Jagnow 1990; Shrestha and Ladha 1996). In Brazil, the approach adopted for sugarcane breeding is to select both local and introduced sugarcane genotypes and evaluate them in the absence of large N inputs, resulting in sugarcane varieties that are able to incorporate up to 70% of their total N from associative bacterial N_2 fixation (Andrews et al. 2003; Baldani et al. 2002). A similarly high N_2 fixing line of the common bean and soybean was developed. Thus, it is evident that traditional and local genotypes derive maximum benefit from autochthonous beneficial microbes.

The development of modern hybrid varieties was based on their evaluation in standardized, high-input systems (agrochemicals) with an emphasis on yield. In such a system, the beneficial interaction between plants and associated soil microorganisms was proven by the provision of large quantities of nutrients in readily available plant form. Furthermore, under such conditions, agromicrobes are faced with an environment that differs substantially from the one in which plant–microbes interaction originally evolved (Drinkwater and Snapp 2007). Thus, it may be difficult to select cultivars for low-input agriculture from the elite germplasm pool of current cultivars. The reintroduction of a gene regulating such adaptive traits from local cultivars into the gene pool of modern varieties may represent the most promising means to improve the nutrient supply system using a microbial approach. In future, both the plant species and the microbial species should be included in breeding programs and such programs should be carried out in a participatory mode under the low-input agricultural systems of poor and marginal farmers. There is an urgent need to bridge the gap between breeding and basic sciences. Information should be generated on gene erosion or gene knockout of microbial regulating traits in modern/hybrid varieties.

4.2.1.3 Tillage

Soil tillage began thousands of years ago. In mechanized farming, deep ploughing using a tractor is a common practice but many poor farmers still practice shallow tillage using a country plough. Soil disturbance by tillage can influence the structure and richness of microbial communities in the rhizosphere and bulk soil (Giller 1996; Lupwayi et al. 1998). Such disturbances may result in a reduction in microbial diversity due to desiccation, mechanical destruction, and disruption of access to food resources; however, reduced tillage has been shown to enhance soil microbial diversity (Wander et al. 1995). Minimum or no tillage were associated with higher root colonization by AMF in subterranean clover, wheat, and maize (Duan et al. 2010), probably because tillage destroyed the hyphal network and hence reduced the infectivity of the soil. Soil disturbance by tillage was identified as a major factor affecting soil biodiversity (Giller 1996). The numerical strength of *Azospirillum* in the maize rhizosphere decreased after deep tillage using a tractor (Verma et al. 2011). The number of nodules, overall nitrogen fixation, P uptake, and mycorrhizal development was greater for plants grown in a reduced or no-tillage system (Douds and Seidel 2012). A strong decline in soil carbon (C) occurred after repeated tillage (Conant et al. 2001), which ultimately reduced the preponderance and diversity of agromicrobes. Apart from reducing the physical disturbance of the soil, the no-tillage system leaves crop residues from the preceding year's growth at the soil surface. In crop rotation, under no-tillage, the litter from several crops in preceding years is likely to result in a greater variety of substrates than deep tillage, where litter does not accumulate. Thus, an additional benefit of no-till might be from the carry-over of the residual population of agromicrobes in the soil. Further, under no-till, the exposure of microorganisms to a greater diversity of substrates probably explains their greater preponderance in this system than modern tillage.

4.2.1.4 Soil Type and Land Use History

Soil type and land use history is an important component in dictating the richness of agromicrobes (Schulter et al. 2004; Steenwerth et al. 2003). A gradient of increasing disturbance intensity represented by a range of land use histories can capture the variation across a landscape, within a landscape unit, and in the physical and chemical characteristics of soil, management, and vegetation, and may provide insights into the complex set of factors that affect soil microbial biomass and community composition. Land use history and its associated management inputs and practices may produce a unique soil environment for which microbes with specific environmental requirements may be selected and supported. Higher values of total soil C, N, and microbial biomass occur in grassland than in cultivated soils (Steenwerth et al. 2003). A given land use type could be identified by the composition of the soil microbial community. In some cases, cultivation history has long-term effects on the microbial community structure in abandoned agricultural fields (Buckley and Schmidt 2001), and the gradients in soil fertility in either grasslands (Grayston et al. 2001) or cultivated sites (Yao et al. 2000) have been shown to influence microbial community composition. Relict and restored stands of long-lived perennial bunchgrass supported microbial communities that differed from those associated with formerly cultivated annual grasslands, while intensive soil disturbance due to current cultivation distinguished the microbial community composition of agricultural sites from grasslands (Steenwerth et al. 2003). Differences in microbial community composition were most highly correlated with soil microbial biomass. The preponderance of *Azospirillum* in the maize rhizosphere of four soil types of different land use history was assessed (Verma et al. 2011). Diara soil, known as the "maize basket" of Bihar (India), having a history of maize cultivation for hundreds of years, had the highest population of *Azospirillum* and noncalcareous soils, with a minimum population in the predominantly rice-growing areas.

We conducted a 5 year field survey of a farmer's field to investigate the effect of agricultural practices on diazotrophs, plant growth–promoting rhizobacteria (PGPR), and AM colonization in the maize rhizosphere. Under no-tillage, traditional maize cultivar and soil organic carbon (>1%) appear to support the preponderance of diazotrophs, PGPR, and AM colonization in maize root. An increase in the preponderance of bacteria, *Azospirillum* population in maize root, AM colonization, microbial biomass, nitrogen mineralization, phosphate-solubilizing microbes, cellulose-hydrolyzing microbes, and an acetylene reduction assay (ARA) of maize root occurred due to the organic fertilization of maize crop (unpublished data). While progress in a number of areas related to agricultural practices has been encouraging, our ability to engineer the agroecosystem in a precise manner remains a challenge. Fundamental issues concerning microbial abundance and diversity in the soil remain unresolved. However, it is probably safer to adopt agricultural practices that preserve or restore the beneficial microbes' preponderance and diversity than to adopt practices that diminish the component of total diversity.

4.2.2 Artificial Inoculation

Artificial inoculation is an ecofriendly strategy to augment the preponderance of efficient beneficial microorganisms in our sustainable agricultural production system. Such artificial introduction of efficient microorganisms is also called biofertilization. Biofertilizer or microbial inoculant is a microbial product containing millions of targeted efficient microorganisms for bionutrient delivery to the crop through seed or soil or root application. The commonly used microbial inoculants for crop productivity are heterotrophic *Rhizobium*, *Azospirillum*, *Acetobacter*, and *Azotobacter*; photosynthetic Cyanobacteria and *Azolla* for bionitrogen acquisition; heterotrophic mycorrhiza and phosphate-solubilizing bacteria/fungi for phosphorus acquisition; and *Pseudomonas* and *Bacillus* for plant growth stimulators. These microbial inoculants can improve soil fertility and crop productivity, augmenting nutrient availability and uptake, stimulating plant growth, controlling and resisting soil-borne diseases, and improving the health and properties of the soil. Recent estimates of the annual inputs of nitrogen of microbial origin in the agricultural ecosystem are about 50–70 million tons annually. Such annual inputs include 12–25 million tons for pasture and fodder legumes, 5 million tons for rice, 0.5 million tons for sugarcane, about 4 million tons for nonlegume crops, and 14 million tons for extensive savannas (Herridge et al. 2008). While *Rhizobium* requires a symbiotic association with RNs of legumes to fix nitrogen, others can fix nitrogen either independently or through loose colonization. Phosphate-solubilizing microbes secrete organic acids that enhance the phosphorus uptake of plants by dissolving rock phosphate, tricalcium phosphate, and unsolubilized phosphate in the soil ecosystem. AMF are the most common and omnipresent phosphorus-mobilizing types. A group of bacteria that enhance the growth of plants through nitrogen fixation, phosphate solubilization, or the production of plant growth–promoting metabolites are known as PGPR. These beneficial microorganisms are artificially inoculated either as single-culture microbial inoculants or in integrated nutrient management (INM) mode or in consortium mode. Under consortium mode, either mixed microbial inoculants or different microbial inoculants are applied singly to a specific crop in its different niches. A microbial inoculant may be more successful under consortium mode because of the increased diversity of the beneficial microbes and the stability that is associated with consortium culture. However, the use of microbial inoculants in consortium mode has not gained widespread acceptance by the agriculture research establishment because it is difficult to demonstrate conclusively which microorganisms are responsible for the observed effects and thus it is difficult to publish the results in reputable scientific journals. An interesting issue with biofertilization is the persistence of biofertilizers after inoculation. If the inoculum potential can be built into agricultural soils, the interval between the application of microbial inoculants can be increased and the cost can be lowered.

4.2.2.1 Nitrogen Acquisition of Microbial Origin

Nitrogen is one of the most commonly deficient nutrients in soils, contributing to reduced agricultural yields throughout the world. Supplying nitrogen through chemical/industrial processes is very energy intensive. However, it can be supplied to crops by BNF, a process that is becoming more important not only for reducing energy costs, but also for seeking more sustainable agricultural production. Diastrophic (nitrogen fixer) microorganisms could therefore be an important component of sustainable agriculture. The physical location of diazotrophs seems to be important in controlling the rates of nitrogen fixation. The interior of plant tissues may be a more favorable niche for N fixation because of the lower partial oxygen pressure and the more direct plant uptake of N fixed by bacteria (James and Olivares 1998). The most interesting type of biofertilizer in the world today is rhizobial biofertilizer. It is used to inoculate legume crops such as soybean, mungbean, chickpea, pigeon pea, pea, clover, lentil, alfalfa, and so on, and can supply them with sufficient amounts of nitrogen using the N_2 fixation process. In association with legumes, rhizobia fix 50–300 kg N/ha/ crop and the percentage of nitrogen derived from nitrogen fixation in plant tissues ranges from 40% to 90% (Franche et al. 2009; Herridge et al. 2008; Peoples et al. 1995). A multilocational trial carried out in India with chickpea, groundnut, rajmah, pigeon pea, and soybean has indicated about a 14%–30% increase in grain yield through rhizobial inoculation technology (Saxena and Tilak 1999). Such gain from *Rhizobium* inoculation depends on cultivars, culture condition, and management regions (Singleton et al. 1997). Nowadays, dual inoculation of *Rhizobium* with vesicular-arbuscular mycorrhiza (VAM)/PGPR/*Azotobacter* is gaining momentum.

The colonization of *Azospirillum* on or in the roots of a wide variety of plants, including those of agronomic importance such as maize, sugarcane, wheat, rice, and sorghum, and several nongraminous species has been observed by many researchers (Bashan et al. 2004; Dobereiner and Pedrosa 1987; Rodrigues et al. 2008; Tarrand et al. 1978). It has been found on the rhizoplane, rhizodermis, apoplast, rhizodermal cells, and root hairs of a number of crops (Rothballer et al. 2003). Agronomical applications of *Azospirillum* have reduced the application of chemical nitrogenous fertilizer by 20%–50% (Bashan et al. 2004). The contribution of *Azospirillum* inoculants to the yield of various crops is equivalent to that of 15–20 N/ha, the magnitude of response varying with location, season, product quality, crop variety, and management practices. Twenty years of data evaluation of field experiments has shown that 60%–70% of all field experiments were successful with a significant yield increase ranging from 5% to 30% (Bashan and Holguin 1997; Dobbelaere et al. 2001). This increase in yield is mainly attributed to nitrogen enrichment and improved root development due to the production of growth-promoting substances and the consequent increased rate of water and mineral uptake (Tilak et al. 2005). Thus, inoculation with *Azospirillum* can also cause phytohormone yield responses. Besides *Azospirillum*, an important diazotrophic species especially for a sugarcane crop is *Gluconacetobacter diazotrophicus*, which colonizes the apoplast and xylem (Dong et al. 1994) in an endophytic manner (Baldani and Baldani 2005). An average of 200 kg N/ ha is fixed by using 30% of the total sugar produced by a sugarcane crop (Dobereiner 1992).

Azotobacter, a free-living diazotroph, contributes 0.026–20 kg N_2/ha/yr of nitrogen to the soil ecosystem. A compilation of 1095 field experiments showed a positive response in yields of cereals and vegetables in 81% of these experiments, but the yield increase was no more than 10% in 47% of the experiments (Wani 1990). Vegetable crops in general, respond better to *Azotobacter* inoculation than other crops. Besides contributing nitrogen, *Azotobacter* is also known to produce indole-3-acetic acid (IAA) and gibberellin-like substances and to exhibit antagonism toward phytopathogens by producing siderosphores, antibiotics, and antifungal compounds (Azcorn and Barea 1975; Pandey and Kumar 1990).

A major question that needs to be answered is: what are the energy, nutritional, or signaling requirements needed to stimulate diazotrophic microbes? It may well be that considerably more carbon is needed than is currently produced in modern nonlegume crop rhizosphere. For example, legumes can fix agronomically significant levels of N but do so by utilizing over 30% of the photosynthate produced by the host plant. Similarly, the cost of AM symbiosis (phosphorus acquisition) is estimated to be 4%–20% of the total C assimilated by host photosynthesis. Besides symbiont, nonsymbiont heterotrophic diazotrophs are also dependent on the C released by the plant roots. Thus, from a plant perspective, C flow into the rhizosphere can be considered a cost for the plant, because a significant amount of C does not contribute to dry matter production. Modern breeding programs that have developed crops under high N levels that emphasize aboveground yield may have selected against a rhizosphere that promotes diazotrophic microorganism independence for both C and N. The artificial inoculation of such photosynthetic Cyanobacteria is an ideal option for mitigating C cost.

Cyanobacterial biofertilizer can create new economic opportunities for small and midsize farms and entrepreneurs in rural development. The commercialization of locally produced, sustainable, cyanobacterial biofertilizer is an important opportunity that could revolutionize twenty-first-century agriculture. Its artificial inoculation in rice resulted in an increase in grain yield of 18%–15%, saving 15–25 kg/ha on chemical nitrogen and maintaining organic carbon as well as a positive nutrient balance (Jha and Prasad 2006). The integration of cyanobacterial inoculants with green manure resulted in a saving of 50 kg N/ha and a significant increase in rice grain yield. The functional relationship (R^2: 83.5%–95.7%) between the different sources of nutrient in rice cultivation under INM mode revealed that the maximum positive contribution of Cyanobacteria was fixed available N (45.2%) and available phosphorus (18.5%). Green manure had the greatest contribution to total N, total P, zinc, iron, and manganese. However, Cyanobacteria had a negative relationship with sodium (−30.19%) (Jha et al. 2013a), which indicates the possibility of using Cyanobacteria as an ameliorating agent for salt-affected soil. The inclusion of Cyanobacteria along with enriched mycostraw and *Azospirillum* as a bionutrient package for rice and its application in consortium mode resulted in a 17% increase in grain yield and a saving of 60 kg N/ha in the research farm's plot. An evaluation of the bionutrient package at the farmer's site showed an increase of 6%–20% and 10%–35% rice grain yield for resource-rich and resource-poor farmers, respectively, over the farmers' conventional practice (Jha et al. 2013b). The low investment and the high expected return probably explain why cyanobacterial inoculation was recommended before the method was full proved.

4.2.2.2 Phosphorus Acquisition through Microbe-Mediated Processes

Phosphate is a critical element in agricultural ecosystems throughout the world. The unavailable phosphates built up in soils are enough to sustain maximum crop yields globally for about 100 years (Goldstein et al. 1993). However, the majority of phosphorus in soils is easily converted into insoluble complexes. The mobility of this element is very slow (10%–15%) in the soil and cannot respond to rapid uptake by plants. This causes the creation and development of phosphorus-depicted zones near the contact area of the root and soil in a rhizosphere. The plant rhizosphere supports a large population of soil microorganisms, including a number of bacteria (*Pseudomonas* and *Bacillus*) and fungi (*Mycorrhiza*, *Aspergillus*, and *Penicillium*) having the ability to increase P availability to crop. Such enhancement of P availability by microbes is due to the solubilization of insoluble phosphate through the production of organic acids, carbonic acids, or H^+ or chelation; the mineralization of organic P by producing extracellular phytases or phosphatases; and increasing the root growth or root hair development by producing phytohormones (Jakobsen et al. 2005). The plants inoculated with phosphate-solubilizing microorganisms (PSMs) showed growth enhancement and increased P content but large variations were found in the effectiveness of PSMs (Wani et al. 2007). According to field experiments, *Penicillium bilaii* and *Bacillus megaterium* are considered the most effective PSMs (Asea et al. 1988). *B. megaterium* has been shown to release P from organic phosphates, but does not solubilize mineral phosphates. However, the results of PSM plant inoculation cannot be conclusively proved because unless there are specific measurements to determine that P deficiency was reduced or eliminated, a growth response may be due to other factors. Nonetheless, a significant number of studies have shown improved P nutrition and yields on a wide range of crops through inoculating with P-enhancing microorganisms (Jakobsen et al. 2005; Leggett et al. 2001). Besides the expected response on P-deficient soils, studies have shown P improvement with P-enhancing microorganisms in the presence of rock phosphate (Baria et al. 2002). Thus, PSM is an important assisting system for plants to provide an available form of phosphorus.

Mycorrhizal fungi, which form a symbiotic, nonpathogenic, permanent association with the roots of land plants, are an appropriate partial substitute for phosphatic fertilizers and promote yield significantly. It is estimated that 250,000 species of plants worldwide, including many arable crops, are capable of symbiosis (Smith and Read 1997). For crop species, endomycorrhizae form arbuscular in close association with host cells; the plant provides the carbohydrate to the fungus. The fungus in turn develops an extensive hyphal network that can transport P and other nutrients to the plant (Smith and Read 1997). The typical zone of uptake for these nutrients for a nonmycorrhizal root is 1 or 2 mm, about the length of a root hair (Li et al. 1991). Further, extraradical hyphae can have a total surface area of several orders of magnitude greater than that of roots alone (Auge 2001). Thus, the same forms of P that are normally unavailable to roots are taken by the extraradical hyphae, where they are synthesized into polyphosphate granules and translocated into the root. There, P is released to the interfacial apoplectic space between the membranes of both partners (Balestrini et al. 2007). Organic acid production by AMF (Lapeyrie 1988) can also solubilize the insoluble mineral phosphate and make it available to the crop. Ectomycorrhizal fungi have been shown to possess the ability to solubilize P. They are capable of utilizing P from inositol phosphates and possess phosphatase activity that could further affect their ability to release P from soil organic matter. Mycorrhizal biofertilizer is extremely beneficial to almost all cultivated plants as it has a broad host range in contrast to other microbial inoculants. It is easy to apply, quite similar to chemical fertilizers. Thus, mycorrhizal biofertilizer has huge potential for widespread use and offers both economic and environmental advantages to farmers and growers, and commercial viability to production units.

The intensive promotion of microbial inoculants can be achieved through intervention at the technical level, the industrial level, and the farmers' level. These interventions include designing new formulation, enhancing shelf life, generating information on agroecology, and educating the farming community. However, the long-term goal should be to create artificial symbiosis in cereals. To achieve such a goal, there is also a need to look beyond *Rhizobium*. Cyanobacteria seem to be ideal candidates due to their ability to form symbiosis in all groups: conclusive evidence of plastid origin from a cyanobacterial symbiont: 14% of gene acquisition from Cyanobacteria is rice; the discovery of the *nif* gene and photosystem-I containing Cyanobacteria; the discovery of a N_2-fixing spheroid body of cyanobacterial origin in *Rhopalodia gibba*; and the detection of new symbiotic associations between unicellular Cyanobacteria and unicellular algae. Much of this information has been generated in microbial ecology and microbial molecular ecology.

4.3 EPILOGUE

Agroecological approaches, especially the utilization of beneficial microorganisms through the promotion of agricultural practices that are benign to agromicrobes or the artificial inoculation of agromicrobes, seem to meet the food requirements of the farming community in a cost-effective and environmentally benign manner. This will facilitate a successful transition from chemical-based farming systems to a more microbe-based sustainable agricultural system. Nevertheless, while agromicrobial technologies have been applied with considerable success in recent years, they have not been widely accepted by the farming community because it is often difficult to consistently reproduce their beneficial effects. Understanding the agroecology of agromicrobial technology, our ability to reliably engineer an agrosystem for a microbial-benign environment remains a challenge. Virtually nothing is known about the attributes that dictate the proliferation of specific microorganisms in an agroecosystem. An understanding of the complex chemical and biological interactions that occur in an agroecosytem is still meager. Fundamental issues concerning the preponderance of agromicrobes in the soil remain unresolved. The complex relationship between different microbes, microbes and plants, and microbes and the agroecosystem remains unresolved because in nature, events never occur in isolation as studied *in vitro*. The situation remains challenging because opportunities in the agroecological domain of agromicrobes are probably being overlooked. A survey of the bibliographic database (CAB abstracts) over the last 29 years (1984–2013) shows that a significant amount of research on beneficial agromicrobes (Cyanobacteria, *Azotobacter, Azospirillum, Rhizobium, Mycorrhiza*, PSM, PGPR) has focused on either the genetics of these microbes (13%–40%) or on their inoculation into crops (15%–42%). In fact, <0.5% of research articles on beneficial agromicrobes have focused on their agroecology. This demonstrates a lack of effort and resources allocation for agroecological research of agromicrobes and underscores the need for basic research on the ecology of agromicrobes. It is imperative that scientific managers and policy makers should support agroecological research on agromicrobes so that the future can benefit from safe, sustainable, and environmentally sound agricultural practices based on microbial-derived nutrients.

REFERENCES

Amman, R., W. Ludwig, and K. H. Schleifer. 1995. Phylogenetic, identification and *in situ* detection of individual microbial cells without cultivation. *FEMS Microbiol Rev* 59:143–169.

Andrews, M., E. K. James, S. P. Cummings, A. A. Zavalin, L. V. Vinogradova, and B. A. McKenzie. 2003. Use of nitrogen fixing bacteria inoculants as a substitute for nitrogen fertilizer for dry land graminaceous crops: Progress made, mechanisms of action and future potential. *Symbiosis* 35:209–229.

Asea, P. E. A., R. M. N. Kucey, and J. W. B. Stewart. 1988. Inorganic phosphate solubilization by two *Penicillium* species solution culture and soil. *Soil Biol Biochem* 20:459–464.

Auge, R. M. 2001. Water relating, drought and vesicular-arbuscular mycorrhizal symbiosis. *Mycorrhiza* 11:3–42.

Azcorn, R. and J. M. Barea. 1975. Synthesis of auxins, gibberellins and cytokininis by *Azotobacter vinelandi* and *Azotobacter beijerinckii* related to effects produced on tomato plants. *Plant Soil* 43:609–619.

Baldani, J. L. and V. L. D. Baldani. 2005. History on the biological nitrogen fixation research in graminaceous plants: Species emphasis on the Brazilian experience. *An Acad Bras Sci* 77:549–579.

Baldani, J. L. V., M. Reis, V. L. D. Baldani, and J. Dobereiner. 2002. A brief story of nitrogen fixation in sugarcane-reasons for success in Brazil. *Funct Plant Biol* 29:417–423.

Balestrini, R., J. Gomez-Ariza, L. Lanfranco, and P. Bonfente. 2007. Laser micro dissection reveals that transcripts for fire plant and one fungal phosphate transporter genes are contemporaneously present in arbusculated cells. *Mol Plant Microb Interac* 20:1055–1062.

Baria, J. M., M. Toro, M. O. Orozco et al. 2002. The application of isotopic (32_p and 15_N) dilution techniques to evaluate the interactive effect of phosphate solubilizing rhizobacteria, mycorrhizal fungi and *Rhizobium* to improve the agronomic efficiency of rock phosphate for legume crops. *Nutr Cycl Agroecosyst* 63:35–42.

Bashan, Y. and G. Holguin. 1997. *Azospirillum* plant relationship environmental and physiological advances (1990–1996). *Can J Microbiol* 43:103–121.

Bashan, Y., G. Holguin, and L. E. De-Bashan. 2004. *Azospirillum* plant relationship physiology, molecular, agricultural and environmental advances (1997–2003). *Can J Microbiol* 50:521–577.

Bloem, J., P. R. Bolhis, M. R. Veninga, and J. Wieinga. 1995. Microbial methods for counting bacteria and fungi in soil. In K. Alef and P. Nannipiery (eds), *Methods in Applied Soil Microbiology and Biochemistry*, pp. 162–173. London: Academic Press.

Bossio, D. A. and K. M. Scow. 1998. Impacts of carbon and flooding on soil microbial communities: Phospholipid fatty acid profiles and substrate utilization patterns. *Microb Ecol* 35:265–278.

Buckley, D. H. and T. M. Schmidt. 2001. The structure of microbial communities in soil and the lasting impact of cultivation. *Microb Ecol* 42:11–21.

Burns, R. G. 1982. Enzyme activity in soil: Location and possible role in microbial ecology. *Soil Biol Biochem* 14:423–427.

Carlsson, G. and K. Huss-Danell. 2003. Nitrogen fixation perennial forage legumes in the field. *Plant Soil* 253:353–372.

Cheshire, M. V., C. N. Bedrock, C. N. Williams, B. O. L. Chapman, S. J. Solntseva, and I. Thomsen. 1999. The immobilization of nitrogen by straw decomposing in soil. *Eur J Soil Sci* 50:329–341.

Cho, J. C. and J. M. Tiedje. 2001. Bacterial species determination from DNA-DNA hybridization by using genome fragments and DNA microarrays. *Appl Environ Microbiol* 67:3677–3682.

Conant, R. T., K. Paustian, and E. T. Elliott. 2001. Grassland management and conversion into grassland: Effects on soil carbon. *Ecol Appl* 11:343–355.

Courtois, S., A. Frostegard, P. Goransson, G. Depret, P. Jeannin, and P. Simonet. 2001. Quantification of bacterial subgroups in soil: Comparison of DNA extracted directly from soil or from cells previously released by density gradient centrifugation. *Environ Microbiol* 3:431–439.

Degens, B. P. 1998a. Microbial functional diversity can be influenced by the addition of simple organic substrates to soil. *Soil Biol Biochem* 30:1981–1988.

Degens, B. P. 1998b. Decrease in microbial functional diversity does not result in corresponding changes in decomposition under different moisture conditions. *Soil Biol Biochem* 30:1989–2000.

Dobbelaere, S., A. Croonenborghs, A. Thys et al. 2001. Responses of agronomically important crops to inoculation with *Azospirillum. Aust J Plant Physiol* 28:871–879.

Dobereiner, J. 1992. History and new perspectives of diazotrophs in association with non-leguminous plants. *Symbiosis* 13:1–13.

Dobereiner, J. and F. O. Pedrosa. 1987. *Nitrogen-Fixing Bacteria in Non-Leguminous Crop Plants.* Berlin: Madisan and Springer.

Dong, Z., M. J. Canny, M. E. Mc Cully et al. 1994. A nitrogen fixing endophyte of sugarcane stems. A new role for the apoplast. *Plant Physiol* 105:1139–1147.

Douds, D. D. and R. Seidel. 2012. The contribution of arbuscular mycorrhizal fungi to the success or failure of agricultural practices. In C. A. Edwards, T. Cheeke, D. Coleman, and D. Wall (eds), *Microbial Ecology in Sustainable Agroecosystems*, pp. 133–152. Boca Raton, FL: CRC Press/Taylor Francis Group.

Drinkwater, L. E. and S. S. Snapp. 2007. Nutrients in agroecosystem: Rethinking the management paradigm. *Adv Agron* 92:163–186.

Duan, T., S. Yuying, E. Facelli, S. E. Smith, and Z. Nan. 2010. New agricultural practices in the Loess Plateau of China do not reduce colonization by arbuscular mycorrhizal or root invading fungi and don't carry a yield penalty. *Plant Soil* 331:265–275.

Ebel, J. and A. Mithofer. 1998. Early events in the elicitation of plant defense. *Planta* 206:335–348.

Egle, K., G. Manse, W. Roemer, and P. L. G. Vlek. 1999. Improved phosphorus efficiency of three new wheat genotypes from CIMMYT in comparison with an older Mexican variety. *J Plant Nutr Soil Sci* 162:353–358.

Engelhard, M., T. Hurek, and B. Reinhold-Hurek. 2000. Preferential occurrence of diazotrophic endophytes, *Azoarcus* spp., in wild rice species and land races of *Oryza sativa* in comparison with modern races. *Environ Microbiol* 2:131–141.

Franche, C., K. Lindstorm, and C. Elemerich. 2009. Nitrogen fixing bacteria associated with leguminous and non-leguminous plant. *Plant Soil* 321:35–59.

Gans, J., M. Wolinsky, and J. Dunbar. 2005. Computational improvements reveal great bacterial diversity and high metal toxicity in soil. *Science* 309:1387–1390.

Garbeva, P. J., A. Vanveen, and J. D. Van Elsas. 2004. Microbial diversity in soil: Selection of microbial population by plant and soil type and implication for disease suppressiveness. *Annu Rev Phytopathol* 42:243–270.

Garland, J. L. 1996a. Analytical approaches to the characterization of samples of microbial communities using patterns of potential C source utilization. *Soil Biol Biochem* 28:213–221.

Garland, J. L. 1996b. Patterns of potential C source utilization by rhizosphere communities. *Soil Biol Biochem* 28:223–230.

Garland, J. L. and A. L. Mills. 1991. Classification and characterization of heterotrophic microbial communities on the basis of patterns of community level sole carbon source utilization. *Appl Environ Microbiol* 57:2351–2359.

Germida, J. J., S. D. Siciliano, J. R. de Freitas, and A. M. Seib. 1998. Diversity of root associated bacteria associated with field grown canola (*Brasissica napus* L.) and wheat (*Triticum aestivum* L.). *FEMS Microbiol Ecol* 26:43–50.

Giller, P. S. 1996. The diversity of soil communities, the "poor man's tropical forest". *Biodiv Conserv* 5:135–168.

Goldstein, A. H., R. D. Roger, and G. Meal. 1993. Mining by microbe. *Nat Biotechnol* 11:1250–1254.

Goodman, R. M., S. B. Bintrim, J. Handelsma et al. 1998. A diary look: Soil microflora and rhizosphere microbiology. In H. E. Flores, J. P. Lynch, and D. Eissenstat (eds), *Radial Biology: Advances and Perspectives on the Function of Plant Roots*, pp. 219–239. Rockville, MD: American Society of Plant Physiologists.

Graham, J. H., N. C. Hodge, and J. B. Morton. 1995. Fatty acid methyl ester profiles for characterization of glomalean fungi and their endomycorrhizae. *Appl Environ Microbiol* 61:58–64.

Grayston, S. J., G. S. Griffith, J. L. Mawdsley, C. D. Campbell, and R. D. Bardgett. 2001. Accounting for variability in soil microbial communities of temperate upland grassland ecosystems. *Soil Boil Biochem* 33:533–551.

Guo, J. H., X. J. Liu, Y. Zhong et al. 2010. Significant acidification in major Chinese croplands. *Science* 327:1008–1010.

Halweil, B. 2002. Farming in the public interest. In C. Flavin, H. French, G. Gardener, S. Dunn et al. (eds), *State of the World*, pp. 51–74. New York: Norton.

Harry, M., N. Jusseaume, B. Gambier, and E. Garnier-Silliam. 2001. Use of RAPD markers for the study of microbial community similarity from termite and tropical soils. *Soil Biol Biochem* 33:417–427.

Hartmann, A., M. Schmid, D. van Tuinen, and G. Berg. 2009. Plant-driven selection of microbes. *Plant Soil* 321:235–257.

Herridge, D. F., M. B. Peoples, and R. M. Boddey. 2008. Global inputs of biological nitrogen fixation in agricultural system. *Plant Soil* 311:1–8.

Hetrick, B. A. D., G. W. T. Wilson, and T. S. Cox. 1993. Mycorrhizal dependence of modern wheat cultivars and ancestors: A synthesis. *Can J Bot* 71:512–518.

Heuer, H., G. Wieland, J. Schonfeld, A. Schnwalder, N. C. M. Gomes, and K. Smalla. 2001. Bacterial community profiling using DGGE or TGGE analysis. In P. A. Rochelle (ed.), *Environmental Molecular Microbiology: Protocols and Applications*, pp. 177–190. Wymondham: Horizon Scientific.

Holben, W. E., J. K. Jansson, B. K. Cheim, and J. M. Tiedje. 1988. DNA probe methods for the detection of specific microorganisms in the soil community. *Appl Environ Microbiol* 54:703–711.

Ibekwe, A. M. and A. C. Kennedy. 1998. Phospholipid fatty acid profiles and carbon utilization patterns for analysis of microbial community structure under field and green house conditions. *FEMS Microbiol Ecol* 26:151–163.

Ingham, R. E., J. A. Trofymo, E. R. Ingham, and D. C. Coleman. 1985. Interactions of bacteria, fungi and their nematode grazers: Effects on nutrient cycling and plant growth. *Ecol Monogr* 55:119–140.

Jagnow, G. 1990. Differences between cereals crop cultivars in root associated nitrogen fixation. Possible causes of variable yield response to seed inoculation. *Plant Soil* 123:255–259.

Jakobsen, I., M. E. Leggett, and A. E. Richardson. 2005. Rhizosphere microorganisms and plant phosphorous uptake. In J. T. Sims and A. N. Sharpley (eds), *Phosphorus: Agriculture and the Environment*, Agronomy Monograph 46, pp. 437–492. Madison, WI: ASA-CSA-SSSA.

James, E. K. and F. L. Olivares. 1998. Infection and colonization of sugarcane and other graminaceous plants by entophytic diazotrophs. *Crit Rev Plant Sci* 17:77–119.

Jha, M. N., S. K. Chourasia, and R. C. Bharti. 2013a. Effect of integrated nutrient management on rice yield, soil nutrient profile, and cyanobacterial nitrogenase activity under rice-wheat cropping system. *Commun Soil Sci Plant Anal* 44(13):1961–1975.

Jha, M. N., S. K. Chourasia, and S. Sinha. 2013b. Microbial consortium for sustainable rice production. *Agroecol Sustain Food Syst* 37:340–362.

Jha, M. N. and A. N. Prasad. 2006. Efficacy of new inexpensive cyanobacterial biofertilizer including its shelf life. *World J Microbiol Biotechnol* 22:73–79.

Johnsen, K., C. S. Jacobsen, V. Torsvik, and J. Sorensen. 2001. Pesticide effects on bacterial diversity in agricultural soils: A review, biology and fertility of soils. *Plant Soil* 170:75–86.

Johnson, N. C., D. L. Rowland, L. Corckidi, L. M. Egerton-Warbourton, and E. B. Allen. 2003. Nitrogen enrichment alters mycorrhizal allocation at five mesic to semiarid grassland. *Ecology* 84:1895–1908.

Kahiluoto, H., E. Ketoja, M. Vestberg, and I. Saarela. 2001. Promotion of AM utilization through reduced P fertilization 2. Field studies. *Plant Soil* 231:65–79.

Killham, K. 1994. *Soil Ecology*. Cambridge: Cambridge University Press.

Kwon, H. Y. and R. J. M. Hudson. 2010. Quantifying management-driven changes in organic matter turnover in an agricultural soil: An inverse modeling approach using historical data and surrogate CENTURY-type model. *Soil Biol Biochem* 42:2241–2253.

Lapeyrie, F. 1988. Oxalate synthesis from soil bicarbonate by fungus *Paxillus involutus*. *Plant Soil* 110:3–8.

Lechevalier, M. P. 1989. Lipids in bacterial taxonomy. In W. O'Leary (ed.), *Practical Handbook of Microbiology*, pp. 57–67. Boca Raton, FL: CRC Press.

Leggett, M. E., S. Gleddie, and G. Helloway. 2001. Phosphate- solubilizing microorganisms and their use. In N. Ae, J. Ariham, K. Okada, and A. Srinivisan (eds), *Plant Nutrient Acquisition: New Perspective*, pp. 299–318. Tokyo: Springer.

Li, X. L., E. George, and H. Marschner. 1991. Extension of the phosphorous depletion zone in VA-mychorrhizal white clover in a calcareous soil. *Plant Soil* 136:41–48.

Liu, W. T., T. L. Marsh, H. Cheng, and L. J. Forney. 1997. Characterization of microbial diversity by determining terminal restriction fragments length polymorphisms of genes encoding 16S rRNA. *Appl Environ Microbiol* 63:4516–4522.

Liu, B., C. Tu, S. Hu, M. L. Gumpertz, and J. B. Ristaino. 2007. Long-term effect of organic and synthetic soils fertility amendments on soil microbial communities and the development of southern blight. *Soil Biol Biochem* 39:2302–2316.

Lundquist, E. J., L. E. Jackson, K. M. Scow, and C. Hsu. 1999. Changes in microbial biomass and community composition and soil carbon and nitrogen pools after incorporation of rye into three California agricultural soils. *Soil Biol Biochem* 31:221–236.

Lupwayi, N. Z., W. A. Rice, and G. W. Clayton. 1998. Soil microbial diversity and community structure under wheat as influenced by tillage and crop sequence. *Soil Biol Biochem* 30:1733–1741.

Mader, P., A. Fliessbach, D. Dubois, L. Gunst, P. Fried, and U. Niggli. 2002. Soil fertility and biodiversity in organic farming. Sc*ience* 296:1694–1697.

Marshell, D., B. Tunali, and L. R. Nelson. 1999. Occurrence of fungal endophytes in species of wild *Triticum*. *Crop Sci* 39:1507–1512.

McCaig, A. E., L. A. Glover, and J. I. Prosser. 1999. Molecular analysis of bacterial community structure and diversity in unimproved upland grass pastures. *Appl Environ Microbiol* 65:1721–1730.

Muyzer, G., E. C. De Wall, and A. G. Uitterlinden. 1993. Profiling of complex microbial population by denaturing gradient gel electrophoresis analysis of polymerase chain reaction-amplified genes coding for 16S rRNA. *Appl Environ Microbiol* 59:695–700.

Muyzer, G. and K. Smalla. 1998. Application of denaturing gradient gel electrophoresis (DGGE) and temperature gradient gel electrophoresis (TGGE). *Int J General Mol Microbiol* 73:127–141.

Nocker, A., M. Burr, and A. K. Camper. 2007. Genotyping microbial community profiling: A critical technical review. *Microb Ecol* 54:276–289.

Ogram, A., G. S. Sayler, and T. Barkay. 1987. The extraction and purification of microbial DNA from sediments. *J Microbiol Methods* 7:57–66.

Pandey, A. and S. Kumar. 1990. Inhibitory effect of *Azotobacter chroococcum* and *Azospirillium brasilense* on arrange of rhizosphere fungi. *Indian J Exp Biol* 28:52–54.

Pankhurst, C. E., S. Yu, B. G. Hawke, and B. D. Harch. 2001. Capacity of fatty acid profiles and substrate utilization patterns to describe differences in soil microbial communities associated with increased salinity and alkalinity at three locations in south Australia. *Biol Fert Soils* 33:204–217.

Peoples, M. B., J. K. Ladha, and D. F. Herridge. 1995. Enhancing legume N_2 fixation through plant and soil management. *Plant Soil* 174:83–101.

Picard, C., E. Baruffa, and M. Bosco. 2008. Enrichment and diversity of plant-probiotic microorganisms in the rhizosphere of hybrid maize during four growth cycles. *Soil Biol Biochem* 40:106–115.

Powlson, D. S., P. C. Brookes, and B. T. Christensen. 1987. Measurement of soil microbial biomass provides an early indication of changes in total organic matter due to straw incorporation. *Soil Biol Biochem* 19:159–164.

Raaijmakers, J., M. C. T. Paulitz, C. Steinberg, C. Alabouvette, and Y. Moenne-Loccoz. 2008. The rhizosphere: A playground and battlefield for soilborne pathogens and beneficial microorganisms. *Plant Soil* 321:341–361.

Rodrigues, E. P., L. S. Rodrigues, A. L. M. de Oliver et al. 2008. *Azospirillum amazonense* inoculation: Effects on growth yield and N_2 fixation on rice (*Oryza sativa* L). *Plant Soil* 302:249–261.

Rondon, M. R., P. R. August, A. D. Bettermann et al. 2000. Cloning the soil metagenome: A strategy for accessing the genetic and functional diversity of uncultured microorganisms. *Appl Environ Microbiol* 66:2541–2547.

Rondon, M. R., R. M. Goodman, and J. Handelsman. 1999. The Earth's bounty: Assessing and accessing soil microbial diversity. *Trends Biotechnol* 17:403–409.

Rothballer, M., M. Schmid, and A. Hartman. 2003. *In situ* localization and PGPR-effect of *Azospirillium brasilense* strains colonizing roots of different wheat varieties. *Symbiosis* 34:261–279.

Russell, A. E., C. A. Cambardella, D. A. Laird, D. B. Jaynes, and D. W. Meek. 2009. Nitrogen fertilizer effects on soil carbon balances in midwestern U.S. agricultural systems. *Ecol Appl* 19:1102–1113.

Saxena, A. K. and K. V. B. R. Tilak. 1999. Potential and prospects of *Rhizobium* biofertilizer. In M. N. Jha, S. Sriram, S. G. Sharma, and G. S. Venkataraman (eds), *Agromicrobes*, pp. 57–78. New Delhi: Today and Tomorrow Publication.

Schulter, M. E., J. M. Sandeno, and R. P. Dick. 2004. Seasonal, soil type and alternative management influences on microbial communities of vegetable cropping system. *Biol Fertil Soils* 345:397–410.

Schwieger, F. and C. C. Tebbe. 1998. A new approach to utilize PCR-single strand conformation polymorphism for 16S r RNA gene based microbial community analysis. *Appl Environ Microbiol* 64:4870–4876.

Shrestha, R. K. and J. K. Ladha. 1996. Genotypic variation in promotion of rice dinitrogen fixation as determined by nitrogen-15 dilution. *Soil Sci Soc AMJ* 60:1815–1821.

Singleton, P. W., N. Boonkerd, T. J. Carr, and J. A. Thompson. 1997. Technical and market constraints limiting legume inoculant use in Asia. In O. P. Rupela, C. Johansen, and D. F. Herridge (eds), *Extending Nitrogen Fixation Research to Farmers Field*, pp. 17–38. New Delhi: ICRISAT.

Smalla, K., G. Wielan, A. Buchhor et al. 2001. Bulk and rhizosphere soil bacterial communities studied by denaturing gradient gel electrophoresis: Plant-dependent enrichment and seasonal shifts revealed. *Appl Environ Microbial* 67:4742–4751.

Smith, S. E. and D. J. Read. 1997. *Mycorrhizal Symbiosis*. London: Academic.

Steenwerth, K. L., L. E. Jackson, F. J. Calderin, M. R. Stromberg, and K. M. Scow. 2003. Soil microbial community composition and land use history in cultivated and grassland ecosystem of coastal California. *Soil Biol Biochem* 35:489–500.

Surekha, K., V. Jhansiilakshmi, N. Somasekhar et al. 2010. Status of organic farming and research experience in rice. *J Rice Res* 3:23–35.

Tarkalson, D. D., G. W. Hergert, and K. G. Cassman. 2006. Long term effects on tillage on soil chemical properties and grain yields of a dry land winter wheat-sorghum/corn-fallow rotation in the Great Plains. *Agron J* 98:26–33.

Tarrand, J. J., N. R. Kreig, and J. Dobereiner. 1978. A taxonomic study of the *Spirillum lipoferum* group with descriptions of a new genus, *Azospirillium* gen nov., and two species, *Azospirillum lipoferum* (Beijerinck) comb. nov. and *Azospirillum brassilense* sp. nov. *Can J Microbiol* 24:967–980.

Tiedje, J. M., J. C. Cho, A. Murray, D. Traves, B. Xia, and J. Zhou. 2001. Soil teeming with life: New frontiers for soil science. In R. M. Rees, B. C. Ball, C. D. Campbell, and C. A. Watson (eds), *Sustainable Management of Soil Organic Matter*, pp. 393–412. Wallingford: CAB International.

Tilak, K. V. B. R., N. Ranganayaki, K. K. Pal et al. 2005. Diversity of plant growth and soil health supporting bacteria. *Curr Sci* 89:136–150.

Torsvik, V., J. Goksoyr, and F. L. Daae. 1990. High diversity in DNA of soil bacteria. *Appl Environ Microbiol* 56:782–787.

Torsvik, V. L., R. Sorheim and J. Goksoyr. 1996. Total bacterial diversity in soil and sediment commodities: A review. *J Ind Micobiol Biotechnol* 17:170–178.

Tunlid, A. and D. C. White. 1992. Biochemical analysis of biomass, community structure, nutritional status and metabolic activity of microbial communities in soil. In G. Stotzky. and J. M. Bollag (eds), *Soil Biochemistry*, pp. 229–262. New York: Marcel Dekker.

Uphoff, N. 2006. Opportunities for overcoming productivity constraints with biologically based approaches. In N. Uphoff, A. S. Ball, E. Fernandes et al. (eds), *Biological Approaches to Sustainable Soil Systems*, pp. 693–713. Boca Raton, FL: CRC Press.

Verma, R., S. K. Chourasia, and M. N. Jha. 2011. Population dynamics and identification of efficient strains of *Azospirillum* in maize ecosystems of Bihar (India). *3 Biotech* 1:247–253.

Vessey, J. K., K. Pawlwosky, and B. Bergman. 2004. Root based N_2 fixing symbioses: Legumes, actinorrhizal plants, *Parasponia* and cycads. *Plant Soil* 266:205–230.

Vestal, R. D. and D. White. 1989. Lipid analysis in microbial ecology. *Bioscience* 39:535–541.

Waldrop, M. P., T. C. Balser, and M. K. Firestone. 2000. Linking microbial community composition to function in a tropical soil. *Soil Biol Biochem* 32:1837–1846.

Wander, M. M., D. S. Hedrick, D. Kaufman et al. 1995. The functional significance of the microbial biomass in organic and conventionally managed soils. *Plant Soil* 170:87–97.

Wani, S. P. 1990. Inoculation with associative nitrogen fixing bacteria, role in cereal grain production improvement. *Indian J Microbiol* 30:363–393.

Wani, P. A., M. S. Khan, and A. Zaidi. 2007. Coinoculation of nitrogen fixing and phosphate solubilizing bacteria to promote growth, yield and nutrient uptake in chickpea. *Acta Agron Hung* 53:315–323.

Yao, H., Z. He, M. J. Wilson, and C.D. Campbell. 2000. Microbial biomass and community and biochemistry structure in a sequence of soils with increasing fertility and changing land use. *Microb Ecol* 40:223–237.

Zelles, L. 1999. Fatty acid partners of phospholipids and lipopolysaccharides in the characterization of microbial communities in soil: A review. *Biol Fertil Soil* 29:111–129.

CHAPTER 5

Management of Rhizosphere Microorganisms in Relation to Plant Nutrition and Health

John Larsen, Miguel Nájera Rincón, Carlos González Esquivel, and Mayra E. Gavito

CONTENTS

5.1   Introduction 103
5.2   Rhizosphere 104
5.3   Functional Groups of Plant-Beneficial Rhizosphere Microorganisms 105
    5.3.1   Arbuscular Mycorrhizal Fungi 105
    5.3.2   *Trichoderma* 106
    5.3.3   Entomopathogenic Fungi 106
    5.3.4   Plant Growth–Promoting Rhizobacteria 107
5.4   Microbial Interactions in the Rhizosphere 108
    5.4.1   Interactions between AMF and *Trichoderma* 108
    5.4.2   Interactions between AMF and PGPR Bacteria 108
    5.4.3   Microbial Rhizosphere Consortia 109
    5.4.4   Nontarget Effects of Microbial Inoculations 109
5.5   Impact of Agricultural Practice on Beneficial Rhizosphere Microorganisms 110
    5.5.1   Crop Genotype 110
    5.5.2   Crop Rotation 110
    5.5.3   Fertilization 111
    5.5.4   Tillage 111
    5.5.5   Pest Management Applications 112
5.6   Strategies to Integrate Beneficial Rhizosphere Microorganisms in Agroecosystems 112
    5.6.1   Inoculation 112
    5.6.2   Conservation 113
5.7   Conclusions and Future Perspectives 114
References 114

5.1 INTRODUCTION

Increasing global food demand is a major challenge for crop production, which should be met with sustainable intensification of existing cropland instead of forest clearing to minimize related environmental impacts (Tilman et al. 2011). However, regardless of future land use strategies to meet the increasing global food demand, integration of biological resources in crop production seems to be an important component in sustainable crop production.

Plant health and nutrition are key elements in agricultural plant production. Conventionally, plant health and nutrition are managed with pesticides and mineral fertilizers, whereas organic farming relies mainly on conservation and the integration of biological resources. Intermediate types of agricultural systems are common, where some agrochemicals are replaced by biological inputs such as inoculation with microbial plant growth promoters (MPGP) and microbial pest control agents (MPCA) and the use of organic fertilizers, both animal and green manure.

Independently of the agricultural production system, integrating biological resources in plant nutrition and health management strategies is receiving more attention worldwide due to increasing prices of agrochemicals and environmental concern. Another driving force for better integrating biological resources in agricultural systems is the future lack of phosphate fertilizers, since the rock phosphate sources are not infinite (Gilbert 2009).

Biological resources, including MPGP and MPCA, both native populations and inoculants, offer a sound strategy for food production, reducing chemical inputs to agroecosystems and consequently involving fewer environmental impacts and human hazards.

The rhizosphere of all crops is naturally inhabited by a broad range of MPGP and MPCA, including arbuscular mycorrhizal fungi (AMF), plant growth–promoting rhizobacteria (PGPR), *Trichoderma*, and entomopathogenic fungi (EPF). However, despite their vital importance, rhizosphere microorganisms are often not considered in agricultural system management, most likely due to their hidden status belowground.

Managing rhizosphere microorganisms in plant production systems to improve plant and soil health requires profound knowledge of possible agricultural impacts on these microorganisms. Larsen et al. (2007) identified main bottlenecks for integrating AMF in horticulture, with fertilization, pest management, crop rotation, crop genotype, and tillage as important factors, all of which should be considered when managing rhizosphere microorganisms.

The main objective of this chapter is to present the state of the art of the management of beneficial rhizosphere microorganisms in agroecosystems.

5.2 RHIZOSPHERE

The rhizosphere is the area directly influenced by the roots of a plant, and, although it extends only a few millimeters beyond the root surface and root hairs, this is considered the hotspot of activity and diversity of living organisms in soil and a dynamic interface of crucial importance for life in terrestrial ecosystems (Hinsinger et al. 2009).

This minute habitat represents the main point of carbon (C) flow from atmosphere to soil and therefore the main entrance for both inorganic and organic forms of C affecting the biogeochemistry and the physical, chemical, and biological nature of the soil (Jones et al. 2009). Carbon fixed in plant leaves has been shown to move progressively from shoots to roots, mycorrhizal fungi, rhizospheric bacteria, and root-feeding animals, and finally to their predators at higher levels through the food web within only a few days (Fitter et al. 2005).

Rhizodeposition is the release of organic C compounds deriving from root-cell loss, death and lysis, C flow to root symbionts, gaseous losses, exudates, and mucilage, in the form of carbohydrates, organic acids, amino acids, phenolics, fatty acids, sterols, enzymes, vitamins, hormones, and nucleosides (Jones et al. 2009). This process induces by far the largest and fastest transformations in soil, and is a research area with great potential to provide alternatives for sustainable management of agroecosystems. The rhizosphere of all crops is naturally inhabited by a broad range of microorganisms accomplishing many ecological functions, including fertility regulation, nutrient cycling, soil formation, and crop protection (Drinkwater and Snapp 2007).

In the following section selected functional groups of rhizosphere microorganisms will be presented.

5.3 FUNCTIONAL GROUPS OF PLANT-BENEFICIAL RHIZOSPHERE MICROORGANISMS

5.3.1 Arbuscular Mycorrhizal Fungi

Arbuscular mycorrhizal fungi (Glomeromycota) are obligate symbionts living inside the roots, usually in mutualistic associations, that involve the delivery of plant C to the fungus and the transfer of immobile nutrients and water from fungus to plant through a specialized exchange interface (Smith and Read 2008). Most crops establish mycorrhizal associations with AMF colonizing the root cortex and also forming an extensive hyphal network (Figure 5.1) proliferating in the surrounding soil, thereby increasing the plant–soil interface and plant nutrient uptake. AMF modulate plant nutrient uptake not only through their own proven capacity to rapidly and efficiently take up, transport, and transfer immobile nutrients such as P (Jakobsen 1994), but also through complex regulation of plant genes involved in nutrient uptake that may down- or upregulate the plant's own nutrient capture mechanisms (Smith and Smith 2012). Moreover, mycorrhizal associations continue to reveal large effects on plant performance under various types of biotic and abiotic stress, but still few advantages have been demonstrated under field conditions, given the difficulty of establishing realistic nonmycorrhizal references or controls in nature (Smith et al. 2010). Although the contribution of AMF to ecosystem services is being increasingly acknowledged (Gianinazzi et al. 2010),

Figure 5.1 (See color insert) External mycelium and young spores of the arbuscular mycorrhizal fungus *Rhizophagus irregularis* growing in soil.

there is a great need for field research to demonstrate the benefits of managing mycorrhizal associations, not only in crop yield, but also in soil fertility, quality, and health, and overall healthy and sustainable agroecosystem functioning.

5.3.2 *Trichoderma*

Trichoderma is a universal fungal genus of saprotrophic soil fungi with strong competitive features (Samuels 1996). Competitive advantages of *Trichoderma* are determined by broad ecological plasticity, favored by rapid growth and sporulation, and a broad enzymatic apparatus allowing growth on a wide variety of simple and complex substrates such as carbohydrates, starch, cellulose, chitin, and pectin, among others (Infante et al. 2009).

Biocontrol and plant growth promotion are important functional traits of *Trichoderma* in relation to plant health and nutrition (Harman et al. 2004). *Trichoderma* is a well-known biocontrol agent against soil-borne root pathogens such as *Pythium* (e.g., Vinale et al. 2008), *Phytophthora* (Bae et al. 2010), *Fusarium* (e.g., Otado et al. 2011), and *Sclerotinia* (e.g., Rabeendran et al. 2006; Figure 5.2).

The mode of action of biocontrol by *Trichoderma* mainly includes competition for space and nutrients, mycoparasitism, antibiosis, and induction of plant defense (Hermosa et al. 2012; Rey et al. 2000). Mycoparasitism is produced by extracellular enzymes, such as cellulases, chitinases, and glucanases, digesting the cell wall of the host (Harman et al. 2004; Infante et al. 2009). Secondary metabolites produced by *Trichoderma* inhibit the development of other microorganisms. These substances include trichodermine, suzukaciline, alameticine, dermadine, trichothecenes, and trichorzianine (Harman et al. 2004; Infante et al. 2009).

More recently, *Trichoderma* has also been reported as a plant growth promoter, enhancing plant nutrient uptake (Ryder et al. 2012; Vinale et al. 2008). Also recently, *Trichoderma* has been reported as an endophytic root symbiont, establishing chemical communication and altering gene expression in plants, improving abiotic stress resistance, nitrogen uptake, pathogen resistance, and photosynthetic efficiency (Friedl and Druzhinina 2012; Hermosa et al. 2012). Growth-promoting hormones produced by *Trichoderma* include auxins and gibberellins (Druzhinina and Kubicek 2005).

5.3.3 Entomopathogenic Fungi

The use of EPF as biological control agents has been widely studied (Jackson and Glare 1992; Lecuona 1996; Zimmermann 2007). Around 700 species of EPF have been identified, including

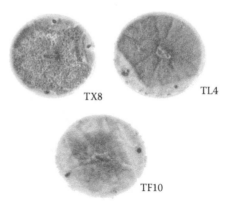

TX8

TL4

TF10

Figure 5.2 In vitro confrontation between *Trichoderma* spp. (TX8, TL4, and TF10) and the plant pathogen *Sclerotinia sclerotiorum*. (Photo: H. García-Núñez.)

several strains of the genera *Beauveria*, *Hirsutella*, *Isaria*, and *Metarhizium* (Ascomycota: Hypocrea). Infections in several species of insects caused by *Metarhizium anisopliae*, *Beauveria bassiana*, and *B. brongniartii* have been reported for over 150 years (Soper and Ward 1981; Figure 5.3).

Several EPF, such as *B. bassiana*, *B. brongniartii*, *Lecanicillium* spp., and *Isaria* spp., have been reported as endophytes of diverse plant groups (Vega 2008). The endophyte behavior of *Neotyphodium* (Clavicipitaceae) is associated with plant growth promotion and host plant stress alleviation against abiotic (drought) and biotic (pest attack) stress. In general, EPF seems to provide multiple functional traits relevant to plant health and nutrition (Goettel 2008; Ownley et al. 2008; Zimmermann 2007).

Improved knowledge of the ecology of EPF, especially in relation to interactions with other soil biota and impacts of agricultural practice, is required to manage EPF in agroecosystems, both in terms of conservation of native populations of EPF (Meyling and Eilenberg 2007) and when inoculating commercial EPF (Krauss et al. 2004; St. Leger 2008).

5.3.4 Plant Growth–Promoting Rhizobacteria

Plant growth–promoting rhizosphere bacteria include a broad range of taxonomic phyla and provide functional traits important to plant growth and nutrition. The most-studied PGPR genera include *Pseudomonas*, *Bacillus*, *Burkholderia*, *Azospirillum*, and *Rhizobium*, with well-known functional traits such as N fixation, P solubilization, production of phytohormones, and biocontrol of root pathogens (Dobbelaere et al. 2003; Lugtenberg and Kamilova 2009; Bhattacharyya and Jha 2012).

Bacterial N fixation occurs in both symbiotic (Herridge et al. 2008) and nonsymbiotic (Kennedy et al. 2004) plant–bacteria associations. Integrating legumes as the main crop, intercrop, or green manure crop is a cornerstone in organic agroecosystems, representing a substantial source of total N input, and when fully integrated it may even be possible to maintain crop production of both cereals and vegetables with biological N fixation from green manure as the sole source of nitrogen (Thorup-Kristensen et al. 2003).

It is well established that symbiotic N fixation can contribute significantly to the N input in agroecosystems. N fixation from Rhizobia in legume crops has been estimated to be in the range of 110–227 kg N/ha/year (Herridge et al. 2008). In nonleguminous crops biological N fixation

Figure 5.3 **(See color insert)** (a) Healthy specimen of *Phyllophaga* sp. (b) Infection with the entomopathogenic fungus *Metarhizium anisopliae*.

from endophytic, associative, and free-living bacteria has been reported to be in the range of 5–25 kg N/ha/year (Herridge et al. 2008).

P solubilization is another important trait of PGPR bacteria (Khan et al. 2007). Plant P nutrition is a key component in plant production due to its scarce bioavailability in the soil. P is present in soil as organic P and inorganic P bound immobilized in soil mineral complexes. P solubilization by PGPR bacteria is provided mainly through production of phosphatases and organic acids, which solubilize organic and mineral P, respectively (Khan et al. 2007). Some fungi are also well-known P solubilizers, such as *Aspergillus* and *Penicillium* (Coutinho et al. 2012) and different soil yeasts (Botha 2011).

Biocontrol of plant diseases is another well-documented trait of PGPR, especially in relation to root diseases caused by fungi, oomycetes, and nematodes (Raaijmakers et al. 2009). The mode of action of biocontrol by PGPR bacteria includes competition for space and nutrients, antibiosis, and induction of plant defense (Lugtenberg and Kamilova 2009).

5.4 MICROBIAL INTERACTIONS IN THE RHIZOSPHERE

Several reviews on microbial interactions in the rhizosphere are available, with a mycorrhiza-centric approach linking rhizosphere interactions to plant nutrition and root health in agroecosystems (Artursson et al. 2006; Barea et al. 2005; Johansson et al. 2004; Linderman 1988; Whipps 2001). In the following section some examples of interactions between different beneficial groups of rhizosphere microorganisms will be presented.

5.4.1 Interactions between AMF and *Trichoderma*

Potential parasitism of mycelium of AM fungi by the fungal biocontrol agent *Trichoderma harzianum* has been demonstrated in monoxenic cultures of transformed carrot roots. Also, mutual inhibition between *T. harzianum* and external mycelium of the AM fungus *G. intraradices* has been demonstrated in root-free soil (Green et al. 1999); mycoparasitism did not seem to be involved in the mode of interaction, whereas competition for nutrients was suggested as a possible mode of interaction. Nevertheless, despite possible mutual inhibition, combining AM fungi and *T. harzianum* has resulted in improved biocontrol against *Fusarium* wilt in tomato (Srivastava et al. 2010) and melon (Martínez-Medina et al. 2009). Similar mutual inhibition between *G. intraradices* and the fungal biocontrol agent *Clonostachys rosea* resulted in synergistic effects on plant growth parameters (Ravnskov et al. 2006).

5.4.2 Interactions between AMF and PGPR Bacteria

Using *Burkholderia cepacia* as a model PGPR bacterium, a series of studies on interactions between *G. intraradices* and PGPR bacteria were examined (Albertsen et al. 2006; Larsen et al. 2003; Ravnskov et al. 2002). Five strains of *B. cepacia* differentially affected hyphal growth of the AM fungus *G. intraradices*, but the hyphal P transport as examined with P isotopes remained unaffected (Ravnskov et al. 2002). The presence of external mycelium of *G. intraradices* reduced *B. cepacia* as examined with the biomarker fatty acid cyclo 17:0 (Ravnskov et al. 2002). The combined effect of the external mycelium of *G. intraradices* and two strains of *B. cepacia* was examined in root-free soil, showing that both types of microbial inoculants reduced the population density of *P. ultimum*, but without improved biocontrol efficacy (Larsen et al. 2003). When studying interactions between AMF and other soil microorganisms, it is important to consider possible effects of organic matter, which could serve as a source of energy for the other soil microorganisms, whereas the AMF obtain their energy from the host plant, and for that reason have a competitive advantage when exploring the soil environment for inorganic nutrients. However, interactions between *G. intraradices* and *B. cepacia* did not seem to be affected by the presence of organic matter (Albertsen et al. 2006).

The population density of the PGPR bacteria *Pseudomonas fluorescens* was also reduced in the rhizosphere and hyphosphere of a cucumber–*G. intraradices* symbiosis (Ravnskov et al. 1999). In another study *P. fluorescens* increased hyphal length of *G. caledonium* but had no effect on AM fungus root colonization and hyphal P transport (Ravnskov and Jakobsen 1999). Barea et al. (1998) also reported improved performance of *Glomus mosseae* in both root colonization and mycelia growth from germinating spores when confronted with *P. fluorescens*. However, clear suppressing effects of the secondary metabolite 2,4-diacetylphloroglucinol produced by *P. fluorescens* against mycelia growth from germinating spores were observed.

Strains of *Paenibacillus polymyxa* and *P. macerans* with PGPR feature differentially affected root colonization of *G. intraradices* and *G. mosseae* (Li et al. 2008). Mutual inhibition between *G. intraradices* and strains of *P. macerans* and *P. polymyxa* resulted in plant growth suppression, but only in soil amended with organic matter (Larsen et al. 2009).

5.4.3 Microbial Rhizosphere Consortia

Combining rhizosphere microorganisms from different functional groups and different life strategies, such as biotrophic AMF and saprotrophic bacteria and fungi, may improve the biocontrol and plant growth promotion efficacy.

Martínez-Medina et al. (2009) reported general improved biocontrol effects against *Fusarium* wilt in melon when combining AMF and *T. harzianum*. Also, combining the AMF fungus *G. intraradices*, a fluorescent *Pseudomonas*, and *T. harzianum* resulted in enhanced biocontrol effects compared with any of the microbial inoculants individually against the tomato wilt pathogen *Fusarium oxysporum* (Srivastava et al. 2010). However, Vestberg et al. (2004) reported no beneficial effects of combining different microbial biocontrol agents against root rot in strawberry.

5.4.4 Nontarget Effects of Microbial Inoculations

When inoculating beneficial rhizosphere microorganisms to agroecosystems, it is important to consider proper risk assessments in relation to human health and environmental impacts (Jensen et al. 2011). Nevertheless, such risk assessment studies are few and should be further addressed (Brimner and Boland 2003). However, in one of the few studies in this area, Vázquez et al. (2000) showed that none of the microbial inoculants *T. harzianum*, *P. fluorescens*, and *Azospirillium brasilense* affected AMF root colonization, but that the background microbial community was altered by both AMF and the other microbial inoculants.

In relation to EPF, most strains have preference for a host or "target" species, but may also infect "nontarget" insects (Hajek 2004). In order to determine the nontarget effects of EPF, it is necessary to stress the difference between the physiological and the ecological host range (Hajek and Butler 2000), which refer to insects that can be infected under laboratory conditions and insects that can be infected under natural conditions, respectively (Hajek and Goettel 2000; Jaronski et al. 2003). Meyling et al. (2005) suggested employing the ecological range approach when examining pathogenicity of *M. anisopliae* in populations of nontarget soil arthropods, ideally in long-term experiments. Furthermore, according to Zimmermann (2007), safety applications of commercial products of EPF should also consider possible effects on native populations of EPF and other soil microorganisms. In two studies in China and Switzerland, exotic strains of *B. bassiana* and *B. brongniartii* were able to coexist with native strains without displacing them (Enkerli et al. 2004; Wang et al. 2004). In contrast, some hyperparasitic and antagonistic fungi such as *Syspastospora parasitica*, *Penicillium urticae*, *Clonostachys* spp., and *Trichoderma* spp. have shown inhibition and displacement by substrate competition against *B. bassiana* and *B. brongniartii* (Krauss et al. 2004; Posada et al. 2004).

5.5 IMPACT OF AGRICULTURAL PRACTICE ON BENEFICIAL RHIZOSPHERE MICROORGANISMS

5.5.1 Crop Genotype

Communities of rhizosphere microorganisms are known to respond to crop genotype, most likely due to differences in root exudation between crop genotypes (Hartmann et al. 2009).

Not only do different plant species harbor specific microbial communities, but also different varieties of the same plant species associate with specific rhizosphere microbial communities (Aira et al. 2010). Such plant genotypic selection of rhizosphere microorganisms should be considered to obtain optimal use of native populations of beneficial rhizosphere microorganisms in relation to plant health and nutrition (Picard and Bosco 2008).

The plant host specificity of beneficial rhizosphere microorganisms seems to be limited, but the functional compatibility between genotypes of beneficial rhizosphere microorganisms and crop genotype seems to be more complex. While the host specificity of AMF is limited, strong differential functional compatibility between crop and AMF genotypes in relation to both plant nutrition (Ravnskov and Jakobsen 1995) and biocontrol traits (Carlsen et al. 2008) has been reported.

Agronomic traits of crop genotypes, such as nutrient uptake efficiency or disease resistance, are most often explained from a phytocentric perspective, and limited information is available on the possible role of rhizosphere-associated microorganisms in the agronomic traits of crop genotypes. Hence, rhizosphere microorganisms are normally not considered when developing plant breeding programs selecting genotypes with desired agronomic traits. Future management of native populations of beneficial rhizosphere microorganisms will depend on their integration in plant breeding programs.

Plant breeding programs are largely directed toward conventional agriculture with high inputs of agrochemicals, marginalizing the role of beneficial rhizosphere microorganisms in relation to crop nutrition and health. Alternative breeding programs directed toward organic farming, exploiting natural environmental adaptation, will most likely better select compatible rhizosphere microorganisms. However, regardless of the breeding approach, the integration of beneficial rhizosphere microorganisms should be considered.

5.5.2 Crop Rotation

Crop rotation, understood either as rotation in the strict sense or as intercropping, has gained acceptance as a positive practice in conventional and alternative agricultural farming during the last decades. The use of legumes to increase soil fertility, in particular, has become widespread, given their capabilities to fix atmospheric nitrogen and mobilize less-labile forms of phosphorus (Nuruzzaman et al. 2005). However, the management of crop diversity requires careful examination of the actions exerted by the different plant species and management practices used to increase fertility and mobilization and storage of nutrients in various pools, since recent evidence suggests that not all plant-rotation-management combination schemes result in positive nutritional balances for crops and soils (Mat Hassan et al. 2012). Evidence of yield decline from short-rotation programs indicates that this practice should be replaced by double-cropping and intercropping for longer periods in order to accomplish the desired target of maintaining yields without jeopardizing soil health and quality (Bennett et al. 2012).

Some crops deserve particular attention, such as those belonging to the Brassicaceae, which release glucosinolates with biocidal potential against many soil organisms and are additionally among the few crops that do not form mycorrhizal associations (Schreiner and Koide 1993). Previous cropping with Brassicaceae plants has shown negative effects on crops highly dependent on mycorrhizal associations and microbial activity to meet their nutrient requirements, especially in early growth stages, such as maize (Gavito and Miller 1998b), leek, or onion (Sørensen et al. 2005, 2008). Managing these crops in rotation also requires good knowledge of the appropriate crops to follow or combine in the same fields to avoid growth depression.

Another important point in designing rotation programs lies in finding the most suitable fertilization regimes that are compatible with the coexistence of a healthy soil biota. Monocropping has been related in some cases to an increase in the abundance of deleterious rhizosphere microorganisms (Bennett et al. 2012), but, since monocropping is nearly always linked to excessive use of fertilizers and tillage that also affect soil biota, it is difficult to determine the contribution of each practice individually. Unfortunately, many farmers who have switched to organic fertilization still maintain the basic scheme of conventional management and carry over the same problems and environmental drawbacks.

5.5.3 Fertilization

Meeting the challenge of feeding several billion people requires adapting a diversity of strategies to improve fertilizer production, nutrient use efficiency, and nutrient recycling for a large continuum of farmers, ranging from those with low resources in low-fertility soils to those with large resources in high-fertility soils (Dawson and Hilton 2011). Despite the current fertilizer costs, the steady increase in fertilizer prices, the "sooner or later" finite amount of the reserves for some nutrients (Elser and Bennett 2011), and the increasing environmental regulations, as a rule and whenever possible, most agricultural fields are fertilized in excess to ensure that yields are not limited by nutrient supply (Powlson et al. 2011). This insurance principle has led to a reduced diversity, function, and efficiency of microbial interactions, and especially microbial symbionts, turning them from predominantly beneficial to either redundant or parasitic relationships (Kiers et al. 2002). Bennett et al. (2012) have suggested that, in many production-oriented agricultural systems, either the C costs derived from the establishment of mycorrhizal associations outweighed any benefits or the type of management had led to selection of inefficient and parasitic microbiota. This is most likely the case for most beneficial microbial and plant–microbial interactions in agricultural scenarios not designed to make use of the contributions of microorganisms to natural fertility regulation and soil quality. Kahiluoto et al. (2009) have shown that rotation, tillage, fertilization, and reincorporation of organic materials can be directed to maximize the benefits of natural fertility mechanisms and microbial activity.

Sustainable agriculture is a scenario in which microbial interactions have better chances to show their benefits, because, although immediate yields will always be important, they should not be the only criterion to define success. The greatest challenge of the numerous new approaches arising as alternatives to conventional agriculture is to maintain crop productivity while improving nutrient balances and soil nutrient reserves (Drinkwater and Snapp 2007).

5.5.4 Tillage

Tillage undoubtedly facilitates agricultural management and may have beneficial effects on soil biota under some circumstances (Treonis et al. 2010). Excessive use of tillage, on the other hand, may lead to compaction of the deeper soil layers, erosion of the upper layers, and eventually an overall reduction in soil quality due to the loss of retention capacity of nutrients, water, and air. Reducing tillage increases microbial biomass and fungi:bacteria ratio (Zhang et al. 2012), but, when considering the overall functional diversity of organisms in soil, changes in soil biota are clearly context specific (van Capelle et al. 2012).

Tillage, like other agricultural practices seeking sustainability, requires careful selection of methods and combination with other practices to succeed in maintaining productivity while improving soil quality (Ludvig et al. 2012). It may be reduced and even eliminated for some crops without yield costs, by increasing the contribution of desirable biological activities such as nitrogen fixation, reducing soil erosion, and improving water and fertilizer use efficiency, but paying attention to other problems such as weed control (Giambalvo et al. 2012) and fertilization (Alvarez and Steinbach 2009; Berner et al. 2008; Triplett and Dick 2008).

Mechanical disturbance of soil damages not only soil structure but also the microbial communities, especially the fungi and arbuscular mycorrhizal fungi in aggregate classes (Helgason et al. 2010; Wang et al. 2010). The hyphal network formed by the association of AMF with the previous crop is fragmented, sometimes causing a reduction in mycorrhizal colonization (Gavito and Miller 1998a) and a delay in the establishment of a functional new hyphal network and thereby in plant P uptake (Gavito and Miller 1998b). Tillage disruption of soil structure and mycorrhizal networks may contribute to crop growth depressions when new networks need to be constructed at the expense of plant carbon, which has led some authors to suggest that mycorrhizal fungi should be eliminated from agricultural fields (Ryan and Graham 2002). Management with excessive tillage and fertilization is a clear example of a context in which mycorrhizal associations may switch from predominantly beneficial to detrimental.

5.5.5 Pest Management Applications

The application of pesticides to control soil-borne pests also affects beneficial rhizosphere microorganisms. Fungicides logically pose a risk to beneficial rhizosphere fungi such as EPF, *Trichoderma*, and AMF. Some fungicides, such as benomyl and copper sulfate, have strong detrimental effects on most true fungi, including AMF (Hagerberg et al. 2010; Larsen et al. 1996). Though fungicides may not be directly toxic to bacteria, indirect effects may occur through fungal-mediated effects. Likewise, insecticides and nematicides are normally not directly toxic to bacteria and fungi, but their target effects on insects and nematodes may alter the soil food web dynamics, hampering the natural soil ecosystem balance, which is why pesticides should ideally be avoided or minimized.

Most types of soil disinfection, including solarization, steaming, and chemical fumigation, adversely affect both target and nontarget soil microorganisms, while biofumigation is less harmful to the soil biota in general (Larsen et al. 2004). However, in situations where it is necessary to employ soil disinfection, the possibility of inoculating beneficial rhizosphere microorganisms should be considered, ideally in combination with organic matter amendments.

Organic matter amendments in terms of animal manure, green manure, compost, and so on offer a sound alternative for the management of soil-borne pests, compatible with beneficial rhizosphere microorganisms.

5.6 STRATEGIES TO INTEGRATE BENEFICIAL RHIZOSPHERE MICROORGANISMS IN AGROECOSYSTEMS

Beneficial rhizosphere microorganisms can be integrated in agroecosystems by either inoculation of commercial products or conservation of native populations.

5.6.1 Inoculation

Commercial products of beneficial rhizosphere microorganisms are available worldwide, basically as biocontrol agents and biofertilizers. Such products most often contain single species, but products with a combination of different microorganisms are also available, mainly either combining microorganisms with similar functional traits or combining AMF, *Trichoderma*, and PGPR bacteria.

Crops with a transplant phase, such as field-grown vegetables, can benefit from preinoculating seedlings with beneficial microorganisms by mixing the microbial inoculum into the growth media. Since mycorrhiza formation is relatively slow compared with other beneficial plant–microbe associations, preinoculation of transplants seems to be ideal in relation to AMF inoculation. AMF-preinoculated transplants of different field-grown vegetables, including onion, leek, and lettuce, have been reported to provide up to 25% higher yield compared with nonmycorrhizal transplants (Larsen et al. 2007).

In general, timely and appropriate application is paramount to the successful introduction of beneficial rhizosphere microorganisms, since they will have to become established and compete with a native, well-adapted microbiota. Introduction is generally carried out using conventional technologies designed for chemical inputs. One of the most common methods is direct inoculation at sowing. Preinoculation of substrates, such as vermiculite or peat, is often used to introduce *Rhizobium* by mixing them with the seed at sowing time. Other PGPR, EPF, and *Trichoderma* can also be applied at sowing time when formulated on cereal grains or as capsules or granules applied with conventional fertilizer spreaders (O'Callaghan 1998). Dressing seeds with N-fixing bacteria (e.g., *Rhizobium*) or with biocontrol agents (e.g., *S. entomophila*) for conventional drilling is now widely used. In every case, from the farmers' point of view, it is important that bioinoculants do not require new equipment or costly technologies for their application.

There is concern about the introduction of exotic microorganisms into agroecosystems, as they could have negative impacts on native communities. To prevent this, every biocontrol programme using exotic inocula should include a detailed risk analysis, besides proving economic feasibility. Exotic inocula should be limited to those cases in which native ones are unavailable or are not efficient in their biocontrol or growth-promoting functions. Shannon et al. (1993) reported that native isolates of EPF were more virulent than exotic ones to control larvae of *Phyllophaga* spp. (Coleoptera: Scarabaeidae).

5.6.2 Conservation

There is a growing body of evidence supporting conservation of native populations of beneficial microbes with agroecological practices (e.g., Meyling and Eilenberg 2007).

In a long-term experiment with maize and wheat, Govaerts et al. (2008) showed that no-tillage combined with residue retention increased the populations of native microorganisms, both beneficial and pathogenic. However, no-tillage with no residue retention actually decreased microbial populations. Other well-recognized benefits of conservation tillage include increased soil organic matter, nutrient and moisture retention, and an increase in mesofauna populations. Raaijmakers et al. (2009) reviews the effect of solarization (using solar energy to heat soil) in reducing pathogen populations and modifying the soil balance to promote the activity and density of beneficial microorganisms such as *Bacillus* and *Pseudomonas*. This increase is caused by both heat and moisture. A combination of solarization and the addition of organic amendments can also increase these benefits. The same authors also mention the effect of crop rotation on decreasing pathogen populations by limiting the host available to the pathogen. It is also possible that crop rotation increases beneficial microbe populations. Xuan et al. (2012) showed that crop rotation resulted in increased composition, abundance, and diversity of microbial populations in rice systems, as well as increasing crop yield.

Plant residues left near the soil surface can suppress pathogens. However, in some cases the opposite effect might be seen. Burying residues with tillage to promote rapid decomposition would benefit the destruction of pathogens, as incorporation displaces them toward deeper layers of the soil, outside their niche. Carbon released through decomposition promotes microbial activity and therefore general pathogen suppression (Raaijmakers et al. 2009).

Studies in soybean agroecosystems in Brazil show that the persistence of EPF (*B. bassiana, M. anisopliae,* and *Isaria fumosorosea*) was higher in zero-tillage systems compared with conventional ones (Moscardi et al. 1985; Sosa-Gómez and Moscardi 1994). Similarly, in a study on crop rotation systems in Mexico, an eightfold increase of *B. bassiana* population density in soil was observed after 15–22 crop cycles under minimum tillage (MT), compared with only three to four cycles. Similar results were obtained when evaluating soil mesofauna, including diverse predators and organic-matter consumers (Nájera-Rincón et al. 2010). However, when the diversity of functional groups of rhizosphere microbes is increased, it is necessary to study the effects on their biology and ecology, as well as the nature of their interactions. This should lead to a more sustainable management of soil microbial diversity, favoring plant nutrition and health (Altieri and Nicholls 2005).

In general, conservation of beneficial rhizosphere microorganisms seems to be favored in organic farming and other low-input farming systems, whereas agricultural production systems with high inputs of agrochemicals marginalize the ecosystem services provided by beneficial rhizosphere microorganisms.

5.7 CONCLUSIONS AND FUTURE PERSPECTIVES

Beneficial rhizosphere microorganisms, including AMF, EPF, *Trichoderma*, and PGPR, provide important ecosystem services to agroecosystems in relation to plant nutrition and health, such as N fixation, P solubilization and transport, and biocontrol of plant pathogens and insect pests. Improved integration of beneficial rhizosphere microorganisms in agroecosystems, in terms of inoculation with commercial products, conservation of native populations of rhizosphere microorganisms, or both, seems important when aiming at providing food security for a growing world population with minimal environmental impacts. The management of beneficial rhizosphere microorganisms relies on agricultural practices favoring their populations, such as low inputs of agrochemicals, crop rotation, conservation tillage, and organic matter amendments. Improved knowledge on the ecology of beneficial rhizosphere microorganisms, especially in relation to interactions between functional groups of microorganisms, will allow the development of microbial consortia with multifunctional traits. Finally, future successful integration of beneficial rhizosphere microorganisms in agroecosystems also depends on plant breeding programs aiming at tailoring plant genotypes selected for beneficial rhizosphere microorganisms.

REFERENCES

Aira, M., M. Gómez-Brandón, C. Lazcano, E. Bååth, and J. Domínguez. 2010. Plant genotype strongly modifies the structure and growth of maize rhizosphere microbial communities. *Soil Biol Biochem* 42:2276–2281.

Albertsen, A., S. Ravnskov, H. Green, D. F. Jensen, and J. Larsen. 2006. Interactions between the external mycelium of the mycorrhizal fungus *Glomus intraradices* and other soil microorganisms as affected by organic matter. *Soil Biol Biochem* 38:1008–1014.

Altieri, M. A. and C. I. Nicholls. 2005. An agroecological basis for insect pest management. In: M. Altieri and C. I. Nicholls (eds), *Agroecology and the Search for a Truly Sustainable Agriculture*, pp. 199–238. Copenhagen: United Nations Environment Programme.

Alvarez, R. and H. S. Steinbach. 2009. A review of the effects of tillage systems on some physical properties, water content, nitrate availability, and crops yield in the Argentinean pampas. *Soil Tillage Res* 104:1–15.

Artursson, V., R. D. Finlay, and J. K. Jansson. 2006. Interactions between arbuscular mycorrhizal fungi and bacteria their potential for stimulating plant growth. *Environ Microbiol* 8:1–10.

Bae, H., D. P. Roberts, H. S. Lim et al. 2010. Endophytic *Trichoderma* isolates from tropical environments delay disease onset and induce resistance against *Phytophthora capsici* in hot pepper using multiple mechanisms. *Mol Plant-Microbe Interact* 24:336–351.

Barea, J. M., G. Andrade, V. Bianciotto et al. 1998. Impact of *Pseudomonas* strains used as inoculants for biocontrol of soil-borne fungal plant pathogens. *Appl Environ Microbiol* 64:2304–2307.

Barea, J. M., M. J. Pozo, R. Azcon, and C. Azcon-Aguilar. 2005. Microbial co-operation in the rhizosphere. *J Exp Bot* 56:761–778.

Bennett, A. J., G. D. Bending, D. Chandler, S. Hilton, and P. Mills. 2012. Meeting the demand for crop production: The challenge of yield decline in crops grown in short rotations. *Biol Rev* 87:52–71.

Berner, A., I. Hildermann, A. Fliessbach, L. Pfiffner, U. Nigli, and P. Mäder. 2008. Crop yield and soil fertility response to reduced tillage under organic management. *Soil Tillage Res* 101:89–96.

Bhattacharyya, P. N and D. K. Jha. 2012. Plant growth-promoting rhizobacteria (PGPR): Emergence in agriculture. *World J Microb Biot* 28:1327–1350.

Botha, A. 2011. The importance and ecology of yeast in soil. *Soil Biol Biochem* 43:1–8.

Brimner, T. A. and G. J. Boland. 2003. A review of the non-target effects of fungi used to biologically control plant diseases. *Agr Ecosyst Environ* 100:3–16.

Carlsen, S. C. K., A. Understrup, I. S. Fomsgaard, A. G. Mortensen, and S. Ravnskov. 2008. Flavonoids in roots of white clover: Interaction of arbuscular mycorrhizal fungi and a pathogenic fungus. *Plant Soil* 302:33–43.

Coutinho, F. P., W. P. Felix, and A. M. Yano-Melo. 2012. Solubilization of phosphates in vitro by *Aspergillus* spp. and *Penicillium* spp. *Ecol Eng* 42:85–89.

Dawson, C. J. and J. Hilton. 2011. Fertiliser availability in a resource limited world: Production and recycling of nitrogen and phosphorus. *Food Policy* 36:14–22.

Dobbelaere, S., J. Vanderleyden, and Y. Okon. 2003. Plant growth-promoting effects of diazotrophs in the rhizosphere. *CRC Crit Rev Plant Sci* 22:107–149.

Drinkwater, L. E. and S. S. Snapp. 2007. Nutrients in agroecosystems: Rethinking the management paradigm. *Adv Agron* 92:163–186.

Druzhinina, S. I. and P. C. Kubicek. 2005. Species concepts and biodiversity in *Trichoderma* and *Hypocrea*: From aggregate species to species clusters? *J Zhejiang Univ Sci B* 6:100–112.

Elser, J and E. Bennett. 2011. A broken biogeochemical cycle. *Nature* 478:29–31.

Enkerli, J., F. Widmer, and S. Keller. 2004. Long-term field persistence of *Beauveria brongniartii* strains applied as biocontrol agents against European cockchafer larvae in Switzerland. *Biol Control* 29:115–123.

Fitter, A. H., C. A. Gilligan, K. Hollingworth et al. 2005. Biodiversity and ecosystem function in soil. *Funct Ecol* 19:369–377.

Friedl, M. A. and I. S. Druzhinina. 2012. Taxon-specific metagenomics of *Trichoderma* reveals a narrow community of opportunistic species that regulate each other's development. *Microbiology* 158:69–83.

Gavito, M. E. and M. H. Miller. 1998a. Changes in mycorrhiza development in maize induced by crop management practices. *Plant Soil* 198:185–192.

Gavito, M. E. and M. H. Miller. 1998b. Early phosphorus nutrition, mycorrhizae development, dry matter partitioning and yield of maize. *Plant Soil* 198:177–186.

Giambalvo, D., P. Ruisi, S. Saia, G. Di Miceli, A. S. Frenda, and G. Amato. 2012. Faba bean grain yield, N_2 fixation, and weed infestation in a long-term tillage experiment under rainfed Mediterranean conditions. *Plant Soil* 360:215–227.

Gianinazzi, S., A. Gollotte, M. N. Binet, D. van Tuinen, D. Redecker, and D. Wipf. 2010. Agroecology: The key role of arbuscular mycorrhizas in ecosystem services. *Mycorrhiza* 20:519–530.

Gilbert, N. 2009. The disappearing nutrient. *Nature* 461:716–718.

Goettel, M. S. 2008. Are entomopathogenic fungi only entomopathogens? A preamble. *J Invertebr Pathol* 98:255.

Govaerts, B., M. Mezzalama, K. D. Sayre et al. 2008. Long-term consequences of tillage, residue management, and crop rotation on selected soil micro-flora groups in the subtropical highlands. *Appl Soil Ecol* 38:197–210.

Green, H., J. Larsen, P. A. Olsson, D. F. Jensen, and I. Jakobsen. 1999. Suppression of the biocontrol agent *Trichoderma harzianum* by mycelium of the arbuscular mycorrhizal fungus *Glomus intraradices* in root-free soil. *Appl Environ Microbiol* 65:1428–1434.

Hagerberg, D., N. Manique, K. K. Brandt, J. Larsen, O. Nybroe, and S. Olsson. 2010. Low concentration of copper inhibits colonization of soil by the arbuscular mycorrhizal fungus *Glomus intraradices* and changes the microbial community structure. *Microb Ecol* 61:844–852.

Hajek, A. 2004. *Natural Enemies. An Introduction to Biological Control.* Cambridge: Cambridge University Press.

Hajek, A. E. and L. Butler. 2000. Predicting the host range of entomopathogenic fungi. In: P. A. Follett and J. J. Duan (eds), *Nontarget Effects of Biological Control*, pp. 263–276. Dordrecht: Kluwer Academic.

Hajek, A. E. and M. S. Goettel. 2000. Guidelines for evaluating effects of entomopathogens on nontarget organisms. In: L. A. Lacey and H. K. Kaya (eds), *Manual of Field Techniques in Insect Pathology*, pp. 847–868. Dordrecht: Kluwer Academic.

Harman, E. G., C. R. Howell, A. Viterbo, I. Chet, and M. Lorito. 2004. *Trichoderma* species opportunistic, avirulent plant symbionts. *Nat Rev Microbiol* 2:43–56.

Hartmann, A., M. Schmid, D. van Tuinen, and G. Berg. 2009. Plant driven selection of microbes. *Plant Soil* 321:235–257.

Helgason, B. L., F. L. Walley, and J. J. Germida. 2010. No-till soil management increases microbial biomass and alters community profiles in soil aggregates. *Appl Soil Ecol* 46:390–397.

Hermosa, R., A. Viterbo, I. Chet, and E. Monte. 2012. Plant-beneficial effects of *Trichoderma* and of its genes. *Microbiology* 158:17–25.

Herridge, D. F., M. B. Peoples, and R. M. Boddey. 2008. Global inputs of biological nitrogen fixation in agricultural systems. *Plant Soil* 311:1–18.

Hinsinger, P., A. G. Bengough, D. Vetterlein, and I. M. Young. 2009. Rhizosphere: Biophysics, biogeochemistry and ecological relevance. *Plant Soil* 321:117–152.

Infante, D., B. Martínez, N. González, and Y. Reyes. 2009. Mecanismos de acción de *Trichoderma* frente a hongos fitopatogenos. *Revista Protección Vegetal* 24:14–21.

Jackson, T. and T. R. Glare. 1992. *Use of Pathogens in Scarab Pest Management*. Andover: Intercept.

Jakobsen, I. 1994. Research approaches to study the functioning of vesicular-arbuscular mycorrhizas in the field. *Plant Soil* 159:141–147.

Jaronski, S. T., M. S. Goettel, and C. J. Lomer. 2003. Regulatory requirements for ecotoxicological assessments of microbial insecticides—How relevant are they? In: H. M. T. Hokkanen and A. E. Hajek (eds), *Environmental Impacts of Microbial Insecticides*, pp. 237–260. Dordrecht: Kluwer Academic Publishers.

Jensen, B., I. M. B. Knudsen, D.F. Jensen et al. 2011. The relative importance of released microbial biocontrol agents and their metabolites in relation to the natural microbiota on strawberries. Pesticides Research No. 128. Denmark: Environmental Protection Agency, Danish Ministry of the Environment.

Johansson, J. F., L. R. Paul, and R. D. Finlay. 2004. Microbial interactions in the mycorrhizosphere and their significance for sustainable agriculture. *FEMS Microbiol Ecol* 48:1–13.

Jones, D. L., C. Nguyen, and R. D. Finlay. 2009. Carbon flow in the rhizosphere: Carbon trading at the root-soil interface. *Plant Soil* 321:25–33.

Kahiluoto, H., E. Ketoja, and M. Vestberg. 2009. Contribution of arbuscular mycorrhiza to soil quality in contrasting cropping systems. *Agr Ecosyst Environ* 134:36–45.

Kennedy, I. R., A. T. M. A. Choudhury, and M. I. Kecskés. 2004. Non-symbiotic bacterial diazotrophs in crop-farming systems: Can their potential for plant growth promotion be better exploited? *Soil Biol Biochem* 36:1229–1244.

Khan, M. S., A. Zaidi, and P. A. Wani. 2007. Role of phosphate-solubilizing microorganisms in sustainable agriculture—A review. *Agron Sustain Dev* 27:29–43.

Kiers, E. T., S. A. West, and R. F. Denison. 2002. Mediating mutualisms: Farm management practices and evolutionary change in symbiont co-operation. *J Appl Ecol* 39:745–754.

Krauss, U., E. Hidalgo, C. Arroyo, and S. R. Piper. 2004. Interaction between the entomopathogens *Beauveria bassiana*, *Metarhizium anisopliae* and *Paecilomyces fumosoroseus* and the mycoparasites *Clonostachys* spp., *Trichoderma harzianum* and *Lecanicillium lecanii*. *Biocontrol Sci Technol* 14:331–346.

Larsen, J., P. Cornejo, and J. M. Barea. 2009. Interactions between the arbuscular mycorrhizal fungus *Glomus intraradices* and the plant growth promoting rhizobacteria *Paenibacillus polymyxa* and *P. macerans* in the mycorrhizosphere of *Cucumis sativus*. *Soil Biol Biochem* 41:286–292.

Larsen, J., S. Ravnskov, and I. Jakobsen. 2003. Combined effects of an AM fungus and BCA bacteria against the root pathogen *Pythium ultimum* in soil. *Folia Geobot* 38:145–154.

Larsen, J., S. Ravnskov, K. Møller, and L. Bødker. 2004. Bæredygtig produktion af småplanter i forstplanteskoler. *Bekæmpelsesmiddelforskning fra Miljøstyrelsen* No. 93.

Larsen, J., S. Ravnskov, and J. N. Sørensen. 2007. Capturing the benefit of mycorrhiza in horticulture. In: C. Hamel and C. Plenchette (eds), *Mycorrhizae and Crop Productivity*, pp. 123–150. New York: Haworth Press.

Larsen, J., I. Thingstrup, I. Jakobsen, and S. Rosendahl. 1996. Benomyl inhibits P transport but not fungal alkaline phosphatase activity in a cucumber–*Glomus* symbiosis. *New Phytol* 132:127–133.

Lecuona, R. 1996. *Microorganismos patógenos empleados en el control microbiano de insectos plaga*. Buenos Aires: Talleres Gráficos Mariano Mas., Argentina.

Li, B., S. Ravnskov, G. Xie, and J. Larsen. 2008. Differential effects of *Paenibacillus* spp. on cucumber mycorrhizas. *Mycol Prog* 7:277–284.

Linderman, R. G. 1988. Mycorrhizal interactions with the rhizosphere microflora: The mycorrhizosphere effect. *Phytopathology* 78:366–370.

Ludvig, B., D. Geisseler, K. Michel et al. 2012. Effects of fertilization and soil management on crop yields and carbon stabilization in soils. *Agron Sustain Dev* 31:361–372.

Lugtenberg, B. and F. Kamilova. 2009. Plant-growth-promoting rhizobacteria. *Annu Rev Microbiol* 63:541–556.

Martínez-Medina, A., J. A. Pascual, E. Lloret, and A. Roldán. 2009. Interactions between arbuscular mycorrhizal fungi and *Trichoderma harzianum* and their effects on *Fusarium* wilt in melon plants grown in seedling nurseries. *J Sci Food Agric* 89:1843–1850.

Mat Hassan, H., P. Marschner, A. McNeil, and C. Tang. 2012. Grain-legume pre-crops and their residues affect the growth, P uptake and size of P pools in the rhizosphere of the following wheat. *Biol Fert Soils* 48:775–785.

Meyling, N. V. and J. Eilenberg. 2007. Ecology of entomopathogenic fungi *Beauveria bassiana* and *Metarhizium anisopliae* in temperate agroecosystems: Potential for conservation biological control. *Biol Control* 43:145–155.

Meyling, N. V., J. Eilenberg, and C. Nielsen. 2005. Non-target effects of insect pathogenic fungi. In: *Insect Pathogens and Insect Parasitic Nematodes "Melolontha", Proceedings of the IOBC/WPRS Meeting*, Innsbruck, 11–13 October, vol. 28, pp. 13–18. IOBC-WPRS Bulletin.

Moscardi, F., B. S. Correa-Ferreira, L. G. Leite, and C. E. O. Zamataro. 1985. Incidência estacional de fungos entomógenos sobre populações de percevejos pragas da soja. Resultados de Pesquisa da Soja 1984/1985. EMBRAPA, Centro Nacional de Pesquisa de Soja, Londrina, PR, Brazil.

Nájera-Rincón, M. B., A. Castro-Ramírez, and A. Aragón-García. 2010. Prácticas culturales y físicas. In: L. A. Rodríguez Del Bosque and M. A. Morón (eds), *Plagas del suelo*, pp. 149–168. Mexico: Mundi Prensa.

Nuruzzaman, M., H. Lambers, M. D. A. Bolland, and E. J. Veneklaas. 2005. Phosphorus benefits of different legume crops to subsequent wheat grown in different soils of Western Australia. *Plant Soil* 271:175–187.

O'Callaghan, M. 1998. Establishment of microbial control agents in soil. In: M. O'Callaghan and T. A. Jackson (eds), *Proceedings of the 4th International Workshop on Microbial Control of Soil Dwelling Pests*, pp. 83–89. Lincoln: AgResearch.

Otado, J. A., S. A. Okoth, J. Ochanda, and J. P. Kahindi. 2011. Assessment of *Trichoderma* isolates for virulence efficacy on *Fusarium oxysporum* F. *sp. Phaseoli*. *Trop Subtrop Agroecosyst* 13:99–107.

Ownley, B. H., M. R. Griffin, W. E. Klingeman, K. D. Gwinn, J. K. Moulton, and M. R. Pereira. 2008. *Beauveria bassiana*: Endophytic colonization and plant disease control. *J Invertebr Pathol* 98:267–270.

Picard, C. and M. Bosco. 2008. Genotypic and phenotypic diversity in populations of plant-probiotic *Pseudomonas* spp. colonizing roots. *Naturwissenschaften* 95:1–16.

Posada, F., F. E. Vega, S. A. Rehner et al. 2004. *Syspastospora parasitica*, a mycoparasite of the fungus *Beauveria bassiana* attacking the Colorado potato beetle *Leptinotarsa decemlineata*: A tritrophic association. *J Insect Sci* 4:24.

Powlson, D. S., P. Gregory, W. R. Whalley et al. 2011. Soil management in relation to sustainable agriculture and ecosystem services. *Food Policy* 36:72–87.

Raaijmakers, J., T. Paulitz, C. Steinberg, C. Alabouvette, and Y. Moënne-Loccoz. 2009. The rhizosphere: A playground and battlefield for soilborne pathogens and beneficial microorganisms. *Plant Soil* 321:341–361.

Rabeendran, N., E. E. Jones, D. J. Moot, and A. Stewart. 2006. Biocontrol of *Sclerotinia* lettuce drop by *Coniothyrium minitans* and *Trichoderma hamatum*. *Biol Control* 39:352–362.

Ravnskov, S. and I. Jakobsen. 1995. Functional compatibility in arbuscular mycorrhizas measured as hyphal P transport to the plant. *New Phytol* 129:611–618.

Ravnskov, S. and I. Jakobsen. 1999. Effects of *Pseudomonas fluorescens* DF57 on growth and P uptake of two arbuscular mycorrhizal fungi in symbiosis with cucumber. *Mycorrhiza* 8:329–334.

Ravnskov, S., B. Jensen, I. M. B. Knudsen et al. 2006. Soil inoculation with the biocontrol agent *Clonostachys rosea* and the mycorrhizal fungus *Glomus intraradices* results in mutual inhibition, plant growth promotion and alteration of soil microbial communities. *Soil Biol Biochem* 38:3453–3462.

Ravnskov, S., J. Larsen, and I. Jakobsen. 2002. Hyphal P transport of the arbsucular mycorrhizal fungus *Glomus intraradices* is not affected by the biocontrol bacterium *Burkholderia cepacia*. *Soil Biol Biochem* 34:1875–1881.

Ravnskov, S., O. Nybroe, and I. Jakobsen. 1999. Influence of an arbuscular mycorrhizal fungus on *Pseudomonas fluorescens* DF57 in rhizosphere and hyphosphere soil. *New Phytol* 142:113–122.

Rey, M., J. Delgado-Jarana, A. M. Rincón, M. C. Limón, and T. Benítez. 2000. Mejora de cepas de *Trichoderma* para su empleo como biofungicidas. *Rev Iberoam Micol* 17:31–36.

Ryan, M. H. and J. H. Graham. 2002. Is there a role of arbuscular mycorrhizal fungi in production agriculture? *Plant Soil* 244:263–271.

Ryder, L. S., B. D. Harris, D. M. Soanes, M. J. Kershaw, N. J. Talbot, and C. R. Thornton. 2012. Saprotrophic competitiveness and biocontrol fitness of a genetically modified strain of the plant growth-promoting fungus *Trichoderma hamatum* GD12. *Microbiology* 158:84–97.

Samuels, G. J. 1996. *Trichoderma*: A review of biology and systematics of the genus. *Mycol Res* 100:923–935.

Schreiner, R. P. and R. T. Koide. 1993. Antifungal compounds from the roots of mycotrophic and non-mycotrophic plant species. *New Phytol* 123:99–105.

Shannon, P. J., S. M. Smith, and E. Hidalgo. 1993. Evaluación en el laboratorio de aislamientos costarricenses y exóticos de *Metarhizium* spp. y *Beauveria* spp. contra larvas de *Phyllophaga* spp. (Coleoptera: Scarabaeidae). In: M. A. Morón (ed.), *Diversidad y manejo de plagas subterráneas*, pp. 203–215. Xalapa, México: Publicación Especial de la Sociedad Mexicana de Entomología e Instituto de Ecología.

Smith, S. E., E. Facelli, S. Pope, and F. A. Smith. 2010. Plant performance in stressful environments: Interpreting new and established knowledge of the roles of arbuscular mycorrhizas. *Plant Soil* 326:3–20.

Smith, S. E. and D. J. Read. 2008. *Mycorrhizal Symbiosis*, 3rd edn. London: Academic Press.

Smith, S. E. and F. A. Smith. 2012. Fresh perspectives on the roles of arbuscular mycorrhizal fungi in plant nutrition and growth. *Mycologia* 104:1–13.

Soper, R. S. and M. G. Ward. 1981. Production, formulation and application of fungi for insect control. In: G. C. Papavizas (ed.), *Biological Control in Crop Production*, BARC Symposium 5, pp. 161–180. Totowa: Allanheld Osmun.

Sørensen, J. N., J. Larsen, and I. Jakobsen. 2005. Mycorrhization and nutrient content of leeks (*Allium porrum* L.) in relation to previous crop and cover crop management on high P soils. *Plant Soil* 273:101–114.

Sørensen, J. N., J. Larsen, and I. Jakobsen. 2008. Pre-inoculation with arbuscular mycorrhizal fungi increases early nutrient concentration and growth of field-grown leeks under high productivity conditions. *Plant Soil* 307:277–287.

Sosa-Gómez, D. R. and F. Moscardi. 1994. Effect of till and no-till soybean cultivation on dynamics of entomopathogenic fungi in the soil. *Fla Entomol* 77:284–287.

Srivastava, R., A. Khalid, U. S. Singh, and A. K. Sharma. 2010. Evaluation of arbuscular mycorrhizal fungus, fluorescent Pseudomonas and *Trichoderma harzianum* formulation against *Fusarium oxysporum* f. sp. Lycopersici for the management of tomato wilt. *Biol Control* 53:24–31.

St. Leger, R. J. 2008. Studies on adaptations of *Metarhizium anisopliae* to life in the soil. *J Invertebr Pathol* 98:271–276.

Thorup-Kristensen, K., J. Magid, and L. Stoumann-Jensen. 2003. Catch crops and green manures as biological tools in nitrogen management in temperate zones. *Adv Agron* 79:227–302.

Tilman, D., C. Balzer, J. Hill, and B. L. Befort. 2011. Global food demand and the sustainable intensification of agriculture. *Proc Natl Acad Sci USA* 108:20260–20264.

Treonis, A., E. E. Austin, J. S. Buyer, J. E. Maul, L. Spicer, and I. Zasada. 2010. Effects of organic amendment and tillage on soil microorganisms and microfauna. *Appl Soil Ecol* 46:103–110.

Triplett, G. B. and W. A. Dick. 2008. No-tillage crop production: A revolution in agriculture! *Agron J* 100:S153–S165.

van Capelle, C., S. Schrader, and J. Brunotte. 2012. Tillage-induced changes in the functional diversity of soil biota. A review with a focus on German data. *Eur J Soil Biol* 50:165–181.

Vázquez, M. M., S. César, R. Azcón, and J. M. Barea. 2000. Interactions between arbuscular mycorrhizal fungi and other microbial inoculants (*Azospirillum, Pseudomonas, Trichoderma*) and their effects on microbial population and enzyme activities in the rhizosphere of maize plants. *Appl Soil Ecol* 15:261–272.

Vega, F. E. 2008. Insect pathology and fungal endophytes. *J Invertebr Pathol* 98:277–279.

Vestberg, M., S. Kukonen, K. Saari et al. 2004. Microbial inoculation for improving the growth and health of micropropagated strawberry. *Appl Soil Ecol* 27:243–258.

Vinale, F., K. Sivasithamparam, E. L. Ghisalberti, R. Marra, S. L. Woo, and M. Lorito. 2008. *Trichoderma*–plant–pathogen interactions. *Soil Biol Biochem* 40:1–10.

Wang, C., M. Fan, Z. Li, and T. M. Butt. 2004. Molecular monitoring and evaluation of the application of the insect-pathogenic fungus *Beauveria bassiana* in southeast China. *J Appl Microbiol* 96:861–870.

Wang, Y., J. Xu, J. Shen, Y. Luo, S. Scheu, and X. Ke. 2010. Tillage, residue burning and crop rotation alter soil fungal community and water-stable aggregation in arable fields. *Soil Tillage Res* 107:71–79.

Whipps, J. M. 2001. Microbial interactions and biocontrol in the rhizosphere. *J Exp Bot* 52:487–511.

Xuan, D., V. Guong, A. Rosling, S. Alström, B. Chai, and N. Högberg. 2012. Different crop rotation systems as drivers of change in soil bacterial community structure and yield of rice, *Oryza sativa*. *Biol Fert Soils* 48:217–225.

Zhang, B., H. B. He, X. L. Ding et al. 2012. Soil microbial community dynamics over a maize (*Zea mays* L.) growing season under conventional and no-tillage practices in a rainfed agroecosystem. *Soil Tillage Res* 124:153–160.

Zimmermann, G. 2007. Review on safety of the entomopathogenic fungi *Beauveria bassiana* and *Beauveria brongniartii*. *Biocontrol Sci Techn* 17:553–596.

CHAPTER 6

Mechanized Rain-Fed Farming and Its Ecological Impact on the Drylands
The Case of Gedarif State, Sudan

Yasin Abdalla Eltayeb Elhadary

CONTENTS

6.1   Introduction 121
6.2   Ecology of Drylands and Sudan 123
6.3   Development of Mechanized Agriculture in Sudan 124
6.4   Fragility of Land Tenure Situation in Sudan 126
6.5   Impact of Semimechanized Farming on Environment and Development 128
        6.5.1   Clearance of Forest 128
        6.5.2   Soil Depletion and Collapse of Yield Production 130
        6.5.3   Notion of Desertification in Sudan 131
6.6   Discussion: Mechanization and Its Socioecological Impact 132
6.7   Conclusions 134
6.8   Summary 134
References 134

6.1 INTRODUCTION

The agricultural sector is still the main economic driver for most developing countries, including African countries, despite the discovery of oil in some, as in the case of Sudan. According to the Food and Agriculture Organization (FAO) (2008), in Africa agriculture generates up to 50% of gross domestic product (GDP) and contributes up to more than 80% of trade in value and more than 50% of raw materials to industries. Besides, its contribution is vital in sustaining food security and slowing down the increase of poverty, especially in rural areas. This implies that agriculture, particularly mechanized farming, is expected to play a major part in solving the food production problem in Africa. It is only through appropriate mechanization that African farmers will be able to feed not only themselves but also the continent's mushrooming urban population (FAO 2008).

Agriculture, the most important potential contributor to economic growth (Abu Raida 2013), has not been given the attention it deserves from policy makers. Currently, this sector faces several constraints that hinder its contribution to the economy of the African countries. These constraints

include, but are not limited to, declining productivity, price fluctuation, and massive land degradation. It is becoming a challenge for planners and decision makers to develop agriculture to meet
the ever-growing demand for food and at the same time to address its negative impact on the environment. Without solving these problems, the capacity of land to secure livelihood and produce
enough food is questionable. The most pressing environmental threat in the twenty-first century is
not global warming, thinning ozone, or depleted energy sources. Rather, it is the decreasing amount
of land reserved for food production (Elhadary et al. 2013).

Like most African countries, Sudan's economy is based heavily on agriculture, although the economic
return of this sector is declining sharply due to the emergence of the oil industry in 2000 (UNEP 2007).
The agricultural sector, with its various sections (traditional, semimechanized, irrigated, and livestock),
plays a vital role in securing a livelihood for 70% of the population, especially in rural areas. It also
contributes around 40% of the country's GDP and more than 90% of the non-oil export earnings. In addition, about 80% of labor is employed in agricultural and related practices (Mustafa 2006), and agriculture
supplies about 60% of the raw materials needed by the manufacturing sector (Elbashir et al. 2004).

Generally, there are two types of rain-fed agriculture in the country: traditional farming and
semimechanized farming. The former comprises mainly subsistence systems, prevalent almost
everywhere in Sudan, accounting for an area of around 10 million hectares. This is how the majority of rural people, mostly found in the Kordofan, Darfur, White Nile, and Blue Nile states, derive
their livelihood (UNEP 2007). The latter is mechanization of rain-fed agriculture, which was
initiated by the British in Gedarif in 1944 to meet the food needs of their army in East Africa
(Elhadary 2010; Eltayeb 1985; UNEP 2007). Since then, it has spread very rapidly, currently covering 6.5 million hectares across the country, and on a large scale in seven states (Elhadary 2007).
These states include El Gedarif, Blue Nile, Kassala, White Nile, Sennar, and Southern Kordofan
(Figure 6.1). The turning point in its expansion was in 1968, when the Mechanized Farming

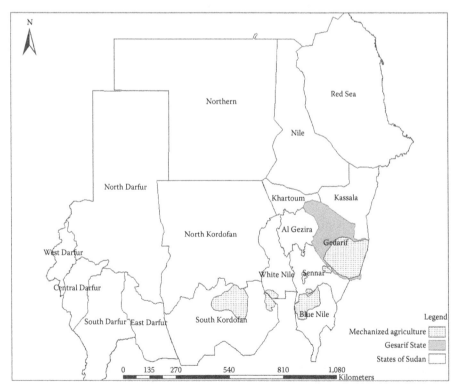

Figure 6.1 Map of Sudan states.

Corporation (MFC) was established by the government on request by the World Bank (Elhadary 2007).

This expansion, particularly where unplanned, has generated severe environmental and socio-economic problems. Symptoms and manifestations of the problems include soil deterioration, declining productivity, removal of trees, land degradation, and destruction of biological habitats. This chapter aims to describe the ecological consequences of mechanized farming on an arid environment, with particular emphasis on Gedarif State in eastern Sudan.

Structurally, this chapter is divided into six sections. The first is about the role of agriculture in the African economy, including Sudan; the second is about the ecological setup of the drylands and the geographical aspect of Sudan; the third focuses on the development of mechanized agriculture in Sudan; the fourth discusses the condition of land tenure in Africa in general and in Sudan in particular; the ecological impact of mechanized farming is discussed in the fifth section, while the sixth section is on the socioecological dimension of mechanical agriculture.

6.2 ECOLOGY OF DRYLANDS AND SUDAN

Generally, dryland is defined as an area characterized by the condition of aridity. Although there is no clear boundary, drylands are considered to be areas where average rainfall is less than the potential moisture losses through evaporation and transpiration (FAO 2004). According to United Nations Convention to Combat Desertification (UNCCD 1999), drylands comprise land within the arid, semiarid, and dry subhumid regions, which are most extensive in Africa (nearly 13 million km^2) and Asia (11 million km^2) (White et al. 2002). Ecologically, this region is characterized by having an immature ecosystem, aridity, variability, water scarcity, and a very short growing season that hardly reaches 200 days (Elhadary 2010). The annual rainfall varies over space and time, ranging from <200 to 800 mm annually (FAO 2004) (Table 6.1). Receiving a lower amount of rainfall over a short period, the world's drylands are in a need of special type of land use management.

Depending on its ecological characteristics, about six billion hectares, which represents 47% of the surface of the earth, can be classified as drylands (UNEP 1992) (Table 6.1). Of this, 3.5 to 4.0 billion hectares (57%–65%) are either desertified or prone to desertification. This implies that almost half of the world's countries (more than 110) are partially or wholly located in dryland environments (FAO 2004). These lands are home to nearly 40% of the world's population (White et al. 2002), and close to one billion people worldwide depend directly on the drylands for their livelihoods (Mwangi and Dohrn 2006). In Africa alone, dryland constitutes 43% of the total area, 40% of the continent's population, and 59% of all ruminant livestock grazing (Scoones 1994). It secures a livelihood for more than 50 million pastoralists and up to 200 million agro-pastoralists (De Jode 2010).

Sudan is located in the eastern part of Africa, between latitudes 8° and 22° north and longitudes 22° and 38° east. The area covered is about 1,882,000 km^2, and the country shares borders with seven countries: Egypt, Libya, Chad, Central Africa Republic, Ethiopia, Eritrea, and Republic of

Table 6.1   Dryland Categories

Classification	Rainfall (mm)	Area (%)	Area ($\times 10^9$ ha)
Hyperarid	<200	7.50	1.00
Arid	<200 (winter) or <400 (summer)	12.1	1.62
Semiarid	200–500 (winter) or 400–600 (summer)	17.7	2.37
Dry subhumid	500–700 (winter) or 600–800 (summer)	9.90	1.32
Total		47.2	6.31

Source: FAO (Food and Agriculture Organization), Carbon Sequestration in Dryland Soils, FAO, Rome, 2004.

South Sudan. The total population of Sudan is estimated to be 39 million, growing at 2.7%, with more than 30 million people living in rural areas (The Sudan Central Bureau of Statistics (CBS) 2011). On July 9, 2011, Sudan was split in two, formalizing the new state of South Sudan (USAID 2012).

Ecologically, the annual rainfall ranges between 75 mm in the extreme north and more than 500 mm in the extreme south (Elhadary 2007). Accordingly, four ecological zones can be identified: desert; semidesert; low-rainfall savannah; and high-rainfall savannah in the flood plain in the south. The main characteristics of these zones remain the same as described by Harrison and Jackson (1958). The desert zone, north of latitude 16° north, represents approximately 29% of Sudan's total area. The average annual rainfall of the desert zone is <100 mm. The semiarid zone encompasses about 19.6% of Sudan's total area and is located between latitudes 14 and 16° north. The annual rainfall of the semiarid zone ranges between 100 and 300 mm. The low-rainfall savannah is located between latitudes 10 and 14° north, with annual rainfall 300–500 mm. This zone covers more than half (51.1%) of Sudan's total area. The high-rainfall savannah, located south of latitude 10° north, has an average annual rainfall <500 mm. This zone constitutes only 10.3% of Sudan's total area. It has been estimated that the semiarid zone of the Sudan encompasses about 70% of the surface area of the country, and 70% of the Sudanese population lives in this zone, with herding and farming as the main sources of livelihood (Elhadary 2007).

Economically, over 80% of Sudan's employment takes place in agriculture and its related activities; the majority of the people are farmers and pastoralists living by subsistence farming and livestock herding in a nomadic way of life (UNEP 2007; Elhadary 2007). Sudan, with its good natural resources (large flat areas, fertile soil, availability of water), is considered a suitable country for securing food, not only for its people but worldwide, particularly for Arab countries. Out of the total area of Sudan, about 81 million hectares (250 million acres), 86 million acres could easily be cultivable, which is more than half the currently cultivated acreage-base of the United States (Abu Raida 2013). Sudan had an estimated gross domestic product (GDP) of US$58 billion in 2008. Agriculture was responsible for 26% of GDP, and industry and services 34% and 40%, respectively (USAID 2012). While agricultural land continues to be an important resource, oil production is decreasing since the secession of South Sudan.

6.3 DEVELOPMENT OF MECHANIZED AGRICULTURE IN SUDAN

Mechanized agriculture is defined simply as the use of mechanical technology to increase the productivity of agriculture, often to achieve results well beyond the capacity of human labor (FAO 2008). In Sudan, tractors and harvesting machines are replacing manual labor, but the technology is not fully applied in agriculture. This explains why people call mechanical agriculture in Gedarif semimechanization, as it partly uses machines. According to Eltayeb (1985) and Elhadary (2010), mechanized farming in Sudan started in the 1940s on a small scale near Gedarif town (Gadmbaliya), and it reached 600 ha in 1945. It was initiated by the British in Gedarif State of eastern Sudan to meet the demands of food for their army in East Africa (UNEP 2007). Until independence, 6000 ha of *Durra* (sorghum), the dominant crop, was cultivated annually, under a sharecropping arrangement between the government and farmers who had been allocated land in the project (Eltayeb and Lewandowski 1983).

The general policy of Sudan, after its independence in 1956, is to increase production of cash crops, whether through irrigation or rain-fed as mechanized farming. This policy is also supported by international organizations such as World Bank in the interest of securing food for the growing population (Elhadary 2007). The bank support for agricultural development during the 1960s and early 1970s was mostly for large-scale development schemes, which often involved mechanization of agriculture, irrigation construction, and rehabilitation (Gibbon et al. 1993). Private sectors and governmental institutions have been encouraged to take up mechanized farming in Gedarif State (UNEP 2007). It important to note that the turning point in expansion was in 1968, when the MFC was established by the government on request from the World Bank (Elhadary 2007). In Sudan, the World Bank sought to introduce modern and off-the-shelf agricultural technologies for food crops, export commodities, and livestock (Gibbon et al. 1993). As a consequence, mechanized farming spread very rapidly in Gedarif as well as in other states in the country. These states include Kassala, Red Sea, Sennar, Blue Nile, White Nile, South Kordofan, and North Kordofan (Figure 6.1). Although demarcation and allocation of land to private investors are the responsibility of the MFC, most of the expansion that has taken place in these areas was not authorized by the MFC. Besides its role in demarcation, the MFC used to play a role in facilitating credit, determining a maximum lease period of 25 years, and lately ensuring that 10% of land leased should be allocated for tree planting around the scheme (Elhadary 2007). This 10% shelterbelt aims to enhance soil fertility and mitigate the impacts of flash floods (UNEP 2007). Technically, the distribution of mechanized schemes to the investors is based on dividing rectangular areas into plots of 420 ha (later increased in places to 630 ha), of which half were leased to farmers and the rest left fallow. After four years, private farmers were to exchange the formerly leased land with adjacent fallow plots to allow the soil to recover.

No one can deny the role of mechanized farming in developing the country in general, and particularly in the states where it is located. Taxes, Islamic *Zakkat* (a proportion of one-tenth of the cultivated crops paid to the government), involvement of subsectors, and generation of jobs are some of its contributions to the economy. However, the unplanned expansion of mechanized farming, coupled with the way in which it has been practiced, has led to negative socioeconomic and ecological impacts in the country. A foremost and concrete example is the violation of the four-year fallow period. This has never been followed in practice, and private farmers illegally seized large areas outside the designated blocks (UNEP 2007). In 1968, more than 750,000 ha were being cultivated in the country, of which it was estimated that more than 200,000 ha constituted unauthorized holdings (Omer 2011). Currently, in Gedarif State, for example, the total area under cultivation amounts to 2.6 million hectares, of which unplanned schemes constitute about 33.8% (Elhadary 2010; UNEP 2007). The same situation is found in other states; for example, in South Kordofan State (Habila region), some 45% of mechanized farms in 1985 were unsanctioned (UNEP 2002). Also in Sennar State, 60% of the two million hectares under rain-fed agriculture are occupied by nonauthorized mechanized schemes, while 30% are under planned mechanization and 10% under traditional agriculture. This unorganized expansion into land belonging to other land users has led to severe environmental degradation and has become one of the major sources of bloody conflict in the country, as in the case of Darfur (Leroy 2009), in Gedarif (Elhadary and Samat 2012a), and in Dali and Mazmum (UNEP 2007).

The MFC, in one way or another, is held responsible for the expansion of unauthorized schemes, as it renews contracts for both authorized and unauthorized schemes (Elhadary 2007; Manger 2009). This led Manger (2009) to describe the MFC as an institution that has full power, using the laws and the system to establish enterprises that benefited the supporters of the state. Several authors have reached the same conclusion, accusing the MFC of allocating large mechanized farming schemes to investors, merchants, or people affiliated to or close to the government (Egemi 2006a; Elhadary 2010; Komey 2009). In Gedarif, for example, 64% of leaseholders of mechanized schemes are considered to be outsiders, and, astonishingly enough, most of them are traders (31%) or retired government officials, including civil servants and army and police officers (48%), with no agricultural background (Ijami 2006). Land grabbing by officials affiliated to the state, the military, private investors, land speculators, religious groups, and urban residents is increasing in Sudan (Pantuliano 2007, 2010). This means that merchants (outsiders), including the Mafia (Assal 2005 in Miller 2005), are the beneficiaries of agricultural "development," a number of them having joined the current regime to maintain their position and privilege. Although it was stated that no farmer should be allowed to "own" more than one scheme (4.2 km^2) as a maximum, the reality showed that one-third (32%) have more than 10 schemes, and in some cases the number reaches 30 schemes each (Elhadary 2010). At a time when land available for traditional producers is reduced or, at best, remains the same, the area under unplanned mechanized farming is increasing very rapidly. This phenomenon is not only confined to Gedarif State; similar cases of acquisition have occurred in other states. According to Manger (2009), in 1997 the grazing area of the Jawamaa and Bidariya tribe in Northern Kordofan state was given to a private company called the Malaysian–African Agriculture Company. While the traditional leadership was not consulted, there were rumors that some village sheikhs of settled farming groups had not objected to the concession, because the company had bribed them.

The seizing of land for mechanized farming has been facilitated by the fragile land tenure situation in the country. Several authors have declared that grabbing of traditional land will continue if no serious action is taken to address the issue of land tenure in the country. Under the guise of compulsory acquisition in the public interest with the rhetoric of providing national development, some agents of the state grab land from ordinary people and, in turn, give land to powerful interest groups such as investors, the rich, and cronies of governments (Elhadary and Obeng-Odoom 2013). More detail on the land tenure system and its role in land grabbing and expanding unplanned mechanized farming is given in the following section.

6.4 FRAGILITY OF LAND TENURE SITUATION IN SUDAN

Despite the effort made by many African governments, including Sudan, to introduce a modern land tenure system, the customary system is still valid, particularly in most African rural areas. For decades, African countries have sought to replace "customary" land tenure systems with a "modern" system of property rights, based on state legislation, on European concepts of ownership, and on land titling and registration (Cotula 2006). According to their opinion, customary land tenure was held not to provide adequate security of tenure, thereby discouraging investment and negatively affecting agricultural productivity (Wilson 1971). Critics claim to have shown that the process is mainly designed to benefit transnational corporations (TNCs) that "grab" land from local people, convert it from farmland, and turn it into investment land (Elhadary and Obeng-Odoom 2013). In addition to this, two other reasons might be that the state was not serious about registering communal land, especially that located in remote areas, because it was of no use at the time, or that the state did so intentionally, aiming to reserve it for use when needed (Elhadary 2010). It is important to note that the full conversion of communal ownership into an official system is not an easy task, and is always surrounded by several constraints These constraints include lack of legal awareness; complicated and lengthy land registration procedures; lack of adequate information; lack of financial resources and of institutional capacity in government agencies; and the difficulty of obtaining exclusive property rights in situations involving complex land use arrangements (Elhadary et al. 2011). These constraints have limited the outreach of state interventions, and, in Sudan, both systems of tenure are still in operation (Cotula 2006). As a result, very little rural land has been registered (across the continent formal tenure covers only between 2% and 10% of the land) (Deininger 2003), and customary land tenure systems continue to be applied in much of rural Africa. Almost 80% of the land in Sudan is held customarily (Babiker 2008).

Prior to the 1970s, most of the land in Sudan was accessed and "owned" through the communal system (unregistered) (Elhadary 2007), a system which has proven to be more effective, especially in rural areas (FAO 2008). This system offers the tribal leaders the power to manage and distribute land resources orally (with no written documents) over all members of their villages (Elhadary et al. 2011). This situation remained without noticeable change even after Sudan's independence in 1956. The turning point was in 1970, with a number of ill-planned initiatives, the consequences of which are still felt today (FAO 2008); among these was the Unregistered Lands Act of 1971.

The act of 1971 decreed that all unregistered land throughout the country, occupied or unoccupied, which was not registered before the commencement of this act, should be registered as government property, and granted the government the legality of disposing of lands as it saw fit (Elhadary 2007; FAO 2008). Not only that, as stated by Egemi (2006b) and Ayoub (2006), the act entitled the government to use force in safeguarding "its land," and this has been further strengthened by the 1991–1993 amendment of the 1984 Civil Transactions Act, which states that no court of law is competent to receive a complaint that goes against the interest of the state. The imposition of the 1971 act was followed by the abolition of the native administration system, a local body that was responsible for organizing access to and use of land in many rural areas. This action was taken by the government to ensure the suppression of communities or individuals that might resist the process of land grabbing and to disable their efforts (Komey 2009). As a result, tribal leaders have lost their historical position and lack the ability to regulate land use and manage inevitable land-related conflicts.

According to Cotula (2006), the central role of the state in land relations in most African countries was to promote agricultural development, on the one hand, and to control a valuable asset and a source of political power, on the other. In Sudan, for example, the act of 1971 was passed by the Nimeiri government (1969–1985) under pressure from the World Bank to serve the purposes of its ambitious plan to make Sudan "the bread basket of the Arab world" (Elhadary et al. 2011; Elhadary and Samat 2012b). For this reason, large-scale investments in both irrigated and mechanized rain-fed agriculture have been adopted in many states of Sudan (Elhadary 2007). As a result, large productive areas in rural lands have been taken and vested to investors, merchants, or people affiliated or close to the government, with no compensation or commitment to the communal right. There are many cases in which no compensation is paid to individuals, families, and communities whose land is taken. The reason is usually that the customary ownership of local land occupants is not legally recognized (Anseeuw et al. 2012). In this way, land as a source of wealth and power remains one of the main differentiating factors between the central and peripheral regions of postcolonial Sudan (Komey 2009). Grabbing land for public and private use under the deceiving pretext of "new development" has completely undermined the rights of rural people and had a negative effect on their livelihood. The rural poor are frequently being dispossessed of land and water resources under customary tenure (Anseeuw et al. 2012). The International Land Coalition estimates that about 83 million hectares of rural land have been taken over by investors in large-scale agriculture. There is evidence that the true owners of these land parcels have been neglected and unlawfully evicted (Molen 2013).

The vulnerability in land tenure is considered as one of the major root causes of many armed conflicts, social disputes, and land degradation in Sudan (Elhadary 2010; Sulieman 2013). This explains clearly why all the peace agreements that have taken place in Sudan (Comprehensive Peace Agreement [CPA] 2005, Darfur Peace Agreement [DPA] 2006, East Peace Agreement [EPA] 2006) have tried to bring to light the issue of land tenure (Elhadary 2010). For example, the CPA calls for the incorporation of customary laws and the establishment of Land Commissions to arbitrate claims, offer compensation, and recommend land reform policies. To date no proper action has been taken to address the land tenure; the agreements addressed several issues, such as the right to self-determination of the peoples of South Sudan, power sharing, oil and non-oil wealth sharing, democracy, and permanent ceasefire and security management, but left the vexed land question to be resolved at a later date (Shanmugaratnam 2008), as if wanting both to benefit from the current situation and to take land whenever there is a need (oil extraction, mechanized, or irrigated schemes) despite the existence of agreements. In this regard, Komey (2009) states that, despite the fact that the CPA provides some mechanisms for settling land-related issues in the postconflict era, the current difficulties facing the implementation of the agreement raised great fear among the local Nuba peoples as to whether their customarily owned land is going to be safeguarded by the agreement or is going to experience further grabbing. This inquiry has been answered by Anseeuw et al.

(2012), who state that, without transparency, accountability, and open debate, decision making over land will continue to be swayed by vested interests at the expense of rural land users.

From what has been said, this study affirms that vulnerability of land tenure has led to, among other things, lack of access to credit, particularly to traditional farmers; environmental degradation, fueling tension between land users, as in Darfur; and an open door to unplanned expansion of mechanized agriculture. In turn, the unauthorized expansion has had a remarkable negative impact on the socioecological condition of the country. More details are given in the following section.

6.5 IMPACT OF SEMIMECHANIZED FARMING ON ENVIRONMENT AND DEVELOPMENT

Land taken by mechanized farming was not vacant; rather, it supported either pastoralism, traditional rain-fed agriculture, or wild habitats, principally open woodlands and treed plains (FAO 2008). Thus, unplanned mechanized farming has deprived many traditional land users from accessing their rights and uses, leading to severe socioecological constraints and, in some cases, bloodshed and conflict. Large areas under capital-intensive land use, such as the mechanized farming schemes in eastern Sudan, have been severely degraded (FAO 2004). Symptoms and manifestations of degradation include the clearance of forests, depletion of palatable vegetation species, soil deterioration, land degradation (desertification), and a collapse in production yield. More details on this are given in the following subsection.

6.5.1 Clearance of Forest

According to Nori (2013), forests in Sudan occupy an area of about 69,949,000 ha, representing about 29.4% of the country's area (including the forested area in South Sudan). Of this area, 20% is classified as primary forest, 71% as naturally regenerated forest, and only 9% as planted forest (FAO 2010). Following the split, the Republic of Sudan has, according to FAO's classification, become a low-forest cover country, with about 11% of its total surface area under forest cover. Yet, the high dependency on forest products and service remains as it was before separation (Gaffar 2011). Forestry elsewhere, despite its role in sustaining the environment, plays a vital role in producing fuel, wood, charcoal, and building materials. Moreover, especially in Sudan, forest is considered as one of the main sources of gum arabic from the Acacia Senegal tree, a product that contributes much to the country's economy. Sudan is the world's largest producer of gum arabic (Abu Raida 2013), which is a source of livelihood for a number of people, particularly in the rural and remote areas of Sudan (Sulieman 2013).

Presently, the forest is being degraded very rapidly, exceeding the average national clearance rate, and the country has gained the reputation of the highest deforestation rate in Africa (FAO 2003). It has been estimated that in 1970 the country's forests and woodlands totaled approximately 915,000 km², or 38.5% of the land area; by 2010, this had decreased to 69,949,000 ha, constituting 28% of the country's land area of 250,581,000 ha (FAO 2008). This implies that Sudan lost 6,432,000 ha between 1990 and 2010, and an average of 643,000 ha annually (see Table 6.2 for more details). This is in line with UNEP (2007) indicating that between 1990 and 2005 the country lost 11.6% of its forest cover.

The degradation of forest cover has occurred in several states in the country, but at a particularly high rate in the Gedarif of the eastern Sudan, a region that witnessed the birth of mechanized farming. According to Sulieman and Ahmed (2013), the area of natural vegetation in eastern Sudan was reduced from 26.1% in 1979 to 12.6% in 1999, and further to 9.4% in 2007. In Gedarif State, for example, the forest area has been degraded from 32% in 1985 to only 15% in 2007 (Elhadary 2007). This is in line with Sulieman (2010), who states that the average natural vegetation clearing rate in

Table 6.2   Area of Forest and Other Wooded Land in Sudan
(North and South) in 2010 Area

Type	1990	2000	2005	2010
Forest	76,381	124,644	67,546	69,949
Other wooded land	58,082	54,153	52,188	50,224
Total land area	237,600	237,600	237,600	237,600
Inland water bodies	12,981	12,981	12,981	12,981
Total area of country	250,581	250,581	250,581	250,581

Source: FAO (Food and Agriculture Organization), Global Forest
Assessment—Sudan, FAO, Rome, 2010.

Note: Values are x1000 ha.

Gedarif was around 0.8% per year, and the most rapid clearing occurred during the 1970s, when the conversion rate to mechanized agriculture increased to about 4.5% per year.

Several factors have been implicated in the reduction of forest area in the country, among which several authors blamed the unplanned expansion of mechanized farming (Elhadary 2007; Sulieman 2013; USAID 2012). According to USAID (2012), forest ecosystems throughout Sudan have been degraded due to drought, fire, uncontrolled grazing, overcutting, and encroachment by agriculture. Arfat (2012) states that the change in forested land is mainly related to small-scale shifting agriculture or land clearing by landlords who want to expand agricultural land. This study is in line with Sulieman and Elagib (2012) and Sulieman (2013), who affirm that the establishment of large-scale mechanized farming has taken place on land that previously used to be covered with natural vegetation (forest and pasture). Mechanized farming has not only invaded forest and rangeland areas but also encroached on legally protected areas like Dinder National Park in Sudan (FAO 2008).

The taking of forestland is not confined to Africa; some Asian countries have witnessed the same grabbing of customary land, but the rate of grabbing is highest in Africa. Millions of hectares of customary forestland are being taken to produce oil palm in Indonesia and parts of Malaysia (Colchester 2011). There is dispute about the accuracy of existing figures on the extent of land taken. It has been reported that total land acquisitions worldwide have reached 134 million hectares (Anseeuw et al. 2012). Approximately 67 million hectares of this are located in Africa, with 43 million hectares reported for Asia, 19 million hectares in Latin America, and 5.4 million in Eastern Europe and Oceania. The high levels of interest in acquiring land in Africa, as well as in Sudan, appear to be driven by a perception that large tracts of land can be acquired from governments with little or no payment (Anseeuw et al. 2012). Land grabbing in Sudan is not a twenty-first-century phenomenon; it goes back to the nineteenth century, it is still going on today, and there are signs that it will continue in the future (Babiker 2011).

Besides the destruction and reduction of forest area, mechanized farming is indirectly responsible for destroying the environment. This is because it generates difficulties for pastoralists in passing along their traditional animal routes, thus forcing them to concentrate in small areas for a longer time or to invade forest land (Elhadary 2007). In Gedarif State, for example, pastoralists have to wait until farmers have harvested their crops, leaving them no option but to bring their livestock into the forest for survival, thus causing serious degradation. This clearly explains the severe degradation of El Rawashda forest in the eastern Sudan, which is the inevitable public grazing for pastoralists during rainy seasons (Elhadary 2007, 2010). Moreover, the use of unsound tillage practices (Eltayeb 1985; Eltayeb and Lewandowski 1983; Sulieman 2013) and cultivating the same monocrops continuously without a fallow period has led to a massive change in the structure of natural vegetation. Some trees and grasses have disappeared, such as *Belepheris edulis* Siha, and some trees that are harmful to the environment have invaded the area, such as *Prosopis glandulosa* (mesquite) and *Striga hermonthica* Boda (Elhadary 2007).

Several strategies have been adopted for renewable natural vegetation in the country. Among these is the requirement to plant a 10% shelterbelt around each mechanized scheme. Inclusion of shelterbelts in the mechanized farming system started in 1994 (Glover 2005). This implies that, of every 420 ha, 10% (42 ha) was expected to be planted with trees. However, these measures, although insufficient to bring back the degraded forest, have not yet been fully enforced, and the application was by and large ignored (FAO 2008). Reasons for this failure include lack of incentive and (at the same time) controversy over rights; farmers believe that trees around the scheme are their "own" private property, while in theory they belong to the government, as the Forests National Corporation (FNC) provided the seeds and technical support (Glover 2005). Moreover, farmers prefer not to have trees on their farms, as trees attract birds and insects and become obstacles to using machines for cultivation (Elhadary 2007). Deforestation, especially in a fragile ecosystem, has become a serious environmental problem in Sudan. The removal of trees for mechanized farming has had a remarkable negative impact on other natural resources, such as the fertility of the soil and the condition of biodiversity. Soil erosion, loss of soil fertility, and salinization are some of the main consequences of deforestation (Cacho 2001).

6.5.2 Soil Depletion and Collapse of Yield Production

Expansion of mechanized cultivation at the expense of forest area is one of the major forces driving land-cover change, not only in Sudan but in the whole Sudano-Sahelian zone of Africa (Babiker 2011). This, coupled with the practice of mechanized agriculture, has increased the rate of soil deterioration, leading to a sharp decrease in crop production. According to Ayoub (1998), about 64 million hectares of soil is degraded in Sudan. Several authors have concluded that soil, mainly in mechanized areas, has been seriously degraded and has lost its fertility (Ayoub 1998; Elhadary 2007; FAO 2008). According to Ayoub (1998), 81% of the total area of degraded soil is in the susceptible drylands (arid, semiarid, and dry subhumid). Most of the degradation (74% of the total degraded soils) is in the arid and semiarid zones, but significant percentages of land are also degraded in the dry subhumid and moist subhumid zones.

In Gedarif, for example, the massive expansion of rain-fed cultivation increases the exposure of soil surfaces to wind and water erosion, leading to more land degradation (Elhadary 2007). Sulieman and Ahmed (2013) affirm that mechanized farming and overgrazing are the major factors behind the increase of bare land in the Butana of Northern Gedarif. For them, mechanical working of the shallow soils, using tractors and wide-level discs, has led to mechanical soil damage, and, in many cultivated areas, gravels appeared on the top soil due to fluctuation of rain. The same conclusion has also been reached by Biro et al. (2013), who state that the chemical and physical properties of soil have deteriorated due to the changes in land use and land cover, mainly by the expansion of mechanized rain-fed agriculture. In the same line, Sulieman (2013) states that, by stripping away the vegetation cover with mechanized cultivation, the soil is laid bare to be carried away by water and wind erosion. The deterioration of soil fertility has been aggravated by the fact that farmers did not use fertilizers or organize crop rotation or fallow systems (FAO 2008). Cropping without appropriate nutrient inputs has degraded about 12 Mha, particularly in small-scale farming on sandy and loamy soils. When these processes of resource mismanagement coincided with the recent recurrent droughts, collapse of the economic base of fragile areas took place (Ayoub 1998).

The inevitable consequences of soil degradation in Sudan and elsewhere in Africa have been the appearance of some parasitic grasses such as *S. hermonthica* (Boda) and a dramatic decline in crop production (Elhadary 2007). Joel et al. (1995) reported that the *Striga* problem in Africa is intimately associated with intensification of land use. This is in line with Glover (2005), who stated that "*Striga* in Gedarif appears to thrive best on heavily used soils of less fertility; hence reduces crop productivity." This implies clearly that there is a clear link between deterioration of soil fertility and the reduction of crop production. Therefore, decline in yields of a crop may be an indicator that soil quality has changed, which in turn may indicate that soil and land degradation are also occurring (Stocking and Murnaghan 2000). This explains why Africa is the only region in the world where agricultural productivity is largely stagnant. Yields of maize and other staple cereals typically remain at about 1 tonne per hectare (1000 kg/ha), which is about one-third of the average achieved in Asia and Latin America (FAO 2008).

As in other African countries, in Sudan crop production is decreasing, and the country has not only lost its status as a breadbasket of the Arab world but has even failed to secure food for its entire population. This is due to large-scale degradation of its rich soil as well as other natural resources, mainly through unsuccessful land use policies and practices (Elhadary 2007, 2010). Glover (2005) affirmed that, since the advent of mechanization, sorghum yields per hectare have been constantly decreasing. In the eastern Sudan crop yields in the past, under the traditional system, were estimated at 10–20 bags of sorghum per hectare (900–1800 kg/ha) and about 750 kg/ha of sesame in the early 1960s. In the 1986/1987 season the sorghum yields had dropped to an average of 4–6 bags/ha (360–540 kg/ha) (Eltayeb 1985). This can be compared with yields from mechanized farming which range from 190 to 980 kg/ha for sorghum and 80–600 kg/ha for sesame (Eltayeb 1985). In Gedarif State, while the area under cultivation is increasing, the production per feddan (equal to 0.42 ha) decreased sharply from four sacks of sorghum in 1980 to 1.1 in 2005 and for sesame from 3.4 gonttar/feddan in 1980 to 1.5 in 2005 (1 gonttar = 100 lb = 44.9 kg).

This section argues that the role of the land tenure situation in soil deterioration should not be ignored. The vulnerability of the land tenure situation led farmers, "owners" or renters, to care about high profit in a short period rather than considering the sustainability of land use, as they lack confidence about having the scheme for the coming years. Without long-term ownership, farmers have little incentive for investment in and protection of natural resources (FAO 2008; Elhadary 2010). This explains why farmers always care about horizontal expansion rather than vertical expansion to maintain their output.

6.5.3 Notion of Desertification in Sudan

The phenomena of desertification affects arid lands all over the world, but tends to be concentrated in Asia and Africa, each of which accounts for 37% of all desertified land (Grainger 1990). The repeated removal of natural vegetation and continuous deterioration in soil fertility are the major factors and, at the same time, indicators of desertification. FAO (2001) reported that, overall, approximately four million hectares of African forests are cleared each year, to the extent that 45% of Africa's original forest cover has disappeared. Shifting cultivation accounts for 70% of the clearing of closed canopy forests and 60% of the cutting of savanna forests, thereby rendering the land vulnerable to desertification/land degradation.

In the case of Sudan, desertification has to do with the deterioration in soil fertility rather than removal of forest. According to Glover (2005), the desertification in the Sudan is more linked to soil degradation than to vegetation degradation. Symptoms of desertification, such as soil degradation, massive runoff, bare land, and reduction in crop production, have been observed in many parts of Sudan. Recently, some signs of desertification on a lower scale have prevailed in states like Gedarif (Elhadary 2007), Kassala, and North Kordofan State (FAO 2008). This is due, among other things, to rapid expansion of mechanized cultivation (Elhadary 2007), the nation's quest for fuel wood (Ayoub 1998), and repeated cultivation with the same crops. However, large-scale mechanized farming has been the main factor contributing to deforestation and consequent land degradation (FAO 2001).

If no serious action is taken to mitigate desertification, a large productive area will be lost, especially in the reality of global warming. According to UNEP (2007), between the mid-1970s and late 2000s, summer rainfall decreased by 15%–20% across parts of Sudan, placing already food-insecure populations at greater risk. Due to climatic changes, crop yields, mainly in Africa, may fall by 10%–20% in 2050 (Jones and Thornton 2003). A climate sensitivity analysis of agriculture concluded that some African countries would virtually lose their entire rain-fed agriculture by 2100 (Mendelsohn et al. 2000). The consequences of desertification are worse, and even beyond human imagination, as they negatively affect all aspects of the socioeconomic and political life of the country. In Sudan, desertification is clearly linked to conflict, as there are strong indications that the hardship caused to pastoralist societies by desertification is one of the underlying causes of the current war in Darfur (UNEP 2007).

6.6 DISCUSSION: MECHANIZATION AND ITS SOCIOECOLOGICAL IMPACT

Agriculture, including mechanized farming, plays a crucial role in shaping the economy of most African countries, securing food and generating jobs for the overwhelmingly rural population. Fifty-eight percent of the active workforce in Africa is employed in agriculture, while 83% of the population depend on farming for their livelihood (FAO 2008). The Sudanese economy is predominantly agricultural (including crop and livestock production, forestry, wildlife, and fisheries, which altogether contribute an average 44.4% of GDP). The agricultural sector contributes more than 85% of Sudan's export earnings and about 90% of the national food requirements. It accommodates more than 80% of the population (Abu Raida 2013). Moreover, in the1980s one–two million laborers moved to the mechanized farming areas, where they could find three or four months' work at weeding and harvest time (UNEP 2007). This situation is no longer sustainable, and the past contribution of agriculture to socioeconomic life ought to be revised. Currently, most land under agricultural mechanization has experienced severe degradation, as highlighted in the previous sections, and agriculture has failed to meet its socioeconomic objectives. In turn, it has become a major factor fuelling land use conflict, especially in a country like Sudan with multiple ethnicities, religions, and races. Two scenarios have been highlighted to evaluate and measure the current situation of mechanized farming. These are the total number of farmers and the farmers–herders relation. The number of farmers and agricultural laborers involved in mechanized farming is decreasing very sharply. Only about 9800 farmers "owned" mechanized schemes in Gedarif in 2007, compared with the total population of 1,336,662 persons (CBS 2011). This implies that only 5% of the total population are benefiting directly from mechanized farming. This, together with the introduction of machines in cultivation, has thrown light on the plight of the seasonal workers. Today, most of the big farmers depend on caretakers (Wakeel) to follow cultivation processes; thus, their role is reduced to funding and remote supervision (Elhadary 2007). Regarding employment, a mechanized scheme of 4200 ha needs only five people as permanent workers and around 35 persons for the process of weeding and harvesting. This figure implies that only a few people depend on mechanized farming for a secure livelihood, and rejects the notion that agriculture employs 80% of the labor force in the country. According to the UNEP (2007), in Sudan 70% depends on traditional rain-fed farming, 12% on irrigated agriculture, and only 0.7% on mechanized agriculture.

Although most of the unplanned mechanized farming expansion has taken place in areas that have traditionally belonged to the local communities, they have not gained much from the utilization of their lands. In southern Gedarif, for example, 84,000 ha have been taken from traditional producers and demarcated for only 200 investors, leaving around 3750 families landless, as only 7%, around 5880 ha, was given to 350 local families (Ijami 2006). Furthermore, mechanized farming is still today creeping into pastoral routes; in Gedarif State, six out of eight traditional pastoral routes have been closed, and their boundaries are not clear due to unorganized expansion (Elhadary 2010; Sulieman 2013). However, the most dangerous environmental impact is that agriculture is now moving into more ecologically marginal areas in the northern Gedarif, exceeding the grazing line: a line that separates agriculture from rangeland areas, as the northern part (arid and semiarid land) of Gedarif is reserved for grazing, while agriculture is allowed in the southern part (humid).

Elhadary (2007) collected evidence that mechanized farming is moving behind the grazing line and has invaded most of the northern area. Astonishingly, nobody is held responsible for this shift, but some rumors have blamed lobbying by big farmers in Gedarif (Elhadary 2010). This illegal push of mechanized agriculture into more ecologically fragile areas means two things: that environmental laws tend to be violated by mechanized famers, and that there is no land left in the south for further horizontal expansion (Sulieman and Ahmed 2013).

Under the pretext of "development," communal lands have been taken and vested to outside investors, without taking into account the historical right of local communities and the nature of drylands. Unequal access to land remains one of the fundamental causes that contribute to the grievance and protracted conflict in most African countries (Elhadary and Samat 2011b). It has been revealed that the reallocation of the grabbed communal land to outsiders remains one of the essential factors behind livelihood insecurity, grievance and conflict, marginalization, and spread of poverty among rural communities. This explains clearly why most pastoral areas, especially in Africa, have recently witnessed severe conflicts and bloodshed. In Sudan, for example, without ignoring the role of ethnicity and environmental problems in escalating the tension, the ongoing conflicts in several parts of the country are political in nature and related in way or another to land rights. This idea is supported by Ayoub (2006), who states that Sudan's conflicts have many causes, but at the root of each conflict are questions over the control and distribution of resources. The most important resource is land: whether used for agriculture, cattle herding, or subterranean resources such as oil or water, land ownership is the key to wealth and power. According to USAID (2012), environmental degradation and competition for limited natural resources has been a contributing cause to conflict in the region. Thus, the unorganized expansion of mechanized farming is considered one of the fundamental reasons why the country has earned the reputation as a home of bloody civil wars, and the country is unlikely to see lasting peace until such issues have been addressed (Sulieman 2013). In view of these challenges, this chapter has pointed out clearly that all the three peace agreements that have taken place in Sudan do not address the issue of communal land right effectively enough. In addition, the law of access to land is still vague and unclear.

It is worth nothing that Sudan has a federal Ministry of Agriculture and a Ministry of Agriculture in each state. These ministries lack technical and financial capacity (UNEP 2007), and they are doing politics rather than fulfilling their objective of formulating policies and developing sustainable agriculture. This explains why most of the agricultural ministries are affiliated to the current regime and do not even have a background in agriculture; the current minister is a medical professional. Furthermore, the linkages between the agricultural ministry and relevant ministries such as the forest, livestock, and environment ministries are weak. These institutional constraints have led to unsound agricultural policy, which has often ignored the nature of drylands and the need for small-scale farmers. The primary focus is on ways of supporting big farmers to obtain loans and machines, while small farmers are completely neglected or left to struggle till they are fed up and give up farming.

Given the nature of drylands, agriculture is not always the best practice in such marginal areas. Other land uses might be more profitable and suitable to the environment. In this regard, White et al. (2002) mentioned that many people have a "tradition" in agriculture that is not always matched by a similar attitude toward other land uses such as forestry, wildlife ranching, or ecotourism, all of which have become profitable enterprises in many dryland regions of the world. The failure to address the needs of the local people and the creation of unprecedented ecological problems are sufficient justification for calling mechanized agriculture an illusion of development rather than sustainable development (Elhadary 2010). Therefore, supporting and encouraging small-scale farming is best suited to the situation of drylands. According to Pearce (2012) cited in Obeng-Odoom (2013), smallholder farmers in either the urban or rural setting in Africa and Asia, and indeed elsewhere, are the answer to the ecological crises in the agrarian sector. Further, he offers evidence on how small-scale farming has led to increased income for farmers and an increase in food production.

In urban areas, Pearce shows how the use of urban agriculture can create greater interaction between the city, the periurban area, and the countryside.

6.7 CONCLUSIONS

This chapter is about the ecological impact of mechanized farming on the drylands of eastern Sudan. The analysis has shown that unplanned mechanized agriculture has expanded very rapidly at the expense of traditional farming, forest, and rangeland. Weak governmental institutions coupled with the vulnerable system of land tenure have accelerated unauthorized expansion and facilitated land grabbing in the country. Desertification, deforestation, soil depletion, and reduction in crops are some of the socioecological consequences of unorganized expansion. In the prevailing climatic changes, these ecological constraints have become a major threat to agriculture and have become one of the fundamental factors that fuel conflicts in several parts of Sudan. Addressing the issue of land tenure, legalizing unplanned mechanized farming, and sustaining agriculture are greatly needed if food insecurity and internal conflicts are to be avoided in the future. This chapter recommended that, in situations where no productive land is left, vertical expansion is essential in achieving food security, poverty alleviation, and overall sustainable development. In the fragile ecosystem of drylands, selection of suitable land use type must be scientifically based, as agriculture is not always the best choice in such a harsh environment. It is time to expand the agribusiness industry and consider agriculture as a coequal partner with industry, rejecting the old views that saw agriculture as a passive agent in socioeconomic development.

6.8 SUMMARY

The Sudanese economy is predominantly agricultural; the sector contributes close to 44.4% of GDP. Fifty-eight percent of the active workforce is employed in agriculture, while 83% of the population depend on farming for securing their livelihood. Mechanized or semimechanized rain-fed farming was initiated in Sudan (Gedarif State) in the 1940s, during British colonization. Since then it has spread very rapidly all over the country, and currently 6.5 million hectares is under mechanized cultivation. Despite its positive role in the economy, mechanized farming has been presented as a serious source of environmental problems in the country. To measure the ecological impact of this sector and trace its development, a literature review has been carried out, and the results are reviewed in detail. The study argues that mechanized farming is not the primary culprit responsible for massive land destruction. Rather, unplanned expansion, seeking high profit, and caring less about sustainability are the essential factors to be blamed. These factors have been exacerbated by the fragile land tenure system in the country. The study concluded that securing land rights; considering the ecological nature of drylands in the scheme's distribution; enforcing all environmental laws, particularly the shelterbelt of 10% around each scheme; bringing back natural vegetation; and creating sound cooperation between stakeholders dealing with natural resources (agriculture, forest, environment) are greatly needed.

REFERENCES

Abu Raida, K. M. M. 2013. Prospects for modernization of agriculture in Sudan with an emphasis to food security. *J Agric Sci Rev* 2:39–50.
Anseeuw, W., L. A. Wily, L. Cotula, and M. Taylor. 2012. *Land Rights and the Rush for Land: Findings of the Global Commercial Pressures on Land Research Project*. Rome: International Land Coalition (ILC).

Arfat, Y. 2012. *Land Use/Land Cover Change Detection and Quantification—A Case Study in Eastern Sudan.* Lund, Sweden: Earth and Ecosystem Sciences, Lund University.

Assal, M. 2005. Economy and politics in Gedaref: A symbiotic encounter. In C. Miller (ed.), *Land, Ethnicity and Political Legitimacy in Eastern Sudan.* Cairo/Khartoum: CEDEJ/DSRC.

Ayoub, M. 1998. Extent, severity and causative factors of land degradation in the Sudan. *J Arid Environ* 38:397–409.

Ayoub, M. 2006. Land and conflict in Sudan. In M. Simmons and P. Dixon (eds), *Peace by Piece Addressing Sudan's Conflicts.* London: Conciliation Resources.

Babiker, M. 2008. Communal land rights and peace-building in North Kordofan: Policy and legislative challenges. Sudan Working Paper. SWP 2008:3. Bergen: Chr. Michelsen Institute.

Babiker, M. 2011. Mobile pastoralism and land grabbing in Sudan: Impacts and responses. In International Conference on Future of Pastoralism. 21–23 May, Addis, Ethiopia. University of Sussex.

Biro, K., B. Pradhan, M. Buchroithner, and F. Makeschin. 2013. Land use/land cover change analysis and its impact on soil properties in the northern part of Gedarif region, Sudan. *Land Degrad Dev* 24:90–102.

Cacho, O. 2001. An analysis of externalities in agroforestry systems in the presence of land degradation. *Ecol Econ* 39:131–143.

CBS (Central Bureau of Statistics Sudan). 2011. Government of Sudan. http://www.cbs.gov.sd/. Retrieved October 2013.

Colchester, M. 2011. Palm oil and indigenous peoples in South East Asia. FPP Contribution to ILC. Collaborative Research Project on Commercial Pressures on Land, Rome.

Cotula, L. 2006. *Changes in "Customary" Land Tenure Systems in Africa.* London: International Institute for Environment and Development (IIED).

Deininger, K. 2003. *Land Policies for Growth and Poverty Reduction World Bank Policy Research.* New York: World Bank.

De Jode, H. 2010. *Modern and Mobile: The Future of Livestock Production in Africa's Drylands.* London: International Institute for Environment and Development (IIED) and SOS Sahel International.

Egemi, O. 2006a. Land tenure in Sudan: Challenges to livelihood security and social peace. In G. E. El Tayeb (ed.), *Land Issue and Peace in Sudan*, p. 13028. Khartoum: Sudan Environment Conservation Society.

Egemi, O. 2006b. Sudan: Land and peace processes in Sudan. In M. Simmons and P. Dixon (eds), *Peace by Piece Addressing Sudan's Conflicts*, pp. 34–54. London: Conciliation Resources.

Elbashir, A., F. A. Siddig, A. Ijaimi, and H. M. Nour. 2004. *Sudan Poverty Reduction and Programs in Agriculture.* Rome: Food and Agriculture Organization (FAO).

Elhadary, Y. A. E. 2007. *Pastoral Adaptation and Socio-Economic Transformations in the Butana Area—Al Gedarif State, Sudan.* Khartoum: University of Khartoum.

Elhadary, Y. A. E. 2010. Challenges facing land tenure system in relation to pastoral livelihood security in Gedarif State, Eastern Sudan. *J Geogr Reg Plann* 3:208–218.

Elhadary, Y. A. E. and N. Samat. 2011a. Political economy and urban poverty in the developing countries: Lessons learned from the Sudanese experience. *J Geogr Geol* 3:63–76.

Elhadary, Y. A. E. and N. Samat. 2011b. Pastoral land rights and protracted conflict in eastern Sudan. *J Pan Afr Stud* 4:74–90.

Elhadary, Y. A. E. and N. Samat. 2012a. Managing scarcity in the dryland of the Eastern Sudan: The role of pastoralists' local knowledge in rangeland management. *Resour Environ* 2(1):55–66.

Elhadary, Y. A. E. and N. Samat. 2012b. Political economy and urban poverty in the developing countries: Lessons learned from Sudan and Malaysia. *J Geogr Geol* 4:212–223.

Elhadary, Y. A. E., N. Samat, and R. Hasni. 2011. Land use changes and its impact on local communities at Seberang Perai Tengah of Penang, Malaysia. Issues in peri-urban regions and ways towards sustainable peri-urban futures. Conference Room, School of Humanities, Penang, Malaysia: Universiti Sains Malaysia.

Elhadary, Y. A. E., N. Samat, and F. Obeng-Odoom. 2013. Development at the peri-urban area and its impact on agricultural activities: An example from the Seberang Perai Region, Penang State, Malaysia. *Agroecol Sustain Food Syst* 37:834–856.

Eltayeb, G. E. 1985. *Environmental Management in the Sudan: Gadarif District Study Area.* Khartoum: Institute of Environmental Studies, University of Khartoum.

Eltayeb, G. E. and A. M. Lewandowski. 1983. Environmental monitoring: Environmental degradation in Gedarif district. In *Proceeding of the National Seminar on Indicators of Environmental Changes and Degradation in Sudan.* Khartoum: University of Khartoum.

FAO (Food and Agriculture Organization). 2001. Global forest resources assessment 2000. Forestry Paper. Rome: FAO.

FAO (Food and Agriculture Organization). 2003. Forest outlook study, sub-regional report, North Africa. Rome: FAO.

FAO (Food and Agriculture Organization). 2004. Carbon sequestration in dryland soils. Rome: FAO.

FAO (Food and Agriculture Organization). 2008. Agricultural mechanization in Africa. Time for action. Rome: FAO.

FAO (Food and Agriculture Organization). 2010. Global forest assessment—Sudan. Rome: FAO.

Gaffar, A. 2011. *Forest Plantation and Woodlots in Sudan*. Nairobi: African Forest Forum.

Gibbon, P., K. J. Havnevik, and K. Hermele. 1993. *A Blighted Harvest: The World Bank and African Agriculture in the 1980s*. Trenton: Africa World Press.

Glover, E. K. 2005. *Tropical Dryland Rehabilitation: Case Study on Participatory Forest Management in Gedarif, Sudan*. Helsinki: Faculty of Agriculture and Forestry, University of Helsinki.

Grainger, A. 1990. *The Threatening Desert: Controlling Desertification*. London: Earthscan Publications.

Harrison, N. and J. Jackson. 1958. *Ecological Classification of the Sudan*. Khartoum: Forest Department, Ministry of Agriculture.

Ijami, A. 2006. Mechanized farming and conflict in Sudan. In G. E. El Tayeb (ed.), *Land Issue and Peace in Sudan*, pp. 25–39. Khartoum: Sudan Environment Conservation Society.

Joel, D. M., J. C. Steffens, and D. E. Matthews. 1995. Germination of weedy root parasites. In J. Kigel and G. Galili (eds), *Seed Development and Germination*, pp. 567–597. New York: Marcel Dekker.

Jones, P. G. and P. K. Thornton. 2003. The potential impacts of climate change in tropical agriculture: The case of maize in Africa and Latin America in 2055. *Glob Environ Change* 13:51–59.

Komey, G. K. 2009. Communal land rights, identities and conflicts in Sudan: The Nuba question. The human rights dimensions of land in the Middle East and North Africa, Cairo, Egypt. 10–12 May.

Leroy, M. 2009. *Environment and Conflict in Africa—Reflections on Darfur*. University for Peace, Africa Programme, Addis Ababa, Ethiopia.

Manger, L. 2009. Resource based conflicts in Western Sudan–Some reflections on the role of the state. In M. Leroy (ed.), *Environment and Conflict in Africa: Reflections on Darfur*, pp. 242–252. Addis Ababa: University of Peace Press.

Mendelsohn, R. and A. Dinar. 2000. Climate change impacts on African agriculture. Mimeo, Yale University.

Miller, C. 2005. Power, land and ethnicity in the Kassala-Gedaref States: An introduction. In C. Miller (ed.), *Land, Ethnicity and Political Legitimacy in Eastern Sudan*, pp. 3–58. Cairo: Centre d'Études et de Documentation Économiques, Development Studies and Research Centre (CEDEJ/DSRC).

Molen, P. V. D. 2013. *How to Stop Land Grabbing in Africa*. Lemmer, The Netherlands: Geoinformation Management GIM, Geomares.

Mustafa, R. H. 2006. *Risk Management in the Rain-Fed Sector of Sudan: Case Study, Gedaref Area Eastern Sudan*. Giessen, Germany: Institute of Agricultural and Food System Management, Faculty of Agricultural Sciences, Home Economics and Environmental Management, Justus Liebig University.

Mwangi, E. and S. Dohrn. 2006. *Biting the Bullet: How to Secure Access to Drylands Resources for Multiple Users?* Washington DC: International Food Policy Research Institute.

Nori, W. M. T. 2013. *Detection of Land Cover Changes in El Rawashda Forest, Sudan: A Systematic Comparison*. Dresden, Germany: Faculty of Forest, Technical University of Dresden.

Obeng-Odoom, F. 2013. The grab of the world's land and water resources. *Braz J Polit Econ* 33:527–537.

Omer, A. M. 2011. Agriculture policy in Sudan. *Agric Sci Res J* 1:1–29.

Pantuliano, S. 2007. *The Land Question: Sudan's Peace Nemesis*. London: Overseas Development Institute.

Pantuliano, S. 2010. *Oil, Land and Conflict: The Decline of Misseriyya Pastoralism in Sudan*. London: Overseas Development Institute.

Pearce, F. 2012. *The Land Grabbers: The New Fight over Who Owns the Earth*. Boston: Beacon Press.

Scoones, I. 1994. *Living with Uncertainty: New Directions in Pastoral Development in Africa*. London: Intermediate Technology Publications.

Shanmugaratnam, N. 2008. Post-War development and the land question in South Sudan. In *International Symposium on Resource under Stress*. Kyoto, Japan: African Centre for Peace and Development, Ryukoku University.

Stocking, M. and N. Murnaghan. 2000. *Land Degradation Guidelines for Field Assessment*. London: Earthscan Publication.

Sulieman, H. M. 2013. Land grabbing along livestock migration routes in Gadarif state, Sudan: Impacts on pastoralism and the environment. Paper # 19. The Hague: The Land Deal Politics Initiative. The International Institute of Social Studies, Erasmus University.

Sulieman, H. M. and A. G. M. Ahmed. 2013. Monitoring changes in pastoral resources in eastern Sudan: A synthesis of remote sensing and local knowledge. *Pastor Res Policy Pract* 3:1–16.

Sulieman, H. M. and N. A. Elagib. 2012. Implications of climate, land-use and land-cover changes for pastoralism in Eastern Sudan. *J Arid Environ* 85:132–141.

UNCCD (United Nations Convention to Combat Desertification). 1999. United Nations Convention to Combat Desertification in those countries experiencing serious drought and/or desertification particularly in Africa. Paris: United Nations.

UNEP (United Nations Environment Programme). 1992. *World Atlas of Desertification*. London: Edward Arnold.

UNEP (United Nations Environment Programme). 2002. Global Environment Outlook 3 (GEO-3). United Nations.

UNEP (United Nations Environment Programme). 2007. *Sudan Post-Conflict Environmental Assessment*. Nairobi: United Nations.

USAID (United States Agency for International Development). 2012. *Sudan Environmental Threats and Opportunities Assessment with Special Focus on Biological Diversity and Tropical Forest*. Washington, DC: Management System International.

White, R. P., D. Tunstall, and N. Henninger. 2002. *An Ecosystem Approach to Drylands: Building Support for New Development Policies*. Washington, DC: World Resources Institute.

Wilson, R. J. A. 1971. Land tenure and economic development—A study of the economic consequences of land registration in Kenya's smallholder areas. *Stat Soc Inquiry Soc Ireland* 22:124–151.

CHAPTER 7

The Paradox of Arable Weeds
Diversity, Conservation, and Ecosystem Services of the Unwanted

Jaime Fagúndez

CONTENTS

7.1   Introduction 139
7.2   Recent Trends in Weed Diversity 140
7.3   Agriculture Intensification and Weed Response 142
7.4   Counteracting the Biodiversity Loss 143
7.5   Conclusions 144
Acknowledgments 144
References 144

7.1 INTRODUCTION

Traditionally, arable weeds have been considered a major problem for agriculture, and strong efforts have been made in their control and eradication. Crop rotation and regular plowing are traditional practices that keep a low density of arable plants but do not eradicate them. In modern times, other tools for weed control have been developed, with the aim of increasing crop yields and raising profits. Chemical weed control turned out to be extraordinarily efficient and traditional practices have been abandoned or reduced in favor of a generalized use of herbicides causing a massive loss of weed abundance. In increasingly industrialized agricultural systems, it is now recognized that a sterile and homogeneous agricultural landscape does not provide the most efficient land use in terms of yields and environmental benefits (Jackson et al. 2007), and a demand for more sustainable land use is emerging.

Noncrop arable plants are responsible for yield reduction by, for example, competition with crops and allelopathy (Dale and Polasky 2007; Zhang et al. 2007). However, new approaches to the study of the weed community have highlighted the benefits obtained from these plants to both the environment and human well-being (Marshall et al. 2003; Sotherton 1998; Storkey et al. 2012). Weeds are a key part of the so-called agroecosystems, which comprise the ecological framework of the human-created environment in agriculture-dominated landscapes. These plants provide a high level of provisioning as well as regulating and supporting ecosystem services (Fiedler et al. 2008;

Firbank et al. 2008; Power 2010; Zhang et al. 2007). The paradox emerges that weeds are valued as an essential element of these agroecosystems, but strategies are needed to maintain density and abundance below a reasonable threshold (Bohan and Haughton 2012).

The weed community contributes a series of functions and services to the agroecosystem such as food for different trophic groups, carbon sequestration, soil stability, and canopy cover (Dale and Polasky 2007; Power 2010). Phytophagous arthropods feed on them (Bàrberi et al. 2010), as do several birds and small mammals (Holland et al. 2006; Orlowski and Czarnecka 2007; Smart et al. 2000; Wilson et al. 1999), thus regulating a biotic system that contributes to biological control and reduces costs. Pollinators are directly affected by changes in the weed community, which affects production in some crop species (Gabriel and Tscharntke 2007; Gibson et al. 2006). The soil biota is closely related to weeds by their root system and canopy cover, but mainly by the persistent seed bank (Dekker 1999; Franke et al. 2009). The weed seed bank plays a key role in the complex interaction of the soil biota, such as food for worms and nematodes. However, the seeds may also act as reservoirs for pathogens and vectors for plant diseases that can be controlled with appropriate crop rotation practices (Franke et al. 2009).

Weeds also contribute to soil stability and nutrient cycling, avoiding erosion, and nutrient leaching and runoff. They participate in regulating the hydric cycle and a proper soil water balance, retaining water and acting as living reservoirs ("green water," Power 2010). Finally, weeds also have an intrinsic value as part of the ecosystem and for their cultural and aesthetic value (Swift et al. 2004). To a certain extent, weeds may also be a positive agent in crop production. The proper management of weeds may raise yields by acting as an alternative source for herbivores and parasites (Fiedler et al. 2008; Schellhorn and Sork 1997), avoiding the intense use of pesticides. Other indirect benefits from arable weeds have been highlighted, such as food and shelter provision for game species (Belda et al. 2011).

A generalized process of agricultural intensification in developed countries has been ongoing in recent decades, especially in the temperate regions of the northern hemisphere such as the European Union (EU) (Benton et al. 2003; Stoate et al. 2001; Storkey et al. 2012). Evidence shows that this process is affecting the stability of agroecosystems, and that this will have strong social and economic effects in our societies (Bohan et al. 2013; Gerowitt et al. 2003; Power 2010; Stoate et al. 2009). Particularly, arable weeds are directly affected by changes in agricultural practices (e.g., tillage abandonment or reduction), but also by other indirect drivers such as loss of habitat heterogeneity (Tscharntke et al. 2005). Here, I summarize new scientific evidence of recent changes in arable plant diversity, focused mainly in the EU, in relation to general trends occurring in agricultural practices, landscape, and new environmental policies that have been established (i.e., Agri-Environment Schemes, AESs). Evidence of diversity loss, species and ecosystem services affected, and future challenges for the development of appropriate policies that may counteract these negative effects are discussed.

7.2 RECENT TRENDS IN WEED DIVERSITY

Several authors have claimed a catastrophic depletion of the weed flora throughout Europe during the last century (Hyvönen 2007; Potts et al. 2010; Robinson and Sutherland 2002; Storkey et al. 2010, 2012), but this requires close inspection and a specific assessment for each region and time period. A generalized loss of arable weeds took place throughout Europe in the first half of the twentieth century (Potts et al. 2010), coinciding with the widespread use of herbicides and the abandonment of other weed-control cultural practices such as crop rotation and regular plowing. Many studies describe a general loss in farmland biodiversity, specifically for plants. Fried et al. (2009) report a loss of 42% of the mean plant species richness across France in 30 years, while Cirujeda et al. (2011) found a significant decrease of the mean number of species in a 30 year study in cereal fields in northeastern Spain. A significant depletion in the mean number of arable weeds per field was found for Germany from 1957 to 2000, but the total number of weed species had increased (Baessler and Klotz 2006). Walker et al. (2009) found that roughly twice as many weed species decreased as increased in a 50 year period since the 1950s in southeastern England. Similarly, a higher percentage of species with lower numbers are found in recent surveys compared to old ones, although some species may increase and others remain unchanged (Robinson and Sutherland 2002; Smart et al. 2000; Wilson et al. 1999). Other cases report a decrease in species abundance rather than number of weed species, with a high number of species becoming rare in 60 years (Sutcliffe and Kay 2000).

On the other hand, a slight recovery has been reported in particular studies for recent decades as a result of the implementation of AESs. For example, recent environmental policies are responsible for the recovery of weeds in Denmark (Andreasen and Stryhn 2008), where higher frequencies for most species were recorded in the 2000s compared to the 1980s. However, the opposite trend has also been reported, as in Ireland, with lower weed diversity values in farms not included in AES areas (Feehan et al. 2005).

Although some authors establish a direct relationship between total species richness and rare species abundance (Sutcliffe and Kay 2000), it is not recommended to assess habitat quality loss using changes in total species richness, whereas measures of the presence and abundance of typical arable weeds are the most accurate indicators for habitat quality in agroecosystems (Albrecht 2003; Hawes et al. 2010). Whenever this approach has been used, there is evidence for a lower frequency and density of the target species in intensified agroecosystems (Fried et al. 2009; Salonen et al. 2013). Thus, the species-specific response to intensification must be addressed individually, since even opposite trends may be found for different species or time periods (Hyvönen and Huusela-Veistola 2008).

Despite general trends in species richness and diversity, markedly different effects have been described over groups of species by ecological requirements, functional types, or rarity (Fried et al. 2009; Pinke et al. 2009). The weed community is strongly dynamic, adapted to a changing environment in which human management plays the leading role. Ruderal environments are commonly open, moderately nitrophilous habitats with low plant cover and unstable soils. Periodic disturbance occurs as plowing or rotation takes place throughout the year, thus the weed seed bank persistence is essential for the survival of the species (Dekker 1999). A short life span and a high production of propagules are common traits of weeds occurring in winter crops, because the aerial part of the plant is rapidly eliminated prior to the sowing period (Albrecht 2003; Lososová et al. 2008). From this original community, agriculture intensification has adjusted the weed community, favoring nitrophilous species (Robinson and Sutherland 2002; Rydberg and Milberg 2000), not specific to the environment (i.e., generalists, Fried et al. 2010), with higher frequencies of perennials and overwintering species (Marshall et al. 2003; Salonen et al. 2013) and neophytes (Pinke et al. 2009; Potts et al. 2010) but not archeophytes (Fried et al. 2009; Potts et al. 2010). Other trends that have been observed include loss of insect-pollinated (Pinke et al. 2009) and bird-food plants (Butler et al. 2010; Donald et al. 2006).

Rare species clearly decline in intensified agroecosystems (Albrecht and Mattheis 1998; Cirujeda et al. 2011; Kleijn and Van der Voort 1997), even in short-term studies of changing farming practices (Albrecht and Mattheis 1998). Rare arable plants are less tolerant to herbicides and are outcompeted by generalist nitrophilous species in fields with higher nutrient availability (Rydberg and Milberg 2000). Moreover, the abandonment of regular plowing promotes the establishment of perennial and winter annual species that outcompete the highly specialized, spring annual arable weeds (Tørresen and Skuterud 2002; Tørresen et al. 2003).

There is also a trend toward a higher number of common species becoming rare (i.e., occurring in <5%–10% of studied fields; Cirujeda et al. 2011; Walker et al. 2007), and the number of arable weeds in Red lists increases throughout the region (Bomanowska 2010; Hulina 2005; Sutcliffe and

Kay 2000; Pinke et al. 2011; Storkey et al. 2012; Türe and Böcük 2008). Consequences of these processes are yet to be fully explored, for example, the genetic erosion of rare weed populations (Brütting et al. 2012).

On the other hand, some new weeds have established themselves in European agricultural areas; several are exotic, and some could be a serious threat to crop production or human well-being such as the invasive *Ambrosia artemisiifolia* (Gladieux et al. 2011; Pinke et al. 2009). This species is strongly allergenic, mainly due to massive pollen production causing hay-fever symptoms (Fumanal et al. 2008). It was introduced from North America and is now frequently found in agricultural areas throughout Europe. A large number of neophytes with high invasive potential, like *A. artemisiifolia*, have been introduced through agricultural systems into European agricultural landscapes (Andreasen and Streibig 2011; Potts et al. 2010).

7.3 AGRICULTURE INTENSIFICATION AND WEED RESPONSE

Agriculture intensification is defined by different trends in management practices such as intense use of herbicides and pesticides, inorganic fertilizers, tillage reduction, seed cleaning, monoculture, and increasing field size and landscape homogenization (Stoate et al. 2001; Storkey et al. 2012). The weed community is strongly affected by these drivers, but the relative effect of each of these factors is difficult to measure since most changes are implemented as a whole, and few controlled experiments have been performed on the subject (Hyvönen and Salonen 2002; José-María et al. 2013; Kleijn and Van der Voort 1997). As an example, crop rotation does not seem to affect species richness, except when combined with other practices such as chemical weed control (Ulber et al. 2009).

Changes in the time of sowing, from spring to winter for cereal, may be a threat for some spring annual weeds and may enhance winter species (Hald 1999; Storkey et al. 2012), while significantly decreasing total species richness. Accordingly, similarities in the floristic composition of several crops were mainly explained by time of sowing in a study in Denmark (Andreasen and Skovgaard 2009). The tillage system has been suggested to be the main factor that explains plant trait changes, such as the increase of perennial versus annual species that takes place when regular plowing is abandoned (Sans et al. 2011; Tørresen and Skuterud 2002; Tørresen et al. 2003). However, a nonsignificant effect from different tillage systems was found in a 23-year study in central Spain (Hernández-Plaza et al. 2011), suggesting that a longer time span is needed to assess the weight of managing factors.

An increase in fertilization was also proved to decrease weed diversity and specifically affect rare weeds (Armengot et al. 2012; Kleijn and Van der Voort 1997), although José-María et al. (2013) found a weak effect of fertilizer addition in fields with poor soils, suggesting that this effect may depend on the soil properties of the region. Improvements in seed cleaning has caused the decline and local extinctions of species such as *Agrostemma githago*, but the enhancement of other species that are harder to discriminate from crop seeds, such as *Polygonum aviculare* (Robinson and Sutherland 2002).

However, the use of herbicides is still the main driver of biodiversity loss in agroecosystems (José-María et al. 2013; Ulber et al. 2009) and the high differences found in field diversity of conventional versus organic farming throughout the region are mainly due to the use of these compounds. The use of herbicides promotes the appearance of resistant clones of common weeds, for example, the widespread nonselective glyphosate (Andreasen and Streibig 2011). This may explain the relative abundance of weeds such as *Lolium rigidum* or *Papaver rhoeas* across the Mediterranean (Bassa et al. 2011; Cirujeda et al. 2011; González-Andújar and Saavedra 2003; Hernández-Plaza et al. 2011; José-María et al. 2010; Ponce et al. 2011), species in which resistance has been reported (Heap 2013). In accordance, the frequency of *L. rigidum*, the most harmful weed in cereal farmland fields in Spain (Loureiro et al. 2010), is lower in organic farming (50%) than in conventional farming (80%) (Ponce et al. 2011).

Agriculture intensification implies a loss of landscape heterogeneity, because small fields are substituted with large ones, and paths, hedgerows, and uncultivated fields are transformed for higher production rates (Gaba et al. 2010; Gabriel et al. 2006). Large fields retain lower species numbers, because the perimeter/area ratio decreases, and field edges are richer in species than the fields themselves (José-María et al. 2010; Tscharntke et al. 2005; Wilson and Aebischer 1995). This results in a biodiversity decrease at the landscape level, but not necessarily at the field level (Bohan and Haughton 2012).

Indeed, there are many cases that demonstrate a positive effect of landscape heterogeneity on weed diversity (Gaba et al. 2010; Gabriel et al. 2005, 2006, Weibull et al. 2003) and a strong interaction with management intensity (Batáry et al. 2012; Winqvist et al. 2011). Agriculture intensification results in a weed diversity loss at the farm, landscape, and region scales (Flohre et al. 2011), but metrics used for assessment of diversity will also influence the conclusions obtained. When diversity is partitioned into mean field species richness (α-diversity), total species number (γ-diversity), and species turnover (β-diversity), each component may show a different trend (Fried et al. 2008).

On a large scale (e.g., the European countries), variation in weed diversity and community assemblages may be explained by environmental (climate, altitude, soil type) or management (crop, tillage intensity, herbicides) factors. Environmental variables, mainly geographical location and mean temperature, accounted for twice the management factors in a study covering all of Hungary (Pinke et al. 2012), but when each factor is treated independently, crop and preceding crop explained the highest variation in community composition across France (Fried et al. 2008), in a broad study across Sweden (Hallgren et al. 1999), and in the weed community of oilseed rape fields across Germany (Hanzlik and Gerowitt 2011). Thus, at large scales, management factors such as those resulting from agriculture intensification may even overcome geographic factors and determine the weed community composition and diversity.

7.4 COUNTERACTING THE BIODIVERSITY LOSS

General concerns about the environmental hazards of agriculture intensification have resulted in the implementation of AESs, including support and promotion of organic farming. In response to a public interest in healthier products, organic farming has increased in importance in the European market since the early 1990s. Organic farming is now regulated by the EU (EC 834/2007) and includes best management practices for the environment such as the strict control of herbicides, pesticides, and synthetic fertilizers. Organic farming has, in general terms, a positive effect on biodiversity (Hole et al. 2005; Bengtsson et al. 2005). An increasing trend for species richness and higher values for habitat quality indicators are found for several groups of organisms such as bats (Wickramasinghe et al. 2004), birds (Smith et al. 2010), butterflies (Rundlöf et al. 2008), bees (Holzschuh et al. 2010), and plants (Clough et al. 2007; Concepción et al. 2012; Gabriel et al. 2006; Hyvönen et al. 2003; José-María et al. 2010; Macfadyen et al. 2009; Ponce et al. 2011; Romero et al. 2008).

Vascular plants are indeed among the organisms more strongly influenced by management type when conventional and organic fields are compared, according to Fuller et al. (2005), because size and isolation of the field does not constrain plant diversity as it does for other taxa such as birds. Rare species benefit from organic farming (Pinke et al. 2009) and ecological networks are reinforced (e.g., plant–pollinators interactions: Holzschuh et al. 2008; mycorrhization: Verbruggen et al. 2010; food webs of plants–herbivores–parasitoids: Macfadyen et al. 2009). Biodiversity may even increase in conventional farming fields when these are surrounded by complex landscapes with organic fields (Roschewitz et al. 2005; Rundlöf et al. 2008). However, some studies have found the adoption of environmental schemes such as switching to organic farming has weak effects, mainly because positive trends strongly depend on the surrounding landscape (Concepción et al. 2008; Gabriel et al. 2010; Gibson et al. 2007; Kleijn and Sutherland 2003).

The occurrence of fallow land, grasslands, margin strips, and hedgerows is very important for biodiversity, and promotion of this diversification is part of the AESs. While other organisms such as arthropods are positively affected, the effects on weeds is not clear (Marshall et al. 2006). Uncropped areas may partially retain the weeds, but the absence of regular plowing will result in profound changes in the community composition in favor of perennial and competitive species (Smith et al. 1999), and ecosystem services such as the indirect effect on birds that feed on herbivore invertebrates may be negatively affected (Storkey et al. 2013).

7.5 CONCLUSIONS

There is a general trend of diversity loss in the weed community across European agroecosystems. This includes a lower frequency and abundance of weeds, and changes in community composition, plant traits, and ecosystem services provided. A general loss of "beneficial" species (Fried et al. 2009) such as insect-pollinated, bird-feeding, and rare species that may become endangered can be inferred. Beneficial species are, in general, poor competitors and highly dependent on agriculture practices such as plowing (Storkey and Westbury 2007). The relationship between biodiversity and ecosystem services is widely accepted (e.g., Isbell et al. 2011), but the direct consequences and thresholds of a generalized process of biodiversity loss in agroecosystems is still under debate (Macfadyen et al. 2009).

Environmental policies applied to agroecosystems may counteract, to a certain extent, the negative impacts of agriculture intensification, but in fact higher diversity values are found in extensive agricultural landscapes with traditional, less intensified management (Concepción et al. 2012; Pinke et al. 2009; Ponce et al. 2011; Šilc et al. 2009) mainly in Eastern and Mediterranean Europe, where extensive cereal farmland provides important sites for biodiversity (Atauri and de Lucio 2001; Oñate et al. 2007). In some of these areas, AESs have proved to be ineffective unless appropriate landscape heterogeneity is available (Benton et al. 2003; Concepción et al. 2008; Roschewitz et al. 2005). Lessons from northern countries should be used to prevent a strong intensification of these regions, where promotion of traditional practices and maintenance of moderately heterogeneous farmland landscapes will be the best policy to be adopted for efficient biodiversity conservation. Incorporating references to the benefits of weeds in agriculture policies will help to meet the countryside's needs (Whittingham 2007) and encourage farmers to implement such environmental policies (De Buck et al. 2001).

ACKNOWLEDGMENTS

M. Sheehy-Skeffington made very useful comments that improved the final text.

REFERENCES

Albrecht, H. 2003. Suitability of arable weeds as indicator organisms to evaluate species conservation effects of management in agricultural ecosystems. *Agr Ecosyst Environ* 98:201–211.

Albrecht, H. and Mattheis, A. 1998. The effects of organic and integrated farming on rare arable weeds on the Forschungsverbund Agrarökosysteme München (FAM) research station in southern Bavaria. *Biol Conserv* 86:347–356.

Andreasen, C. and Skovgaard, I. M. 2009. Crop and soil factors of importance for the distribution of plant species on arable fields in Denmark. *Agr Ecosyst Environ* 133:61–67.

Andreasen, C. and Streibig, J. C. 2011. Evaluation of changes in weed flora in arable fields of Nordic countries—Based on Danish long-term surveys. *Weed Res* 51:214–226.

Andreasen, C. and Stryhn, H. 2008. Increasing weed flora in Danish arable fields and its importance for biodiversity. *Weed Res* 48:1–9.

Armengot, L., Sans, F. S., Fischer, C., Flohre, A., José-María, L., Tscharntke, T., and Thies, C. 2012. The b-diversity of arable weed communities on organic and conventional cereal farms in two contrasting regions. *Appl Veget Sci* 15:571–579.

Atauri, J. A. and de Lucio, J. V. 2001. The role of landscape structure in species richness distribution of birds, amphibians, reptiles and lepidopterans in Mediterranean landscapes. *Landsc Ecol* 16:147–159.

Baessler, C. and Klotz, S. 2006. Effects of changes in agricultural land-use on landscape structure and arable weed vegetation over the last 50 years. *Agr Ecosyst Environ* 115:43–50.

Bàrberi, P., Burgio, G., Dinelli, G., Moonen, A. C., Otto, S., Vazzana, C., and Zanin, G. 2010. Functional biodiversity in the agricultural landscape: Relationships between weeds and arthropod fauna. *Weed Res* 50:388–401.

Bassa, M., Boutin, C., Chamorro, L., and Sans, F. X. 2011. Effects of farming management and landscape heterogeneity on plant species composition of Mediterranean field boundaries. *Agr Ecosyst Environ* 141:455–460.

Batáry, P., Holzschuh, A., Orci, K. M., Samu, F., and Tscharntke, T. 2012. Responses of plant, insect and spider biodiversity to local and landscape scale management intensity in cereal crops and grasslands. *Agr Ecosyst Environ* 146:130–136.

Belda, A., Martinez-Perez, J. E., Peiro, V., Seva, E., and Arques, J. 2011. Main landscape metrics affecting abundance and diversity of game species in a semi-arid agroecosystem in the Mediterranean region. *Sp J Agr Res* 9:1197–1212.

Bengtsson, J., Ahnström, J., and Weibull, A.-C. 2005. The effects of organic agriculture on biodiversity and abundance: A meta-analysis. *J Appl Ecol* 42:261–269.

Benton, T. G., Vickery, J. A., and Wilson, J. D. 2003. Farmland biodiversity: Is habitat heterogeneity the key? *Trends Ecol Evol* 18:182–188.

Bohan, D. A. and Haughton, A. J. 2012. Effects of local landscape richness on in-field weed metrics across the Great Britain scale. *Agr Ecosyst Environ* 158:208–215.

Bohan, D. A., Raybould, A., Mulder, C., Woodward, G., Tamaddoni-Nezhad, A., Bluthgen, N., Pocock, M. J. O., et al. 2013. Networking agroecology: Integrating the diversity of agroecosystem interactions. *Adv Ecol Res* 49:1–67.

Bomanowska, A. 2010. Threat to arable weeds in Poland in the light of national and regional red lists. *Plant Breed Seed Sci* 61:55–74.

Brütting, C., Wesche, K., Meyer, S., and Hensen, I. 2012. Genetic diversity of six arable plants in relation to their Red List status. *Biodiv Conserv* 21:745–761.

Butler, S. J., Boccaccio, L., Gregory, R. D., Vorisek, P., and Norris, K. 2010. Quantifying the impact of land-use change to European farmland bird populations. *Agr Ecosyst Environ* 137:348–357.

Cirujeda, A., Aibar, J., and Zaragoza, C. 2011. Remarkable changes of weed species in Spanish cereal fields from 1976 to 2007. *Agr Sust Dev* 31:675–688.

Clough, Y., Holzschuh, A., Gabriel, D., Purtauf, T., Kleijn, D., Kruess, A., Steffan-Dewenter, I., and Tscharntke, T. 2007. Alpha and beta diversity of arthropods and plants in organically and conventionally managed wheat fields. *J Appl Ecol* 44:804–812.

Concepción, E. D., Díaz, M., and Baquero, R. A. 2008. Effects of landscape complexity on the ecological effectiveness of agri-environment schemes. *Landsc Ecol* 23:135–148.

Concepción, E.D., Fernández-González, F., and Díaz, M. 2012. Plant diversity partitioning in Mediterranean croplands: Effects of farming intensity, field edge, and landscape context. *Ecol Appl* 22:972–981.

Dale, V. H. and Polasky, S. 2007. Measures of the effects of agricultural practices on ecosystem services. *Ecol Econ* 64:286–296.

De Buck, A. J., Van Rijn, I., Roling, N. G., and Wossink, G. A. A. 2001. Farmers' reasons for changing or not changing to more sustainable practices: An exploratory study of arable farming in the Netherlands. *J Agr Edu Ext* 7:153–166.

Dekker, J. 1999. Soil weed seed banks and weed management. *J Crop Prod* 2:139–166.

Donald, P. F., Sanderson, F. J., Burfield, I. J., and van Bommel, F. P. J. 2006. Further evidence of continentwide impacts of agricultural intensification on European farmland birds, 1990–2000. *Agr Ecosyst Environ* 116:189–196.

Feehan, J., Gillmor, D. A., and Culleton, N. 2005. Effects of an agri-environment scheme on farmland biodiversity in Ireland. *Agr Ecosyst Environ* 107:275–286.

Fiedler, A. K., Landis, D. A., and Wratten, S. D. 2008. Maximizing ecosystem services from conservation biological control: The role of habitat management. *Biol Control* 45:254–271.

Firbank, L. G., Petit, S., Smart, S., Blain, A., and Fuller, R. J. 2008. Assessing the impacts of agricultural intensification on biodiversity: A British perspective. *Philos Trans R Soc Lond B Biol Sci* 363:777–787.

Flohre, A., Fischer, C., Aavik, T., Bengtsson, J., Berendse, F., Bommarco, R., Ceryngier, P., et al. 2011. Agricultural intensification and biodiversity partitioning in European landscapes comparing plants, carabids, and birds. *Ecol Appl* 21:1772–1781.

Franke, A. C., Lotz, L. A. P., Van Der Burg, W. J., and Van Overbeek, L. 2009. The role of arable weed seeds for agroecosystem functioning. *Weed Res* 49:131–141.

Fried, G., Norton, L. R., and Reboud, X. 2008. Environmental and management factors determining weed species composition and diversity in France. *Agr Ecosyst Environ* 128:68–76.

Fried, G., Petit, S., Dessaint, F., and Reboud, X. 2009. Arable weed decline in Northern France: Crop edges as refugia for weed conservation? *Biol Conserv* 142:238–243.

Fried, G., Petit, S., and Reboud, X. 2010. A specialist-generalist classification of the arable flora and its response to changes in agricultural practices. *BMC Ecol* 10:20.

Fuller, R. J., Norton, L. R., Feber, R. E., Johnson, P. J., Chamberlain, D. E., Joys, A. C., Mathews, F., et al. 2005. Benefits of organic farming to biodiversity vary among taxa. *Biol Lett* 1:431–434.

Fumanal, B., Girod, C., Fried, G., Bretagnolle, F., and Chauvel, B. 2008. Can the large ecological amplitude of *Ambrosia artemisiifolia* explain its invasive success in France? *Weed Res* 48:349–359.

Gaba, S., Chauvel, B., Dessaint, F., Bretagnolle, V., and Petit, S. 2010. Weed species richness in winter wheat increases with landscape heterogeneity. *Agr Ecosyst Environ* 138:318–323.

Gabriel, D., Roschewitz, I., Tscharntke, T., and Thies, C. 2006. Beta diversity at different spatial scales: Plant communities in organic and conventional agriculture. *Ecol Appl* 16:2011–2021.

Gabriel, D., Sait, S. M., Hodgson, J. A., Schmutz, U., Kunin, W. E., and Benton, T. G. 2010. Scale matters: The impact of organic farming on biodiversity at different spatial scales. *Ecol Lett* 13:858–869.

Gabriel, D., Thies, C., and Tscharntke, T. 2005. Local diversity of arable weeds increases with landscape complexity. *Perspect Plant Ecol Evol Syst* 7:85–93.

Gabriel, D. and Tscharntke, T. 2007. Insect pollinated plants benefit from organic farming. *Agr Ecosyst Environ* 118:43–48.

Gerowitt, B., Bertke, E., Hespelt, S.-K., and Tute, C. 2003. Towards multifunctional agriculture—Weeds as ecological goods? *Weed Res* 43:227–235.

Gibson, R. H., Nelson, I. L., Hopkins, G. W., Hamlett, B. J., and Memmott, J. 2006. Pollinator webs, plant communities and the conservation of rare plants: Arable weeds as a case study. *J Appl Ecol* 43:246–257.

Gibson, R. H., Pearce, S., Morris, R. J., Symondson, W. O. C., and Memmott, J. 2007. Plant diversity and land use under organic and conventional agriculture: A whole-farm approach. *J Appl Ecol* 44:792–803.

Gladieux, P., Giraud, T., Kiss, L., Genton, B. J., Jonot, O., and Shykoff, J. A. 2011. Distinct invasion sources of common ragweed (*Ambrosia artemisiifolia*) in Eastern and Western Europe. *Biol Invasions* 13:933–944.

González-Andújar, J. L. and Saavedra, M. 2003. Spatial distribution of annual grass weed populations in winter cereals. *Crop Prot* 22:629–633.

Hald, A. B. 1999. The impact of changing the season in which cereals are sown on the diversity of the weed flora in rotational fields in Denmark. *J Appl Ecol* 36:24–32.

Hallgren, E., Palmer, M. W., and Milberg, P. 1999. Data diving with cross-validation: An investigation of broad-scale gradients in Swedish weed communities. *J Ecol* 87:1037–1051.

Hanzlik, K. and Gerowitt, B. 2011. The importance of climate, site and management on weed vegetation in oilseed rape in Germany. *Agr Ecosyst Environ* 141:323–331.

Hawes, C., Squire, G. R., Hallett, P. D., Watson, C. A., and Young, M. W. 2010. Arable plant communities as indicators of farming practice. *Agr Ecosyst Environ* 138:17–26.

Heap, I. 2013. The international survey of herbicide resistant weeds. Available at: http://www.weedscience.org (accessed December 2013).

Hernández-Plaza, E., Kozak, M., Navarrete, L., and González-Andújar, J. L. 2011. Tillage system did not affect weed diversity in a 23-year experiment in Mediterranean dryland. *Agr Ecosyst Environ* 140:102–105.

Hole, D. G., Perkins, A. J., Wilson, J. D., Alexander, I. H., Grice, P. V., and Evans, A. D. 2005. Does organic farming benefit biodiversity? *Biol Conserv* 122:113–130.

Holland, J. M., Hutchison, M. A. S., Smith, B., and Aebischer, N. J. 2006. A review of invertebrates and seed-bearing plants as food for farmland birds in Europe. *Ann Appl Biol* 148:49–71.

Holzschuh, A., Steffan-Dewenter, I., and Tscharntke, T. 2008. Agricultural landscapes with organic crops support higher pollinator diversity. *Oikos* 117:354–361.

Holzschuh, A., Steffan-Dewenter, I., and Tscharntke, T. 2010. How do landscape composition and configuration, organic farming and fallow strips affect the diversity of bees, wasps and their parasitoids? *J Ann Ecol* 79:491–500.

Hulina, N. 2005. List of threatened weeds in the continental part of Croatia and their possible conservation. *Agr Conspect Sci* 70:37–42.

Hyvönen, T. 2007. Can conversion to organic farming restore the species composition of arable weed communities? *Biol Conserv* 137:382–390.

Hyvönen, T. and Huusela-Veistola, E. 2008. Arable weeds as indicators of agricultural intensity—A case study from Finland. *Biol Conserv* 141:2857–2864.

Hyvönen, T., Ketoja, T., Salonen, J., Jalli, H., and Tiainen, J. 2003. Weed species diversity and community composition in organic and conventional cropping of spring cereals. *Agr Ecosyst Environ* 97:131–149.

Hyvönen, T. and Salonen, J. 2002. Weed species diversity and community composition in cropping practices at two intensity levels—A six-year experiment. *Plant Ecol* 154:73–81.

Isbell, F., Calcagno, V., Hector, A., Connolly, J., Harpole, W. S., Reich, P. B., Scherer-Lorenzen, M., et al. 2011. High plant diversity is needed to maintain ecosystem services. *Nature* 477:199–202.

Jackson, L. E., Pascual, U., and Hodgkin, T. 2007. Utilizing and conserving agrobiodiversity in agricultural landscapes. *Agr Ecosyst Environ* 121:196–210.

José-María, L., Armengot, L., Blanco-Moreno, J. M., Bassa, M., and Sans, F. X. 2010. Effects of agricultural intensification on plant diversity in Mediterranean dryland cereal fields. *J Appl Ecol* 47:832–840.

José-María, L., Armengot, L., Chamorro, L., and Sans, F. X. 2013. The conservation of arable weeds at crop edges of barley fields in northeast Spain. *Ann Appl Biol* 163:47–55.

Kleijn, D. and Sutherland, W. J. 2003 How effective are agrienvironment schemes in maintaining and conserving biodiversity? *J Appl Ecol* 40:947–969.

Kleijn, D. and Van der Voort, L. A. C. 1997. Conservation headlands for rare arable weeds: The effects of fertilizer application and light penetration on plant growth. *Biol Conserv* 81:57–67.

Lososová, Z., Chytrý, M., and Kühn, I. 2008. Plant attributes determining the regional abundance of weeds on central European arable land. *J Biogeogr* 35:177–187.

Loureiro, I., Rodríguez-García, E., Escorial, C., García-Baudín, J. M., González-Andújar, J. L., and Chueca, M. C. 2010. Distribution and frequency of resistance to four herbicide modes of action in *Lolium rigidum* Gaud. accessions randomly collected in winter cereal fields in Spain. *Crop Prot* 29:1248–1256.

Macfadyen, S., Gibson, R. H., Polaszek, A., Morris, R. J., Craze, P. G., Planque, R., Symondson, W. O. C., and Memmot, J. 2009. Do differences in food web structure between organic and conventional farms affect the ecosystem service of pest control? *Ecol Lett* 12:229–238.

Marshall, E. J. P., Brown, V. K., Boatman, N. D., Lutman, P. J. W., Squire, G. R., and Ward, L. K. 2003. The role of weeds in supporting biological diversity within crop fields. *Weed Res* 43:77–89.

Marshall, E. J. P., West, T. M., and Kleijn, D. 2006. Impacts of an agri-environment field margin prescription on the flora and fauna of arable farmland in different landscapes. *Agr Ecosyst Environ* 113:36–44.

Oñate, J. J., Atance, I., Bardají, I., and Llusia, D. 2007. Modelling the effects of alternative CAP policies for the Spanish high-nature value cereal-steppe systems. *Agr Syst* 94:247–260.

Orlowski, G. and Czarnecka, J. 2007. Winter diet of reed bunting *Emberiza schoeniclus* in fallow and stubble fields. *Agr Ecosyst Environ* 118:244–248.

Pinke, G., Karácsony, P., Czúcz, B., Botta-Dukát, Z., and Lengyel, A. 2012. The influence of environment, management and site context on species composition of summer arable weed vegetation in Hungary. *Appl Veg Sci* 15:136–144.

Pinke, G., Király, G., Barina, Z., Mesterházy, A., Balogh, L., Csiky, J., Schmotzer, A., Molnár, A. V., and Pál, R. W. 2011. Assessment of endangered synanthropic plants of Hungary with special attention to arable weeds. *Plant Biosyst* 145:426–435.

Pinke, G., Pál, R., Botta-Dukát, Z., and Chytrý, M. 2009. Weed vegetation and its conservation value in three management systems of Hungarian winter cereals on base-rich soils. *Weed Res* 49:544–551.

Ponce, C., Bravo, C., de León, D. G., Magana, M., and Alonso, J. C. 2011. Effects of organic farming on plant and arthropod communities: A case study in Mediterranean dryland cereal. *Agr Ecosyst Environ* 141:193–201.

Potts, G. R., Ewald, J. A., and Aebischer, N. J. 2010. Long-term changes in the flora of the cereal ecosystem on the Sussex Downs, England, focusing on the years 1968–2005. *J Appl Ecol* 47:215–226.

Power, A. G. 2010. Ecosystem services and agriculture: Tradeoffs and synergies. *Philos Trans Royal Soc B* 365:2959–2971.

Robinson, R. A. and Sutherland, W. J. 2002. Post-war changes in arable farming and biodiversity in Great Britain. *J Appl Ecol* 39:157–176.

Romero, A., Chamorro, L., and Sans, F. X. 2008. Weed diversity in crop edges and inner fields of organic and conventional dryland winter cereal crops in NE Spain. *Agr Ecosyst Environ* 124:97–104.

Roschewitz, I., Gabriel, D., Tscharntke, T., and Thies, C. 2005. The effects of landscape complexity on arable weed species diversity in organic and conventional farming. *J Appl Ecol* 42:873–882.

Rundlöf, M., Bengtsson, J., and Smith, H. G. 2008. Local and landscape effects of organic farming on butterfly species richness and abundance. *J Appl Ecol* 45:813–820.

Rydberg, N. T. and Milberg, P. 2000. A survey of weeds in organic farming in Sweden. *Biol Agr Hort* 18:175–185.

Salonen, J., Hyvonen, T., Kaseva, J., and Jalli, H. 2013. Impact of changed cropping practices on weed occurrence in spring cereals in Finland—A comparison of surveys in 1997–1999 and 2007–2009. *Weed Res* 53:110–120.

Sans, F. X., Berner, A., Armengot, L., and Mäder, P. 2011. Tillage effects on weed communities in an organic winter wheat–sunflower–spelt cropping sequence. *Weed Res* 51:413–421.

Schellhorn, N. A. and Sork, V. L. 1997. The impact of weed diversity on insect population dynamics and crop yield in collards, *Brassica oleracea* (Brassicaceae). *Oecologia* 111:233–240.

Šilc, U., Vrbničanin, S., Božić, D., Čarni, A., and Stevanović, Z. D. 2009. Weed vegetation in the north-western Balkans: Diversity and species composition. *Weed Res* 49:602–612.

Smart, S. M., Firbank, L. G., Bunce, R. G. H., and Watkins, J. W. 2000. Quantifying changes in the abundance of food plants for butterfly larvae and farmland birds. *J Appl Ecol* 37:398–414.

Smith, H., Firbank, L. G., and Macdonald, D. W. 1999. Uncropped edges of arable fields managed for biodiversity do not increase weed occurrence in adjacent crops. *Biol Conserv* 89:107–111.

Smith, H. G., Dänhardt, J., Lindström, Å., and Rundlöf, M. 2010. Consequences of organic farming and landscape heterogeneity for species richness and abundance of farmland birds. *Oecologia* 162:1071–1079.

Sotherton, N. W. 1998. Land use changes and the decline of farmland wildlife: An appraisal of the set-aside approach. *Biol Conserv* 83:259–268.

Stoate, C., Báldi, A., Beja, P., Boatman, N. D., Herzon, I., van Doorn, A., de Snoo, G. R., Rakosy, L., and Ramwell, C. 2009. Ecological impacts of early 21st century agricultural change in Europe—A review. *J Environ Manage* 91:22–46.

Stoate, C., Boatman, N. D., Borralho, R. J., Carvalho, C. R., de Snoo, G. R., and Eden, P. 2001. Ecological impacts of arable intensification in Europe. *J Environ Manage* 63:337–365.

Storkey, J., Brooks, D., Haughton, A., Hawes, C., Smith, B. M., and Holland, J. M. 2013. Using functional traits to quantify the value of plant communities to invertebrate ecosystem service providers in arable landscapes. *J Ecol* 101:38–46.

Storkey, J., Meyer, S., Still, K. S., and Leuschner, C. 2012. The impact of agricultural intensification and land-use change on the European arable flora. *Pro Royal Soc B* 279:1421–1429.

Storkey, J., Moss, S. R., and Cussans, J. W. 2010. Using assembly theory to explain changes in a weed flora in response to agricultural intensification. *Weed Sci* 58:39–46.

Storkey, J. and Westbury, D. B. 2007. Managing arable weeds for biodiversity. *Pest Manag Sci* 63:517–523.

Sutcliffe, O. L., and Kay, Q. O. 2000. Changes in the arable flora of central southern England since the 1960s. *Biol Conserv* 93:1–8.

Swift, M. J., Izac, A.-M. N., and van Noordwijk, M. 2004. Biodiversity and ecosystem services in agricultural landscapes—Are we asking the right questions? *Agr Ecosyst Environ* 104:113–134.

Tørresen, K. S. and Skuterud, R. 2002. Plant protection in spring cereal production with reduced tillage. IV.: Changes in the weed flora and weed seedbank. *Crop Prot* 21:179–193.

Tørresen, K. S., Skuterud, R., Tandsaether, H. J., and Hagemo, M. B. 2003. Long-term experiments with reduced tillage in spring cereals. I. Effects on weed flora, weed seedbank and grain yield. *Crop Prot* 22:185–200.

Tscharntke, T., Klein, A. M., Kruess, A., Steffan-Dewenter, I., and Thies, C. 2005. Landscape perspectives on agricultural intensification and biodiversity—Ecosystem service management. *Ecol Lett* 8:857–874.

Türe, C. and Böcük, H. 2008. Investigation of threatened arable weeds and their conservation status in Turkey. *Weed Res* 48:289–296.

Ulber, L., Steinmann, H. H., Klimek, S., and Isselstein, J. 2009. An on-farm approach to investigate the impact of diversified crop rotations on weed species richness and composition in winter wheat. *Weed Res* 49:534–543.

Verbruggen, E., Röling, W. F. M., Gamper, H. A., Kowalchuk, G. A., Verhoef, H. A., and van der Heijden, M. G. A. 2010. Positive effects of organic farming on below-ground mutualists: Large-scale comparison of mycorrhizal fungal communities in agricultural soils. *New Phytol* 186:968–979.

Walker, K. J., Critchley, C. N. R., Sherwood, A. J., Large, R., Nuttall, P., Hulmes, S., and Mountford, J. O. 2007. The conservation of arable plants on cereal field margins: An assessment of new agri-environment scheme options in England, UK. *Biol Conserv* 136:260–270.

Walker, K. J., Preston, C. D., and Boon, C. R. 2009. Fifty years of change in an area of intensive agriculture: Plant trait responses to habitat modification and conservation, Bedfordshire, England. *Biodiv Conserv* 18:3597–3613.

Weibull, A. C., Östman, Ö., and Granqvist, Å. 2003. Species richness in agroecosystems: The effect of landscape, habitat and farm management. *Biodiv Conserv* 12:1335–1355.

Whittingham, M. J. 2007. Will agri-environment schemes deliver substantial biodiversity gain, and if not why not? *J Appl Ecol* 44:1–5.

Wickramasinghe, L., Harris, S., Jones, G., and Vaughan, N. 2004. Abundance and species richness of nocturnal insects on organic and conventional farms: Effects of agricultural intensification on bat foraging. *Conserv Biol* 18:1–10.

Wilson, J. D., Morris, A. J., Arroyo, B. E., Clark, S. C., and Bradbury, R. B. 1999. A review of the abundance and diversity of invertebrate and plant foods of granivorous birds in northern Europe in relation to agricultural change. *Agr Ecosyst Environ* 75:13–30.

Wilson, P. J. and Aebischer, N. J. 1995. The distribution of dicotyledonous arable weeds in relation to distance from the field edge. *J Appl Ecol* 32:295–310.

Winqvist, C., Bengtsson, J., Aavik, T., Berendse, F., Clement, L. W., Eggers, S., Fischer, C., et al. 2011. Mixed effects of organic farming and landscape complexity on farmland biodiversity and biological control potential across Europe. *J Appl Ecol* 48:570–579.

Zhang, W., Ricketts, T. H., Kremen, C., Carney, K., and Swinton, S. M. 2007. Ecosystem services and disservices to agriculture. *Ecol Econ* 64:253–260.

CHAPTER **8**

Proteomics Potential and Its Contribution toward Sustainable Agriculture

Abhijit Sarkar, Md. Tofazzal Islam, Sajad Majeed Zargar, Vivek Dogra, Sun Tae Kim, Ravi Gupta, Renu Deswal, Ganesh Bagler, Yelam Sreenivasulu, Rungaroon Waditee-Sirisattha, Sophon Sirisattha, Jai Singh Rohila, Manish Raorane, Ajay Kohli, Dea-Wook Kim, Kyoungwon Cho, Abdiani Attiq Saidajan, Ganesh Kumar Agrawal, and Randeep Rakwal

CONTENTS

8.1   Climate Change, Biodiversity, and Sustainable Agriculture: An Introduction 152
8.2   Sustainable Agriculture: Past and Existing Issues.. 154
     8.2.1   Issues That Brought the First Green Revolution... 154
     8.2.2   The "Second" Green Revolution and Climate Change 155
         8.2.2.1   How Genetically Engineered Plants (Enhanced Salt Tolerance in
                  Plants by Ion Homeostasis and Osmoprotectant) Have Been Integrated
                  into Agriculture and Their Contribution to Sustainable Agriculture.......... 156
         8.2.2.2   Plant Probiotics as Biofertilizers and Biopesticides 157
8.3   High-Throughput Approaches: Necessity in Expediting Sustainable Agriculture............... 160
     8.3.1   Proteomics Approach and Its Importance among Omics Technologies 160
     8.3.2   Molecular Breeding for Crop Improvement 160
8.4   Proteomics Contribution in Cereal Crops: Rice and Wheat... 162
8.5   Proteomics Contribution in Model Legume Species: A Case Study in Seeds 164
     8.5.1   Proteomic Studies in Soybean 165
     8.5.2   Proteomic Studies in Peanut 167
8.6   Proteomics Contribution in Horticultural Crops 167
     8.6.1   Strawberry 167
     8.6.2   Grape 168
     8.6.3   Pear 169
8.7   Proteomics in Combination with Other Omics Approaches toward
     Next-Generation Crops 170
     8.7.1   Genomes, Proteomes, Transcriptomes: Application of Computational
         Approaches for Reconstruction, Modeling, and Analysis of -Omes 170
Acknowledgments.. 171
References 171

8.1 CLIMATE CHANGE, BIODIVERSITY, AND SUSTAINABLE AGRICULTURE: AN INTRODUCTION

Sustainable agriculture has run its course during decades gone by, and "food security" issues are not what they used to be in the past. Due to the ever-changing global climate and population growth, food security is now one of the hottest topics of discussion around the world. To overcome food shortages, hunger, and starvation, a breakthrough in cutting-edge research and translation of the research findings to the field are desperately needed. Technologies including high-yielding varieties, intensive use of agrochemicals, and so on, used in the green revolution, helped to feed billions of people. However, they have caused significant environmental degradation and seem unable to meet the future demand for food security. The impacts of global climate and environmental change mean that crops are prone to damage on an unprecedented scale. Therefore, there is an increased demand for research efforts and scientific collaboration to address the need for enhanced food production by minimizing the impacts of climate change in crop production. In this chapter, we not only discuss the past and future of sustainable agriculture but also try to bring to the forefront the necessity of expediting research using high-throughput omics technologies such as genomics, proteomics, and metabolomics. Among these, proteomics is the main focus of this review, in which we describe its use in crops, giving examples from cereals to legumes and fruits. We also give an insight into potential biomarkers and their exploitation in screening natural genetic resources, alone or in combination with other technology-derived results, for generating the next-generation crop plants for the twenty-first century and beyond.

In human history, the Industrial Revolution starting in the eighteenth century, followed by the green revolution in the mid-twentieth century, did not only dramatically increase human capacity but also accelerated world population growth. However, both revolutions have impacted on acceleration of changes in the global environment, biodiversity, and climate. Although the green revolution helped to feed billions of people, this heavy input-based technology seems unable to meet the future demands for food supply, especially in developing countries, due to deterioration of soil health and environmental pollution. It has been estimated that, over the last century, the atmospheric concentrations of carbon dioxide (CO_2), nitrous oxide (N_2O), and methane (CH_4) have increased by 25%, 16%, and 100%, respectively (Houghton et al. 2001; Hoffert et al. 2002; IPCC AR4 WG1 2007). As a result, the temperature on the earth's surface has increased by about 0.76°C (IPCC 2007a). However, it is now known that over the last 50 years the rate of warming (0.13°C ± 0.03°C per decade) has increased nearly twofold compared with the last 100 years, which poses a serious threat to biodiversity, agricultural productivity, and food security in many countries. For example, it has been reported that global warming is causing receding and thinning of glaciers in the Himalayas at an accelerated rate (Hasnain 2002; IPCC 2007b). The Himalayan glaciers form a unique reservoir supporting the perennial rivers Indus, Ganga, and Brahmaputra, which are considered the lifeline of millions of people in the South Asian countries of Pakistan, Nepal, Bhutan, India, and Bangladesh.

Over the past three decades, and especially, since the 1990s climate change has become one of the most heavily researched subjects in science (Lamb 1995). The theoretical analysis of most of the modelling approaches to climate change prediction does not consider potential evolutionary changes of living organisms. Recently published work has shown that many species are capable of relatively rapid genetic changes, which enhances their ability to invade new areas in response to anthropogenic ecosystem modification (Clements and Ditommaso 2011). Therefore, climate change represents a major challenge to maintenance of biodiversity and richness of plant species (Sommer et al. 2010). Soil microbes, including probiotic bacteria, play a key role in the ecosystem by driving major biogeochemical processes directly contributing to the maintenance of plant productivity. Climate change generally affects organisms either directly via physiological stress or indirectly via changing relationships among species (Harley 2011). As mentioned by Harley

(2011), there is some evidence suggesting that anthropogenic climate change can alter interspecific interactions and produce unexpected changes in species distributions, community structure, and diversity. The changed climate will create some common problems for wildlife, such as loss of habitat due to inundation, shortage of food due to breaks in the food chain, loss of breeding grounds, altered breeding patterns, and so on. In this situation many wild species may find it difficult to maintain their existence. Therefore, it is important to understand whether and how climate change impacts on the biodiversity of organisms, including microorganisms, and how such changes might feed back to influence changes in plant productivity (Thuiller 2007; van der Heijden et al. 2008). As well as biodiversity and plant productivity, climate change is also posing a challenge to the interconnected development goals such as supply of food, energy, and water (Beddington 2009; Godfray et al. 2010).

It is now well accepted that climate change is posing a threat to agricultural production systems worldwide, and more so severely in tropical and subtropical countries (Rosenzweig and Parry 1994). Although several studies suggest that doubling of the atmospheric CO_2 concentration will lead to only a small decrease in total global crop production, developing countries are likely to bear the brunt of the problem, including a considerable decrease in crop production (Rosenzweig and Parry 1994). Recently, Lobell et al. (2011) analyzed the climate trends and global crop production since 1980. Those authors found that, in the cropping regions and growing seasons of most countries, temperature trends from 1980 to 2008 exceeded one standard deviation of historic year-to-year variability. Further, their analysis indicates that global maize and wheat production has declined by 3.8% and 5.5%, respectively, relative to a counterfactual without climate trends (Lobell et al. 2011). Rice production has also decreased, but to a lesser extent than the other cereals. It has been noted that the climate trends were large enough in some countries to offset a significant portion of the increases in average yields that arose from technology, CO_2 fertilization, and other factors (Lobell et al. 2011).

It must be emphasized that climate change has already led to altered distributions of species, phenotypic variation, and allele frequencies (Bradshaw and Holzapfel 2001; Franks et al. 2007; Lobell et al. 2011; Lynch and Lande 1993; Umina et al. 2005), and that the impact of changing climate is expected to intensify (Hancock et al. 2011). Hoffmann and Sgrò (2011) have reported that the capacity to respond to changing climate is likely to vary widely as a consequence of variation among species in their degree of phenotypic plasticity and their potential for genetic adaptation, which in turn depends on the amount of standing genetic variation and the rate at which new genetic variation arises. Understanding the genetic basis and modes of adaptation to current climatic conditions will be essential to accurately predict responses to future environmental change.

Arabidopsis thaliana (hereafter called *Arabidopsis*) is an excellent model for investigating the genetic basis and mode of adaptation to climate. One reason is the extensive climatic variation across its native range. The other reason is the availability of genome-wide single-nucleotide polymorphism data among a geographically diverse collection. Recently, Hancock et al. (2011) conducted a genome-wide scan to identify climate-adaptive genetic loci and pathways in *Arabidopsis*. Amino acid-changing variants were significantly enriched among the loci strongly correlated with climate, suggesting that our scan effectively detects adaptive alleles. Their findings predicted relative fitness among a set of geographically diverse *Arabidopsis* accessions when grown together in a common environment and provide a set of candidates for dissecting the molecular bases of climate adaptations, as well as insights about the prevalence of selective sweeps, which has implications for predicting the rate of adaptation. Findings of their study also suggested that species such as *Arabidopsis* may reach adaptive limits under rapid climate change, due to the constraints imposed by waiting for new mutations. In another study on *Arabidopsis*, Fournier-Level et al. (2011) have identified candidate loci for local adaptation from a genome-wide association study of lifetime fitness in geographically diverse accessions. The fitness-associated loci were found to exhibit both geographic and climatic signatures of local adaptation. Relative to genomic controls, high-fitness alleles were generally distributed closer to the site where they increased fitness, occupying specific and distinct climate spaces. Independent loci with different molecular functions contributed most strongly to fitness variation in each site, which suggested that independent local adaptation by distinct genetic mechanisms may facilitate a flexible evolutionary response to changing environment across a species range.

Rice (*Oryza sativa* L.) is the cereal crop and monocot genome model (Goff et al. 2002; Yu et al. 2002), and, as it is a food for life, research into rice biology has progressed greatly over the years since its genome sequence became available in 2002. In particular, proteomics has progressed at an unexpected pace since 2000, establishing and understanding the proteomes of tissues, organs, and organelles under normal and adverse environmental conditions (for review see Agrawal and Rakwal 2011, and references therein). Established proteomes have also helped in reannotating the rice genome and revealing new roles of previously known proteins. Proteomics-based discoveries in rice are likely to be translated into improving crop plants, and vice versa, against ever-changing environmental factors. A major role of rice proteomics is envisioned in addressing the basic global problem of food security under threat from climate change and to meet the demands of the human population, which is projected to reach 6–9 billion by 2040. Moreover, rice, together with wheat and maize, provides 50% of the total calories consumed by the human population worldwide (Maclean et al. 2002). Thus, rice is a crop that requires our utmost attention in both fundamental and applied research.

More research is required to estimate the likely scale and timing of climate change impacts on different sectors of the economy, for informed planning of future investment strategies. Advanced technologies such as biotechnology and genetic engineering should be applied to develop cropping systems that are resilient to climate change (including crop varieties tolerant of flooding, drought, and salinity, and also those based on indigenous cultivars and other varieties suited to the needs of resource-poor farmers), fisheries, and livestock systems to ensure food security. Applying emerging genetic techniques holds great promise for understanding how biodiversity in soil and environments will respond to global changes (Chakraborty et al. 2008; Roesch et al. 2007).

8.2 SUSTAINABLE AGRICULTURE: PAST AND EXISTING ISSUES

8.2.1 Issues That Brought the First Green Revolution

Increasing population and issues such as environmental pollution and climate change brought the need for enhancing agricultural production and food security (Borlaug and Dowswell 2004). In the 1940s, people in Mexico, under the leadership of the great scientist Norman Borlaug adopted agricultural practices that combined Borlaug's wheat varieties with new mechanized agricultural technologies. This resulted in Mexico producing more wheat than was needed by its own citizens, and becoming an exporter of wheat by the 1960s. Due to the success of this improved practice, very soon agricultural practitioners around the world adapted this practice for increasing their respective agricultural productions. This movement, including a series of research, development, and technology transfer initiatives in agriculture for increasing agriculture production around the world during the 1960s, was named the *green revolution* by former United States Agency for International Development (USAID) director, William Gaud, in 1968 (Gaud 1968). The initiatives, led by Norman Borlaug, also called the "father of the green revolution," were credited with saving over a billion people from starvation. They involved the development of high-yielding varieties of cereal grains; expansion of irrigation infrastructure; modernization of management techniques; and distribution of hybridized seeds, synthetic fertilizers, and pesticides to farmers. Borlaug was later awarded the Nobel Peace Prize.

The green revolution spread technologies that had already existed before, but had not been widely used outside the industrialized nations. These technologies include modern irrigation projects, pesticides, synthetic nitrogen fertilizer, and improved crop varieties developed through the conventional, science-based methods available. The technological advancements that led to the success of the green revolution were based largely on molecular breeding and genetic engineering, which produced high-yielding varieties/cultivars of cereal crops, especially wheat (Norin 10, Japanese semi-dwarf wheat cultivar) (De Datta et al. 1968), rice (IR8) (IRRI 1962), Golden Rice (Ye et al. 2000), maize (Bt corn), and so on. It should be noted that these high-yielding varieties were able to significantly outperform traditional varieties in the presence of adequate irrigation, pesticides, and fertilizers. Thus, production of crops increased dramatically around the world during 1961–1985.

8.2.2 The "Second" Green Revolution and Climate Change

Since the 1960s, world per capita agricultural production has increased by 25% and per capita food production in Asia has increased by 76%. That was an era when we needed to ensure food for all, but now the scenario has changed. We had developed strategies to achieve food security for all. The green revolution played a central role in this achievement. However, this success has been at the expense of the natural resource base, by the overuse of natural resources or through their use as a sink for pollution. These practices eroded crop biodiversity and gave rise to other ecological problems, such as soil infertility, chemical pollution of land and water resources, pesticide poisoning, pest infestation, and so on. Also, due to the ever-increasing population, global demand for associated agricultural products is projected to rise by at least 50% over the next two decades (UN Millennium Project 2005). The key challenges now are assurance of quality food and sustainability where the green revolution failed, despite the challenges of increasing population, decreasing fertility, pollution, and, most of all, climate change.

Climate change is an ongoing process, but the pace of this change is very slow. Issues such as global warming and dimming resulting from environmental pollution, and other human-induced as well as natural factors, have increased the rate of climate change. As global warming increases, we can see that species and their habitats will be or are on the decrease. This will affect the chances for ecosystems to adapt naturally, leading to an imbalance in biodiversity. So, in the present scenario, a major challenge for agricultural production will be to manage biotic and abiotic stresses with perpetually rising numbers of pests and pathogens due to the changing climate. Thus, climate change has become a major issue, and an intensive drive will be needed to overcome its effects (Adams et al. 1998).

Significant effort to assess the impending impacts of climate change on agriculture began in 1978, when the National Defense University assembled an international group of climate experts to predict the probabilities of various climate change events and the resulting impacts on agriculture. In 1988, the Intergovernmental Panel on Climate Change (IPCC) was created by the United Nations Environment Programme (UNEP) and the World Meteorological Organization (WMO) to assess the scientific knowledge on global warming. The IPCC concluded in 1990 that there was broad international consensus that climate change was anthropogenic. That report led to an international convention for climate change, the United Nations Framework Convention on Climate Change (UNFCCC), signed by over 150 countries at the Rio Earth Summit in 1992. Since then, more structured scientific studies have resulted in a growing consensus on the interactions between climate change and agriculture (Pachauri and Reisinger 2007), culminating in the 2007 Fourth Assessment Report of the IPCC. Environmental pollution, deforestation, and increased emission of greenhouse gases, especially CO_2, have resulted in its accumulation in the atmosphere, leading to global warming and ultimately climate change. Some of the harmful effects of elevated levels of CO_2 have been observed recently, such as reduction in the sea ice in the Arctic and acidification of oceans, both leading to a decline in biodiversity (Yamamoto et al. 2012). Therefore, we need to tackle climate change and warming by introducing the use of carbon sinks to soak up CO_2, reforestation, and at the same time adapting sustainable agriculture practices.

Humankind is on the verge of achieving a second green revolution to ensure food security to the ever-increasing population, but at the same time climate change has an important issue, which stresses the need for sustainability. Therefore, it is necessary to introduce agricultural practices that can fulfill the increasing demand for food and at the same time protect the climate. The scientific community has responded remarkably well and has taken initiatives to reduce the pace and the effects of climate change using modern biotechnological techniques and better equipment. Newer molecular biology disciplines, such as genomics, transcriptomics, metabolomics, and proteomics, along with systems biology, are being utilized today to address issues in agriculture and its sustainability.

8.2.2.1 How Genetically Engineered Plants (Enhanced Salt Tolerance in Plants by Ion Homeostasis and Osmoprotectant) Have Been Integrated into Agriculture and Their Contribution to Sustainable Agriculture

Adverse environmental conditions such as high salinity, water deficit, and temperature extremes are found in many agricultural areas, limiting crop productivity worldwide. Plants have evolved various responses to these stresses at the molecular and cellular levels (Hasegawa et al. 2000; Shinozaki et al. 2003; Zhu 2002; Hakeem et al. 2012). So far, numerous genes related to plant response to abiotic stress have been identified and characterized (Gaxiola et al. 1999; Shi et al. 2000; Song et al. 2004; Waditee et al. 2003). Complete information is available on various plant genomes and advances in omics technologies are proceeding at a great pace, thus providing opportunities for the identification of transcriptional, translational, and posttranslational mechanisms, including signalling pathways that regulate the plant stress response.

To breed or genetically engineer plants to enhance abiotic stress tolerance would be part of a comprehensive strategy for sustainable agriculture. Manipulation of genes that affect specific targets, such as expression of genes encoding enzymes associated with the accumulation of compatible solutes, ion transport, stress proteins, and enzymes involved in scavenging oxygen radicals, has been used to generate transgenic plants (Pardo 2010; Umezawa et al. 2006; Yang et al. 2010). It has been shown that various compatible solutes enable plants to tolerate abiotic stress, and glycine betaine is one of the most potent compatible solutes found in nature. Many transgenic lines that overexpressed genes for the biosynthesis of glycine betaine have shown enhanced abiotic stress tolerance (see review, Chen and Murata 2011): for example, codA-transgenic rice (Sakamoto and Murata 2001) and ApGSMT-DMT-transgenic Arabidopsis (Waditee et al. 2005). However, all transgenic plants accumulate only low levels of glycine betaine compared with natural accumulators of glycine betaine.

Manipulation of genes encoding ion transport for improving plant tolerance to salinity stress has been reported in both model and crop plants. In Arabidopsis, physiological roles of the genes AtNHX1, AtNHX7 (SOS1), AtCHX17, AtCHX23, and AtHKT, which contribute to salt tolerance, have been described (Apse et al. 1999, 2003; Berthomieu et al. 2003; Cellier et al. 2004; Shi et al. 2003; Song et al. 2004). Numerous investigations have led to the conclusion that these genes play important roles in expelling Na^+ across the plasma membrane, sequestrating Na^+ into a tonoplast, recirculating or removing large amounts of Na^+ from shoots, or a combination of these. Overexpression of an Arabidopsis vacuolar Na^+/H^+ antiporter, AtNHX1, in Arabidopsis (Apse et al. 1999) and its ectopic expression in Brassica (Zhang et al. 2001), tomato (Zhang et al. 2001), cotton (He et al. 2005), wheat (Xue et al. 2004), and Beta vulgaris (Yang et al. 2005) resulted in improved salt tolerance. In addition, overexpression of the rice ortholog, OsNHX1, in rice and its ectopic expression in maize enhanced salt stress tolerance of transgenic rice cells and plants (Apse et al. 2003; Chen et al. 2007). Recently, ectopic expression of the Arabidopsis AtNHX5 resulted in enhanced salt and drought tolerance in rice seedlings (Bassil et al. 2011).

High-altitude plant habitats are very specific because of their niche area of extreme environmental conditions. Their genes and proteins are being used as molecular tools for engineering crop and other plants for better stress tolerance and adaptability against the present scenarios of climate change. Ectopic expression of copper zinc superoxide dismutase (CuZnSOD/PaSOD) from a high-altitude plant, *Potentilla astrungunea*, improved copper and salt stress tolerance in *Arabidopsis* (Gill et al. 2010b, 2012). Later, it was also found that the ectopic expression of *PaSOD* improved stress tolerance in *Arabidopsis* because of overaccumulation of lignin in vascular bundles (Gill et al. 2010a). Apparently, overexpression of the same *SOD* in potato also enhanced photosynthetic performance under drought stress (Pal et al. 2012). These studies confirmed that the antioxidant genes from specific niche areas can be successfully utilized for engineering abiotic stress tolerance in crop plants without disturbing the native physiology of the plants. Recently, Dogra et al. (2013) and Rana and Sreenivasulu (2013) reported the role of cell wall hydrolases in seed germination of the high-altitude plants *Podophyllum hexandrum* and *Aconitum heterophyllum*. These cell wall hydrolases, due to their comparatively smaller size and enzymatic activity over a broad temperature range, can be utilized for manipulating seed germination problems. β-1,3-glucanase, a germination-related cell wall–degrading enzyme from *P. hexandrum*, reduced mean germination time and induced early seed germination even at low temperature (5°C) in *Arabidopsis* (unpublished data, personal communication with Vivek Dogra).

The results described above suggest that engineering plants by overexpressing genes encoding the accumulation of compatible solutes and ion transport appears to be a useful technology for various kinds of agriculturally important crops. To date, it has been reported that over 30 genetically engineered crops are being grown in 25 countries (Clive 2009). These efforts would be part of a comprehensive strategy for sustainable global agriculture that can meet the need for food production.

8.2.2.2 *Plant Probiotics as Biofertilizers and Biopesticides*

Probiotics are living microorganisms, which, when administered in adequate amounts, confer a health benefit on the host (Lilly and Stillwell 1965; Parker 1974). The term *probiotics* is widely used for bacteria such as *Lactobacillus* and *Bifidobacterium* that pass through the gastrointestinal tract of animals and humans and might prevent, or even cure, diarrhea and other gastrointestinal diseases (Haas and Defago 2005). Similarly, the term *plant probiotics* has been used to describe plant-associated elite bacteria that, when applied, promote the growth of the host plant (Maheshwari 2012). The promotion of plant growth by beneficial bacteria is achieved in several ways, including better nutrition, disease suppression, and hormonal activity (Borriss 2011; Haas and Defago 2005; Islam 2011). Understanding the mechanisms of interaction between plant probiotics and their host is considered important for their large-scale application in agriculture.

Our ability to provide adequate plant nutrition and pest control to increase crop yields and reduce land requirements will not be able to keep pace with the growing demand of the world's population using conventional high input-based agriculture. Hence, this has led to higher chemical inputs to promote plant nutrition and control insect pests and plant diseases. Consequently, an undue burden is placed on the planet's ecosystems, along with serious pressure on the depleting natural resources used for making agrochemicals. The costs associated with the rapid development of modern fertilizers and pesticides include environmental pollution (Zaidi et al. 2009), unpredicted human health consequences, and deleterious effects on wildlife and other nontargeted organisms in the food chain. The realization that a sustainable agricultural system must be compatible with environmental concerns has developed into the philosophy of green agriculture. Microbial biofertilizers (Islam and Hossain 2012) and biopesticides (Haas and Defago 2005; Islam et al. 2005; Islam and Hossain 2012) represent an alternative to hazardous synthetic chemicals for sustainable green agriculture. Several lines of evidence suggest that probiotic bacteria can be used as alternatives to synthetic fertilizers such as nitrogenous (e.g., urea) and phosphatic fertilizers (e.g., super phosphates) (Borriss 2011; Islam and Hossain 2012; Maheshwari 2012). This section reviews the current knowledge and trends of using plant probiotics as biofertilizers and biopesticides in low-input sustainable agriculture.

From the global viewpoint, agriculture is confronted with huge problems resulting from deterioration of the environment and depletion of natural resources. For example, we must be concerned about depletion of phosphate rock and energy sources (fossil fuels, natural gas, etc.), which are the important raw materials for production of phosphorus and nitrogenous fertilizers (Zaidi et al. 2009). Low levels of soluble soil phosphorus (P) are a serious constraint to crop production in tropical and subtropical soils, as most of the inorganic soil P forms complexes of iron, aluminum, and calcium phosphates that are adsorbed on clay particles. The solubility of these inorganic P compounds, as well as organic P, is extremely low, and only very small amounts of soil P are in solution at any one time. Probiotic bacteria that are known to enhance the solubilization of fixed soil P and applied phosphate fertilizer, resulting in better P nutrition in crop plants, are known as phosphate-solubilizing bacteria (PSB) (Abd-Alla 1994; Islam and Hossain 2012). An important trait in plant growth-promoting bacteria is the ability of PSB to convert insoluble forms of phosphorus to an accessible form for increasing plant yields (Islam et al. 2007; Richardson and Simpson 2011; Islam and Hossain 2012).

Bacteria from diverse taxonomic groups such as *Pseudomonas*, *Bacillus*, *Klebsiella*, *Rhizobium*, *Acinetobacter*, and so on have been shown to solubilize soil-insoluble P, and some of them have shown high promise for P nutrition in crop plants (Bianco and Defez 2010; Borriss 2011; Islam et al. 2007; Islam and Hossain 2012; Naik et al. 2008; Richardson and Simpson 2011). The widely recognized mechanisms of phosphate solubilization mediated by these plant-associated bacteria are production of organic acids (such as gluconic, citric, oxalic), secretion of hydrolytic enzymes (such as phytases, phosphatases), or both (Abd-Alla 1994; Islam and Hossain 2012). The production of organic acids by PSB appears to be independent of their genetic relatedness, and each strain has its own ability to produce organic acids during solubilization of inorganic phosphates (Vyas and Gulati 2009). Despite their potential as low-input practical agents for plant P nutrition, application of PSB has been hampered by their inconsistent performance in the field, which is usually attributed to their rhizosphere competence (Islam et al. 2007; Richardson and Simpson 2011). Hence, the full potential of PSB for P nutrition in the production of major crops has not yet been achieved (Islam and Hossain 2012; Lugtenberg and Kamilova 2009). In fact, our knowledge of bacterial quorum sensing and ongoing complex molecular cross-talk within the rhizosphere after inoculation of a certain PSB is limited (Naik et al. 2008; Richardson and Simpson 2011). Application of plant probiotics such as PSB for improving P nutrition in crop plants is strongly connected to our better understanding of bacterial diversity, host specificity, mode of action, appropriate formulation, and method of application (Bianco and Defez 2010; Islam and Hossain 2012; Richardson and Simpson 2011). Recent advances on whole genome sequencing of several important crop plants and PSB strains will provide a future basis for better understanding of PSB–plant interactions and development of improved strains as effective biophosphorus fertilizer for eco-friendly low-input sustainable agriculture. Several comprehensive reviews have recently been published on plant growth-promoting microorganisms (Haas and Defago 2005; Lugtenberg and Kamilova 2009) or soil microorganisms mediating phosphorus availability (Islam and Hossain 2012; Richardson and Simpson 2011).

Similarly to P nutrition, a large body of literature reveals that bacteria from diverse genera such as *Acetobacter*, *Azoarcus*, *Azospirillum*, *Bacillus*, *Burkholderia*, *Herbaspirillum*, *Klebsiella*, *Rhizobium*, and *Pseudomonas* are capable of N_2 fixation from the atmosphere (Dixon and Kahn 2004; Doty 2011). The process of N_2 fixation by bacteria is known as biological nitrogen fixation (BNF). BNF by a variety of symbiotic, associative, and free-living microorganisms has tremendous importance for the environment and for world agriculture. Nitrogen fixation is considered one of the key steps of the nitrogen cycle, as it replenishes the overall nitrogen content of the biosphere and compensates for the losses that are incurred due to denitrification. Application of diazotrophic bacteria substantially supplements the N requirement and promotes the growth of crop plants. The fixed N_2 that is provided by BNF is less prone to leaching and volatilization as it is utilized *in situ*.

Therefore, this biological process contributes as an important and sustainable input into agriculture (Dixon and Kahn 2004).

Although symbiotic nitrogen fixation by *Rhizobium* spp. and their molecular cross-talk with leguminous plants have been well studied, nitrogen fixation by nonsymbiotic free-living or endophytic bacteria and their interactions with plants are poorly understood (Dixon and Kahn 2004; Doty 2011; Vyas and Gulati 2009). The discovery of symbiosis between N_2-fixing bacteria and legumes raises the eventual question of whether such a relationship is possible for nonleguminous plants (Ladha et al. 1997). More research is needed to discover the molecular mechanisms of BNF in nonleguminous crop species based on our understanding of nitrogen fixation biology in legumes (Godfray et al. 2010). Recently, Markmann et al. (2008) have found that several genes, including the so-called "symbiosis-receptor-kinase-gene" (SYMRK), are involved in a genetic program that links arbuscular mycorrhiza (AM) and one form of bacterial nodule symbiosis. The analysis of SYMRK in several species of plant provided striking evidence that most plants have a short version of SMYRK, which is required for AM symbiosis, while a longer variant was found only in plants involved in symbiotic relationships with nitrogen-fixing bacteria. This finding can be considered as an important step toward understanding the evolution of nitrogen fixation in plants, and even whether plants that do not form a symbiosis with nitrogen-fixing bacteria could be engineered to do so, thus increasing their N nutrition to ensure higher productivity. Making crop plants capable of fixing their own nitrogen via a close interaction with diazotrophic bacteria may be used as an alternative strategy for solving nitrogen nutrition in economically important crops such as rice, wheat, maize, and so on (Ladha et al. 1997). Although providing nitrogen nutrition to nonleguminous crops through BNF is a novel approach, its potential has a considerable payoff in term of increasing the production of these crops, which would not only help resource-poor farmers reduce the cost of production but also significantly reduce environmental pollution (Ladha et al. 1997). Several good reviews have recently been published on diazotrophic bacteria in nitrogen nutrition in plants (Beatty and Good 2011; Borriss 2011; Dixon and Kahn 2004; Doty 2011; Mia et al. 2012). In concert with plant nutrition, some elite strains exert multiple effects to promote plant growth through phytohormone production, enzyme activity, siderophore production, antagonisms against phytopathogens, or a combination of these (Bianco and Defez 2010; Duarah et al. 2011; Islam 2011; Islam and Hossain 2012; Naik et al. 2008; Zaidi et al. 2009).

Biological control involves the use of organisms, their products, genes, or gene products to control undesirable organisms (pests) and favor desirable organisms, such as crops, trees, beneficial organisms, and insects. The organism that suppresses the disease or pathogen is referred to as the biological control agent (BCA). In the last three decades, we have witnessed a dramatic development in research on biological control of plant diseases caused by fungi, bacteria, nematodes, and peronosporomycetes (Cook 1993; Islam et al. 2005; Islam 2011; Islam and Hossain 2013). Plant pathologists have been fascinated by the perception that disease-suppressing soil or antagonistic plant-associated bacteria could be used as environment-friendly BCA (Haas and Defago 2005). The concept of biological control by plant probiotic bacteria is becoming popular because of not only increasing public concern about the use of hazardous chemical pesticides, but also the uncertainty or inefficiency of current disease control strategies against phytopathogens (Cook 1993; Islam et al. 2005). Biological control strategies attempt to enhance the activities of BCA either by introducing high populations of a specific bacterium or by enhancing the conditions that enable the bacterium in its natural habitat to suppress the diseases (Nelson 2004). In fact, bacteria are easy to deliver, increase biomass production and yield, and improve soil and plant health (Islam and Hossain 2013). Isolation and characterization of new potential bacteria and understanding their ecology, behavior, and mode of action are considered as major foci in current biocontrol research (Cook 1993; Islam et al. 2005, 2011). The rapid development of convenient techniques in molecular biology has revolutionized this field by facilitating the identification of the underlying molecular mechanism of pathogen suppression (Islam et al. 2005, 2011; Islam 2008; Islam and von

Tiedemann 2011), and by providing means for construction of "superior" bacteria through genetic engineering (Bainton et al. 2004).

A large body of literature indicates that bacterial antagonists can significantly suppress disease caused by peronosporomycete and fungal phytopathogens and increase the yield of crops. Bacterial antagonists commonly studied and deployed for the control of peronosporomycete diseases include *Pseudomonas, Bacillus, Burkholderia, Lysobacter, Actinobacter, Enterobacter, Paenibacillus,* and *Streptomyces* (Cook 1993; Haas and Defago 2005; Islam et al. 2005, 2011; Islam 2011; Islam and Hossain 2013). Suppression of pathogens or diseases by the biocontrol agents is accomplished in several ways, such as production of antibiotics or lytic enzymes (Islam et al. 2005, 2011; Islam 2008; Islam and von Tiedemann 2011; Osburn et al. 1995; Perneel et al. 2008), competition for specific nutrients (e.g., iron or carbon) (Lee et al. 2008), induction of systemic resistance in the host plants (Yan et al. 2002), and parasitizing the pathogen's hyphae (Tu 1978), reproductive structures, or both (Islam et al. 2005; Khan et al. 1997). Several good reviews on biocontrol of plant diseases by plant probiotic bacteria have been published (Borriss 2011; Haas and Defago 2005; Islam 2011; Islam and Hossain 2013; McSpadden Gardener and Fravel 2002).

8.3 HIGH-THROUGHPUT APPROACHES: NECESSITY IN EXPEDITING SUSTAINABLE AGRICULTURE

The twenty-first century has seen the unravelling of the genomes of various plant species, including the most important food crops (Feuillet et al. 2010). It is these crop genomes that are bringing a paradigm shift in the approach to plant biology and crop breeding to meet future global food demand through crop improvement (Flavell 2010). With this genomic information, scientists are able to efficiently utilize the high-throughput technologies—transcriptomics (expression of genes genome-wide), proteomics (expression of proteins), and metabolomics (metabolites)—in order to systematically reveal the function of each gene in the genome (Fukushima et al. 2009; Weckwerth 2011). These technologies will not only address fundamental biological questions but also help create new resources in terms of plants that will be able to withstand adverse climatic conditions, which is one of the main goals of plant biologists and breeders in particular.

8.3.1 Proteomics Approach and Its Importance among Omics Technologies

In the functional genomics era, proteomics is undoubtedly one of the rapidly emerging and expanding fields of study, dealing with large-scale and systematic study of the protein population in the cell. Proteomics is absolutely necessary, and the statement of Watson and coworkers presents the reason—"as we seek to better understand the gene function and to study the holistic biology of systems, it is inevitable that we study the proteome"—highlighting the power of proteomics to address important physiological questions (Watson et al. 2003). Plant proteomics followed the advances in proteomics of mammalian systems, and readers are referred to the comprehensive book *Plant Proteomics: Technologies, Strategies, and Applications* for an in-depth reading on the subject (Agrawal and Rakwal 2008). Examples of proteomics research in plants/crops are given in Section 8.4.

8.3.2 Molecular Breeding for Crop Improvement

Plant breeding describes methods for the creation, selection, and fixation of superior plant phenotypes in the development of improved cultivars suited to the needs of farmers and consumers. The primary goal of plant breeding is to improve yield, nutritional qualities, and other traits of commercial value. The integration of advances in biotechnology, genomic research, and molecular marker applications with conventional plant breeding practices has created the foundation for molecular plant breeding. Methods for marker-assisted backcrossing were developed rapidly for the introgression of desirable traits and reduction of linkage drag, by which molecular markers were used in genome scans to select those individuals that contained both the transgene and the greatest proportion of favorable alleles from the recurrent plant genome (Ragot et al. 1995). The continued development and application of plant biotechnology, molecular markers, and genomics has established new tools for the creation, analysis, and manipulation of genetic variation and the development of improved cultivars during the past 25 years (Collard and Mackill 2008; Sharma et al. 2002; Varshney et al. 2006). Although molecular markers and other genomic applications have been successful in characterizing existing genetic variation within species, plant biotechnology generates new genetic diversity that often extends beyond species boundaries (Gepts 2002). For genes that cannot be transferred through crossing, molecular tools have made it possible to create an essentially infinite pool of novel genetic variation. Genes may be acquired from existing genomes spanning all kingdoms of life, or designed and assembled de novo in the laboratory. Biotechnology also facilitates the molecular stacking of transgenes that control a trait or suite of traits into a single locus haplotype defined by a transgenic event. Examples include development of Golden Rice (Ye et al. 2000), or the combination of transgenes that simultaneously increase synthesis and decrease catabolism of lysine in maize seeds (Frizzi et al. 2008). The combination of phenotypic data and molecular marker scores increase selection gains for maize grain yield and resistance to European corn borer (Johnson 2004). A breeding population of 250 corn lines obtained from the use of multiple trait indices and marker-assisted selection (MAS) showed that the use of molecular markers increased breeding efficiency approximately twofold relative to phenotypic selection alone (Eathington et al. 2007). MAS can also significantly enhance genetic gain for traits where the phenotype is difficult to evaluate because of its expense or its dependence on specific environmental conditions. The probability of identifying truly superior genotypes has been made possible by the use of molecular markers (Knapp 1998). Molecular markers for identification of plant resistance to soybean cyst nematode (Young 1999), resistance to cereal diseases (Varshney et al. 2006), and drought tolerance in maize (Tuberosa et al. 2007) have been used successfully. Molecular markers are being employed for detection and exploitation of naturally occurring DNA sequence polymorphisms. A number of molecular markers have been developed and used for the detection of polymorphisms. This started with the use of restriction endonuclease digestion of total genomic DNA followed by hybridization with a radioactively labelled probe, revealing differently sized hybridized fragments. This type of polymorphism, termed restriction fragment length polymorphism (RFLP), has been used extensively for genetic studies. However with the development of polymerase chain reaction (PCR), the field of molecular biology was revolutionized. A number of PCR-based molecular marker techniques have been developed, beginning with randomly amplified polymorphic DNAs (RAPDs) and simple sequence repeats (SSRs). These PCR-based procedures have the advantages that they are technically simple, are quick to perform, require only small amounts of DNA, and do not involve radioactivity (Waugh and Powell 1992). However, the gel-based assays that are needed for most molecular markers are time-consuming and expensive, limiting their utility at times. The new-generation molecular markers, single-nucleotide polymorphisms (SNPs), do not always need these gel-based assays. They are the most abundant of all marker systems known so far. A beginning has been made in the development and use of SNPs in higher plants, including some crops and tree species. Several approaches can be used for the discovery of new SNPs, and about a dozen different methods are also suitable for automation and high-throughput approaches. These methods, in principle, make a distinction between a perfect match and a mismatch (at the SNP site) between a probe of known sequence and the target DNA containing the SNP site. The target DNA in most of the methods is a PCR product, except in some cases such as "invasive cleavage assay" and "reduced representation shotgun (RRS)" (Gupta et al. 2001). SNPs are the only new-generation molecular markers for individual genotyping needed for MAS. There is some evidence that the stability of SNPs and, therefore, the relative fidelity of their inheritance are higher than those of the other molecular systems such as SSRs and AFLPs. SNPs at a particular site in the DNA molecule should, in principle, involve four possible nucleotides, but in practice only two of these four possibilities have been observed at a specific site in a population. Consequently, SNPs are biallelic. However, the extraordinary abundance of SNPs largely offsets the disadvantage of their being biallelic and makes them the most attractive molecular marker developed so far. According to recent estimates, one SNP occurs every 100–300 bp in any genome, thus making SNPs the most abundant molecular markers known so far. One can only hope that SNPs will be developed expeditiously in all major crops in a variety of crop improvement programs, although the nonavailability of adequate sequence data may limit this activity.

The other high-throughput technique, diversity arrays technology (DArT), is an upcoming technology that does not require sequence information. DArT discovers a large number of markers in parallel and does not require further development of an assay once markers are discovered. DArT was developed to provide a practical and cost-effective whole genome fingerprinting tool (Jaccoud et al. 2001). Development of DArT starts with assembling a group of DNA samples representative of the germplasm anticipated to be analyzed with the technology. This group of samples usually corresponds to the primary gene pool of a crop species, but can be restricted to the two parents of a cross or expanded to secondary or even tertiary gene pools. The DNA mixture representing the gene pool of interest is processed using a complexity reduction method—a process by which a defined fraction of genomic fragments, called representation, is then used to construct a library in *Escherichia coli*. The inserts from individual clones are amplified and used as molecular probes on DArT assays. DArT markers are biallelic markers that are either dominant or hemidominant. DArT has been successfully developed for 16 plant species and two species of plant pathogenic fungi. Rice was used for the initial proof-of-concept work of DArT (Jaccoud et al. 2001).

The choice of molecular marker technology depends strongly on the intended application. It is likely that the science of genomics in plants will follow the path of human genetics: increasing the resolution of analysis through increased marker density.

8.4 PROTEOMICS CONTRIBUTION IN CEREAL CROPS: RICE AND WHEAT

Although rice, like other plants, faces numerous challenges from abnormal environmental factors during its growth and development toward the final product (seed), it is more prone to abiotic factors such as high and low temperatures, drought, flooding, ozone (O_3), and so on (for review see Agrawal et al. 2006, 2009; Agrawal and Rakwal 2006, 2011; Rakwal and Agrawal 2003). These major abiotic stress factors also cause numerous changes in plants at the level of the proteome (for review see Kosová et al. 2011). Physiology, molecular biology, and genetics have greatly improved our understanding of the responsiveness of rice to stress. Recently a lot of effort has been applied to proteomic analysis in rice, and systematic studies have been performed on the functional identification of proteins present in different tissues at different developmental stages. In the last couple of years some interesting high-throughput proteomic approaches have been undertaken to understand the proteomic response against drought, high and low temperatures, and salinity. Rice, being a crop that requires excellent water management to achieve the desired yields, is particularly susceptible to water shortage or drought conditions. Drought, a meteorological event causing absence of rainfall for a period of time, can cause the soil to be depleted of moisture, and the water deficit results in a decrease of water potential in plant tissue (Hadiarto and Tran 2010; Mitra 2001; Zhang 2007). A recent paper by the group of Paul Haynes at Macquarie University, New South Wales, Australia, has recently assessed changes in the physiology and proteome of rice (cv. Nipponbare) leaf due to drought stress (Mirzaei et al. 2012a). These authors used a controlled regime of drought lasting for two weeks, which was then followed by rewatering and rapid recovery in order to understand the gene-level events as plants experienced drought and recovery. Proteins were identified from leaf samples after moderate drought, extreme drought, and three and six days of rewatering by a combination of label-free quantitative shotgun proteomic and spectral counting using normalized spectral abundance factors, which resulted in the identification of 1548 nonredundant proteins (Mirzaei et al. 2012a). Results revealed that proteins were downregulated in the early stages of drought, and, as the drought became severe, proteins were found to be upregulated. When the stressed plants were rewatered, proteins were downregulated, suggesting that stress-related proteins were being degraded. The dominant identified proteins were found to be involved in signalling and transport under severe drought and decreased on rewatering. Mirzaei and coworkers speculated that water transport and drought signalling are critical elements of the overall response to drought in rice and might be the key to biotechnological approaches to drought tolerance (Mirzaei et al. 2012a). Similar studies in rice plants grown in split root systems to analyze long-distance drought signalling within root systems indicated a general upregulation of pathogenesis-related (PR), heat shock proteins (HSPs), and oxidation–reduction proteins in drought-exposed roots (Mirzaei et al. 2012b). PR proteins are known to play an important role in general adaptation to stress, along with HSPs, which act as molecular chaperones. Several peroxidases, SODs and catalases were also identified, which perhaps helped to suppress the levels of reactive oxygen species (ROS) in response to drought. Interestingly, plant microtubules and chitinases were also shown to be affected by transmissible defense-inducing signals in response to drought, thus suggesting that there is no single class of proteins unaffected by stress. Quantitative tandem mass tag proteomic analysis of drought-stressed parental and drought-tolerant near isogenic lines (NILs) for a rice quantitative trait locus (QTL) for yield under drought were also studied to understand the proteomic response to severe reproductive-stage drought stress. Systematic analysis of protein profiling in three different tissues of the same plant (flag leaf, panicles, and roots), suggested effective carbon–nitrogen remobilization in flag leaf, panicle, and roots and enhanced detoxification of ROS species during stress as the key contributing factor to the remarkable performance of NILs (unpublished data, personal communication with Ajay Kohli). The proteomic response of rice plants to other abiotic stress factors such as high and low temperature has also been studied. Quantitative label-free shotgun proteomic analysis of 850 stress-responsive proteins from cultured rice cells exposed to high and low-temperature stress provided molecular insights into thermal stress response in plants (Gammulla et al. 2010). This study mainly concluded that there was a higher abundance of proteins involved in stress response, carbohydrate metabolism, lipid metabolism, cell redox homeostasis, cell wall modification, and cell division in response to high-temperature stress. Low-temperature stress elicited proteins involved in protein metabolism and cellular component organization. The same authors conducted a similar successor study to investigate the effect of high and low-temperature stress on the leaves of rice seedlings. Contrasting responses were observed in leaves and in studies involving suspension-cultured cells Proteins involved in protein metabolism, such as eukaryotic initiation factors and elongation factors, were seen to be upregulated in leaves more than in cell culture (Gammulla et al. 2011). The authors hypothesized that this difference was due to the rapid cell turnover and coexistence of dividing and senescent cells in the case of cell cultures. This study also led to the identification of 20 novel stress-response proteins (Gammulla et al. 2011). They also provided an initial functional annotation to these novel proteins. The study also highlighted for the first time the presence of chloroplast ribosomal proteins and translation releasing factors in response to cold stress (Gammulla et al. 2011). Differentially expressed ubiquitinated proteins were identified in rice roots by nanospray liquid chromatography/tandem mass spectrometry in response to the initial phase of salt stress responses in rice seedlings. The expression of ubiquitination on pyruvate phosphate dikinase 1, heat shock protein 81–1, probable aldehyde oxidase 3, plasma membrane ATPase, cellulose synthase A catalytic subunit 4 (UDP-forming), and cyclin-C1-1 was identified and analyzed before and after salt treatment (Liu et al. 2012). Anther proteomic patterns for two contrasting rice genotypes under salt stress were also compared using matrix-assisted laser desorption/ionization time-of-flight/time-of-flight (MALDI-TOF/TOF) mass spectrometry (MS) analysis to understand the basis of salt tolerance. Several proteins were identified that might increase plant adaptation to salt stress by modulating important metabolic or biochemical processes such as carbohydrate/energy metabolism, anther and pollen wall remodelling and metabolism, and protein synthesis and assembly (Sarhadi et al. 2012).

Knowledge of protein alterations under biotic and abiotic stresses should help us understand the molecular mechanism of stress tolerance in rice at the translational/posttranslational instead of the transcriptional level. This is imperative for understanding complex environmental signalling responses and engineering cultivars tolerant to stress, further leading toward sustainable agriculture.

In wheat, let us look at one of the less investigated components of climate change, O_3 stress, which is also a major environmental pollutant affecting plant growth and productivity (Cho et al. 2011). A recent study examining for the first time proteomics aspects of O_3-exposed wheat plants was carried out by Sarkar and coworkers at Banaras Hindu University, Varanasi, India (Sarkar et al. 2010). This study was designed to evaluate the impact of elevated concentrations of O_3 on phenotypical, physiological, and biochemical traits in two high-yielding cultivars of wheat, and also to analyze the leaf proteome using one/two-dimensional gel electrophoresis (1-/2-DGE) in conjunction with immunoblotting and MS analyses under near-natural conditions using open-top chambers. The O_3 exposure caused specific foliar injury in both the wheat cultivars. Results also showed that O_3 significantly decreased photosynthetic rate, stomatal conductance, and chlorophyll fluorescence kinetics (Fv/Fm) in test cultivars. Biochemical evaluations also showed a greater loss of photosynthetic pigments. A significantly induced antioxidant system under elevated O_3 concentrations indicates the ability of O_3 to generate oxidative stress. 1-DGE analysis showed drastic reductions in the abundantly present ribulose-1,5-bisphosphate carboxylase/oxygenase (RuBisCO) large and small subunits. Immunoblotting with specific antibodies confirmed induced accumulation of antioxidative enzymes such as SOD and ascorbate peroxidase protein(s) and common defense/stress-related thaumatin-like protein(s). 2-DGE analysis revealed a total of 38 differentially expressed protein spots, common to both the wheat cultivars. It was found that some leaf photosynthetic proteins (including RuBisCO and RuBisCO activase) and important energy metabolism proteins (including ATP synthase, aldolase, and phosphoglycerate kinase) were drastically reduced. On the other hand, some stress/defense-related proteins (such as harpin-binding protein and germin-like protein) were found to be induced. The study reveals an intimate molecular network, provoked by O_3, affecting photosynthesis and triggering antioxidative defense and stress-related proteins, culminating in accelerated foliar injury in wheat plants (Sarkar et al. 2010).

8.5 PROTEOMICS CONTRIBUTION IN MODEL LEGUME SPECIES: A CASE STUDY IN SEEDS

Legumes, including *Medicago truncatula*, soybean, and peanut are among the most important crops worldwide, playing a major role in agriculture, the environment, and human and animal nutrition and health (Graham and Vance 2003). Legume seeds provide a valuable source of edible oils and proteins for feeding both animals and humans (Mitchell et al. 2009), although the levels of protein, oils, and starch vary between legume species. The protein content of legume seeds ranges from 20% to 40%, depending on the species. Seeds of soybean and *M. truncatula* exhibit high protein content, whereas the starch content is very low.

Seed development is a key step of the plant life cycle and is a complex process that determines the nutrient value of seeds for proteins and fatty acids. Particularly during the seed filling period, drastic changes in protein and oil composition occur. Due to the importance of seed filling, systematic studies of this phase of seed development are beginning in legumes. Comparative analysis during seed development is also being studied, with considerable attention to understanding protein and allergen accumulation and metabolic regulation at the level of the proteome.

Proteomic analyses provide a powerful tool to address biochemical and physiological aspects, not only of plant responses to abiotic and biotic stresses, but in various tissues at different developmental stages (leaf senescence, root symbiosis, seed development and germination). In addition, advances in 2-DGE coupled with MALDI-TOF-MS and nano-electrospray ionization liquid chromatography-tandem mass spectrometry (nESI-LC-MS/MS) and bioinformatics for protein identification, has been applied to understand gene function and characteristics of various genotypes in many legume species, including *Glycine max* (soybean), peanut (*Arachis hypogaea*), *Lotus japonicus* (Japanese trefoil) and *Medicago truncatula* (barrel medic). During the past decade, several proteomic studies during seed development or in mature seeds have been carried out for different legume species, such as *L. japonicus* (Dam et al. 2009), *M. truncatula* (Gallardo et al. 2003, 2007), *G. max* (Agrawal et al. 2008; Hajduch et al. 2005; Krishnan et al. 2009), and *A. hypogaea* (Kottapalli et al. 2008). These data, along with the reports on other legume proteomics, will be useful for future analysis of legume seed proteins. Furthermore, an interactive web database for proteomics data for seed filling in soybean and other legume oilseeds studied has been established (http://oilseedproteomics.missouri.edu/), which is freely available on publication of individual datasets.

8.5.1 Proteomic Studies in Soybean

Soybeans are responsible for approximately $12 billion in annual crop value and more than $5 billion in annual export value to the U.S. economy (Gunstone 2001). Soybean is also one of the important sources of fatty acids and proteins for human and animal nutrition (for review, see Thelen and Ohlrogge 2002), as well as other legumes. As mentioned above, seed filling during seed development is a period that largely determines the relative levels of storage reserves in seeds in accordance with significant changes in protein and oil composition. Particularly at this stage, soybean seed accumulates 40% storage reserves, mainly comprising the two abundant seed storage proteins, glycinin and β-conglycinin (Hill and Breidenbach 1974). So far, almost 80 storage proteins have been identified in soybean seeds. These results suggested that the relative proportion of storage components in seeds should vary dramatically among different plant species. Therefore, research on mechanisms of soybean development is not only useful for acquiring a basic understanding of plant developmental biology, but also valuable for improving agronomic traits, such as improvements in nutritional quality and nonallergenic proteins, that could influence other traits of agronomic value.

To gain insight into the complex process of seed development, identification of proteins and their expression profiles was comprehensively investigated, resulting in the establishment of metabolic biosynthetic flow during seed filling. An investigation on soybean (var. Maverick) seed filling (Hajduch et al. 2005) successfully profiled 679 2-DGE-separated protein spots at five sequential developmental stages (2–6 weeks after flowering). Analysis of each of these protein spot groups by MALDI-TOF-MS yielded the identity of 422 of these proteins, representing 216 protein activities. These proteins were classified into 14 functional classes, according to a revised classification scheme originally used for the *Arabidopsis* genome. Overall, metabolism-related proteins were decreased, while proteins associated with destination and storage were increased. The accumulation of unknown proteins, sucrose transport and cleavage enzymes, cysteine and methionine biosynthesis enzymes, 14-3-3-like proteins, lipoxygenases, storage proteins, and allergenic proteins during seed filling is also identified and discussed to elucidate whether they are involved in seed development. Agrawal et al. (2008) also performed an in-depth investigation of proteins expressed during seed filling in soybean using 2-DGE and semicontinuous multidimensional protein identification technology (MudPIT) (Sec-MudPIT) in combination with nESI-LC-MS/MS. 2-DGE and Sec-MudPIT analyses were conducted on five sequential seed stages (two–six weeks after flowering) and identified 478 proteins out of 675 protein spots on high-resolution 2-D gel reference maps. The identified proteins were mainly protein components of metabolic biosynthetic pathways of carbohydrates, fatty acids, and proteins in soybean, and they were compared with a parallel study of rapeseed using high-quality and integrated datasets. These studies provide knowledge about the importance of metabolic pathways involved in legume seed development, especially the synthesis of storage globulins. Furthermore, these results suggest that a detailed analysis of the proteomes of seeds could contribute to efforts aimed at increasing the nutritional value of seeds marketed for human consumption.

Seed storage proteins have been detected on 2-D gels having different positions and amounts depending on the variation in different genotypes (genetically determined quantitative variations). Several proteomic studies have demonstrated that seed storage protein content and composition (glycine subunits, trypsin inhibitors, major seed allergens, low abundant metabolic proteins) in soybean vary in accordance with genetic variability (Mahmoud et al. 2006; Natarajan et al. 2007a,b, 2013; Xu et al. 2007). For example, Natarajan et al. (2007a) characterized that the wild genotypes have a higher number of glycinin protein spots (G1, G2, G3, G4, and G5), which are high-abundant seed storage proteins, when compared with the other three genotypes. Major variation was observed in acidic polypeptides of G3, G4, and G5 compared with G1 and G2, and minor variation was observed in basic polypeptides of all subunits. These data indicated that there are major variations of glycinin subunits between wild and cultivated genotypes rather than within the same groups. Thus, approaches combining plant proteomics and plant genetics can help us elucidate the genetic control of seed protein composition, which can be exploited for legume crop improvement. These results give us important knowledge about the variability of protein and protein subunit accumulation among various cultivars, which is useful for improving both quantity (high-yield proteins) and quality (allergen-free proteins) of soybean seed proteins, which will benefit the overall utilization of soybean in the food and feed industries.

There have been technical advances in the fractionation of high-abundant seed proteins and high-resolution 2-D gels of soybean seeds. Technical advances related to 2-DGE and MS for protein identification have improved the sensitivity, reproducibility, and accuracy of proteome analysis (Rabilloud and Lelong 2011). It is possible to characterize various complex protein samples, since the 2-DGE-MS strategy has proven to be a reliable and efficient means of proteome analysis. However, this 2-DGE-MS strategy has its own difficulties in resolving low-abundant proteins (LAPs) due to the limited loading capacity of isoelectric focusing (IEF) gels (Rothemund et al. 2003). For instance, proteomic assessment of the LAPs within soybean seed is difficult when the overwhelming majority, sometimes 60%–80%, is made up of storage proteins (Krishnan et al. 2009). Basically, cultivated soybean seed storage proteins consist primarily of two major storage protein complexes, glycinin and β-conglycinin. Glycinin, accounting for roughly 40%–60% of the total seed protein, is a hexameric protein, ranging from 320 to 375 kDa. There have been attempts to establish an extraction method for storage proteins, including Ca^{2+} (Krishnan et al. 2009) and isopropanol fractionation (Natrajan et al. 2009). More recently, a simplified extraction method to improve the detection of LAPs by a protamine sulfate precipitation (PSP) method (unpublished data, personal communication with SunTae Kim) has been developed. These fractionation methods are simple, fast, economic, and reproducible for seed storage protein fractionation and suitable for downstream in-depth proteomics analysis.

Soybean possesses about 15 proteins recognized by IgEs from soy-sensitive people. So far, it has been reported that soybean seed contains 16 known protein allergens, including the seed storage proteins glycinin, beta conglycinin and alpha subunit, Gly m Bd 30K (vacuolar thiol protease), and Kunitz trypsin inhibitor (KTI), with differing degrees of severity (Thelen 2009). Glycinin, the most abundant seed storage protein in soybean seed, consists of five subunits, G1, G2, G3, G4, and G5. Among these subunits, only G2 and acidic polypeptides of G1 are reported to be allergens (Ogawa et al. 2000). The well-known soybean allergenic protein is P34, a member of the papain superfamily of cysteine proteases, and more than 65% of soy-sensitive patients react to the P34 protein (Wilson et al. 2005). Xu et al. (2007) have conducted comparative studies of allergen proteins in 16 cultivated and wild soybean genotypes, revealing that considerable heterogeneity of the α subunit of β-conglycinin distribution exists among these 16 soybean genotypes. The data may serve as a 2-D reference map for comparison of soybean allergenic proteins in various soybean genotypes, and may also be useful for the modification of soybean to improve its nutritional value.

8.5.2 Proteomic Studies in Peanut

Peanut is an important food legume and is also one of the most important leguminous crops in the world, both for vegetable oil and as a protein source. Several research groups have reported studies of peanut proteomics. To date, proteomics has enabled the identification and characterization of methionine-rich protein (MRP) from cultivated peanut (Basha and Pancholy 1981; Sathanoori and Basha 1996), the establishment of genetic variation among peanut cultivars (Kottapalli et al. 2008), and the discovery of protein markers that are able to distinguish given subspecies (Liang et al. 2006). Kottapalli et al. (2008) first profiled total seed proteins isolated from mature seeds of four peanut cultivars (New Mexico Valencia C (NM Valencia C), Tamspan 90, Georgia Green, and NC-7) using 2-DGE coupled with nESI-LC–MS/MS. Among 20 abundant protein spots showing differences in relative abundance among these cultivars, 14 nonredundant proteins were identified by nESI-LC–MS/MS, suggesting that the major proteins belong to arachin (glycinin and Arah3/4) and conarachin seed storage proteins, as well as other allergenic proteins. Some of these proteins showed cultivar-specific expression patterns. For example, New Mexico Valencia C showed low levels of the antinutritive proteins, such as lysyl oxidase (LOX) and galactose-binding lectin. Conversely, Arah3/h4, an allergen with decreased allergenic properties, was highly abundant in Tamspan 90, suggesting that the identified proteins might serve as potential markers for cultivar differentiation. It may be implied that no single cultivar has all the desirable traits for breeding a cultivar to increase seed quality in hypoallergenic peanut lines.

8.6 PROTEOMICS CONTRIBUTION IN HORTICULTURAL CROPS

8.6.1 Strawberry

Strawberries are a rich source of vitamins and have immense economic value due to their unique flavor. The strawberry proteome was analyzed to investigate the developmental changes during berry ripening (Bianco et al. 2009). Two-dimensional difference gel electrophoresis (2-D-DIGE) gels of three different stages of berry development (immature, turning, and red) showed 568, 622, and 520 spots, respectively. Alternatively, a shotgun approach was also followed to identify the proteins involved in berry development. The identified proteins mainly belonged to the defense, energy, and secondary metabolism categories. Citric acid is an important constituent of the strawberry. Interestingly, many enzymes of the citric acid cycle, including citrate synthase, aconitase, isocitrate dehydrogenase, 2-oxoglutarate dehydrogenase complex, succinyl-CoA synthetase, succinate dehydrogenase, fumarase, and malate dehydrogenase, were identified. In addition, ripening-related proteins showed accumulation from immature to red berries. Four allergens were also detected in the strawberry proteome, suggesting a plausible explanation of strawberry allergy in some people. A comparison of protein sequences of strawberry Fra a 1 allergen with birch pollen allergen Bet v 1 and corresponding apple allergen showed 54% and 77% identity, respectively. Decreased abundance of allergens and enzymes involved in red pigment synthesis was observed in white strawberries, further confirming why red strawberry-allergic persons are not allergic to white strawberries (Hjernø et al. 2006). A detailed comparison of proteome profiles of different varieties of strawberries (red and white) showed that the existence of the allergen Fra a 1 varied between varieties, suggesting why some people are allergic to a particular variety of strawberry and not to others (Alm et al. 2007).

8.6.2 Grape

Wine grapes are an important horticultural crop due to their high nutrient content and high demand in the wine industry. Although analysis of the grape berry proteome is relatively difficult due to the high content of secondary metabolites and sugars that interferes with the protein extraction and separation methods, a number of reports have been published regarding the proteome analysis of the grape berry.

The proteomics of berry development has been studied in all parts of the berry, including skin (Deytieux et al. 2007; Negri et al. 2008a), pulp/flesh (Giribaldi et al. 2007; Martinez-Esteso et al. 2011), and whole berries (Sarry et al. 2004). In addition, grape berry proteome has been analyzed during postharvest withering (Carli et al. 2010). Results obtained from these studies reveal the accumulation of PR proteins such as thaumatin-like protein, different isoforms of chitinase, β-1,3-glucanase, abscisic stress ripening protein, polyphenol oxidase, and so on, in all the berry parts during berry maturation and withering. Increased activity of chitinase and β-1,3-glucanase was also reported in berry skin as the berry ripens. However, decreased abundance of HSPs and energy and general metabolism-related proteins was observed in berry skin from onset of ripening to color change. Besides, proteins associated with anthocyanin biosynthesis and cell wall loosening showed increased abundance during berry maturation (Deytieux et al. 2007).

In berry flesh, a majority of the identified proteins were associated with sugar and organic acid metabolism, indicating a vast reprogramming of sugar and acid content of the berries during ripening. In contrast to berry skin, proteins related to energy metabolism showed increased abundance during ripening. In addition, enzymes of protein synthesis and regulation also showed higher expression in flesh, suggesting a significant modulation of the flesh proteome during ripening (Martinez-Esteso et al. 2011). During maturation and withering, dehydration of berries takes place, and wax deposition in berries is observed to minimize this dehydration. An increased abundance of lipid metabolism-related proteins such as dienelactone hydrolase and 3-ketoacyl-(acyl-carrier-protein)-reductase was observed in matured and withered berries. The increased abundance of these enzymes can be correlated with wax synthesis, which minimizes water loss. Proteins associated with the cytoskeleton showed accumulation during ripening (Carli et al. 2010; Giribaldi et al. 2007) and remained similar up to withering (Carli et al. 2010).

The effect of water-deficit stress was analyzed in grape pericarp, seeds, and skin. Water deficit affected the abundance of approximately 7% of the grape pericarp proteins. In skin, water-deficit stress increases the abundance of proteosome subunit proteins and proteases, ROS-detoxification enzymes, and selected enzymes involved in flavonoid biosynthesis, whereas the pulp showed increases in isoflavone reductase, glutamate decarboxylase, and an endochitinase. In contrast, the seed proteome was least affected by water deficit, and mainly comprised seed storage, maturation, and late embryogenesis abundant (LEA) proteins. Analysis of the metabolome in these tissues under water-deficit conditions showed accumulation of caffeic acid, proline, shikimate, and gluconate in skin and alanine, catechine, myo-inositol, shikimate, and sucrose in pulp (Grimplet et al. 2009).

Identification of herbicide (flumioxazin)-treated grape wine berry proteins showed degradation of RuBisCO, suggesting a negative effect of the herbicide on photosynthesis. However, several isoforms of PR-10 showed increased abundance, indicating their pivotal roles in decreasing the negative effects of herbicide stress. In addition, accumulation of enzymes involved in the photorespiration and antioxidant system was observed, suggesting activation of photorespiration and ROS as a result of herbicide stress (Castro et al. 2005).

The effect of abscisic acid (ABA) was analyzed before veraison in deseeded berries and at veraison in berry flesh and skin. ABA treatment affects the abundance of 60 protein spots of the berries, out of which 40 (15 from whole berries treated before veraison, 9 from berry flesh, and 16 from berry skins of berries treated at veraison) were identified by LC-MS/MS. Results showed that mainly ripening-related proteins, such as vacuolar invertase and NADP-dependent malic enzyme, were affected by ABA treatment. An increased abundance of oxidative stress-related proteins such as ascorbate peroxidase was also identified before and at veraison (Giribaldi et al. 2010).

The effect of methylated cyclodextrins (MBCD) and methyl jasmonate (MeJA) elicitors was analyzed in grapevine cell suspension culture. Out of 233 spots detected in Coomassie Brilliant Blue (CBB)-stained gel, 39 were differentially modulated upon elicitation. Peptide mass finger-printing (PMF) and MS/MS identification of 25 differentially modulated spots showed that they were involved in plant defense. The proteins included class III secretory basic peroxidase, class III chitinase, β-1,3-glucanase, thaumatin-like protein, and so on. Interestingly, some of the identified proteins are involved in systemic acquired resistance (SAR), suggesting that elicitation with MBCD mimics the effect of SAR in plants (Martinez-Esteso et al. 2009). In another similar study in which grapevine cell suspension culture was elicited with MBCD and MeJA, a total of 1031 spots were detected in the 2D-DIGE gel, of which 67 showed altered abundance upon elicitation. The enzymes involved in the trans-resveratrol (tR) biosynthetic pathway showed increased abundance with either MBCD or combined MBCD and MeJA but not with MeJA alone. In addition, accumulation of stil-benoids was observed in response to elicitation (Martinez-Esteso et al. 2011).

For analysis of the cell wall proteome of skin and seeds of the grape berry, four different meth-ods for cell wall protein extraction were tried. Out of the three methods tried for protein extraction (extraction in $LiCl_2$ (lithium chloride), $CaCl_2$ (calcium chloride), and phenol), the phenol method showed the best results, as seen by the highest number of detectable spots and greatest spot resolu-tion. A total of 904 spots were observed in the cell wall proteome extracted by the phenol method, of which 47 were identified. The identified proteins included β-1,3-glucanase, several isoforms of chitinase, and ROS-detoxifying enzymes such as catalase and CuZnSOD. Several proteins related to carbohydrate metabolism were also observed in the cell wall fraction, which was devoid of signal peptide. These proteins included phosphoglyceromutase, glyceraldehyde-3-phosphate dehydroge-nase, fructose-biphosphate aldolase, and so on, and might be involved in cell wall biosynthesis (Negri et al. 2008b).

The effect of the developmental stage was also analyzed in the plasma membrane of the grape berry. Silver-stained 2-D gels of the pre-veraison, veraison, and post-veraison berries showed 119, 98, and 86 spots respectively, of which 12 showed developmental stage-specific expressions. MALDI-TOF-MS identification of the differentially modulated proteins showed their involvement in transportation, metabolism, protein synthesis, and signalling. An increased abundance of ubiq-uitin proteolysis and cytoskeletal proteins at the veraison and post-veraison stages indicates protein degradation during berry ripening, which could be correlated with the decreased number of spots in these stages (Zhang et al. 2008).

8.6.3 Pear

After *Vitis* and *Malus*, pear is the third most economically important fruit produced in the temperate regions. A proteome analysis of pear fruit was conducted to analyze the effect of core breakdown disorder (Pedreschi et al. 2007), extreme gas conditions (Pedreschi et al. 2009), storage in a controlled atmosphere (Pedreschi et al. 2008), and bagging treatment (Feng et al. 2011). To analyze the effect of core breakdown disorder, proteins were extracted from healthy, sound, and brown tissue and were separated on 2-D gels. The amount of protein was highest in the healthy tis-sue, followed by sound and brown tissue, suggesting protein degradation during core breakdown. Identification of the spots showing differential abundances in healthy and brown tissue by MS/MS showed their involvement in energy metabolism, ROS detoxification, and ethylene biosynthesis. Proteins associated with energy metabolism and defense showed increased abundance in the brown tissue (Pedreschi et al. 2007). DIGE technology was employed to analyze the effect of anoxia and air on pear fruit slices. Upregulation of pentose phosphate pathway (PPP) enzymes was observed dur-ing anoxic conditions, suggesting that PPP is activated as an alternate source of energy production during low-oxygen conditions. In addition, accumulation of PR proteins was also observed, indicating their involvement in overcoming the effect of anoxia in addition to other abiotic and biotic stress conditions (Pedreschi et al. 2009). Proteome analysis of the skin tissue of natural and bagged pear showed decreased abundance of photosynthesis; signalling; and protein, carbohydrate, and acidity metabolism. Regarding photosynthesis, targets related to PS II and RuBisCO were identified, suggesting their probable degradation during fruit bagging (Feng et al. 2011).

8.7 PROTEOMICS IN COMBINATION WITH OTHER OMICS APPROACHES TOWARD NEXT-GENERATION CROPS

8.7.1 Genomes, Proteomes, Transcriptomes: Application of Computational Approaches for Reconstruction, Modelling, and Analysis of -Omes

Plants are an important source of food and various multipurpose products for humans. Traditional studies have focused on reductionist approaches to characterize plant functions and to probe their molecular basis. The complexity of plant functions and responses is not adequately addressed by focusing on the functional details of DNA, RNA, proteins, and metabolites alone. For better understanding of plant functions, one must invoke the complex and subtle details of interactions among these constituent molecules. Lately, in addition to proteomics, there has been a tremendous rise in the exploration of plant systems-level genomes (AGI 2000; IRGSP 2005; Paterson et al. 2009; Schnable et al. 2009), protein interactomes (Dreze et al. 2011; Lee et al. 2010; Uhrig 2006), and transcriptomes (Yamada et al. 2003). The understanding of the plant protein–protein interaction network and other interactomes is expected to provide crucial insights into the regulation of plant developmental, physiological, and pathological processes (Lin et al. 2011; Morsy et al. 2008; Umezawa et al. 2006; Zhang et al. 2010).

Analysis of complex biological systems and their macromolecular interaction networks may be central in characterizing genotype-to-phenotype relationships in plants. Advances in graph theoretical analysis provide methods of functional interpretation of plant protein interactomes (Uhrig 2006). Identification of network modules as putative cellular modularity and network motifs that represent cores of functional modules could lead us to better identification and control of systems-level features specifying various biological processes (Uhrig 2006). It has been indicated that evolutionary processes have left their imprint on protein interactomes (Dreze et al. 2011). Network analysis has also been used to show how pathogens may exploit protein interactions to manipulate a plant's cellular machinery (Mukhtar et al. 2011).

Thus, macromolecular interaction networks help us draw detailed maps of cellular networks reflecting the architecture and dynamic interplay of cellular dynamics. These interaction network models could be used to infer functionally important proteins and regulatory motifs that specify plant processes such as biotic and abiotic stresses, plant defense, plant–pathogen interactions, and so on, which have a strong bearing on sustainable agriculture. As a theoretically oriented question, which could lead to ways of controlling molecular mechanisms, it would be enriching to construct models of evolution of plant interactomes that address evolutionary mechanisms and features specific to plants.

Constructing comprehensive plant–protein interaction maps is a vast challenge, given the estimated number of proteins and interactions among them. There are three chief types of approaches used for construction of plant protein interactomes: *in vitro*, *in vivo*, and *in silico*. *In vitro* approaches consist of experimental methods such as protein microarrays, surface plasmon resonance, immuno-purification, and so on, and are limited by the need for sophisticated instruments and procedures. *In vivo* approaches consist of experimental methods such as yeast two hybrid (Y2H) and protoplast Y2H. A high false positive rate is one of the major shortcomings of these methods. Interactions could also be predicted by *in silico* analysis methods. One of the methods used is that of predicted orthologs. Predicted interologs identification has been used as a predictor of protein interaction on the premise that orthologous proteins, which are known to interact in one organism, can interact in the organism under study.

Several protein interaction databases have been compiled, including IntAct (Kerrien et al. 2012), Molecular Interaction Database (MINT) (Zanzoni et al. 2002), *Arabidopsis thaliana* Protein Interaction Networks (AtPID) (Cui et al. 2007), Database of Interacting Proteins (DIP) (Xenarios et al. 2000), Biomolecular Interaction Network Database (BIND) (Bader et al. 2001), and Biological General Repository for Interaction Datasets (BioGRID) (Stark et al. 2006).

So, in this era of integrative sciences, systems biology is being used for analysis and better understanding of complex interactome data generated through wet lab experiments. It has been inferred that using novel combinatorial approaches employing computational expertise, along with knowledge gained from genomics, transcriptomics, metabolomics, and proteomics-based studies, would lead to identification of targets for resolving the issues in agriculture and sustainability.

ACKNOWLEDGMENTS

Randeep Rakwal acknowledges the great support of Professor Yoshihiro Shiraiwa (Provost, Faculty of Life and Environmental Sciences, University of Tsukuba), Professor Koji Nomura (Organization for Educational Initiatives, University of Tsukuba), and Professor Seiji Shioda and Dr. Tetsuo Ogawa (Department of Anatomy I, Showa University School of Medicine) in promoting interdisciplinary research and unselfish encouragement. M. T. Islam acknowledges funding from World Bank to CP#2071 under HEQEP. The authors acknowledge the International Plant Proteomics Organization (INPPO) (www.inppo.com) platform for this initiative in bringing together scientists of different disciplines in constructing this review. Finally, given the vast amount of research in these disciplines and space limitations, many works could not be cited and discussed in this review.

REFERENCES

Abd-Alla, M. H. 1994. Phosphatases and the utilization of organic phosphorus by *Rhizobium leguminosarum* biovar *viceae*. *Lett Appl Microbiol* 18:294–296.

Adams, R. M., B. H. Hurd, S. Lenhart, and N. Leary. 1998. Effects of global climate change on agriculture: An interpretative review. *Clim Res* 11:19–30.

AGI (The Arabidopsis Genome Initiative). 2000. Analysis of the genome sequence of the flowering plant *Arabidopsis thaliana*. *Nature* 408:796–815.

Agrawal, G. K., M. Hajduch, K. Graham, and J. J. Thelen. 2008. In-depth investigation of the soybean seed-filling proteome and comparison with a parallel study of rapeseed. *Plant Physiol* 148:504–518.

Agrawal, G. K., N. S. Jwa, Y. Iwahashi, M. Yonekura, H. Iwahashi, and R. Rakwal. 2006. Rejuvenating rice proteomics: Facts, challenges, and visions. *Proteomics* 6:5549–5576.

Agrawal, G. K., N. S. Jwa, and R. Rakwal. 2009. Rice proteomics: End of phase I and beginning of phase II. *Proteomics* 9:935–963.

Agrawal, G. K. and R. Rakwal. 2006. Rice proteomics: A cornerstone for cereal food crop proteomes. *Mass Spectrom Rev* 25:1–53.

Agrawal, G. K. and R. Rakwal. 2008. *Plant Proteomics: Technologies, Strategies, and Applications*. Hoboken, NJ: Wiley.

Agrawal, G. K. and R. Rakwal. 2011. Rice proteomics: A move toward expanded proteome coverage to comparative and functional proteomics uncovers the mysteries of rice and plant biology. *Proteomics* 11:1630–1649.

Alm, R., A. Ekefjärd, M. Krogh, J. Häkkinen, and C. Emanuelsson. 2007. Proteomic variation is as large within as between strawberry varieties. *J Proteome Res* 6:3011–3020.

Apse, M. P., G. S. Aharon, W. A. Snedden, and E. Blumwald. 1999. Salt tolerance conferred by overexpression of a vacuolar Na$^+$/H$^+$ antiport in *Arabidopsis*. *Science* 285:1256–1258.

Apse, M. P., J. B. Sottosanto, and E. Blumwald. 2003. Vacuolar cation/H$^+$ exchange, ion homeostasis, and leaf development are altered in a T-DNA insertional mutant of AtNHX1, the *Arabidopsis* vacuolar Na$^+$/H$^+$ antiporter. *Plant J* 36:229–239.

Bader, G. D., I. Donaldson, C. Wolting, B. F. Ouellette, T. Pawson, and C. W. Hogue. 2001. BIND: The biomolecular interaction network database. *Nucleic Acids Res* 29:242–245.

Bainton, N. J., J. M. Lynch, D. Naseby, and J. A. Way. 2004. Survival and ecological fitness of *Pseudomonas fluorescens* genetically engineered with dual biocontrol mechanisms. *Microbiol Ecol* 48:349–357.

Basha, S. M. M. and S. K. Pancholy. 1981. Identification of methionine-rich polypeptides in peanut leaf. *J Agric Food Chem* 29:331–335.

Bassil, E., M. A. Ohto, T. Esumi et al. 2011. The *Arabidopsis* intracellular Na$^+$/H$^+$ antiporters NHX5 and NHX6 are endosome associated and necessary for plant growth and development. *Plant Cell* 23:224–239.

Beatty, P. H. and A. G. Good. 2011. Future prospects for cereals that fix nitrogen. *Science* 333:416–417.

Beddington, J. 2009. *Food, Energy, Water and the Climate: A Perfect Storm of Global Events?* London, UK: Government Office for Science.

Berthomieu, P., G. Conéjéro, A. Nublat et al. 2003. Functional analysis of AtHKT1 in *Arabidopsis* shows that Na$^+$ recirculation by the phloem is crucial for salt tolerance. *EMBO J* 22:2004–2014.

Bianco, C. and R. Defez. 2010. Improvement of phosphate solubilization and *Medicago* plant yield by an indole-3-acetic acid-overproducing strain of *Sinorhizobium meliloti*. *Appl Environ Microbiol* 76:4626–4632.

Bianco, L., L. Lopez, A. G. Scalone et al. 2009. Strawberry proteome characterization and its regulation during fruit ripening and in different genotypes. *J Proteomics* 72:586–607.

Borlaug, N. E. and C. R. Dowswell. 2004. Prospects for world agriculture in the twenty-first century. In: R. La, P. R. Hobbs, N. Uphoff, and D. O. Hansen (eds), *Sustainable Agriculture and the International Rice-Wheat System*, pp. 1–18. Madison, WI: Marcel Dekker.

Borriss, R. 2011. Use of plant-associated *Bacillus* strains as biofertilizers and biocontrol agents in agriculture. In: D. K. Maheshwari (ed.), *Bacteria in Agrobiology: Plant Growth Responses*, pp. 41–76. Berlin: Springer-Verlag.

Bradshaw, W. E. and C. M. Holzapfel. 2001. Genetic shift in photoperiodic response correlated with global warming. *Proc Natl Acad Sci USA* 98:14509–14511.

Carli, M. D., A. Zamboni, M. E. Pe et al. 2010. Two-dimensional differential in gel electrophoresis (2D-DIGE) analysis of grape berry proteome during postharvest withering. *J Proteome Res* 10:429–446.

Castro, A. J., C. Carapito, N. Zorn et al. 2005. Proteomic analysis of grapevine (*Vitis vinifera* L.) tissues subjected to herbicide stress. *J Exp Bot* 56:2783–2795.

Cellier, F., G. Conejero, L. Ricaud et al. 2004. Characterization of AtCHX17, a member of the cation/H$^+$ exchangers, CHX family, from *Arabidopsis thaliana* suggests a role in K$^+$ homeostasis. *Plant J* 39:834–846.

Chakraborty, S., J. Luck, G. Hollaway et al. 2008. Impacts of global change on diseases of agricultural crops and forest trees. *CAB Rev Perspect Agr Vet Sci Nutr Nat Res* 3:1–15.

Chen, M., Q. J. Chen, X. G. Niu et al. 2007. Expression of OsNHX1 gene in maize confers salt tolerance and promotes plant growth in the field. *Plant Soil Environ* 11(53):490–498.

Chen, T. H. and N. Murata. 2011. Glycinebetaine protects plants against abiotic stress: Mechanisms and biotechnological applications. *Plant Cell Environ* 34:1–20.

Cho, K., S. Tiwari, S. B. Agrawal et al. 2011. Troposphere ozone and plants: Absorption, responses and consequences. *Rev Environ Contam Toxicol* 212:61–111.

Clements, D. R. and A. Ditommaso. 2011. Climate change and weed adaptation: Can evolution of invasive plants lead to greater range expansion than forecasted? *Weed Res* 51:227–240.

Clive, J. 2009. Global Status of Commercialized Biotech/GM Crops. Ithaca, NY: The International Service for the Acquisition of Agri-biotech Applications (ISAAA) Brief No.41.

Collard, B. C. Y. and D. J. Mackill. 2008. Marker-assisted selection: An approach for precision plant breeding in the twenty-first century. *Philos Trans R Soc Lond Ser B Biol Sci* 363:557–572.

Cook, R. J. 1993. Making greater use of introduced microorganisms for biological control of plant pathogens. *Annu Rev Phytopathol* 31:53–80.

Cui, J., P. Li, G. Li et al. 2007. AtPID: *Arabidopsis thaliana* protein interactome database—An integrative platform for plant systems biology. *Nucleic Acids Res* 36:D999–D1008.

Dam, S., B. S. Laursen, J. H. Ornfelt et al. 2009. The proteome of seed development in the model legume *Lotus japonicus*. *Plant Physiol* 149:1325–1340.

Datta, S. K., A. C. Tauro, and S. N. Balaoing. 1968. Effect of plant type and nitrogen level on growth characteristics and grain yield of indica rice in the tropics. *Agron J* 60:643–647.

Deytieux, C., L. Geny, D. Lapaillerie, S. Claverol, M. Bonneu, and B. Doneche. 2007. Proteome analysis of grape skins during ripening. *J Exp Bot* 58:1851–1862.

Dixon, R. and D. Kahn. 2004. Genetic regulation of biological nitrogen fixation. *Nat Rev Microbiol* 2:621–631.

Dogra, V., P. S. Ahuja, and Y. Sreenivasulu. 2013. Change in protein content during seed germination of a high altitude plant *Podophyllum hexandrum* Royle. *J Proteomics* 78:26–38.

Doty, S. L. 2011. Nitrogen-fixing endophytic bacteria for improved plant growth. In: D. K. Maheshwari (ed.), *Bacteria in Agrobiology: Plant Growth Responses*, pp. 183–199. Berlin: Springer-Verlag.

Dreze, M., A. R. Carvunis, B. Charloteaux et al. 2011. Evidence for network evolution in an *Arabidopsis* interactome map. *Science* 333:601–607.

Duarah, I., M. Deka, N. Saikia, and H. P. Deka Boruah. 2011. Phosphate solubilizers enhance NPK fertilizer use efficiency in rice and legume cultivation. *3 Biotech* 1(4):227–238.

Eathington, S. R., T. M. Crosbie, M. D. Edwards, R. S. Reiter, and J. K. Bull. 2007. Molecular markers in a commercial breeding program. *Crop Sci* 47:S154–S163.

Feng, S., X. Chen, Y. Zhang et al. 2011. Differential expression of proteins in red pear following fruit bagging treatment. *Protein J* 30:194–200.

Feuillet, C., J. E. Leach, J. Rogers, P. S. Schnable, and K. Eversole. 2010. Crop genome sequencing: Lessons and rationales. *Trends Plant Sci* 16:77–88.

Flavell, R. 2010. From genomics to crop breeding. *Nat Biotechnol* 28:144–145.

Fournier-Level, A., A. Korte, M. D. Cooper, M. Nordborg, J. Schmitt, and A. M. Wilczek. 2011. A map of local adaptation in *Arabidopsis thaliana*. *Science* 334:86–89.

Franks, S. J., S. Sim, and A. E. Weis. 2007. Rapid evolution of flowering time by an annual plant in response to a climate fluctuation. *Proc Natl Acad Sci USA* 104:1278–1282.

Frizzi, A., S. Huang, L. A. Gilbertson, T. A. Armstrong, M. H. Luethy, and T. M. Malvar. 2008. Modifying lysine biosynthesis and catabolism in corn with a single bifunctional expression/silencing transgene cassette. *Plant Biotechnol J* 6:13–21.

Fukushima, A., M. Kusano, H. Redestig, M. Arita, and K. Saito. 2009. Integrated omics approaches in plant systems biology. *Curr Opin Chem Biol* 13:532–538.

Gallardo, K., C. Firnhaber, H. Zuber et al. 2007. A combined proteome and transcriptome analysis of developing *Medicago truncatula* seeds: Evidence for metabolic specialization of maternal and filial tissues. *Mol Cell Proteomics* 6:2165–2179.

Gallardo, K., C. L. Signor, J. Vandekerckhove, R. D. Thompson, and J. Burstin. 2003. Proteomics of *Medicago truncatula* seed development establishes the time frame of diverse metabolic processes related to reserve accumulation. *Plant Physiol* 133:664–682.

Gammulla, C. G., D. Pascovici, P. A. Hayne, and B. J. Atwell. 2010. Differential metabolic response of cultured rice (*Oryza sativa*) cells exposed to high- and low-temperature stress. *Proteomics* 10:3001–3019.

Gammulla, C. G., D. Pascovici, P. A. Hayne, and B. J. Atwell. 2011. Differential proteomic response of rice (*Oryza sativa*) leaves exposed to high- and low-temperature stress. *Proteomics* 11:2839–2850.

Gaud, W.S. 1968. The green revolution: Accomplishments and apprehensions. AgBioWorld. http://www.agbioworld.org/biotech-info/topics/borlaug/borlaug-green.html. Retrieved 8 August 2011.

Gaxiola, R. A., R. Rao, A. Sherman, P. Grisafi, S. L. Alper, and G. R. Fink. 1999. The *Arabidopsis thaliana* proton transporters, *AtNhx1* and *Avp1*, can function in cation detoxification in yeast. *Proc Natl Acad Sci USA* 96:1480–1485.

Gepts, P. 2002. A comparison between crop domestication, classical plant breeding, and genetic engineering. *Crop Sci* 42:1780–1790.

Gill, T., V. Dogra, S. Kumar, P. S. Ahuja, and Y. Sreenivasulu. 2012. Protein dynamics during seed germination under copper stress in *Arabidopsis* over-expressing *Potentilla* superoxide dismutase. *J Plant Res* 125:165–172.

Gill, T., S. Kumar, P. S. Ahuja, and Y. Sreenivasulu. 2010a. Over-expression of superoxide dismutase exhibits lignifications of vascular structures in *Arabidopsis thaliana*. *J Plant Physiol* 167:757–760.

Gill, T., S. Kumar, P. S. Ahuja, and Y. Sreenivasulu. 2010b. Over-expression of *Potentilla* superoxide dismutase improves salt stress tolerance during germination and growth in *Arabidopsis thaliana*. *J Plant Genet Transgenics* 1:1–10.

Giribaldi, M., L. Geny, S. Delrot, and A. Schubert. 2010. Proteomic analysis of the effects of ABA treatments on ripening *Vitis vinifera* berries. *J Exp Bot* 61:2447–2458.

Giribaldi, M., I. Perugini, F. X. Sauvage, and A. Schubert. 2007. Analysis of protein changes during grape berry ripening by 2-DE and MALDI-TOF. *Proteomics* 7:3154–3170.

Godfray, H. C. J., J. R. Beddington, I. R. Crute et al. 2010. Food security: The challenge of feeding 9 billion people. *Science* 327:812–818.

Goff, S. A., D. Ricke, T. H. Lan et al. 2002. A draft sequence of the rice genome (*Oryza sativa* L. ssp. Japonica). *Science* 296:92–100.

Graham, P. H. and C. P. Vance. 2003. Legumes: Importance and constraints to greater use. *Plant Physiol* 131:872–877.

Grimplet, J., M. D. Wheatley, H. B. Jouira, L. G. Deluc, G. R. Cramer, and J. C. Cushman. 2009. Proteomic and selected metabolite analysis of grape berry tissues under well-watered and water-deficit stress conditions. *Proteomics* 9:2503–2528.

Gunstone, F. D. 2001. Soybeans pace boost in oilseed production. *Inform* 11:1287–1289.

Gupta, P. K., J. K. Roy, and M. Prasad. 2001. Single nucleotide polymorphisms: A new paradigm for molecular marker technology and DNA polymorphism detection with emphasis on their use in plants. *Curr Sci* 80:524–535.

Haas, D. and G. Defago. 2005. Biological control of soil-borne pathogens by fluorescent pseudomonads. *Nat Rev Microbiol* 3:307–319.

Hadiarto, T. and L. S. P. Tran. 2010. Progress studies of drought-responsive studies in rice. *Plant Cell Rep* 30:297–310.

Hajduch, M., A. Ganapathy, J. W. Stein, and J. J. Thelen. 2005. A systematic proteomic study of seed filling in soybean. Establishment of high-resolution two-dimensional reference maps, expression profiles, and an interactive proteome database. *Plant Physiol* 137:1379–1419.

Hakeem, K. R., R. Chandna, P. Ahmad, M. Iqbal, and M. Ozturk. 2012. Relevance of proteomic investigations in plant abiotic stress physiology. *OMICS* 16:621–635.

Hancock, A. M., B. Brachi, N. Faure et al. 2011. Adaptation to climate across the *Arabidopsis thaliana* genome. *Science* 334:83–86.

Harley, C. D. G. 2011. Climate change, keystone predation, and biodiversity loss. *Science* 334:1124–1127.

Hasegawa, P. M., R. A. Bressan, J. K. Zhu, and H. J. Bohnert. 2000. Plant cellular and molecular responses to high salinity. *Annu Rev Plant Physiol Plant Mol Biol* 51:463–499.

Hasnain, S. I. 2002. Himalayan glaciers meltdown: Impacts on South Asian rivers. In: H. A. J. van Lanen and S. Demuth (eds), *FRIEND 2002-Regional Hydrology: Bridging the Gap Between Research and Practice*, pp. 417–423. Wallingford: IAHS Publications.

He, C., J. Yan, G. Shen et al. 2005. Expression of an *Arabidopsis* vacuolar sodium/proton antiporter gene in cotton improves photosynthetic performance under salt conditions and increases fiber yield in the field. *Plant Cell Physiol* 46:1848–1854.

Hill, J. E. and R. W. Breidenbach. 1974. Proteins of soybean seeds. II. Accumulation of the major protein components during seed development and maturation. *Plant Physiol* 53:747–751.

Hjernø, K., R. Alm, B. Canback et al. 2006. Down-regulation of the strawberry Bet v 1-homologous allergen in concert with the flavonoid biosynthesis pathway in colourless strawberry mutant. *Proteomics* 6:1574–1587.

Hoffert, M. I., K. Caldeira, G. Benford et al. 2002. Advanced technology paths to global climate stability: Energy for a greenhouse planet. *Science* 298:981–987.

Hoffmann, A. A. and C. M. Sgrò. 2011. Climate change and evolutionary adaptation. *Nature* 470:479–485.

Houghton, J. T., Y. Ding, D. J. Griggs, M. Noguer, P. J. Linden, and D. Xiaosu. 2001. Climate change 2001: The scientific basis. In: *Contribution of Working Group to the Third Assessment Report of the Intergovernmental Panel on Climatic Change (IPCC)*, p. 944, Cambridge University Press, Cambridge, UK.

IPCC AR4 WG1. 2007. Climate change 2007: The physical science basis. In: S. Solomon, D. Qin, M. Manning et al. (eds), *Contribution of Working Group I to the Fourth Assessment Report of the Intergovernmental Panel on Climate Change*, Cambridge, United Kingdom: Cambridge University Press.

IPCC (Intergovernmental Panel on Climate Change). 2007a. Summary for policymakers. In: S. Solomon, D. Qin, M. Manning et al. (eds), *Climate Change 2007: The Physical Science Basis, Contribution of Working Group I to the Fourth Assessment Report of the Intergovernmental Panel on Climate Change*, Cambridge, United Kingdom: Cambridge University Press.

IPCC (Intergovernmental Panel on Climate Change). 2007b. Summary for policymakers. In: M. L. Parry, O. F. Canziani, J. P. Palutikof et al. (eds), *Climate Change 2007: Impacts, Adaptation and Vulnerability, Contribution of Working Group II to the Fourth Assessment Report of the Intergovernmental Panel on Climate Change*, Cambridge, United Kingdom: Cambridge University Press.

IRGSP (International Rice Genome Sequencing Project). 2005. The map-based sequence of the rice genome. *Nature* 436:793–800.

IRRI (International Rice Research Institute). 1962. The long road. *Rice Today* 7(4).

Islam, M. T. 2008. Disruption of ultrastructure and cytoskeletal network is involved with biocontrol of damping-off pathogen *Aphanomyces cochlioides* by *Lysobacter* sp. SB-K88. *Biol Control* 46:312–321.

Islam, M. T. 2011. Potentials for biological control of plant diseases by *Lysobacter* spp., with special reference to strain SB-K88. In: D. K. Maheshwari (ed.), *Bacteria in Agrobiology: Plant Growth Responses*, pp. 335–363. Berlin: Springer.

Islam, M. T., A. Deora, Y. Hashidoko, T. Itoa, and S. Tahara. 2005. Suppression of damping-off disease in host plants by the rhizoplane bacterium *Lysobacter* sp. strain SB-K88 is linked to plant colonization and antibiosis against soil borne *Peronosporomycetes*. *Appl Environ Microbiol* 71:3776–3786.

Islam, M. T., A. Deora, Y. Hashidoko, A. Rahmana, T. Itoa, and S. Tahara. 2007. Isolation and identification of potential phosphate solubilizing bacteria from the rhizoplane of *Oryza sativa* L. cv. BR29 of Bangladesh. *Z Naturforsch* 62:103–110.

Islam, M. T. and M. M. Hossain. 2012. Plant probiotics in phosphorus nutrition in crops, with special reference to rice. In: D. K. Maheshwari (ed.), *Bacteria in Agrobiology: Plant Probiotics*. pp. 325–363. Berlin: Springer.

Islam, M. T. and M. M. Hossain. 2013. Biological control of *Peronosporomycete* phytopathogens by bacterial antagonists. In: D. K. Maheshwari (eds), *Bacteria in Agrobiology: Plant Protection*. pp. 167–218. Berlin: Springer.

Islam, M. T. and A. von Tiedemann. 2011. 2,4-Diacetylphloroglucinol suppresses zoosporogenesis and impairs motility of *Peronosporomycete* zoospores. *World J Microbiol Biotechnol* 27:2071–2079.

Islam, M. T., A. von Tiedemann, and H. Laatsch. 2011. Protein kinase C is likely to be involved in zoosporogenesis and maintenance of flagellar motility in the Peronosporomycete zoospores. *Mol Plant Microbe Interact* 24:938–947.

Jaccoud, D., K. Peng, D. Feinstein, and A. Kilian. 2001. Diversity arrays: A solid state technology for sequence information independent genotyping. *Nucleic Acids Res* 29:e25.

Johnson, G. R. 2004. Marker assisted selection. In: J. Janick (ed.), *Plant Breeding Reviews, Long Term Selection: Maize*, vol. 24, pp. 293–309. Hoboken, NJ: Wiley.

Kerrien, S., B. Aranda, L. Breuza et al. 2012. The IntAct molecular interaction database in 2012. *Nucleic Acids Res* 40:D841–D846.

Khan, N. I, A. B. Filonow, and L. L. Singleton. 1997. Augmentation of soil with sporangia of *Actinoplanes* spp. for biological control of *Pythium* damping-off. *Biocontrol Sci Technol* 7:11–22.

Knapp, S. 1998. Marker-assisted selection as a strategy for increasing the probability of selecting superior genotypes. *Crop Sci* 38:1164–1174.

Kottapalli, K. R., P. Payton, R. Rakwal et al. 2008. Proteomics analysis of mature seed of four peanut cultivars using two-dimensional gel electrophoresis reveals distinct differential expression of storage, antinutritional, and allergenic proteins. *Plant Sci* 175:321–329.

Kosová, K., P. Vítámvás, I. T. Prášil, and J. Renaut. 2011. Plant proteome changes under abiotic stress—contribution of proteomics studies to understanding plant stress response. *J Proteomics* 74:1301–1322.

Krishnan, H. B., N. W. Oehrle, and S. S. Natarajan. 2009. A rapid and simple procedure for the depletion of abundant storage proteins from legume seeds to advance proteome analysis: A case study using *Glycine max*. *Proteomics* 9:3174–3188.

Ladha, J. K., F. J. de Bruijin, and K. A. Malik. 1997. Assessing opportunities for nitrogen fixation in rice: A frontier project. *Plant Soil* 194:1–10.

Lamb, H. H. 1995. *Climate, History and the Modern World*. London: Routledge.

Lee, K., D. Thorneycroft, P. Achuthan, H. Hermjakob, and T. Ideker. 2010. Mapping plant interactomes using literature curated and predicted protein–protein interaction data sets. *Plant Cell* 22:997–1005.

Lee, K. J., S. Kamala-Kannan, H. S. Sub, C. K. Seong, and G. W. Lee. 2008. Biological control of *Phytophthora* blight in red pepper (*Capsicum annuum* L.) using *Bacillus subtilis*. *World J Microb Biot* 24:1139–1145.

Liang, X. Q., M. Luo, C. C. Holbrook, and B. Z. Guo. 2006. Storage protein profiles in Spanish and runner market type peanuts and potential markers. *BMC Plant Biol* 6:24.

Lilly, D. H. and R. H. Stillwell. 1965. Probiotics. Growth promoting factors produced by micro-organisms. *Science* 147:747–748.

Lin, M., X. Zhou, X. Shen, C. Mao, and X. Chen. 2011. The predicted *Arabidopsis* interactome resource and network topology-based systems biology analyses. *Plant Cell* 23:911–922.

Liu, C.-W., Y. K. Hsu, Y. H. Cheng, H. C. Yen, Y. P. Wu, C. H. Wang, and C. C. Lai. 2012. Proteomic analysis of salt-responsive ubiquitin-related proteins in rice roots. *Rapid Commun Mass Spectrom* 26:1649–1660.

Lobell, D. B., W. Schlenker, and J. Costa-Roberts. 2011. Climate trends and global crop production since 1980. *Science* 333:616–620.

Lugtenberg, B. and F. Kamilova. 2009. Plant-growth-promoting rhizobacteria. *Annu Rev Microbiol* 63:541–556.

Lynch, M. and R. Lande. 1993. Evolution and extinction in response to environmental change. In: P. M. Kareiva, J. G. Kingsolver, and R. B. Huey (eds), *Biotic Interactions and Global Change*, pp. 234–250. Sunderland, MA: Sinauer Associates.

Maclean, J. L., D. C. Dave, B. Hardy, and G. P. Hettel. 2002. *Rice Almanac*. Wallingford: CAB International.

Maheshwari, D. K. 2012. *Bacteria in Agrobiology: Plant Probiotics*. Berlin: Springer.

Mahmoud, A. A., S. S. Natarajan, J. O. Bennett, T. P. Mawhinney, W. J. Wiebold, and H. B. Krishnan. 2006. Effect of six decades of selective breeding on soybean protein composition and quality: A biochemical and molecular analysis. *J Agric Food Chem* 54:3916–3922.

Markmann, K., G. Giczey, and M. Parniske. 2008. Functional adaptation of a plant receptor-kinase paved the way for the evolution of intracellular root symbioses with bacteria. *PLoS Biol* 6:e68.

Martinez-Esteso, M. J., S. Selles-Marchart, J. C. Vera-Urbina, M. A. Pedreno, and R. Bru-Martinez. 2009. The extracellular proteome of grapevine (*Vitis vinifera* cv Gamay) elicited cell culture reveals specifically induced defense gene products. *J Proteomics* 73:331–341.

Martinez-Esteso, M. J., S. Selles-Marchart, J. C. Vera-Urbina, M. A. Pedreno, and R. Bru-Martinez. 2011. DIGE analysis of proteome changes accompanying large resveratrol production by grapevine (*Vitis vinifera* cv. Gamay) cell cultures in response to methyl-β-cyclodextrin and methyl jasmonate elicitors. *J Proteomics* 74:1421–1436.

McSpadden Gardener, B. B. and D. R. Fravel. 2002. Biological control of plant pathogens: Research, commercialization, and application in the USA. Online. *Plant Health Progress*. doi: 10.1094/PHP-2002-0510-01-RV.

Mia, M. A. B., M. M. Hossain, Z. H. Shamsuddin, and M. T. Islam. 2012. Plant-associated bacteria in nitrogen nutrition in crops, with special reference to rice and banana. In: D. K. Maheshwari (ed.), *Bacteria in Agrobiology: Plant Nutrition*. pp. 97–126. Berlin: Springer.

Mirzaei, M., D. Pascovici, B. J. Atwell, and P. A. Haynes. 2012a. Differential regulation of aquaporins, small GTPases and V-ATPases proteins in rice leaves subjected to drought stress and recovery. *Proteomics* 12:864–877.

Mirzaei, M., N. Soltani, E. Sarhadi, D. Pascovici, T. Keighley, G. H. Salekdeh, P. A. Haynes, and B. J. Atwell. 2012b. Shotgun proteomic analysis of long-distance drought signaling in rice roots. *J Proteome Res* 11:348–358.

Mitchell, D. C, F. R. Lawrence, T. J. Hartman, and J. M. Curran. 2009. Consumption of dry beans, peas, and lentils could improve diet quality in the US population. *J Am Diet Assoc* 109:909–913.

Mitra, J. 2001. Genetics and genetic improvement of drought resistance in crop plants. *Curr Sci* 80:758–763.

Morsy, M., S. Gouthu, S. Orchard et al. 2008. Charting plant interactomes: Possibilities and challenges. *Trends Plant Sci* 13:183–191.

Mukhtar, M. S., R. Carvunis, M. Dreze et al. 2011. Independently evolved virulence effectors converge onto hubs in a plant immune system network. *Science* 333:596–601.

Naik, P. R., G. Raman, K. B. Narayanan, and N. Sakthivel. 2008. Assessment of genetic and functional diversity of phosphate solubilizing fluorescent pseudomonads isolated from rhizospheric soil. *BMC Microbiol* 8:230–243.

Natarajan, S., D. Luthria, H. Bae, D. Lakshman, and A. Mitra. 2013. Transgenic soybeans and soybean protein analysis: An overview. *J Agric Food Chem* 61:11736–11743.

Natarajan, S., C. Xu, H. Bae, B. A. Bailey, P. Cregan, T. J. Caperna, W. M. Garrett, and D. Luthria. 2007a. Proteomic and genetic analysis of glycinin subunits of sixteen soybean genotypes. *Plant Physiol Biochem* 45:436–444.

Natarajan, S., C. Xu, H. Bae, B. A. Bailey. 2007b. Proteomic and genomic characterization of Kunitz trypsin inhibitors in wild and cultivated soybean genotypes. *J Plant Physiol* 164:756–763.

Natarajan, S. S., H. B. Krishnan, S. Lakshman, and W. M. Garrett. 2009. An efficient extraction method to enhance analysis of low abundant proteins from soybean seed. *Anal Biochem* 394:259–268.

Negri, A. S., B. Prinsi, A. Scienza, S. Morgutti, M. Cocucci, and L. Espen. 2008a. Analysis of grape berry cell wall proteome: A comparative evaluation of extraction methods. *J Plant Physiol* 165:1379–1389.

Negri, A. S., B. Prinsi, A. Scienza, M. Rossoni, F. Osvaldo, M. Cocucci, and L. Espen. 2008b. Proteome changes in the skin of the grape cultivar Barbera among different stages of ripening. *BMC Genomics* 9:378.

Nelson, L. M. 2004. Plant growth promoting rhizobacteria (PGPR): Prospects for new inoculants. *Crop Management*, doi:10.1094/CH-2004-0301-05-RV.

Ogawa, T., M. Samoto, and K. Takahashi. 2000. Soybean allergens and hypoallergenic soybean products. *J Nutr Sci Vitaminol* 46:271–279.

Osburn, R. M., J. L. Milner, E. S. Oplinger, R. S. Smith, and J. Handelsman. 1995. Effect of *Bacillus cereus* UW85 on the yield of soybean at two field sites in Wisconsin. *Plant Dis* 79:551–556.

Pachauri, R. K. and A. Reisinger. 2007. Contribution of working groups I, II and III to the fourth assessment report of the intergovernmental panel on climate change. In: *Climate Change 2007: Synthesis Report (IPCC Fourth Assessment Report (AR4))*, p. 104. Geneva, Switzerland: IPCC.

Pal, A. K., K. Acharya, S. K. Vats, S. Kumar, and P. S. Ahuja. 2012. Over-expression of *PaSOD* in transgenic potato enhances photosynthesis performance under drought. *Biol Plantarum* 57:359–364.

Pardo, J. M. 2010. Biotechnology of water and salinity stress tolerance. *Curr Opin Plant Biol* 21:185–196.

Parker, R. B. 1974. Probiotics, the other half of the antibiotic story. *Anim Nutr Health* 29:4–8.

Pedreschi, R., M. Hertog, J. Robben et al. 2008. Physiological implications of controlled atmosphere storage of 'Conference' pears (*Pyrus communis* L.): A proteomic approach. *Postharv Biol Technol* 50:110–116.

Paterson, A. H., J. E. Bowers, R. Bruggmann et al. 2009. The *Sorghum bicolor* genome and the diversification of grasses. *Nature* 457:551–556.

Pedreschi, R., M. Hertog, J. Robben et al. 2009. Gel-based proteomics approach to the study of metabolic changes in pear tissue during storage. *J Agric Food Chem* 57:6997–7004.

Pedreschi, R., E. Vanstreels, S. Carpentier et al. 2007. Proteomic analysis of core breakdown disorder in Conference pears (*Pyrus communis* L.). *Proteomics* 7:2083–2099.

Perneel, M., L. D'Hondt, K. De Maeyer, A. Adiobo, K. Rabaey, and M. Hofte. 2008. Phenazines and biosurfactants interact in the biological control of soil-borne diseases caused by *Pythium* spp. *Environ Microbiol* 10:778–788.

Rabilloud, T. and C. Lelong. 2011. Two-dimensional gel electrophoresis in proteomics: A tutorial. *J Proteomics* 74:1829–1841.

Ragot, M., M. Biasiolli, M. F. Delbut et al. 1995. Marker-assisted backcrossing: A practical example. In: A. Berville and M. Tersac (eds), *Techniques et utilisations des marqueurs moléculaires*, pp. 45–56. Paris: INRA.

Rakwal, R. and G. K. Agrawal. 2003. Rice proteomics: current status and future perspectives. *Electrophoresis* 24:3378–3389.

Rana, B. and Y. Sreenivasulu. 2013. Protein changes during ethanol induced seed germination in *Aconitum heterophyllum*. *Plant Sci* 198:27–38.

Richardson, A. E. and R. J. Simpson. 2011. Soil microorganisms mediating phosphorus availability update on microbial phosphorus. *Plant Physiol* 156:989–996.

Roesch, L. F. W., R. R. Fulthorpe, A. Riva et al. 2007. Pyrosequencing enumerates and contrasts soil microbial diversity. *ISME J* 1:283–290.

Rosenzweig, C. and M. L. Parry. 1994. Potential impact of climate change on world food supply. *Nature* 367:133–138.

Rothemund, D. L., V. L. Locke, A. Liew, T. M. Thomas, V. Wasinger, and D. B. Rylatt. 2003. Depletion of the highly abundant protein albumin from human plasma using the Gradiflow. *Proteomics* 3:279–287.

Sakamoto, A. and N. Murata. 2001. The use of bacterial choline oxidase, a glycinebetaine-synthesizing enzyme, to create stress-resistant transgenic plants. *Plant Physiol* 125:180–188.

Sarhadi, E., M. M. Bazargani, A. G. Sajise, S. Abdolahi, N. A. Vispo, M. Arceta, G. M. Nejad, R. K. Singh, and G. H. Salekdeh. 2012. Proteomic analysis of rice anthers under salt stress. *Plant Physiol Biochem* 58:280–287.

Sarkar, A., R. Rakwal, S. B. Agrawal et al. 2010. Investigating the impact of elevated levels of ozone on tropical wheat using integrated phenotypical, physiological, biochemical and proteomics approaches. *J Proteome Res* 9:4565–4584.

Sarry, J. E., N. Sommerer, F. X. Sauvage et al. 2004. Grape berry biochemistry revisited upon proteomic analysis of the mesocarp. *Proteomics* 4:201–215.

Sathanoori, R. S. and S. M. Basha. 1996. Methionine content of the polypeptides of methionine-rich protein from peanut. *J Agric Food Chem* 44:2134–2136.

Schnable, P. S., D. Ware, R. S. Fulton et al. 2009. The B73 maize genome: Complexity, diversity, and dynamics. *Science* 326:1112–1115.

Sharma, H. C., J. H. Crouch, K. K. Sharma, N. Seetharama, and C. T. Hash. 2002. Applications of biotechnology for crop improvement: Prospects and constraints. *Plant Sci* 163:381–395.

Shi, H., M. Ishitani, C. Kim, and J. K. Zhu. 2000. The *Arabidopsis thaliana* salt tolerance gene SOS1 encodes a putative Na$^+$/H$^+$ antiporter. *Proc Natl Acad Sci USA* 97:6896–6901.

Shi, H., B. H. Lee, S. J. Wu, and J. K. Zhu. 2003. Overexpression of a plasma membrane Na$^+$/H$^+$ antiporter gene improves salt tolerance in *Arabidopsis thaliana*. *Nat Biotechnol* 21:81–85.

Shinozaki, K., K. Yamaguchi-Shinozaki, and M. Seki. 2003. Regulatory network of gene expression in the drought and cold stress responses. *Curr Opin Plant Biol* 6:410–417.

Sommer, J. H., H. Kreft, G. Kier, W. Jetz, J. Mutke, and W. Barthlott. 2010. Projected impacts of climate change on regional capacities for global plant species richness. *Proc Royal Soc B: Biol Sci* 277:2271–2280.

Song, C. P., Y. Guo, Q. Qiu et al. 2004. A probable Na$^+$(K$^+$)/H$^+$ exchanger on the chloroplast envelope functions in pH homeostasis and chloroplast development in *Arabidopsis thaliana*. *Proc Natl Acad Sci USA* 101:10211–10216.

Stark, C., B. J. Breitkreutz, T. Reguly, L. Boucher, A. Breitkreutz, and M. Tyers. 2006. BioGRID: A general repository for interaction datasets. *Nucleic Acids Res* 34:D535–D539.

Thelen, J. J. 2009. Proteomics tools and resources for investigating protein allergens in oil seeds. *Regul Toxicol Pharm* 54:S41–S45.

Thelen, J. J. and J. Ohlrogge. 2002. Metabolic engineering of fatty acid biosynthesis in plants. *Metab Eng* 4:12–21.

Thuiller, W. 2007. Biodiversity: Climate change and the ecologist. *Nature* 448:550–552.

Tu, J. C. 1978. Protection of soybean from severe *Phytophthora* root rot by *Rhizobium*. *Physiol Plant Pathol* 12:237–240.

Tuberosa, R., S. Salvi, S. Giuliani et al. 2007. Genome-wide approaches to investigate and improve maize response to drought. *Crop Sci* 47:S120–S141.

Uhrig, J. F. 2006. Protein interaction networks in plants. *Planta* 224:771–781.

Umezawa, T., M. Fujita, Y. Fujita, K. Yamaguchi-Shinozaki, and K. Shinozaki. 2006. Engineering drought tolerance in plants: Discovering and tailoring genes to unlock the future. *Curr Opin Biotechnol* 17:113–122.

Umina, P. A., A. R. Weeks, M. R. Kearney, S. W. McKechnie, and A. A. Hoffmann. 2005. A rapid shift in a classic clinal pattern in *Drosophila* reflecting climate change. *Science* 308:691–693.

van der Heijden, M. G. A., R. D. Bardgett, and M. van Straalen Nico. 2008. The unseen majority: Soil microbes as drivers of plant diversity and productivity in terrestrial ecosystems. *Ecol Lett* 11:296–310.

Varshney, R. V., D. A. Hoisington, and A. K. Tyagi. 2006. Advances in cereal genomics and applications in crop breeding. *Trends Biotechnol* 24:490–499.

Vyas, P. and A. Gulati. 2009. Organic acid production in vitro and plant growth promotion in maize under controlled environment by phosphate-solubilizing fluorescent *Pseudomonas*. *BMC Microbiol* 9:174–188.

Waditee, R., M. N. H. Bhuiyan, V. Rai et al. 2005. Genes for direct methylation of glycine provide high levels of glycinebetaine and abiotic-stress tolerance in *Synechococcus* and *Arabidopsis*. *Proc Natl Acad Sci USA* 102:1318–1323.

Waditee, R., Y. Tanaka, K. Aoki et al. 2003. Isolation and functional characterization of N-methyltransferases that catalyze betaine synthesis from glycine in a halotolerant photosynthetic organism *Aphanothece halophytica*. *J Biol Chem* 278:4932–4942.

Watson, B. S., V. S. Asirvatham, L. Wang, and L. W. Sumner. 2003. Mapping the proteome of barrel medic (*Medicago truncatula*). *Plant Physiol* 131:1104–1123.

Waugh, R. and W. Powell. 1992. Using RAPD markers for crop improvement. *Trends Biotechnol* 10:186–191.

Weckwerth, W. 2011. Green systems biology—From single genomes, proteomes and metabolomes to ecosystems research and biotechnology. *J Proteomics* 75:284–305.

Wilson, S., K. Blaschek, E. Gonzalez, and D. Mejia. 2005. Allergenic proteins in soybean: Processing and reduction of P34 allergenicity. *Nutr Rev* 63:47–58.

Xenarios, I., D. W. Rice, L. Salwinski, M. K. Baron, E. M. Marcotte, and D. Eisenberg. 2000. DIP: The database of interacting proteins. *Nucleic Acids Res* 28:289–291.

Xu, C., T. J. Caperna, W. M. Garrett et al. 2007. Proteomic analysis of the distribution of the major seed allergens in wild, landrace, ancestral, and modern soybean genotypes. *J Sci Food Agric* 87:2511–2518.

Xue, Z. Y., D. Y. Zhi, G. P. Xue, H. Zhang, Y. X. Zhao, and G. M. Xia. 2004. Enhanced salt tolerance of transgenic wheat (*Tritivum aestivum* L.) expressing a vacuolar Na^+/H^+ antiporter gene with improved grain yields in saline soils in the field and a reduced level of leaf Na^+. *Plant Sci* 167:849–859.

Yamada, K., J. Lim, J. M. Dale et al. 2003. Empirical analysis of transcriptional activity in the *Arabidopsis* genome. *Science* 302:842–846.

Yamamoto, A., M. Kawamiya, A. Ishida, Y. Yamanaka, and S. Watanabe. 2012. Impact of rapid sea-ice reduction in the Arctic Ocean on the rate of ocean acidification. *Biogeosciences* 9:2365–2375.

Yan, Z., M. S. Reddy, C. M. Ryu, J. A. McInroy, M. Wilson, and J. W. Kloepper. 2002. Induced systemic protection against tomato late blight elicited by plant growth-promoting rhizobacteria. *Phytopathology* 92:1329–1333.

Yang, A. F., X. G. Duan, X. F. Gu, F. Gao, and J. R. Zhang. 2005. Efficient transformation of beet (*Beta vulgaris*) and production of plants with improved salt-tolerance. *Plant Cell Tissue Org* 83:259–270.

Yang, S., B. Vanderbeld, J. Wan, and Y. Huang. 2010. Narrowing down the targets: Towards successful genetic engineering of drought-tolerant crops. *Mol Plant* 3:469–490.

Ye, X., S. Al-Babili, A. Kloti et al. 2000. Engineering the provitamin A (beta-carotene) biosynthetic pathway into (carotenoid-free) rice endosperm. *Science* 287:303–305.

Young, N. D. 1999. A cautiously optimistic vision for marker-assisted breeding. *Mol Breed* 5:505–510.

Yu, J., S. Hu, J. Wang et al. 2002. A draft sequence of the rice genome (*Oryza sativa* L. *ssp.* Indica). *Science* 296:79–92.

Zaidi, A., M. S. Khan, M. Ahemad, M. Oves, and P. A. Wani. 2009. Recent advances in plant growth promotion by phosphate-solubilizing microbes. In: M. S. Khan et al. (eds), *Microbial Strategies for Crop Improvement*, pp. 23. Berlin: Springer-Verlag.

Zanzoni, A., L. Montecchi-Palazzi, M. Quondam, G. Ausiello, M. Helmer-Citterich, and G. Cesareni. 2002. MINT: A Molecular INTeraction database. *FEBS Lett* 513:135–140.

Zhang, H. X., J. N. Hodson, J. P. Williams, and E. Blumwald. 2001. Engineering salt-tolerant *Brassica* plants: Characterization of yield and seed oil quality in transgenic plants with increased vacuolar sodium accumulation. *Proc Natl Acad Sci USA* 98:12832–12836.

Zhang, J., H. Ma, J. Feng, L. Zeng, Z. Wang, and S. Chen. 2008. Grape berry plasma membrane proteome analysis and its differential expression during ripening. *J Exp Bot* 59:2979–2990.

Zhang, Q. 2007. Strategies for developing green super rice. *Proc Natl Acad Sci USA* 104:16402–16409.

Zhang, Y., P. Gao, and J. S. Yuan. 2010. Plant protein–protein interaction network and interactome. *Curr Genomics* 11:40–46.

Zhu, J. K. 2002. Salt and drought stress signal transduction in plants. *Annu Rev Plant Biol* 53:247–273.

CHAPTER 9

The Food System Approach in Agroecology Supported by Natural and Social Sciences
Topics, Concepts, Applications

Alexander Wezel, Philippe Fleury, Christophe David, and Patrick Mundler

CONTENTS

9.1   Introduction 181
9.2   Natural and Social Sciences Approaches to Agroecology of the Food System 183
      9.2.1   Natural Sciences Approach 183
      9.2.2   Social Sciences Approach.. 184
9.3   Examples of Agroecology of Food System Research and Natural and Social Sciences
      Approaches 186
      9.3.1   Organic Grain Production: From Wheat Production at Field Scale to Wheat Baking....186
      9.3.2   Community-Supported Agriculture: The Case of the French AMAP System......... 189
      9.3.3   Organic Agriculture and Drinking Water Catchments 191
9.4   Discussion and Conclusion 194
References 195

9.1 INTRODUCTION

The expected growth of the world population for at least the next four decades will demand not only increased food production worldwide, but also improved food availability and fair production and distribution of food at local, national, and global scales. This demands a concept of sustainable agricultural and food systems that not only addresses quantitative production issues, but also increasingly considers environmental issues of agricultural food production, such water and air pollution, biodiversity loss, and land degradation. In addition, social and economic aspects have to be taken into account, such as economic viability of farmers, organization and efficiency of supply chains and markets, communication and coordination among stakeholders, and farmer–consumer relationships. These aspects can no longer be analyzed in isolation if we wish to establish sustainable agricultural and food systems. Thus, global and holistic approaches are required. Such approaches have been, for example, developed in recent years in the framework of agroecology, applied to the food system, whereby agronomic, ecological, economic, and social dimensions are taken simultaneously into account at different scales (Francis et al. 2003; Gliessman 2007; Wezel and David 2012).

Although agroecology as a scientific discipline has existed for many decades, the food systems approach in agroecology has only recently been discussed (Wezel and Soldat 2009; Wezel and Jauneau 2011). Current research is focused on outlining more clearly and precisely the concepts, topics, and applications of the food system approach, as well as defining specific models and methods. Besides agroecology as a scientific discipline, other interpretations, such as agroecology as a practice or as a movement, are present to different degrees, depending on the geographical and institutional context (Wezel et al. 2009).

The scale and dimension of research in agroecology have been enlarged from (1) the plot, field, or animal scale to (2) the farm or agroecosystem scale, and (3) finally, in the last years, to the dimension of the food system, which is increasingly cited as a major outlook for agroecology (Wezel and Soldat 2009).

At the plot/field/animal scale, agroecological research aims to develop new farming practices, for example, in improving nutrient cycling, in more efficiently using natural resources, and in enhancing diversity of soil organisms, crops, and livestock to provide healthier systems. At this scale, research does not really consider the interactions and implications of these techniques for the agroecosystem, or the environment at a larger scale, or the food system.

The agroecosystem approach is a second major approach in agroecology. Here, ongoing research focuses on the agroecosystem scale, including exchange with, and impact on, the surrounding environment. Agroecological analyses focus, for example, on plant and animal communities and food web interactions for biological control, biodiversity and nature conservation, and drinking water pollution and land degradation in agricultural landscapes and agroecosystems. Within the agroecosystem approach the definitions and concepts might vary, depending on the delimitation of an agroecosystem. Sometimes, the farm is seen as equivalent to an agroecosystem where the relations between farmers' practices and natural resources are analyzed (Conway 1987). For others an agroecosystem is larger, that is, a local or regional landscape where relations between different types of agriculture and the natural resources of the landscape are investigated.

The food systems approach is the most recent and broadest approach in agroecology. It was first defined by Francis et al. (2003) as "the integrative study of the ecology of the entire food systems, encompassing ecological, economic and social dimensions," or more simply "the ecology of food systems." Gliessman (2007) defined agroecology as "the science of applying ecological concepts and principles to the design and management of sustainable food systems." These two definitions are based on former definitions of Altieri (1989, 1995, 2002).

Since the beginning of the 2000s, several authors have demanded that agriculture be analyzed in a holistic manner. For example, Robertson et al. (2004) stated that agricultural research needs long-term, system-level research at multiple scales, and that natural and social sciences must be better integrated. Gliessman (2007) stated that "to recognize the influence of social, economic, cultural, and political factors on agriculture, we must eventually shift our focus from sustainability of agroecosystems to the sustainability of our food systems." Nevertheless, it is still difficult to outline clear concepts, theoretical models, and methods that specify and translate these demands, and in particular the expanded definition of agroecology of the food system, into concrete cases and applications. So far, very few papers have applied agroecological concepts and theory to the food system; examples include Francis and Rickerl (2004) and Wezel and David (2012).

In the field of social sciences, numerous research works have dealt with alternative food systems to conventional agribusiness systems. An increasing number of research topics on food systems, local food networks, and alternative food systems such as direct selling, farm stores and short supply chain, farmers' markets, community-supported agriculture, consumer cooperatives, or quality labels for local or regional products have been developed since the mid-1980s (Feenstra 1997; Jarosz 2008; Marsden et al. 2000; Maye et al. 2007; Renting et al. 2003; Tregear 2011). Other works have dealt with change of farmers' practices and production of knowledge allowing changes toward more sustainable systems (e.g., conversion to organic farming, adoption of no-tillage practices, commitment to agri-environmental measures) (Hall and Mogyorody 2001; Lamine and Bellon 2009;

Walford 2003), as well as with social relationships in food systems (Guptill and Wilkins 2002) or consumers' practices (O'Hara and Stagle 2002). These research topics were mainly developed in the field of social sciences, and have been increasingly linked to agroecology in recent years.

Today, a broad diversity of food system topics is covered in literature dealing explicitly with agroecology. The topics include food sovereignty (Altieri and Nicholls 2008; Altieri and Toledo 2011; Altieri et al. 2012; Cohn et al. 2006; Reardon and Pérez 2010; Rosset et al. 2011), alternative and local food networks (Gliessman 2007), social agricultural networks (Warner 2005, 2007a,b), food crisis (Gliessman 2011), food security (Chappell and LaValle 2011), right to food (De Schutter 2012), agri-food systems (Thompson and Scoones 2009), food markets (Lockie and Carpenter 2010), and consumers (Stassart and Claes 2010).

In this chapter we will first provide a conceptual analysis of the food system approaches in the natural (ecology and agronomy) and social sciences, emphasizing fields of research topics and scientific disciplines involved. Second, we will take three varied examples to illustrate specificities and common aspects of natural and social sciences approaches to analyzing and assessing food system questions and topics. The first example will illustrate a research program that was initially dominated by a natural science approach (organic grain production), the second will show an example dominated by a social science approach (Associations pour le Maintien de l'Agriculture Paysanne (AMAP)—a community-supported agriculture system), and the third will illustrate a simultaneous natural and social sciences approach (organic agriculture and drinking water catchments). Finally, we will discuss to what degree the different approaches enrich the concept of agroecology of the food system, and will look more specifically at the added value of an interdisciplinary approach.

9.2 NATURAL AND SOCIAL SCIENCES APPROACHES TO AGROECOLOGY OF THE FOOD SYSTEM

In recent years, there has been an important ongoing debate in agroecology about the interest and the value of the food system approach of agroecology. The major questions deal with how to approach food system issues, what research concepts have to be used, and what methodology has to be developed. It is broadly acknowledged that for food system issues a global and systemic approach is needed (Francis et al. 2003; Gliessman 2007). According to Wezel and David (2012), the concepts of holism with a systemic approach including different scales and interdisciplinarity exist already, so they can now form the basis for research and analyses for the agroecology of the food system. Nevertheless, research approaches to the agroecology of the food system show themselves to be quite different when driven by either the natural or the social sciences. In addition, the specific research subjects and topics dealt with at the different scales may vary considerably for the two groups of disciplines.

9.2.1 Natural Sciences Approach

With the natural sciences approach, food-related research questions are often first dealt with at the scale of the field, and in fewer cases directly at the farm/agroecosystem scale (Figure 9.1). The increasing scales used for the food system approach of the natural sciences have caused the involvement of an increasing number of disciplines to deal with the increasing complexity of research questions and objects. The basic disciplines for the natural sciences approach to agroecology are agronomy and ecology, but currently it is still necessary to design concepts to handle the level of the food system as well as to deal with the high complexity of research questions at this scale. In general, the natural sciences research approach is based on an upscaling approach from the field level to the food system. A typical example of research in agroecology driven by a natural sciences approach will be illustrated by the first example in the second part of this chapter.

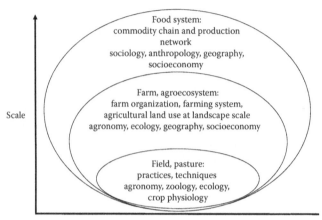

Figure 9.1 The natural sciences food system approach in agroecology illustrated for different scales, the major research subjects, and the disciplines involved. Agronomy and ecology are the basic disciplines for the natural sciences approach. With increasing scale, other disciplines are integrated in order to be able to deal with the increasing complexity of research questions.

For the natural sciences approach, the major research subjects and topics generally dealt with at the field/pasture scale are farmers' practices and techniques, crop or animal performance, and input and output fluxes. More specifically, this may include nutritional or protein content of crops in relation to agricultural practices, or the effect of vegetation composition and fodder quality on animal growth or milk quality. At the farm/agroecosystem scale, research topics are mainly farm organization, inputs and outputs at the farming system scale, or agricultural land use at the territory or landscape scales. More specifically, the questions of diversification of food production at the farm level as well as food transformation on-farm are analyzed. Finally, at the food system scale, the main research subjects are the analysis of the commodity chain and production networks of farmers and other stakeholders, although these food system topics are still seldom integrated into agroecological research initiated by natural scientists.

9.2.2 Social Sciences Approach

Here we look more generally at the social sciences approaches to food system topics and research. So far, we cannot explicitly speak of social sciences approaches in agroecology, because the increase of social sciences topics in agroecology research has only recently begun. However, an extensive corpus of literature on food systems topics and their recent developments can be revealed, some of which are mobilized today in agroecology. In contrast to the natural sciences approaches, the social sciences approaches do not follow a progressive enlargement of scales and subjects, but are centered on different research questions and topics (illustrated in the second example). The evolution of research questions and subjects can be quite complex, even if we limit this to the disciplines and approaches we could frame with agricultural sciences in the larger sense, with economic, social, technical, and biological knowledge creation in respect to agriculture. Three levels or scales can be distinguished: the farmer and farm, the local level, and the global level (Figure 9.2). At the farm level, research topics are mainly farm organization or decision making, economic performance, and farmers' social representations and knowledge. At the local level, topics dealing with collective organization of farmers, short supply chains, and relationships between farmers and consumers are the more frequent, whereas at the global level analyses on long supply chains and markets trends prevail. The basic disciplines for all levels are economy, sociology, and geography.

Scale

Global level:
long supply chains, agricultural
policy, markets, consumer behaviour,
economy, geography, sociology

Local level:
short supply chains, collective organization
of farmers, stakeholder-consumer
relationships, rural development
economy, geography, sociology,
anthropology, agronomy, ecology

Farmer and farm:
farm organization, economic performance,
farmers' social representation and knowledge
economy, geography, sociology,
anthropology, agronomy, ecology

Integration of disciplines and topics

Figure 9.2 The social sciences food system approaches illustrated for different scales, the major research subjects, and the disciplines involved. Economy, sociology, and geography are the basic disciplines at all scales. The social sciences approach does not follow a progressive enlargement of scales and subjects, but is more constructed around certain research questions and topics.

As the social sciences approaches to the food system are more centered on different research questions and topics and less on scales, we will present here the three main approaches. The first refers to the analysis of supply chains and commodity chains either at the local (short supply chains) or at the global scale (long supply chains). A supply chain is centered on a basic agricultural product and its successive complete or partial transformations (Fraval 2000). It is, thus, an approach by fluxes: linkage between vertical flux of products, money, or both between the different stakeholders of the supply chain. This can start from the production and production factors toward the consumption, or the reverse.

The second approach focuses on stakeholders and their networks. It analyses the strategies and behaviors of stakeholders, and the relationships between these stakeholders as they are integrated into a larger collectivity or network (Frayssignes 2005). The vertical supply chain approach is here enlarged by also studying the horizontal relations (Mikkola and Seppänen 2006; Murdoch 2000), or even taking into account the connections with other supply or commodity chains, as well as with other stakeholders not directly linked to the supply chain. Here, the actor-network theory (Granovetter 2000; Law 1992) is often used to analyze the shape and composition of a network, which is given not simply by its socioeconomic components, but by all the linkages between all the enrolled entities (Murdoch 2000). In this context, the contribution of local agricultural products to territorial development is studied in relation to stakeholders' networks. This contribution can be twofold (Tregear et al. 2007): (i) favoring the constitution of an active network of producers, which allows increasing employment and income in the interior of the network; and (ii) creation of relations between biophysical (landscapes, local flora and fauna), cultural (techniques, local know-how), and economic (employment) resources, thus obtaining supplementary added value.

The third approach concentrates on the contribution of food systems to the economic and social development of local or regional areas. For this, connections between the rural agro-industry and the territory are investigated in analyzing the organization of productions and services (farm unities, food production enterprises, commercialization enterprises, catering enterprises) and their links to the characteristics and the management of a specific area. It refers to areas where the relations between the environment, people, products, enterprises, traditions, and consumption habits are handled in a holistic way and developed to produce a specific and localized organization of the food systems. This refers to the concept of a localized agro-food system (Muchnik and de Sainte Marie 2010). These systems are often seen as a potential alternative to globalized and standardized food systems.

9.3 EXAMPLES OF AGROECOLOGY OF FOOD SYSTEM RESEARCH AND NATURAL AND SOCIAL SCIENCES APPROACHES

In the second part of this chapter, we will use three research examples dealing with different food system topics to illustrate how the natural and social sciences approached the topics differently, and which scientific disciplines have been involved. In addition, we will evaluate how the different approaches allow us to consider the holistic approach to the agroecology of the food system, but also which constraints still remain with the different approaches. The three chosen examples are:

1. *Organic grain production*: applied research program initiated with a natural sciences approach focused on field level, with a progressive extension toward farm and food system issues.
2. *The AMAP system in France—an example of community-supported agriculture*: a social sciences approach to analyze different topics of a local food system at different scales.
3. *Organic agriculture and drinking water catchments*: an interdisciplinary initiative crossing social and natural sciences approaches to consider simultaneously agronomic, ecological, social, and food system topics from the plot/field to the food systems scale.

For the three examples we will describe the scale of analyses and the contributions from the interdisciplinary approaches carried out, but also their limits.

9.3.1 Organic Grain Production: From Wheat Production at Field Scale to Wheat Baking

This first example, previously published in Wezel and David (2012), illustrates a research project on organic grain production, where the central question was to improve nitrogen management of organic wheat and the baking quality of wheat flour.

The on-farm research program on organic wheat was initiated to identify the main limiting factors (David et al. 2004) and improve N management (David et al. 2005a) and weed management (Casagrande et al. 2009), to increase wheat performance (Figure 9.3). This first phase had been set up on 17 farms in two different agroecosystems in France (Diois, Plain de Valence), first, to take into account a wider range of growing conditions; second, to benefit from farmers' expert knowledge when research on organic grains systems was still very limited; and, finally, to

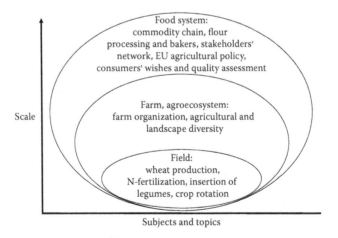

Figure 9.3   Main research topics and subjects of the organic wheat production and commodity chain example. Initially a natural sciences approach was carried out at the field scale, which was subsequently complemented with social sciences topics at larger scales.

consider the entire farm system and its socioeconomic parameters. Nitrogen and weed management were investigated on more than 40 organic fields from 1993 to 1998 by testing various techniques and equipment (field scale, agronomy). From 1998 to 2004, management of N fertilization had also been studied on experimental farms under controlled conditions, to produce references for N nutrition of organic and low-input wheat from off-farm organic N sources (David et al. 2004). This research also allowed the development of a decision support system to manage N fertilization of organic wheat (David et al. 2005b; David and Jeuffroy 2009) to improve grain yield and grain protein content. During the second phase of the program, research went beyond the restricted field scale analyses in integrating farm management aspects. A multivariate analysis of quantitative and qualitative data such as grain yield, protein content, crop management, and farming system management from 97 organic farms demonstrated, for example, the effects of crop management conditions, such as cultivar, preceding crop, N fertilization, and weed control, but also the effects of soil and climatic conditions, such as water deficit and temperature, on grain yield and protein content (field and agroecosystem scale; agronomy). Furthermore, interviews with farmers, which had been started in the first phase, increased in the second phase and have now become a key element of the research program, enabling a more complete study of farm management (field, farm, and food system scale; agronomy, economy, and sociology). In parallel, the analysis of the wheat-flour food chain allowed determination of the interactions between producers, collectors, processors, and consumers (David and Joud 2008). Also, a structured organic food chain supported by cooperatives and bakers improved the economic viability of farms.

The major results so far are that the agroecosystem characteristics of the studied areas strongly influence the farming systems, but also the food system. In general, different factors limiting organic crop production, such as weed and pest infestation, soil compaction, or climatic conditions like water stress and hot temperatures, could be determined (Casagrande et al. 2009; David et al. 2005a).

The Diois agroecosystem consists of limited areas with fertile soil in the Drome Valley, where cereals are produced in a long-term and diversified crop rotation of 8–11 years, surrounded by large areas with low soil fertility occupied by vineyards, lavender fields, permanent pastures, and (semi) natural ecosystems (Figure 9.4). The agricultural productivity is limited in this area. In contrast, the high agricultural diversity, together with the Drome River and the adjacent Vercors Mountains, make it a beautiful landscape and give it high value for tourism, for which farmers produce local

Figure 9.4 **(See color insert)** Diois agroecosystem with cereal production and viticulture, southeastern France.

food, wine, and lavender as well as offering accommodation. Conversion to organic production allowed the economic value of low-input agricultural productions such as wine, grains, and aromatic plants to be maintained. Moreover, the marketing of these organic products, promoted by cooperatives, is associated with identity and origin, and supported by traditional varieties and specific products, for instance by the Clairette de Die, a famous sweet sparkling wine produced exclusively in this area.

As the agroecosystem of the Plain of Valence consists of a large fertile plain, the yield performance of dominant grain production is much higher compared with the Diois. Organic grain systems differ only slightly from conventional systems. Cropping systems are based on a balanced proportion of spring crops, mostly irrigated, such as maize and soybean, associated in a crop rotation of four–six years with winter cereals such as wheat, barley, or triticale. The organic grains are collected by conventional cooperatives, where a limited organic sector has been developed to answer farmers' requirements. Tourism is very limited in the Plain of Valence area; thus, direct selling, provision of local food products, and accommodation on farms are rare.

The agroecosystem characteristics of the two subareas do not only influence the farming systems, but also involve differences in the food systems. For instance, in the Diois, the wheat-flour–bread chain is essentially based on a small niche market for traditional organic bakers or organic retailers looking for a specific flavor obtained with ancient varieties, but also providing identity as originating from the area. On the contrary, the wheat-flour food chain in the Plain of Valence is essentially based on standardized quality requirements, for example, protein content over the conventional threshold of 11.5 g per 100 g and no mycotoxins, applied to mass distribution or enterprises (David and Joud 2008). In general, it can be concluded that diversification of farm production and activities, off-farm employment, and professional and social networking contributed significantly to farm viability (David et al. 2010).

The ongoing research project now integrates many different scientific disciplines, such as agronomy, food technology, economy, and sociology, and works simultaneously at different scales, namely the field, the farm, and the food system levels, to develop a more holistic approach. Thus, the present research objectives are to improve nitrogen management, not only from additional fertilizers but also through innovations within crop rotations. Intercropping or undersowing systems with leguminous species, and also reduced tillage, have been tested to improve wheat production, as well as baking quality and nutritional value, and avoid mycotoxin contamination. Recently, agronomical and technological methods to improve the technological quality and safety of wheat have been determined (David et al. 2011). Further research questions are how local and regional processing, marketing, distribution, and selling enterprises in the region can be established or better implemented in the region, considering the increasing requirements from processors for quality and safety of organic wheat, as well as the demand from the regional and national organic food markets to decrease the variation in offer and quality as well as to limit price instability. And, last but not least, how can the organic farmers become better integrated in this food chain network, also considering the different support payment systems at national and European levels for organic agriculture?

This example shows the evolution of a natural sciences-oriented research program in which research objectives and methodology have slowly developed from technical questions on nitrogen management of organic wheat, supported by agronomists and applied at field scale, to overall agroecological questions around organic grain producers in which the social sciences (economists, sociologists, and food technologists) have been increasingly integrated, focusing on the wheat-flour food chain, stakeholder networks, and quality and safety issues of wheat, applied at the farm and food system scales. Finally, after several years, a more holistic approach to the agroecology of the food system has been established. Nevertheless, the program is still dominated by natural scientists, with the result that some more social sciences-oriented topics, such as stakeholder networks and marketing of organic food products, are somewhat neglected in achieving a balanced holistic approach.

9.3.2 Community-Supported Agriculture: The Case of the French AMAP System

In France, modes of direct selling with food boxes have strongly developed in recent years. Among different systems, the AMAP system is the best known. URGENCI (2012) describes this system as "local solidarity-based partnerships between producers and consumers." AMAP is an association based on mutual engagement between a group of consumers and a farmer. Each consumer signs a contract at the beginning of the season with a commitment to buy part of the production, which is then periodically delivered to him at a constant price. The engagement of the consumers normally also involves a contribution to the association, such as organizing the locality for distribution or distributing information, but can go as far as participation in practical work on the farm. On his side, the farmer makes a commitment to deliver quality products and to respect the rules of the AMAP charter (Alliance Provenance 2003).

The principles were born in Japan at the end of the 1950s (Amemiya 2007). At the same time, similar systems emerged in Switzerland and Germany, and since 1985 also in the United States (Cone and Myhre 2000; Cooley and Lass 1998; Fieldhouse 1996), called "Community-Supported Agriculture" (CSA). Similar systems are today also found in Canada, England, Australia, New Zealand, Brazil, Sweden, and Norway. In France, this system is more recent, but is strongly developing (Lamine 2008; Mundler 2007). According to MIRAMAP (2012), the number of AMAP systems reached 1600 at the end of 2011, with more than 66,000 involved consumers and a global turnover of 40 million euros. Even if the management of the CSA systems may be different from country to country, all these systems claim to offer an alternative to the conventional industrialized agri-food system (Fieldhouse 1996; Hinrichs 2000), and benefits such as education and social capital have been widely demonstrated (Cooley and Lass 1998; Sharp et al. 2002).

As the AMAP systems can vary significantly, they cannot be analyzed without taking into account the diversity of social and geographical contexts. In the following we will summarize research results about these systems in the Rhône-Alpes region, looking at different research topics mainly analyzed with a social sciences approach. Our research objective was to assess different components of the sustainability of the AMAP systems: their stability and persistence, their internal management, the economic viability of farmers, their capacity to produce quality products to all types of consumers having different incomes, and the environmental efficiency of the delivery of the products.

The Rhône-Alpes region is the second of 22 regions in France in terms of economic weight, numbering slightly more than 6.2 million inhabitants. The region is characterized by larger areas of the high mountains of the French Alps, mid-altitude mountainous and hilly areas, some urban agglomerations, and some smaller areas of plains dominated by agriculture. Its agriculture is distinguished by smaller farms (37 ha in average) compared with the French national average (55 ha), a large diversity of agricultural production, an important rate of production with quality labels, and a shared concern for environmental issues in agriculture (Mundler 2008). The AMAP systems have shown a strong development since 2004 in the Rhône-Alpes region. According to MIRAMAP (2012), AMAPs delivered to more than 10,000 families in 2010.

Different research questions were analyzed by economists and sociologists, mainly at farmer and consumer levels, but also at the scale of the AMAP system (Figure 9.5). Those farm enterprises that process their products and deliver them in short supply chains created more work units per farm: 2.1 annual work units for direct-selling farms compared with 1.3 work units for traditional farms (Capt and Dussol 2004). Direct selling with food boxes (Figure 9.6) allows family farms to be maintained, although income remains modest, and can be lower than in conventional supply chains (Mundler et al. 2008). The perception of risk for famers also changed. Even though the contracts of engagement are signed for 6–12 months, the farmers feel less subject to fluctuating market prices for standard agricultural commodities.

On the social level, the subjects and topics analyzed are mainly focusing on the collective and the territorial scales. It can be shown that the AMAP systems are competitive in comparison with

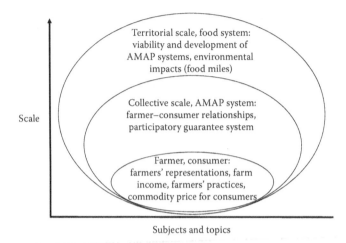

Figure 9.5   Subjects and research topics at different scales for community-supported agriculture illustrated with the case of the AMAP system in France. The research approach was dominated by a social sciences approach.

Figure 9.6   Food baskets and boxes and their distribution with an AMAP system. (Courtesy of Alliance PEC Rhône-Alpes. With permission.)

other modes of distribution (Mundler and Audras 2010). The consumers of the AMAPs are a population with strong social and cultural values, but are not distinguished by higher income. Those consumers who do not favor buying their food products in an AMAP system are more characterized by having a social and cultural distance from this kind of system, but not adhering because of a question of price. Another important fact is the social and professional recognition for farmers in short supply chains (including the AMAP). This recognition concerns satisfaction with their work and better socialization due to their relationships with clients, but also better financial recognition (Dufour et al. 2011). The consumers engaged in the AMAP system also insist on the educational role implicated in the supply of the food boxes: education of taste and the link between health and alimentation. Finally, the support and engagement of consumers for AMAP farmers have contributed to the implementation of different types of actions by territory-based organizations of stakeholders to protect land tenure of farmers and to favor short supply chains (Mundler et al. 2008).

If we look at environment-related questions, the farmer and farm scale and the territorial food system scale are important. Here, the dominant social sciences approach is complemented by the disciplines of agronomy and ecology. The charter of the AMAP system (Alliance Provenance 2003) mentions that the products from AMAP should "respect nature, environment, biodiversity, and soil fertility, and the production has to be managed without synthetic pesticides and fertilizers and responsible water use." In fact, the majority of AMAP farmers are certified for organic agriculture.

But the AMAP system even wants to be enrolled in an evolutionary progress approach that goes beyond organic farming in the sense of a social, economic, and ecological practice. This approach is opposed to the conception of organic farming as being too technical and instrumental, and exclusively defined by the production guidelines (Sylvander et al. 2006). In this conception, the AMAP systems have implemented a participatory guarantee system that associates consumers and farmers in the evaluation of the applied practices in the partner farm enterprises of the AMAP (Mundler and Bellon 2011). In some cases, this evaluation goes beyond the question of farmers' practices by also mutually evaluating other social and ethical questions, with the final objective of guaranteeing close relationships between the stakeholders.

Another important point is that the AMAP system favors local agriculture (Mundler 2009) and the consumption of seasonal products. Stakeholders see this as an advantage, based on the assumption, widely discussed in the literature, that these systems provide environmental benefits in reducing food miles, the distance travelled by a food product between the field and the plate. Several recent publications have questioned these benefits, showing the low efficiency of transport of low quantities (Coley et al. 2009; Edwards-Jones et al. 2008; Schlich and Fleissner 2005). In contrast, when the practices of the stakeholders in AMAP systems in France were analyzed and compared with other types of food distribution, it was shown that the AMAP systems had good efficiency (Mundler and Rumpus 2012).

In conclusion, the AMAP example represents a typical food system research topic that is often driven by a social sciences approach. For economic and socially related questions a lot of information is available already; however, less information is so far available on the impact of AMAP systems on certain practices of farmers, as well as on environmental issues such as biodiversity conservation or management (Maréchal 2008). In our example, this fact is mainly due to a low involvement of natural sciences approaches. We assume that the willingness to provide the possible largest diversity of products to consumers might induce agricultural practices favoring cultivated biodiversity. We also assume that receiving consumer members of an AMAP on the farm might promote the establishment of landscape elements such as hedges or herbaceous vegetation strips, thus favoring natural biodiversity. Nevertheless, these assumptions still have to be verified with interdisciplinary reseach involving agronomists, sociologists, economists, and ecologists.

9.3.3 Organic Agriculture and Drinking Water Catchments

In the last decade, different initiatives by national, regional, or local authorities in many countries in the world have been started to improve water quality in catchments for drinking water. The proposed solutions applied for catchments where agricultural land use is dominant are (i) to limit pesticide and nutrient inputs from conventional agricultural practices, (ii) to purchase agricultural land by the institution managing the catchment and to lend it to farmers with fixed rules for agricultural practices, (iii) to conduct reforestation, or (iv) to convert partially or completely to organic agriculture (Fleury and Vincent 2011).

In relation to organic farming, different approaches are carried out to favor conversion in drinking water catchment areas. In general, replacing conventional with organic farming practices normally results in a significant decrease of nitrate leaching and contamination with synthetic pesticides, as they are forbidden in organic agriculture (Haas 2010; Wilbois et al. 2007). Other, larger-scale approaches are (i) to promote collective catering with organic food for schools or other local catering facilities (as in the case of Munich, Germany (Stadtwerke München 2012) and Lons-le-Saunier (Martin 2010)), or (ii) to promote the establishment of local and regional market structures for organic raw products as well as to provide more information and establish better exchanges with farmers showing an interest in converting to organic agriculture (Haas 2010; Hermanowski et al. 2008; Wilbois et al. 2007). Often a combination of different options is favored in the different areas concerned.

This third research example deals with the issue of drinking water catchments and the potential contribution of organic agriculture to improving water quality. In contrast to the two preceding cases, it is an example in which a combined natural and social sciences approach was carried out from the beginning of the research project, with scientists from the fields of agronomy, ecology, hydrology, geography, sociology, and economics. In the project, initiated by the regional water agency, three drinking water catchments in southeastern France are analyzed and evaluated simultaneously at the field, farm, agroecosystem/local area, and food system scales. Different stakeholders were integrated from the beginning of the research project. The project analyzes and evaluates the contribution of organic agriculture to improving water quality in three territories where problems exist with nitrate and pesticide concentrations in the groundwater (Figure 9.7). Each territory includes a drinking water catchment, but is considered as a much larger area without specific delimitation to take into account different food system issues.

At the field scale, the different cropping and livestock production practices are analyzed to assess their potential risk for groundwater pollution (Figure 9.8). The project also investigates how similar they are to organic farming practices, which are generally considered to be less polluting, and evaluates the feasibility of changing or adapting certain practices. The initial results show that in one catchment certain crop rotations dominated by maize rely on intensive use of fertilizers,

Figure 9.7 **(See color insert)** Area dominated by agriculture in a drinking water catchment in southeastern France.

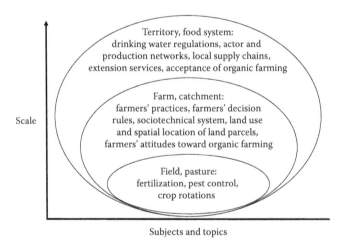

Figure 9.8 Different subjects and research topics are dealt with in a simultaneous social and natural sciences approach for analyzing and evaluating the potential contribution of organic agriculture to improving water quality in drinking water catchments.

pesticides, and irrigation. Also, some farmers seem to be already close to organic agriculture practices, but have not yet considered or decided to convert to organic agriculture. Many farmers indicated in interviews that they (i) will face major technical constraints to changing their practices, (ii) fear that their incomes will decrease, (iii) cannot imagine changing their farming system to organic, and (iv) would need technical and financial assistance when converting to organic agriculture.

At the scale of the drinking water catchments, the spatial location of farms and their land parcels and their respective practices are analyzed to evaluate zoning effects, such as concentration of certain "polluting" practices in certain zones, and the possibility of exchanging land parcels or modifying crop rotations, or even farming systems in certain zones (Figure 9.8). The field and water catchment scale analyses are led by natural sciences investigations, but also take into account social sciences aspects such as farmers' motivations for respective farming systems or practices and their decision rules for fertilization or pest control practices.

In interviews with farmers, the analysis of their farming practices and their spatial arrangement showed that especially intensive crop rotations in one catchment are located in zones with a high risk of transfer to the groundwater, and that farmers would be reluctant to locate them elsewhere because of difference in soil types, access to irrigation water, and distance from the farm buildings. In the second catchment, farmers have already adjusted their N-fertilization of crops in the core zone of the catchment relatively well, for example with fractionated application, which makes it difficult for them to improve this type of fertilization further. In addition, these fertilization practices are managed differently depending on which commune the farm is located in, as certain communes impose stronger regulations than others. For some mixed crop–livestock farms in the same catchment, it also became evident that they could not easily relocate their temporary pastures within their systems because of restricted access with their livestock to potential other land parcels, or because of remoteness of land parcels from the farm stable.

Simultaneously with the abovementioned mainly natural sciences analyses, a social sciences approach is carried out, analyzing stakeholder networks, sociotechnical networks, existing organic food commodity chains and production networks, but also farmers' flexibility and willingness to convert to organic agriculture. The latter are linked to the farm and water catchment scale, but the other analyses focus on the "territory" of stakeholder networks or even go beyond this when dealing with commodity chains or production networks.

Many farmers indicated in interviews that, besides technical problems and their low willingness to change to organic agriculture mentioned above, they would not know where they could place their products in the commodity supply chain if they converted to organic agriculture. The supply chain analyses have shown so far that short supply chains are more or less nonexistent and would need to be established first. Also, the development of collective catering with organic food in one of the study territories is in an initial state. Although longer supply chains for organic products exist on a regional scale, they are still under development. Interviews and meetings with stakeholders showed that so far there is little cooperation or coordination among many stakeholders, and that farmers are often not aware or well informed about the issue of drinking water quality and the different courses of action of certain stakeholders, such as the local or regional water management institutions.

This third example clearly illustrates the benefits of involving different scientific disciplines from the beginning of a research project in a simultaneous natural and social sciences approach, and at different scales. Without this, there would probably have been a risk of favoring certain more technical short-term solutions to improve farmers' practices at the field or the catchment level, without adequately taking into account the fact that a conversion to organic agriculture (or to conventional systems with practices close to organic practices) is not currently feasible, as the necessary network of collectors, raw product processors, retailers, technical advisors, and local to national markets for their organic products is presently underdeveloped or even nonexistent for the studied territories.

9.4 DISCUSSION AND CONCLUSION

Different authors demand that agroecology of the food system requires a holistic approach to guarantee that the different facets, subjects, and topics of the food systems are sufficiently considered. Nevertheless, two major questions remain: when is a holistic approach accomplished, and how much integration is needed from either natural or social sciences?

In our opinion, a holistic approach can be assured in considering at least the first three of the four prerequisites for food system research in agroecology, as mentioned in Wezel and David (2012):

1. Agroecological research has to be carried out simultaneously at different scales.
2. Agroecological research has to integrate different scientific disciplines as well as stakeholders from the different food system networks.
3. The potential environmental, social, and economic impacts of the expected research results have to be anticipated during the development of a research project and its hypothesis.
4. The recommendations from agroecological research have to be impact assessment-driven for the different scales.

The fourth prerequisite goes somewhat beyond the pure research approach, as it also defines the use of research results. Thus, it might not always be applied by the different research communities, as this last step is not automatically integrated or planned in many research projects.

In practice, this means that when starting to build a research program the focus is often on special subjects and topics, for example, analysis of short supply chains, and research is often led by only one or two scientific disciplines. But, if the intention is to build an agroecology program with a holistic approach, a second step is needed, in which additional subjects and topics are also taken into account. In the case of the short supply chains, these could include, for example, impact on farmers' income, farmers' practices, family work organization, food quality, food miles, and presence of (agro)biodiversity elements on farms. In adding these topics, the question of scales will appear automatically, as well as which other disciplines have to be integrated to analyze the additional topics adequately.

The second question, "how much integration is needed from either natural or social sciences to achieve a holistic agroecology of the food system approach?", probably cannot be answered satisfactorily.

Indeed, the status of agroecology from an interdisciplinary viewpoint remains ambiguous. For many social sciences researchers, agroecology is not yet a research object with clearly outlined frontiers or scales. Therefore, it remains an object with vague contours; the concept expresses an intention, a horizon, a process in constant evolution, claimed by a diversity of actors, which themselves can be in strong disagreement. From this viewpoint, the debates around sustainable development have to be reconsidered for agroecology and social sciences. For example, should agroecology be restricted only to alternative approaches to the agro-industry? Which agricultural practices should be considered, or not, for agroecology? Numerous questions remain. Sustainable development can be defined as "a vector of desirable social objectives, that is a list of attributes which society seeks to achieve or maximize" (Pearce et al. 1990); therefore, agroecology could be handled as a set of technical, social, environmental, and economic practices aiming at favoring a permanent transformation of agriculture, with the objective of better taking into account ethics of exchanges, a social and economic cohesion, and the preservation of natural resources adapted to a certain territory. In this sense, agroecology is a movement, not only in the sense of a social movement, but in the sense of flexibility and a permanent adaptation of technical and social practices in each particular context.

Can agroecology be seen as an interdiscipline or a metadiscipline that is able to associate and integrate natural and social sciences? In our opinion, yes, if it stimulates research associating researchers from different disciplines and practitioners in a permanent mutual learning process, allowing them to preserve natural resources, to improve the efficiency of practices in reducing external inputs, favoring control of pests and maladies with biological control, and to redesign agroecosystems based on ecosystem services and adapted to social and environmental conditions specific to each territory.

The required interdisciplinarity for agroecology research needs, therefore, an evolution in the way to design programs and research objects. As shown with the third research example, there exist disjunctions between the ecological and agronomic functional areas (a drinking water catchment and farmers' fields), the management area (the farm territory, the collecting area for an agricultural commodity, or even the territory of a local interest group of actors), and the areas used by the citizens. It is, therefore, necessary to establish coherence and a common perspective for the varied knowledge of different origins that concerns the objectives, the sectors, and the different scales. Ecology and agronomy, historically the basic disciplines of agroecology, contribute to a holistic approach with their capacity to analyze changes of farmers' practices and their production and environmental consequences, as well as applying the concepts and methods of ecology for food chain analyses, such as matter fluxes and transformation, to the evaluation of food systems, something the social sciences are not able to carry out. The social sciences have historically developed and continue to develop theories, concepts, and methods to deal with different issues in agriculture and food systems. The individual in interaction and exchange with other actors is, therefore, at the center of analyses, and increasingly these analyses are carried out in a participatory way. To progress toward a global and systemic vision of agroecology of the food system, therefore, both interdisciplinary and participatory research is needed (Petit et al. 2011).

In this regard, the third example, about organic agriculture and drinking water catchments, is an example in which all the prerequisites for research into the agroecology of the food system have been considered from the beginning of the research program, and thus a holistic approach has been established. This, of course, does not guarantee that the final research results will automatically provide satisfactory solutions for stakeholders and farmers, but it provides assurance that relevant research topics from natural and social sciences have been considered in order to analyze the whole food system at different scales.

In contrast, the first example, about organic wheat production and wheat baking, clearly shows that a holistic approach was not accomplished in the first years of the project, as it was guided only by the natural sciences. Although social sciences topics were gradually integrated thereafter, a bias still exists, with natural sciences questions dominating. We could call this a "biased holistic approach," in which a broad variety of relevant research topics were integrated, but with a remaining weakness and insufficient depth in the analyses of certain social sciences topics.

With the second example, about the AMAP system, we find an even stronger bias. Here, an almost complete dominance of social sciences topics and subjects prevails. Natural sciences topics are so far only marginally integrated. Therefore, a holistic approach to the agroecology of the food system is not yet achieved.

To conclude, the challenging questions of today's agriculture and food systems can only be adequately solved if interdisciplinary research is carried out from the beginning, integrating natural and social sciences, as well as concerned stakeholders. For this, the holistic approach to the agroecology of the food system seems very promising, as it is based on interdisciplinarity, multiscale analyses, stakeholder integration, and anticipation of environmental, social, and economical impacts from the expected research results.

REFERENCES

Alliance Provence (Réseau régional des AMAP). 2003. Charte d'Alliance Provence sur les AMAP. http://www.allianceprovence.org/La-nouvelle-Charte-des-AMAP.html (accessed: 5 February 2014).

Altieri, M. A. 1989. Agroecology: A new research and development paradigm for world agriculture. *Agr Ecosyst Environ* 27:37–46.

Altieri, M. A. 1995. Agroecology: *The Science of Sustainable Agriculture*. Boulder: Westview Press.

Altieri, M. A. 2002. Agroecology: The science of natural resource management for poor farmers in marginal environments. *Agr Ecosyst Environ* 93:1–24.

Altieri, M. A., F. R. Funes-Monzote, and P. Petersen. 2012. Agroecologically efficient agricultural systems for smallholder farmers: Contributions to food sovereignty. *Agron Sustain Dev* 32:1–13.

Altieri, M. A. and C. I. Nicholls. 2008. Scaling up agroecological approaches for food sovereignty in Latin America. *Development* 51:472–480.

Altieri, M. A. and V. M. Toledo. 2011. The agroecological revolution in Latin America: Rescuing nature, ensuring food sovereignty and empowering peasants. *J Peasant Stud* 38:587–612.

Amemiya, H. 2007. Le Teikei: La reference japonaise de la vente directe de produits fermiers locaux. In Amemiya, H. (ed.), *L'agriculture participative. Dynamiques bretonnes de la vente directe*, pp. 21–48. France: Presses Universitaires de Rennes.

Capt, D. and A. M. Dussol. 2004. Exploitations diversifiées: Un contenu en emploi plus élevé. *Agreste Cahiers* 2:11–18.

Casagrande, M., C. David, M. Valantin-Morison, D. Makowski, and M. H. Jeuffroy. 2009. Factors limiting the grain protein content of organic winter wheat in south-eastern France: A mixed-model approach. *Agron Sustain Dev* 29:565–574.

Chappell, M. J. and L. A. LaValle. 2011. Food security and biodiversity: Can we have both? An agroecological analysis. *Agr Hum Values* 28:3–26.

Cohn, A., J. Cook, M. Fernandez, R. Reider, and C. Steward. 2006. Agroecology and struggle for food security in the Americas. London, England: International Institute for Environment and Development (IIED), Yale School of Forestry and Environmental Studies (Yale F&ES), IUCN Commission on Environmental, Economic and Social Policy (CEESP). http://pubs.iied.org/pdfs/14506IIED.pdf (accessed: 17 March 2012).

Coley, D., M. Howard, and M. Winter. 2009. Local food, food miles and carbon emissions: A comparison of farm shop and mass distribution approaches. *Food Policy* 34:150–155.

Cone, C. A. and A. Myhre. 2000. Community-supported agriculture: A sustainable alternative to industrial agriculture? *Hum Organ* 59:187–197.

Conway, G. R. 1987. The properties of agroecosystems. *Agr Syst* 24:95–117.

Cooley, P. J. and D. A. Lass. 1998. Consumer benefits from community supported agriculture membership. *Rev Agr Econ* 20:227–237.

David, C., F. Celette, J. Abecassis, et al. 2011. AGronomical and TEChnological methods to improve ORGanic wheat quality (AGTEC-Org). EU-Project no. ERAC-CT-2004-011716. Final report, p. 53. http://www.coreorganic.org/research/projects/agtec-org/index.html (accessed: 3 June 2012).

David, C. and M. H. Jeuffroy. 2009. A sequential approach for improving Azodyn crop model under conventional and low-input conditions. *Eur J Agron* 31:177–182.

David, C., M. H. Jeuffroy, J. Henning, and J. M. Meynard. 2005a. Yield variation of organic winter wheat: A crop diagnosis on a field network in south-east of France. *Agron Sustain Dev* 25:1–11.

David, C., M. H. Jeuffroy, F. Laurent, M. Mangin, and J. M. Meynard. 2005b. The assessment of a decision making tool for managing the nitrogen fertilization of organic winter wheat. *Eur J Agron* 23:225–242.

David, C., M. H. Jeuffroy, S. Recous, and F. Dorsainvil. 2004. Adaptation and assessment of the Azodyn model managing the nitrogen fertilization of organic winter wheat. *Eur J Agron* 21:249–266.

David, C. and S. Joud. 2008. Etats des lieux de la collecte du blé biologique panifiable en France. *Industries des Céréales* 159:21–26.

David, C., P. Mundler, O. Demarle, and S. Ingrand. 2010. Long-term strategy and flexibility of organic farmers in Southeastern France. *Int J Agr Sustain* 8:305–318.

De Schutter, O. 2012. Agroecology, a tool for the realization of the right to food. In Lichtfouse, E. (ed.), *Agroecology and Strategies for Climate Change, Sustainable Agriculture Reviews*, vol. 8, pp. 1–16. Dordrecht: Springer Verlag.

Dufour, A., C. Herault-Fournier, E. Lanciano, and N. Pennec. 2011. L'herbe est-elle plus verte dans le panier? Satisfaction au travail et intégration professionnelle de maraîchers qui commercialisent sous forme de paniers. In Traversac, J. B. (ed.), *Circuits courts. Contribution au développement regional*, pp. 71–85. Dijon: Editions Educagri.

Edwards-Jones, G., L. Milai Canals, N. Hounsome, et al. 2008. Testing the assertion that "local food is best": The challenges of an evidence-based approach. *Trends Food Sci Tech* 19:265–274.

Feenstra, G. 1997. Local food systems and sustainable communities. *Am J Alternative Agr* 12:28–36.

Fieldhouse, P. 1996. Community shared agriculture. *Agr Hum Values* 13(3):43–47.

Fleury, Ph. and A. Vincent. 2011. Water quality protection and organic farming development in France. In *3rd Scientific Conference of ISOFAR*, vol. 2, pp. 51–54. IFOAM, Seoul, South Korea, 27–30 October 2011.

Francis, C., G. Lieblein, S. Gliessman, et al. 2003. Agroecology: The ecology of food systems. *J Sustain Agr* 22:99–118.

Francis, C. and D. Rickerl (eds). 2004. Ecology of food systems: Visions for the future. In *Agroecosystems Analysis*. Agronomy 43, pp. 177–197. Madison, WI: American Society of Agronomy.

Fraval, P. 2000. Éléments pour l'analyse économique des filières agricoles en Afrique sub-saharienne. Bureau des politiques agricoles et de la sécurité alimentaire DCT/EPS, Ministère des Affaires étrangères. DGCID, p. 100.

Frayssignes, J. 2005. Les AOC dans le développement territorial. Une analyse en terme d'ancrage appliqué aux cas français des filières fromagères. Thèse de doctorat en géographie. Institut National Polytechnique de Toulouse.

Gliessman, S. R. 2007. *Agroecology: The Ecology of Sustainable Food Systems*, p. 384. New York: CRC Press, Taylor & Francis.

Gliessman, S. R. 2011. Agroecology and the growing food crisis. *J Sustain Agr* 35:697–698.

Granovetter, M. 2000. *Le marché autrement*. Essais de Mark Granovetter. Paris, Desclée de Brouwer, p. 238.

Guptill, A. and J.-L. Wilkins. 2002. Buying into the food system: Trends in food retailing in the US and implications for local foods. *Agr Hum Values* 19:39–51.

Haas, G. 2010. Wasserschutz im Ökologischen Landbau: Leitfaden für Land- und Wasserwirtschaft. Report, Bundesprogramm Ökologischer Landbau, p. 61. http://orgprints.org/16897/1/16897-06OE175-agraring-enieurbuero-haas-2010-wasserschutz.pdf (accessed: 7 February 2012).

Hall, A. and V. Mogyorody. 2001. Organic farmers in Ontario: An examination of the conventionalization argument. *Sociol Ruralis* 41:399–322.

Hermanowski, R., A. Bauer, B. Schwab, and D. Pfennigwerth. 2008. Wenn Markt und Ökologie an Einem Strang Ziehen. *Ökologie & Landbau* 147:47–49. http://orgprints.org/8451/ (accessed: 7 February 2012).

Hinrichs, C. C. 2000. Embeddedness and local food systems: Notes on two types of direct agricultural market. *J Rural Stud* 16:295–303.

Jarosz, L. 2008. The city in the country: Growing alternative food networks in metropolitan areas. *J Rural Stud* 24:231–244.

Lamine, C. 2008. *Les AMAP: Un nouveau pacte entre producteurs et consommateurs?* p. 163. Paris: Editions Yves Michel.

Lamine, C. and S. Bellon. 2009. Conversion to organic farming: A multidimensional research object at the crossroads of agricultural and social sciences. A review. *Agron Sustain Dev* 29:97–112.

Law, J. 1992. Notes on the theory of the actor-network: Ordering, strategy and heterogeneity. *Syst Pract* 5:379–393.

Lockie, S. and D. Carpenter. 2010. *Agriculture, Biodiversity and Markets: Livelihoods and Agroecology in Comparative Perspective*. London, England: Earthscan.

Maréchal, G. 2008. *Les circuits courts alimentaires: Bien manger dans les territoires*. Dijon, France: Educagri.

Marsden, T., J. Banks, and G. Bristow. 2000. Food supply chain approaches: Exploring their role in rural development. *Sociol Ruralis* 40(4):424–438.

Martin, L. 2010. Agriculture et qualité de l'eau: Le dispositif de Lons-le-Saunier, Evolution des jeux d'acteurs. Unpublished master thesis, ESITPA, Mont-Saint-Aignan, France.

Maye, D., L. Holloway, and M. Kneafsey (eds). 2007. *Alternative Food Geographies: Representation and Practice*. Bingley, UK: Emerald Group.

Mikkola, M. and L. Seppänen. 2006. Farmers' new participation in food supply chains: Making horizontal and vertical progress by networking. In Langeveld, H. and N. Röling (eds), *Changing European Farming Systems for a Better Future: New Visions for Rural Areas*. The Netherlands: Wageningen Academic Publishers.

MIRAMAP. 2012. Mouvement Interrégional des AMAP. http://miramap.org/spip.php (accessed: 7 February 2012).

Muchnik, J. and C. de Sainte Marie (eds). 2010. *Le temps des Syal: Techniques, vivres et territoires*. Versailles: éditions Quae.

Mundler, P. 2007. Les AMAP en Rhône-Alpes, entre marché et solidarité. *Ruralia* 20:185–215.

Mundler, P. 2008. L'agriculture en Rhône-Alpes, une multifonctionnalité évidente? In Cornu, P. and J.-L. Mayaud (eds), *Nouvelles questions agraires: Exploitants, fonctions et territoires*, pp. 275–298. Paris: Editions Boutique de l'Histoire, coll. Mondes ruraux contemporains.

Mundler, P. 2009. Les associations pour le maintien de l'agriculture paysanne: Solidarité, circuits courts et relocalisation de l'agriculture. *Pour* 201:155–162.

Mundler, P. and A. Audras. 2010. Le prix des paniers, analyse de la formation des prix dans 7 AMAP en Rhône-Alpes. Colloque INRA-SFER-CIRAD, Rennes, 9–10 December 2010.

Mundler, P. and S. Bellon. 2011. Les systèmes participatifs de garantie. Une alternative à la certification par organismes tiers? *Pour* 212:57–65.

Mundler, P., J. M. Ferrero, A. Jan, and R. Thomas. 2008. Petites exploitations diversifiées en circuits courts. Soutenabilité sociale et économique. Research report. Lyon, France: Isara.

Mundler, P. and L. Rumpus. 2012. The energy efficiency of local food systems: A comparison between different modes of distribution. *Food Policy* 37:609–615.

Murdoch, J. 2000. Networks—A new paradigm of rural development? *J Rural Stud* 16:407–419.

O'Hara, S.-U. and S. Stagle. 2002. Endogenous preferences and sustainable development. *J Soc Econ* 31:511–527.

Pearce, D. W., E. B. Barbier, and A. Markandya. 1990. *Sustainable Development*. London: Earthscan.

Petit, S., C. Mougenot, and Ph. Fleury. 2011. Stories on research, research on stories. *J Rural Stud* 27:394–402.

Reardon, J. A. S. and R. A. Pérez. 2010. Agroecology and the development of indicators of food sovereignty in Cuban food systems. *J Sustain Agr* 34:907–922.

Renting, H., T. K. Marsden, and J. Banks. 2003. Understanding alternative food networks: Exploring the role of short food supply chains in rural development. *Environ Plann* 35:393–411.

Robertson, G. P., J. C. Broome, E. Chornesky, J. R. Frankenberger, P. Johnson, M. Lipson, J. A. Miranowski, E. Owens, D. Pimentel, and L. A. Thrupp. 2004. Rethinking the vision for environmental research in US agriculture. *Bioscience* 54:61–65.

Rosset, P. M., B. M. Sosa, A. M. R. Jaime, and D. R. A. Lozano. 2011. The Campesino-to-Campesino agroecology movement of ANAP in Cuba: Social process methodology in the construction of sustainable peasant agriculture and food sovereignty. *J Peasant Stud* 38:161–191.

Schlich, E. and U. Fleissner. 2005. The ecology of scale: Assessment of regional energy turnover and comparison with global food. *Int J Life Cycle Ass* 10:219–223.

Sharp, J., E. Imerman, and G. Peters. 2002. Community Supported Agriculture (CSA): Building community among farmers and non-farmers. *J Extension* 40. Online. www.joe.org/joe/2002june/a3.html (retrieved March 2012).

Stadtwerke München. 2012. Trinkwasser-Schutz und Organischer Landbau. http://www.swm.de/privatkunden/m-wasser/gewinnung/wasserschutz.html (accessed: 7 February 2012).

Stassart, P. M. and C. Claes. 2010. Agroécologie: Le chainon manquant. Rôle de consommateurs et d'ONG dans les processus émergents d'apprentissages. In *Proceeding of ISDA* 2010. Montpellier, 28–30 June 2010. http://hal.archives-ouvertes.fr/docs/00/52/14/06/PDF/STASSART-CLAES_AgroA_cologie_Le_chainon_manquant_Final.pdf (accessed: 7 February 2012).

Sylvander, B., S. Bellon, and M. Benoît. 2006. Facing the organic reality: The diversity of development models and their consequences on research policies. In *Proceedings of European Joint Organic Congress "Organic Farming and European Rural Development"*, pp. 58–61. Odense, Denmark, 30–31 May 2006. http://orgprints.org/8247/ (accessed: 5 May 2012).

Tregear, A. 2011. Progressing knowledge in alternative and local food networks: Critical reflections and a research agenda. *J Rural Stud* 27:419–430.

Tregear, A., F. Arfini, G. Belletti, and A. Marescotti. 2007. Regional foods and rural development: The role of product qualification. *J Rural Stud* 23:12–22.

Thompson, J. and I. Scoones. 2009. Addressing the dynamics of agri-food systems: An emerging agenda for social science research. *Environ Sci Policy* 12:386–397.

URGENCI. 2012. The international network of community supported agriculture. www.urgenci.net (accessed: 7 February 2012).

Walford, N. 2003. Productivism is allegedly dead, long live productivism: Evidence of continued productivist attitudes and decision-making in south-east England. *J Rural Stud* 19:491–502.

Warner, K. D. 2005. Extending agroecology: Grower participation in partnerships is key social learning. *Renew Agr Food Syst* 21:84–94.

Warner, K. D. 2007a. *Agroecology in Action: Extending Alternative Agriculture through Social Networks.* USA: MIT Press.

Warner, K. D. 2007b. The quality of sustainability: Agroecological partnerships and the geographic branding of California winegrapes. *J Rural Stud* 23:142–155.

Wezel, A., S. Bellon, T. Doré, C. Francis, D. Vallod, and C. David. 2009. Agroecology as a science, a movement or a practice. A review. *Agron Sustain Dev* 29:503–515.

Wezel, A. and C. David. 2012. Agroecology and the food system. In Lichtfouse, E. (ed.), *Agroecology and Strategies for Climate Change*, Sustainable Agriculture Reviews, vol. 8, pp. 17–34. Dordrecht: Springer.

Wezel, A. and J. C. Jauneau. 2011. Agroecology—Interpretations, approaches and their links to nature conservation, rural development and ecotourism. In Campbell, W. B. and S. López Ortiz (eds), *Integrating Agriculture, Conservation and Ecotourism: Examples from the Field.* Issues in Agroecology—Present Status and Future Prospectus, vol. 1, pp. 1–25. Dordrecht: Springer.

Wezel, A. and V. Soldat. 2009. A quantitative and qualitative historical analysis of the scientific discipline agroecology. *Int J Agr Sust* 7:3–18.

Wilbois, K.-P., M. Szerencsits, and R. Hermanowski. 2007. Eignung des Ökologischen Landbaus zur Minimierung des Nitrataustrags ins Grundwasser. FiBL Deutschland. http://orgprints.org/13270/ (accessed: 7 February 2012).

CHAPTER 10

Agroecology Applications in Tropical Agriculture Systems

Noureddine Benkeblia and Charles A. Francis

CONTENTS

10.1 Introduction 201
10.2 Agroecosystems..203
    10.2.1 What Is an Agroecosystem? 203
    10.2.2 Types of Agroecosystems 203
        10.2.2.1 Seasonally Cropped Agroecosystems (SCA)...203
        10.2.2.2 Permanently Cropped Agroecosystems (PCA)...204
        10.2.2.3 Forestry Agroecosystems (FAE) 205
10.3 Tropical Agroecosystems 205
    10.3.1 Humid Tropics 205
    10.3.2 Wet–Dry Tropics..206
10.4 Tropical Agroecosystem Biodiversity..206
10.5 Managed Agroecosystems..207
    10.5.1 Crops and Varieties..207
    10.5.2 Cropping Systems 208
10.6 High-Tech and Alternative Agricultural Production Systems in the Tropics......................210
10.7 Soil Fertility in Tropical Agroecosystems...212
10.8 Pest Management with Chemicals and Alternatives 213
10.9 Human Influences on Tropical Agroecosystems 214
10.10 Agroecology and Applications of New Technologies 214
10.11 Conclusions and Perspectives 215
References 215

10.1 INTRODUCTION

Current research shows that national per capita incomes are related to ecological wealth of countries, including their fertile soils and adequate rainfall, and other natural resources that provide strong indicators of the wealth or poverty of a nation. Most economies in tropical regions are generally poor, while those in temperate ecoregions are rich (Sachs 2001). Although tropical countries have temperatures that may permit year-round possibilities for crop growth, seasonal rainfall and highly weathered soils reduce that potential in many areas. These important factors limit the productivity of many current tropical agroecosystems.

Soils and water resources, weather and climate, and native and cultivated plants are among the most valuable natural resources of any country. Soil in particular is the foundation for successful and sustainable cropping systems. Several different sectors including agriculture and others closely allied can sustain economic viability and quality of life for human populations and communities only if the soil resources, water supply, and plant and animal biodiversity can be preserved and regenerated. Therefore, it is fundamental to understand both inherent potentials and limitations of soils for designing management strategies for sustained tropical agriculture and an ecological balance (Gajbhiye and Mandal 2000).

Starting in the twentieth century, the world population has increased as never before. It is becoming ever more urgent to grow food in sufficient quantities and of sufficient quality and to promote equity in access in this growing population. Concurrently, we recognize that intensive high-tech cropping systems with all their external fossil fuel needs are depleting nonrenewable resources. These systems may also seriously affect our environment, and our contemporary challenge is to supply people's needs for food, preserve our environment and a sustainable agriculture, and maintain ecological systems and services (Charlton 1987; Daily 1997) in ways that do not reduce the options for future generations (Brundtland 1988).

To assure that we can accomplish these goals, it is essential to understand and manage ever more complex and diverse agricultural and ecological systems. This challenge is particularly acute for developing countries in the tropics and subtropics, where the population is increasing rapidly and resources are often limited.

Late in the twentieth century, we began to appreciate that the major challenges include more than designing and managing cropping systems, but also include social, cultural, economic, and particularly environmental concerns. It is now obvious that crop production issues cannot be resolved without considering resource and social concerns. For example, the distribution of benefits from the agricultural sector, including access to food, has become skewed toward wealthy people in many small and low-income countries, including those in the tropics. There are large and persistent sectoral income and wage discrepancies. Widespread mobility of land and labor from agriculture to tourism and related uses, for example in many island countries, further complicates the food system. There have been massive changes in ownership and control of agricultural lands away from farmers to people and corporations in cities and in other countries (Kugelman and Levenstein 2013). However, near the end of the century, new sets of policies and practices have been developed to reverse agricultural decline and preserve environmental stability (McElroy and de Albuquerque 1990).

Based on principles of agroecology, a new approach to technology use in development has emerged that has the potential to provide the increasing world population with their food and other needs without depleting the limited natural resource base (IAASTD 2008). One goal of agroecology is to satisfy contemporary needs through the design of farming systems that are based on natural processes and renewable resource use. This discipline also must develop broad performance criteria that include ecological sustainability, as well as food security, economic viability, resource conservation, and social equity, in addition to the traditional goal of increased production (Altieri et al. 1984). Traditional agroecosystems represent a strategy that ensures diverse diets and income sources, stable production, minimum risk, efficient use of land resources, and enhanced ecological integrity. The legacy of traditional agriculture developed over centuries demonstrates that a combination of stable and diverse production, internally generated and renewable inputs, favorable energy output/input ratios, and incorporation with both subsistence and market needs can be an effective approach to achieve food security, income generation, and environmental conservation. Traditional approaches represent multiple resource use strategies that enhance the multifunctional nature of agriculture, an important feature for the health of rural regions in the next century (Altieri 2000). The challenges are to integrate science with appropriate traditional systems and technologies to increase productivity to meet human food needs while preserving other species and quality of life.

10.2 AGROECOSYSTEMS

10.2.1 What Is an Agroecosystem?

Many definitions have been used to define agroecosystems. A general concept and definition have emerged to describe a human activity system that provides specific services and possesses certain characteristics such as a defined biodiversity, a determined ecological succession, and a promotion of specific food webs and nutrient cycles. To achieve human goals, we have developed intensive, high-tech, and industrial agriculture, increased deforestation and drainage of natural wetlands, and appropriated a large share of natural resources to produce food and other products. Thus, an agroecosystem differs from a natural ecosystem because it lacks species biodiversity, relies on focused plant and animal breeding, depends on infusions of fossil fuels, and often has high inputs and loss levels of both nutrients and energy (Figure 10.1) (Sheaffer and Moncada 2009). An agroecosystem is also extractive, unlike natural ecosystems.

Focusing on biological and resource use interactions to meet human goals leads to a definition of an agroecosystem as "a biological and natural resource system managed by humans for the primary purpose of producing food as well as other socially valuable nonfood goods and environmental services" (Wood et al. 2000). Most natural ecosystems have been greatly changed by human activities, and very few are still in their natural state. Through the practice of agriculture and other activities, intentional agroecosystems have been developed that are now highly modified ecosystems playing the main role in the production of foods and other commodities. In the process, these agroecosystems have caused major changes in our environment, affecting soil, surface water and groundwater, and biodiversity (Sheaffer and Moncada 2009; Wood et al. 2000). Rapid changes in land use are considered a major factor accelerating soil erosion, including overgrazing, deforestation due to shortage of farmland, and cropping practices that affect soil quality (Lal 1985). In recent years, increased widespread cultivation of soybeans, cotton, oil palm, and biofuel crops has been fueled by global demands for agricultural commodities, and some of this activity has expanded from tropical to subtropical regions. Export of raw commodities to the wealthy global North as a result of this agricultural transition is likely the most important major change since the green revolution (Nepstad and Stickler 2008).

10.2.2 Types of Agroecosystems

The several definitions of agroecosytems do not differentiate among different types of production systems, whether based on plants, animals, or both. From an ecological point of view, it is difficult to distinguish different types of agroecosystems based only on region or climate, the associated human community, or the predominant agricultural practices. One possible classification of agroecosystems (INSEDA 2012) is utilitarian and based on what is included in the system and the predominant natural conditions: types of crops, types of animals, and natural conditions/predominant vegetation. This functional classification mixes components with geography and climate and seems less useful than one based more on agroecozones. We choose a more robust system to describe three major agroecosystems commonly found in the tropics, based on the choices of farmers to include different categories of crop and animal species.

10.2.2.1 Seasonally Cropped Agroecosystems (SCA)

SCAs can be defined as agroecological systems cultivated seasonally that include plants (generally annuals) that require one season or one cropping year from sowing to harvest. The major component plants in SCAs are annual herbaceous vegetables, oilseeds, cereals, and other plants such as herbs, spices, and medicinal plants. SCAs may also include livestock feed production including diverse pastures for grazing. These systems are characterized by seasonal agricultural practices

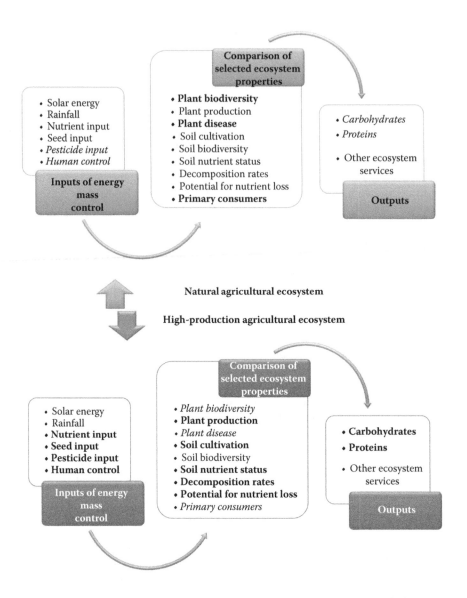

Figure 10.1   Similarities and differences between typical natural and high-production agricultural ecosystems. Inputs of energy, mass, and control (*left*), comparison of selected ecosystem properties (*center*), and outputs (*right*). (**Bold**, markedly higher value than in the other ecosystem type; *italic*, very low). (Redrawn from *Ecosystem Ecology*, Andrén, O. and Kätteter, T., Agricultural systems, pp. 145–156, (2008), with permission from Elsevier.)

adapted to available rainfall, short but intense tillage operations, and nutrient export that depends on the nature of what is harvested and sold. SCA soil erosion is difficult to generalize due to the different timing of crops in the field, types of tillage practices, local soil types and topographic features, and climatic conditions of the regions, especially frequency and intensity of rainfall.

10.2.2.2 *Permanently Cropped Agroecosystems (PCA)*

PCAs are established perennial cropping systems that include different perennial plants such as herbaceous (pineapple), industrial (palm for oil), and fruit and nut trees. Other plants cropped in these systems include coconut, cocoa, coffee, and other minor tropical and subtropical fruits. Because of the perennial nature of the crops and minimal soil disturbance, PCAs are considered more stable biologically than SCAs. They present the advantage of including two or more stories to better utilize light and other resources, and many annual herbaceous crops can be grown in the inter-row spaces. Once established, they provide some resilience due to deep root systems that can withstand periodic droughts and are promoted by high levels of biodiversity, but PCAs are tied to specific markets and do not allow short-term changes in crop species, which are possible in annual systems.

10.2.2.3 Forestry Agroecosystems (FAE)

Although they may not be considered cropping systems, forests are fundamental and vital components in many niches in the global system, especially in the tropics. Most countries in the tropics have forests and these constitute one-third of the world's total forest area. Tropical forests are known for their richness in species and invaluable genetic resources of plants for food, fiber, and many other products used by human populations (Adedire 2002; Pirarda and Karsenty 2009). FAEs are sometimes used for growing pastures for animal grazing. Tropical FAEs are also used to raise shaded coffee, as in Jamaica, where this system is predominant in coffee plantations, or cacao, as in West Africa. Complex multicrop systems in the humid forest zone of West Africa may include up to 18 different harvested species per hectare, providing biological and economic resilience (Francis 1986). Large areas of forest land have been cleared for timber exploitation, and not only for agricultural crops and grazing, but also for many other social, economical, and political reasons including oil palm for production of biofuel. Deforestation causes the loss of 10–25 Mha in the tropics each year. This could lead to catastrophic global environmental changes including impacts on agricultural systems (Adedire 2002; Brown and Pearce 1996; Fortner 1992), while the clearing and planting of crops and pasture systems is seen as vital to provide food for a growing human population.

10.3 TROPICAL AGROECOSYSTEMS

Tropical agroecosystems are located in regions between the latitudes of 22.5° north and 22.5° south and are characterized by the following climatic factors:

- High level of solar energy received because of latitude
- Low to moderate elevation of most areas, some with proximity to mountains
- Relatively high and consistent temperatures throughout the year
- Nearness to vast areas of water causing moderation of extreme temperatures
- High number of storm systems as cyclones, hurricanes, and thunderstorms
- Air pressure distribution producing varying patterns of wind and air-mass movement

Categories of tropical agroecosystems are consistent with general ecoregions because they are based on temperature patterns and humid or arid periods through the year that determine the choice of crops, growth rates of crops, and predominant cropping systems. An empirical classification, arguably the most known and widely used, was developed by Wladimir Köppen at the beginning of the last century, who defined tropical climates as those having average temperatures higher than 18°C (Köppen 1900a,b). According to his criteria, two main tropical agroecosystems fall into this classification:

10.3.1 Humid Tropics

Humid tropical areas are constantly warm and receive high annual rainfall. Their high humidity throughout the year supports dense tropical rain forests, while high rainfall and good soil moisture maintain evergreen forests. Agroecosystems in the humid tropics are well adapted due to their wide range of tree crops, tilled crops, and forages where soil erosion is controlled and soil fertility maintained, since the main cropping systems are perennially cropped or forest systems that are more stable than the seasonally cropped areas described in Section 10.2.2.1.

10.3.2 Wet–Dry Tropics

Wet–dry tropics are characterized by greater variation and contrasting rainfall in specific seasons during the year. Summer is generally hot, wet, and rainy, while winter is cooler and dry. The agroecosystems in wet–dry tropics are generally adapted to dry seasons of 2.5–5 months' duration, with rain in the rest of the year, and some well-suited annual crops, forages, and tree fruits. Among the several regions where agriculture is practiced around the globe, tropical agroecosystems in the wet–dry tropics are considered the most fragile and most in need of novel production approaches.

In general, tropical agricultural systems have not benefitted as much as others have from improved agricultural technologies. Major research investments and applications of new technologies have been more effective in boosting productivity in temperate systems, although some notable advances were achieved with production packages using appropriate technologies in tropical rice production during the green revolution. There have been substantial research investments in extensive monoculture "plantation crops" such as rubber, sugar cane, pineapple, and oil palm for export. High rainfall and elevated temperatures are two factors in the tropics that favor growth and reproduction of weeds, insects, and pathogens as well as promoting rapid nutrient cycling and leaching. These problems often plague plantations of perennial monocrops and annual crop agroecosystems that are currently found in tropical regions; there are fewer such problems in highly diverse and resilient cropping systems (MCSs) (Beets 1990; Francis 1986).

10.4 TROPICAL AGROECOSYSTEM BIODIVERSITY

Although many traditional tropical farming systems managed by small farmers are highly diverse, most tropical regions also include extensive zones where high-tech monoculture agriculture systems have greatly simplified structure and function compared to natural ecosystems. Tropical regions are characterized by a range of climate conditions, but share the characteristics of high temperatures, high humidity, continuous or seasonal rainfall, and relatively consistent photoperiod. As a result, plant growth varies from one place to another, and agroecosystems and practices must be adapted to each set of environmental conditions. Based on success in temperate regions, conventional thinking in the design of "improved" tropical agricultural systems has led to replacing nature's diversity and traditional farmer systems with a small number of crop varieties and limiting genetic diversity across the landscape by planting monocultures. Although this strategy may be economically feasible for the short term, with ample fossil fuels and production inputs derived from them, the long-term viability of such systems is under discussion. A singular strategy of monoculture that homogenizes the agricultural landscape also does not take into account the uniqueness of place and the potentials that presents for local adaptation of unique cropping combinations to exploit the internal, renewable resources of each place.

In spite of notable success with rice, oil palm, banana, and other plantation crops, many researchers have warned about the vulnerability associated with this genetic uniformity, rightly claiming that "ecological homogeneity in agriculture is closely linked to pest invasions" (Adams et al. 1971; Robinson 1996). Farmers who adopted these high-tech systems have tried to mitigate biotic constraints by using chemical pesticides and to overcome nutrient leaching losses with chemical fertilizers. Nevertheless, with the exception of a few crops such as rice, this approach has led to negative biological and social impacts and found limited adoption by most farmers due to ecological, environmental, and health issues, as well as high costs of production inputs. We agree with Conway (1997) that there has been a lack of adequate research on integrated systems for tropical environments.

There is potential for cultivation of a wide range of plants that can be grown under these conditions:

- Trees—including a large diversity of evergreen and deciduous species
- Shrubs—including a huge selection of foliage and flowering shrubs
- Palms, ferns, and cycads
- Climbers and succulents
- Root and tuber crops
- Crops with bulbs and rhizomes
- Local or "traditional" stone fruits and vegetables
- Herbs and spices
- Ornamentals including flowers

In tropical agriculture, biodiversity of crops leads to structural diversity of systems, and these have important functions and benefits that include sustainability and food security, especially for small farmers. Yet entrepreneurs motivated by the potential for short-term economic gains, facilitated by their access to land, power, and export markets, have caused declines in the planting of many traditional crops. Their monoculture systems are contributing to loss of biodiversity in agriculture and in the rural landscape. To mitigate these impacts, including enhancing farming systems, pursuit of development of new smaller-scale and biodiverse principles and practices are suggested, as well as the political will to support this strategy. Such an initiative will combine experience in traditional farming systems with some advantages offered by modern technologies and scientific findings in environmental science, ecosystems, and agroecology. This strategy would be valuable for establishing agroecosystems with enhanced farming methods, for example organic farming, permaculture, and integrated ecological pest and soil management, that could build in greater biodiversity and sustainability (Heyd 2010; Grant 2006; Ishwaran 2010; Small 2011; Thrupp 2004).

10.5 MANAGED AGROECOSYSTEMS

10.5.1 Crops and Varieties

In the tropics, the natural vegetation in the lowlands is often lush rainforest, and many crops are grown successfully under these humid conditions. Major species include maize, tuber crops, rice, plantains, cassava, and tree crops such as oil palm and coconut (Table 10.1). In addition, other important crops such as wheat, barley, grain legumes, potatoes, and high-quality coffee and tea are cultivated in tropical regions at higher elevations (Aune 1998).

Tropical crops have played a major role in the agricultural and economic development of human society and have become major food sources in the temperate zone as well. Among the most important crops in term of civilization's origin and food cultures are rice, maize, and potato (Buddenshagen 1977). Today, irrigated rice in the lowland deltas and some terraced mountain areas, plus rain-fed rice in Asia and Africa, are the staple cereals in many diets. Likewise, maize is grown from near sea level to over 3000 m elevation in the tropics and often is intercropped with a range of legumes and other lower-story crops. Potato likewise is grown from sea level to high elevations in the tropics and is a staple crop for many highland people. Rice that evolved in Asia and maize and potato that evolved in the Americas are now grown across a wide range of ecoregions and provide the staple foundation of diets for people living far from the crops' centers of origin.

Several major cereal, legume, and root crops that originated in the tropics—including rice, maize, grain sorghum, beans, cassava, cowpea—have received priority research attention from the International Agricultural Research Centers (IARCs) and national research programs in the tropics. Yet the potential for improving the genetics of other important food crops has not been explored

Table 10.1   Major Crops Cultivated in Tropical Regions

Legumes	African yam bean, bambarra groundnut, black-eyed pea, chickpea, cowpea, faba bean, groundnut, jack bean, kidney bean, lathyrus pea, lentils, lima bean, mung bean, pigeon pea, winged bean
Vegetables	Amaranth, arugula (rocket), bell pepper, cabbage, Ceylon spinach, chard (silverbeet), Chinese cabbage, chilli pepper, cucumber, eggplant (aubergine), endive, kang kong (water spinach), lettuce, luffa, okra, pepper, pumpkin, radish, squash, tomatoes, water chestnut, zucchini
Roots and tubers	Cassava, sweet potato, taro, yams
Fruits	Acerola (West Indian cherry), avocado, banana, bilimbi, carambola, calamondin, canistel, capulin, Jamaica cherry, cashew, chempedak, citrus (citron, lime, grapefruit, king mandarin, naranjita, mandarin, pommelo, sour orange sweet orange), custard apple (soursop, sugar apple, sweetsop, atemoya), durian, guava, inyam, jackfruit, jambolan, jujube, lychee, mango, mangosteen, naseberry, papaya, passionfruit, persimmon, pineapple, pomegranate, rambutan, ribena, otaheite apple, star apple, tamarind water apple, wax jambu
Beverage crops	Coffee, tea, cocoa
Cereals and grains	Amaranth, barley, corn (maize), millet, oat, rice, sorghum, wheat

with anywhere near the same intensity, especially in the IARCs, namely banana and platano, yams and sweet potato, taro, and a wide array of other fruit, root, and tuber crops that are often specific to small farm cropping systems. Many of these are not as important to the global food export and import business, with the exception of banana and some specialty crops that have found markets in the North. Since many of the major cereal and legume crops have been researched extensively and are apparently nearing their genetic yield potential plateaus, it is likely that a number of species that have received less research attention will be a key to improving food production in the tropics in the future. In addition, there is great potential for improving cropping systems.

10.5.2 Cropping Systems

The twentieth century was marked by an unprecedented human demographic explosion and a resulting acceleration in demand for food. Although part of this need for food was met in the short term by genetic and agronomic advances of the green revolution, there are just as many people today who lack access to sufficient food for normal human development and activity. Increased population and pressures on the land in tropical regions have led to shorter fallow periods and fewer nutrients available for crops following fallow. Soil erosion caused by rain has increased as farmers changed to more annual crops to meet market demand and in many areas abandoned their long-term and established mixed annual/perennial MCSs. With annual crops in monoculture, there is less cover on the fields during more months of the year, and one result of the reduction in crop diversity is greater soil loss. Erosion has been mitigated in some areas by planting fruit trees (citrus) or expanding small plantations such as mango, coconut, papaya, and other minor tropical fruits. Further research is needed to determine the advantages of agroecosystems that combine trees with crops, and crops with animals, because these systems are increasingly recognized and promoted to improve sustainable use of tropical lands (Nicholas 1988). Currently, the majority of small-farm tropical agroecosystems still display a wide array of agricultural practices that are appropriate to the range of crops that can be grown in a series of diverse environments, as described above.

Tropical systems have historical roots, some of which are relevant to our quest to improve systems for the future. In the tropics, and other regions as well, agroecosystems have undergone many changes over the centuries, and early preagricultural systems were established to supply crops through harvesting natural species that were valuable for human nutrition and some that could be stored in preparation for periods of the year when food was less available. The "slash and burn"

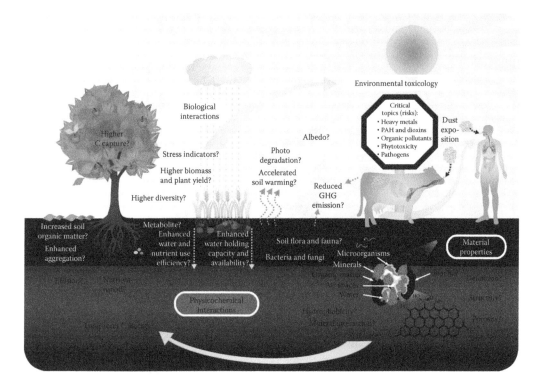

Figure 1.25 Most important unknowns on the interaction of biochar within ecosystem components. An integrative approach covers the whole range from the molecular to the ecosystem level and could be divided into four thematic modules: material properties (green), physicochemical interaction with soil and water (blue), biological interaction (orange), and environmental toxicology (black).

Figure 3.3 Typical soil profile of humid tropical soil with no biochar exhibiting low SOM levels and low soil fertility (*left*) due to intensive weathering and a terra preta with biochar (*right*) showing exactly the opposite; namely, high SOM and nutrient contents and thus high soil fertility.

Figure 5.1 External mycelium and young spores of the arbuscular mycorrhizal fungus *Rhizophagus irregularis* growing in soil.

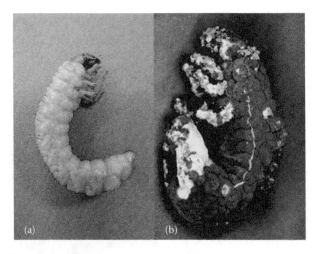

Figure 5.3 (a) Healthy specimen of *Phyllophaga* sp. (b) Infection with the entomopathogenic fungus *Metarhizium anisopliae*.

Figure 9.4 Diois agroecosystem with cereal production and viticulture, southeastern France.

Figure 9.7 Area dominated by agriculture in a drinking water catchment in southeastern France.

Silviculture with trees
planted in a pasture field
with livestock

Intercropping of trees
with row crops

Mixed-use forests with
multiple crop types
integrated

Figure 11.1 Examples of different types of agroforestry systems with different levels of complexity. The different types of systems are implemented in different regions of the world based on the selection of crops and the climate regime.

Figure 11.5 Belowground carbon storage is highly affected by site-specific conditions, but root mass and mulch from plant material contribute heavily to soil organic matter and long-term belowground carbon storage.

Figure 12.5 Black mangrove pneumatophores. (Courtesy of M. K. Webber.)

Figure 12.7 Leaves of the black mangrove with salt crystals on the upper surface, scale bar = 1 cm. (Courtesy of M. K. Webber.)

Figure 12.10 White mangrove pneumatophores, which can be wider at the tip than at the base, scale bar = 1 cm. (Courtesy of M. K. Webber.)

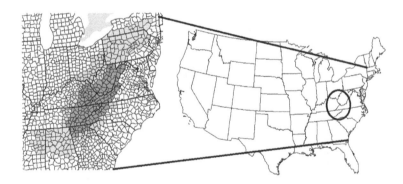

Figure 14.1. The Appalachian region in the eastern United States according to the Appalachian Regional Commission (*yellow*) and Williams (2002) (*orange*). The focus of this analysis is on the central Appalachian area surrounded by the oval. (Source of national and county maps: Wiki Commons, Creative Commons.) (Data from Williams, J.A., *Appalachia: A History*, University of North Carolina Press, Chapel Hill, 2002.)

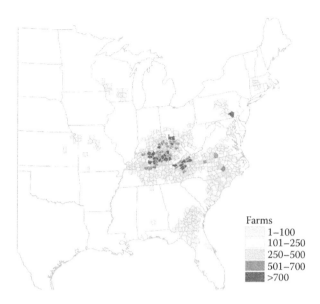

Figure 14.3 The number of tobacco farms per county. The concentration of tobacco production, particularly burley tobacco, in the Appalachian region has helped to maintain the abundance of small farms. There is uncertainty about what, if anything, will replace tobacco in the region. (From Capehart, T., Trends in US tobacco farming, Economic Research Service, US Department of Agriculture, 2004, using data from the 2002 Census of Agriculture.)

Figure 14.5 Changes in the number of farms between the agriculture censuses of 2002 and 2007. The decrease in farm numbers has been particularly acute in the central Appalachian region. (From USDA, National Agricultural Statistics Service, http://www.agcensus.usda.gov/Publications/2007/Online_Highlights/Ag_Atlas_Maps/Farms/Size/07-M004-RGBDot2-largetext.pdf.)

Figure 16.2 Photograph of existing vegetable field in a floodplain (*top*) and the proposed condition with hoop houses, crop diversification, and wetlands (*bottom*).

Figure 16.3 Photograph of an existing cornfield on a riverbank (*top*) and the proposed condition with riparian buffer (*bottom*).

agricultural system no doubt evolved from people observing what happened in nature following a fire, with renewed vegetation and available nutrients. With increasing population density and pressure on land resources, this system reached limits in feeding the growing human population. Length of fallow periods was reduced due to pressure for more land. With more frequent burning, soils become more impoverished, and this system was shown to be appropriate only when human population density is low. The fallow systems in many places developed into continuous plantings and a permanent agriculture, dependent on fossil fuel–based production inputs, and an increase in farming intensity and agricultural mechanization (Aune 1998). There is increasing debate about the long-term potentials of such high-tech monocultures to produce enough food and especially to contribute to the distribution of benefits needed to lift many rural families from poverty and help them provide for their own adequate nutrition. Agroecology and improvement of traditional systems have been recommended for solving this food dilemma (Altieri 1999, 2000; IAASTD 2008).

Highly diverse, small-farm systems have survived the green revolution in many parts of the tropics, and they hold one key to potential improved production systems in the future with dispersed benefits and focus on local food systems. In tropical agroecosystems, a wide range of crop production systems that include leafy vegetables, cereals, legumes, roots, tubers, and fruit trees, as well as other components such as forage crops and forest species, can be preserved or established because many of these food crops are adapted to high temperature and humidity in the lowland regions (Table 10.1). It is also well recognized that traditional tropical farmers cultivate a great variety or diversity of crops in order to maximize harvest security within the limited areas available to them, as most farms are still small. Typically, tropical agroecosystems are a mix of grains, roots and tubers, vegetables, legumes, and even tree fruits such as banana and papaya in cropping systems with a high level of interspecific diversity (Clawson 1985). Generically, these systems are known as *multiple cropping systems* (MCSs), which are defined as a "multiple cropping by growing two or more crops on the same field in a year" (Clawson 1985; Francis 1986). MCS are as old as agriculture itself, and these intensive farming systems persist and are most prevalent in areas of high rainfall in the tropics, as well as in many other areas of the world where small farmer agriculture predominates. MCSs are used to maximize land productivity per unit area, per cropping season, and per unit of inputs because in tropical regions temperatures and moisture are often favorable for year-round crop production.

Furthermore, in most tropical regions, farmers interact closely with their ecological surroundings, as they transform land use patterns but may ignore the ways farming decisions, intensities of systems, and use of inputs may affect the nearby natural systems. Farmers must continuously make choices about what to plant, how to plant, when to plant, and what types of technologies are accessible and perceived as useful to meet their multiple goals of food production, sustainable income, lowest possible risk, and meeting social obligations in the family and community. Their decisions each season depend on natural, geographical, cultural, and economic conditions. Continuing changes in these conditions give rise to a continuous evolution of agricultural systems, which often involves the intensification of production systems and may reflect both short-term needs as well as long-term concerns about resilience and sustainability (Aune 1998).

As examples, in many tropical agroecosystems aroids (in particular *Colocasia* and *Xanthosoma* spp.), cassava (*Manihot esculenta*), sweet potato (*Ipomoea batatas*), and yams (*Dioscorea* spp.) are staple foods and considered the major sources of carbohydrates. These root and tuber crops are often grown in small-scale farming systems, and with the exception of cassava, there has been relatively little research on these crops. Current agroecological research includes studies of productivity under mixed cropping systems, soil resource utilization (water, nutrients), resistance to abiotic stresses, pest resistance, and cultural practices. The development of good agroecological practices for productivity improvement of this category of crops in the tropics depends on how well underlying agroecological principles are understood (Valenzuela and Defrank 1995). Research to date on productivity of cassava (*M. esculenta*), maize (*Zea mays*), and yams (*Dioscorea* spp.) in tropical agroecological systems has shown that labor and asset endowments of producers were positively correlated with level of output. Farm size, financial assets, and use of inorganic fertilizer were all significant factors associated with productivity (Adeyemo and Akinola 2010).

Another agroecosystem that is frequently practiced is "mixed farming," generally considered one of the most environmentally friendly agricultural production strategies since it can be designed as a relatively closed system. In crop rotations, the residues of one combination of crops in one season are broken down and used by a succeeding set of crops in another season, which returns its own residues back to the first set of crops, and the cycle continues. This system has the advantages of increasing resource recycling opportunities that could contribute to the success of organic farming and other systems that depend on internal resources (Christiaensen et al. 1995; Pimental et al. 1992, 1995).

In summary, these mixed farming systems have been designed by farmers to

- Maintain soil fertility (recycling organic residues, crop rotation)
- Maintain soil biodiversity
- Manage insects, weeds, pathogens, and other pests
- Minimize soil erosion and maximize water and nutrient storage
- Make possible an intensification of farming without external inputs

These are all attributes of a sustainable, resilient, long-term farming strategy that can provide for family needs as well as some production for the market, enhancing the nutritional and financial well-being of small-farm families. Yet many such systems are being replaced in favorable zones by high-tech monocultures.

10.6 HIGH-TECH AND ALTERNATIVE AGRICULTURAL PRODUCTION SYSTEMS IN THE TROPICS

Although highly diverse and sustainable at modest levels of yield, some traditional tropical agroecosystems based on slash and burn with limited time for recovery have created degraded soils. In contrast, high-tech agroecosystems and monoculture crops can be grown where soils are relatively fertile, farmers have resources and access to inputs, and stable markets are available. With the challenge to feed the world's increasing population and provide some surplus for export markets to stimulate national economies, there has been pressure to adopt high-input agroecosystems. Some people claim that these contribute to the preservation of the biologically diverse native vegetation by preventing cultivation of marginal lands, thus ensuring the sustainability of such cropping systems (Boddey et al. 2003). The success of this strategy depends entirely on future availability of fossil fuels, of course.

In spite of the positive aspects of increased food production in the green revolution, there is a down side to this singular, monoculture strategy. During the twentieth century, agriculture that moved to more industrialized agrosystems has exacted a large toll on surrounding environments, polluting soils and water, reducing biodiversity, modifying natural habitats, directly or indirectly contaminating the food chain, endangering public health via pesticide exposure, and contributing to climate change, which in turn affects the true sustainability of these cropping systems (Diaz and Rosenberg 2008; Foley et al. 2011; Horrigan et al. 2002; Kremen et al. 2012; Marks et al. 2010; Newman et al. 2007; Sinclair et al. 2005; Tilman et al. 2002).

Thoughtful establishment of truly sustainable tropical agroecosystems should take into account the impacts of the transition from use of the land for agriculture and forestry based on land clearing, cultivation for some years, and then moving on to new areas. More sustainable use of land and preservation of soil, biodiversity, and water quality will require new models. To understand how to design durable and resilient agricultural systems, agroecological principles should be considered as they contribute to creating a more sustainable and secure global food system (Cribb 2010; Childers et al. 2011). Among the different modified agroecosystems that have been conceptualized, some of the most developed have been:

- Organic farming systems
- Multiple cropping systems
- Ecologically intensive agricultural systems
- Biodynamic farming systems
- Permaculture
- Crop/animal integrated systems

These systems are not mutually exclusive, for example, an organic integrated crop/animal system is what actually prevails on the majority of small farms throughout the tropics, in spite of the large push from international lending organizations and development programs at the national level to adopt technologies that have been successful in Western Europe and North America. We discuss later how specific practices in maintaining soil fertility and protecting against pests have been developed using these smaller-scale models.

It is important to explore how farming systems innovations, often designed and practiced by farmers without the benefit of formal research, relate to time-honored, traditional farming systems. The systems that have developed and prevailed over centuries are based on use of the land, rainfall, seeds, tillage methods, and power sources available to produce what is possible within the natural environment. Traditional technologies including equipment are used to till the land, plant seeds selected by farmers, maintain fertility, protect plants from pests, and harvest. These systems are often characterized by poor productivity because they depend on natural soil fertility and unpredictable weather conditions, and they often lack appropriate technology and management, which results in slow growth of productivity. There is often limited access to markets and availability of credit as well as limited government investment in needed infrastructure (Motes 2012).

Conventional agriculture has sometimes encountered economic problems associated with overproduction of crops in some regions, increased costs of energy-based inputs, and wide fluctuations in farm incomes. A primary goal is to increase food production in the tropics by increasing yields without prejudicing biodiversity, soils, and the resource base for future generations (Boddey et al. 1997). Both traditional and conventional types of agriculture have produced ecological problems such as poor ecological diversity, soil and water pollution, and soil erosion. It is important to develop sustainable production systems that require low inputs of fertilizer and pesticides and provide incentives for selection of high-yield and resistant crops that can help alleviate these ecological and economical problems. By understanding the interactions among fertilizers, pesticides, rotations, and other cultural practices, and how they influence growth and yields of crops, appropriate systems can satisfy human needs for food production and also mitigate the impacts of the inputs on the environment. Well-designed systems can contribute to biodiversity preservation as well. Moreover, other alternatives can contribute to the success of these systems such as crop rotation with legumes, use of organic waste either from plants or animals, integrated pest management (IPM), biological and cultural pest control; use of organic and mechanical weed control; and conservation tillage, intercropping or double-row cropping, strip cropping, and undersowing (Edwards 1989). Consequently, focus on a diverse sustainable agriculture has led to the development of a new class of modern agricultural systems, such as those recently described by Kremen et al. (2012) as *Diversified Farming Systems* (DFS), which are defined as "farming practices and landscapes that intentionally include functional biodiversity at multiple spatial and/or temporal scales in order to maintain ecosystem services that provide critical inputs to agriculture, such as soil fertility, pest and disease control, water-use efficiency, and pollination."

Although high-tech agricultural systems features are clearly understood by large farmers and agricultural professionals, the distinctions between these "modern" and more traditional systems with multiple goals do not easily fall into specific categories, but rather are found along a spectrum in technology use (Motes 2012). As reported by many authors, high-tech systems put the farmers in central roles, since they apply technologies and current information to dominate or control most of the production environment, and the success of these systems depends on resource access, information and communication technologies, management of inputs and costs, investment, markets, and supportive government policies. Soil fertility is achieved by applying chemical fertilizer, machines are used to create ideal soil conditions for optimal plant growth and minimal soil loss, transgenic seed varieties are used to simplify management, and these technologies combine to achieve the success of high-tech systems. Efficient irrigation systems, mechanized harvesting and handling practices, and appropriate storage and transportation to maintain product shelf life and reduce losses also contribute to the success of these systems. Further, capital investment, equipment, information, physical facilities, well-functioning commercial markets and financial systems, and supportive policies are required for the success of high-tech agricultural systems (Altieri 1999; Pretty et al. 2003; Reddy and Ankaiah 2005; Subbarao et al. 2006; Vanloqueren and Baret 2009; Welch and Graham 1999). Many of the assumptions on stability of institutions that support these systems as well as those related to fossil fuel and fresh water availability are under discussion, and the alternative systems described later provide options for what may be needed in the future for a more sustainable tropical food system that serves to improve people's nutrition and not just the profits of large farmers and multinational corporations.

10.7 SOIL FERTILITY IN TROPICAL AGROECOSYSTEMS

Soil fertility is the foundation for sustainable agriculture, and the agricultural sector must find ways to produce food for humans and feed for livestock without impairing soil fertility. In tropical agroecosystems, it is essential to design cropping systems that can ensure long-term soil fertility. Knowledge of different physical and chemical attributes and the relationships between soil and plants is essential to understand the soil system and then to inform the design of cropping systems. Besides the often-measured physical and biological attributes of soils, it is important to understand how fertility can be maintained in tropical soils (Pilatti et al. 2003). Soil fertility is an attribute that requires continuous management, and that management begins well before the crops are planted each season. Systems should be designed to provide optimum nutrient supplies when the crops need them. Thus, it is obvious that efficient soil fertility management is the key to sustainable agriculture and agroecosystems success.

Soil fertility is primarily concerned with the essential plant nutrients, including their levels in soils, their availability to crops, their behaviors and interactions in soils, and all the complex factors affecting their long-term maintenance. In conventional systems that depend on fertilizers, only the simple chemical nutrients known to be absolutely essential for plant growth are commonly available commercially. Plant nutrition depends on soil type and rainfall. In moist climates with sandy loams, plant nutrients are generally more accessible than in silt or clay soils, although in general clay soils contain higher native fertility resources. In very wet or very dry soils, nutrients become restricted or unavailable for physical reasons such as a lack of root absorption by most crop species (Cakmak 2002; Lynch 1998; Tisdale et al. 1985). Whereas hydrogen (H), oxygen (O), and carbon (C) are copiously available in water and carbon dioxide, the macronutrients nitrogen (N), phosphorus (P), and sulfur (S) are less present and available. The elements H, O, C, silicon (Si), aluminum (Al), and iron (Fe) are abundant and contribute to soil solid matter, water, and carbon dioxide, while the remaining ones are more important to the chemistry of plants (Deevey 1970).

In the tropics, the myth of high fertility of soils was dispelled when the forest was cut and crops planted, and yield levels were low and rapidly declining; it was quickly learned that soil fertility in the tropics was uniformly low and easily lost depending on the cultivation process, type of crop, and cropping system established (Ewel et al. 1991; Jacks and Whyte 1939). In fact, when farmers burn the vegetation and cultivate the land for some years, the effect of burning is to add nutrients, increase soil pH, and reduce weed infestation. However, crop yields often decline during the first years of cultivation due to increased weed infestation and loss of soil fertility, which negatively affect yields. Then the land is returned to fallow for a period of time, then cleared and burned again, and this is the pattern in traditional "slash and burn" agroecosystems, which are still practiced with some success in the tropics where human population densities are low and there is a long enough cycle for nutrients to accumulate (Aune 1998).

In general, lack of complexity of crop species mix in agroecosystems negatively affects soil fertility and simplifies many biological processes in agricultural soils; in contrast, increasing the complexity of agroecosystems may positively affect agricultural soil fertility and improve the stability of belowground systems, thereby contributing to the ecological sustainability of agroecosystems (Ewel 1999; Hu and Zhang 2004; Krebs et al. 1974; Lynch 1998). Given the rapid conversion of tropical forests to crop and pasture land, the high-tech agroecosystems tend to reduce the negative ecological impacts of both fertilizers and pesticides in the short term and improve the management of organic inputs and soil organic matter (SOM) dynamics. SOM dynamics are affected by manipulation of the soil environment (tillage, mulching, application of organic or inorganic fertilizers) and varying the quantity and quality of organic inputs including the placement and timing of application. By manipulating SOM via management practices, soil conservation is promoted, ensuring the sustainable productivity of agroecosystems and increasing the capacity of tropical soils to act as a nutrient sink rather than only a source. Nevertheless, the management of SOM dynamics in tropical agroecosystems is constrained by the lack of appropriate practices to take advantage of SOM pools that are responsive to management, because SOM tightly controls many soil properties and major biogeochemical cycles. Therefore, SOM status is often taken as a strong indicator of fertility and land degradation (Fernandes et al. 1997; Manlay et al. 2007; Nair 1997).

10.8 PEST MANAGEMENT WITH CHEMICALS AND ALTERNATIVES

In agriculture, high yields of crops have been attributed in part to the use of pesticides, a practice that has provided high productivity at a relatively low price, when negative "externalities" such as environmental contamination and negative effects on human health are ignored. However, recognition of negative environmental impacts and consumer concerns are demanding strategies to stop or at least reduce pesticide use. In the tropics, several major environmental factors such as wind, rainfall, solar radiation intensity, and soil moisture affect the presence and/or disappearance of pesticides in the soil (Arias-Estévez et al. 2008; Lalah et al. 2001). The efficacy and fate of pesticides in soil strongly depend on sorption reversibility, which is known to decrease with increasing contact time (or aging) (Laabs and Amelung 2005), although soil characteristics such as pH, organic carbon concentration, texture, and microbial activity also influence the distribution and degradation of pesticides in soil (Lalah et al. 2001; Oliver et al. 2005). In the tropics, as a substitute for burning, herbicides are also used on occasion for bush clearing, although this use should not be encouraged; herbicides are also used to control perennial weeds, protect crops, and reduce the seasonal peak of labor demand (Ogborn 1969). However, Bernard et al. (2005) indicated that the potential leaching of herbicides is lower during the dry season, while during the cyclonic period the probability is much higher, and herbicides in soil present a high risk of potential contamination of ground water resources.

There is current interest in a middle way between highly efficient practices using less pesticides and smaller-scale organic farming systems. One intermediate scheme is development and application of IPM, and retailers and consumers have a part to play in encouraging growers to consider such schemes. Farmers should be educated in the value of alternatives to conventional agricultural practices, both utilizing less pesticide and developing organic growing methods (Spriegel 1992). The design of cropping systems with multiple species, appropriate crop rotations, and thoughtful planting dates can greatly reduce or eliminate many problem pests, and IPM is applied to include these options as well as using varieties that are resistant or tolerant to pests.

10.9 HUMAN INFLUENCES ON TROPICAL AGROECOSYSTEMS

Disturbances of natural ecosystems by human activities last a long time and greatly influence the structure and function of the systems (Binford et al. 1987; Robock and Graf 1994). Human activities are the major factors that are putting at risk ecologically and culturally rich lands and natural resources in the tropics, especially the Amazon rainforest, the Cerrado woodland complex, African savannas, and many other tropical regions. Strategies for reducing the negative ecological and social impacts are needed (García-Montiel and Scatena 1994; Nepstad and Stickler 2008). Indeed, human disturbances are normally widespread and progressive rather than discrete and invariably have adverse impacts on ecosystem stability (García-Montiel and Scatena 1994). In the twentieth century, agriculture substantially altered the face of the rural landscape, and croplands have replaced natural vegetation over large areas. In order to grow their food crops, humans have exploited the advantages provided by climate in the tropics such as favorable temperature and rainfall; however, clearing of natural ecosystems to establish croplands is still contributing to changes in the global climate (Ramankutty et al. 2006; Tinker et al. 1996), the biogeochemistry of soils (Jennerjahn et al. 2004; Li et al. 2013), and the extent and health of forest ecosystems (Abdullah and Nakagoshi 2007; Chazdon 2003; Walters 1997).

In many tropical regions, swidden agriculture has been important because farmers often have little or insecure access to investment and market opportunities, or lands are preserved as a strategy to adapt to current ecological, economic, and political circumstances. In some areas, swidden practice still remains important because intensive cropping systems are not a viable option (van Vliet et al. 2012). The transition from swidden systems to more intensive cropping systems has increased, but has led to negative effects on natural, social, and human capital, while contributing to permanent deforestation, loss of biodiversity, increasing weed pressure, declines in soil fertility, and accelerating soil erosion (van Vliet et al. 2012). However, in the past 10–15 years, swidden agriculture has been decreasing in tropical forest agriculture, especially with access to local, national, and international markets that encourages cash cropping, the development of conservation policies that restrict forest clearing, and incentives to encourage commercial agriculture (van Vliet et al. 2012).

10.10 AGROECOLOGY AND APPLICATIONS OF NEW TECHNOLOGIES

The improvement and sustainability of agroecosystems depends on the use of existing technologies and understanding synergies among soils, plants, climate, and management practices. However, many systems are characterized by a lack of understanding of the structure of their biological and ecological relationships that are driving resource dynamics (Duru 2013). Therefore, the development of "appropriate technologies" that build on tropical production potentials in designing sustainable agriculture and modern, resource-efficient agoecosystems is the central idea. A number of agroecological research and development strategies such as modern farming systems, development of research and extension, agroecosystems analysis, and extensive and improved small-scale farming systems are suggested in order to achieve this target (Altieri 1989).

Modern technologies can contribute to the development of alternative agroecosystems in the tropics, especially when combined with farmer wisdom. This development could be achieved in part by creating or transferring new and appropriate technologies including small-scale mechanical tools, farmer-based as well as science-based genetic improvements, and carefully chosen chemical inputs (MacIntyre 1987;

Palladino 1996; Perkins 1982; Richards 1985). Progress could also include encouraging development of an enlightened attitude, appreciation of criteria to be used in choosing appropriate technologies, and confirmation of the validity and sustainability of newly designed systems (Richards 1985).

Learning from the green revolution, we should appreciate that introducing technological innovations is essential, but these changes should be undertaken keeping agricultural and socioeconomic issues and the importance of policies in mind. The complex long-term interactions among resources, people, and their environment should be understood well, and to attain this understanding, agriculture must be conceived of as an ecological system as well as a human-dominated socioeconomic system. Therefore, a new interdisciplinary framework to integrate the omics technologies, nanotechnologies, and biophysical sciences is indispensable.

10.11 CONCLUSIONS AND PERSPECTIVES

The last century has been marked by the rising pressures of population and demands on resource use in the world, particularly in tropical regions, and these pressures are constraining humanity from establishing and maintaining productive capacity. All signals indicate that an agroecosystems approach is ever more necessary. Although expansion and intensification of crop production using high-yielding varieties, chemical fertilization, irrigation systems, and pesticides have contributed substantially to the tremendous increases in food production over the past half century, they have also contributed to the degradation of the agricultural systems, pollution, soil erosion, land conversion, and deforestation. These actions resulted in altering the biotic interactions and patterns of resource availability in ecosystems and led to serious negative local, regional, and global environmental consequences.

The use of ecologically based management strategies and practices that were adopted during the last two decades have reduced some off-site consequences and promoted more sustainable agricultural production in some areas. Although a range of agricultural practices offering an improved level of outputs on a sustained basis are already well recognized, there are still many practices that should be developed such as fallow-management effectiveness, green fertilization and manuring practices, improved tillage practices, intercropping, and development of agroforestry systems. The scientific community has achieved much in tropical regions, yet the potential contributions of all the stakeholders including farmers have been undervalued and should be recognized. We believe that further progress depends on their close integration into the research and development process.

The development of sustainable production systems will continue to face many ecological constraints and also many economic, social, and political challenges. The development of industrial agrosystems slows the strengthening of sustainable agriculture and strategies that would generate more efficient and small-scale strategies. Effective development of ecologically sound agriculture in tropical regions requires inclusion of political and socioeconomic perspectives as part of an interdisciplinary approach to this crucial problem to manage resources in tropical countries, particularly the small tropical islands. Additionally, conservation agriculture is nowadays promoted as a technology tool to reduce soil degradation, mitigate drought effects, increase productivity, and reduce inputs. This technology relies on minimizing tillage, maintaining a permanent soil cover, and diversifying profitable crop rotation.

REFERENCES

Abdullah, S. A. and N. Nakagoshi. 2007. Forest fragmentation and its correlation to human land use change in the state of Selangor, peninsular Malaysia. *Forest Ecol Manag* 241:39–48.

Adams, M. W., A. H. Ellingboe, and E. C. Rossman. 1971. Biological uniformity and disease epidemics. *BioScience* 21:1067–1070.

Adedire, M. O. 2002. Environmental implications of tropical deforestation. *Int J Sust Dev World* 9:33–40.

Adeyemo, R. and A. A. Akinola. 2010. Productivity of cassava, yam, and maize under tropical conditions. *Int J Veg Sci* 16:118–127.

Altieri, M. A. 2000. Multifunctional dimensions of ecologically-based agriculture in Latin America. *Int J Sust Dev World* 7:62–75.

Altieri, M. A. 1999. The ecological role of biodiversity in agroecosystems. *Agr Ecosyst Environ* 74:19–31.

Altieri, M. A. 1989. Agroecology: A new research and development paradigm for world agriculture. *Agr Ecosyst Environ* 27:37–46.

Altieri, M. A., D. K. Letourneau, and J. R. Davis. 1984. The requirements of sustainable agroecosystems. In G. K. Douglass (ed.), *Agricultural Sustainability in a Changing World*, pp. 175–189. Boulder, CO: Westview Press.

Andrén, O. and T. Kätteter. 2008. Agricultural systems. In S. E. Jørgensen (ed.), *Ecosystem Ecology*, pp. 145–156. Amsterdam: Academic Press.

Arias-Estévez, M., E. López-Periago, E. Martínez-Carballo, J. Simal-Gándara, J.-C. Mejuto, and L. García-Río. 2008. The mobility and degradation of pesticides in soils and the pollution of groundwater resources. *Agr Ecosyst Environ* 123:247–260.

Aune, J. 1998. Introduction to agricultural systems in developing countries. In C. K. Eicher and J. M. Staatz (eds), *International Agricultural Development*, pp. 3–42. Baltimore: Johns Hopkins University Press.

Beets, W. C. 1990. *Raising and Sustaining Productivity of Smallholder Farming Systems in the Tropics*. Amsterdam: AgBe Publishing.

Bernard, H., P. F. Chabalier, J. L. Chopart, B. Legube, and M. Vauclin. 2005. Assessment of herbicide leaching risk in two tropical soils of Reunion Island (France). *J Environ Qual* 34:534–543.

Binford, M. W., M. Brenner, T. J. Whitmore, A. Higuera-Gundy, E. S. Deevey, and B. Leyden. 1987. Ecosystems, paleoecology and human disturbance in subtropical and tropical America. *Quat Sci Rev* 6:115–128.

Boddey, R. M., D. F. Xavier, and B. J. R. Alves. 2003. Brazilian agriculture: The transition to sustainability. *J Crop Prod* 9:591–621.

Boddey, R. M., J. C. De Moraes Sá, B. J. R. Alves, and S. Urquiaga. 1997. The contribution of biological nitrogen fixation for sustainable agricultural systems in the tropics. *Soil Biol Biochem* 29:787–799.

Brown, K. and D. W. Pearce. 1996. *The Causes of Tropical Deforestation*. London: UCL Press.

Brundtland, G. H. 1988. *Our Common Future. U.N. World Commission on Environment and Development*. Wallingford: Oxford Press.

Buddenshagen, I. W. 1977. Resistance and vulnerability of tropical crops in relation to their evolution and breeding. *Ann NY Acad Sci* 287:309–326.

Cakmak, I. 2002. Plant nutrition research: Priorities to meet human needs for food in sustainable ways. *Plant Soil* 247:3–24.

Charlton, C. A. 1987. Problems and prospects for sustainable agricultural systems in the humid tropics. *Appl Geogr* 7:153–174.

Chazdon, R. L. 2003. Tropical forest recovery: Legacies of human impact and natural disturbances. *Perspect Plant Ecol* 6:51–71.

Childers, D., J. Corman, M. Edwards, and J. Elser. 2011. Sustainability challenges of phosphorus and food: Solutions from closing the human phosphorus cycle. *Bioscience* 61:117–118.

Christiaensen, L., E. Tollens, and C. Ezedinma. 1995. Development patterns under population pressure: Agricultural development and the cassava-livestock interaction in smallholder farming systems in sub-Saharan Africa. *Agr Syst* 48:51–72.

Clawson, D. L. 1985. Harvest security and intraspecific diversity in traditional tropical agriculture. *Econ Bot* 39:56–67.

Cribb, J. 2010. *The Coming Famine: The Global Food Crisis and What We Can Do to Avoid It*. Berkeley: University of California Press.

Conway, G. R. 1997. *The Doubly Green Revolution*. London: Penguin Books.

Daily, G. C. 1997. *Nature's Services: Societal Dependence on Natural Ecosystems*. New York: Island Press.

Deevey, E. S. 1970. Mineral cycles. *Sci Am* 223:149–158.

Diaz, R. J. and R. Rosenberg. 2008. Spreading dead zones and consequences for marine ecosystems. *Science* 321:926–929.

Duru, M. 2013. Combining agroecology and management science to design field tools under high agrosystem structural or process uncertainty: Lessons from two case studies of grassland management. *Agr Syst* 114:84–94.

Edwards, C. A. 1989. The importance of integration in sustainable agricultural systems. *Agr Ecosyst Environ* 27:25–35.

Ewel, J. J. 1999. Natural systems as models for the design of sustainable systems of land use. *Agroforest Syst* 45:1–21.

Ewel, J. J., M. J. Mazzarino, and C. W. Berish. 1991. Tropical soil fertility changes under monocultures and successional communities of different structure. *Ecol Appl* 1:289–302.

Fernandes, E. C. M., P. P. Motavalli, C. Castilla, and L. Mukurumbira. 1997. Management control of soil organic matter dynamics in tropical land-use systems. *Geoderma* 79:49–67.

Foley, J., N. Ramankutty, K. Brauman et al. 2011. Solutions for a cultivated planet. *Nature* 478:337–342.

Fortner, R. 1992. Visualizing the impacts of deforestation. *Sci Activities: Classroom Projects Curric Ideas* 29:25–30.

Francis, C. A. 1986. *Multiple Cropping Systems*. New York: Macmillan.

Gajbhiye, K. S. and C. Mandal. 2000. Agro-ecological zones, their soil resource and cropping systems. Status of farm mechanization in India. Mumbai: Center for Education and Documentation. Available at: http://www.indiawaterportal.org/sites/indiawaterportal.org/files/01jan00sfm1.pdf (accessed: January 17, 2014).

García-Montiel, D. C. and F. N. Scatena. 1994. The effect of human activity on the structure and composition of a tropical forest in Puerto Rico. *Forest Ecol Manag* 63:57–78.

Grant, S. M. 2006. The importance of biodiversity in crop sustainability: A look at monoculture. *J Hunger Environ Nutr* 1:101–109.

Heyd, D. 2010. Cultural diversity and biodiversity: A tempting analogy. *Crit Rev Int Soc Polit Philos* 13:159–179.

Horrigan, L., R. Lawrence, and P. Walker. 2002. How sustainable agriculture can address the environmental and human health harms of industrial agriculture. *Environ Health Persp* 10:445–456.

Hu, S. and W. Zhang. 2004. Impact of global change on biological processes in soil. *J Crop Improv* 12:289–314.

IAASTD. 2008. Agriculture at a crossroads: Food for survival. Global Report. New York: United Nations.

INSEDA. 2012. Inseda biodiesel at a glance. Inseda. Available at: http://www.inseda.eu/SiteBiodiesel/English/index.html (accessed: January 14, 2013).

Ishwaran, N. 2010. Biodiversity, people and places. *Aust J Environ Manag* 17:215–222.

Jacks, G. V. and R. O. Whyte. 1939. *The Rape of the Earth—A World Survey of Soil Erosion*. London: Faber and Faber.

Jennerjahn, T. C., V. Ittekkot, S. Klöpper et al. 2004. Biogeochemistry of a tropical river affected by human activities in its catchment: Brantas River estuary and coastal waters of Madura Strait, Java, Indonesia. *Estuar Coast Shelf S* 60:503–514.

Köppen, W. 1900a. Versuch einer Klassifikation der Klimate, Vorzugsweise nach ihren Beziehungen zur Pflanzenwelt. *Geogr Zeitschr* 6:593–611.

Köppen, W. 1900b. Versuch einer Klassifikation der Klimate, Vorzugsweise nach ihren Beziehungen zur Pflanzenwelt. *Geogr Zeitschr* 6:657–679.

Krebs, J. E., K. H. Tan, and F. B. Golley. 1974. A comparative study on chemical characteristics of tropical soils from volcanic material under forest and agriculture. *Commun Soil Sci Plan* 5:579–596.

Kremen, C., A. Iles, and C. Bacon. 2012. Diversified farming systems: An agroecological, systems-based alternative to modern industrial agriculture. *Ecol Soc* 17:44. http://dx.doi.org/10.5751/ES-05103-170444.

Kugelman, M. and S. L. Levenstein. 2013. *The Global Farms Race. Land Grabs, Agricultural Investment, and the Scramble for Food Security*. Washington, D.C.: Island Press.

Laabs, V. and W. Amelung. 2005. Sorption and aging of corn and soybean pesticides in tropical soils of Brazil. *J Agr Food Chem* 53:7184–7192.

Lal, R. 1985. Soil erosion and sediment transport research in tropical. *Africa Hydrol Sci J* 30:239–256.

Lalah, J. O., P. N. Kaigwara, Z. Getenga, J. M. Mghenyi, and S. O. Wandiga. 2001. The major environmental factors that influence rapid disappearance of pesticides from tropical soils in Kenya. *Toxicol Environ Chem* 81:161–197.

Li, R., S. Liu, G. Zhang, J. Ren, and J. Zhang. 2013. Biogeochemistry of nutrients in an estuary affected by human activities: The Wanquan River estuary, eastern Hainan Island, China. *Cont Shelf Res* 57:18–31.

Lynch, J. 1998. The role of nutrient-efficient crops in modern agriculture. *J Crop Prod* 1:241–264.

MacIntyre, A. A. 1987. Why pesticides received extensive use in America: A political economy of agricultural pest management since 1970. *Nat Resour J* 27:533–578.

Manlay, R. J., C. Feller, and M. J. Swift. 2007. Historical evolution of soil organic matter concepts and their relationships with the fertility and sustainability of cropping systems. *Agr Ecosyst Environ* 119:217–233.

Marks, A. R., K. Harley, A. Bradman et al. 2010. Organophosphate pesticide exposure and attention in young Mexican-American children: The CHAMACOS Study. *Environ Health Persp* 118:1768–1774.

McElroy, J. L. and K. de Albuquerque. 1990. Sustainable small-scale agriculture in small Caribbean Islands. *Soc Nat Res Int J* 3:109–129.

Motes, W. C. 2012. Modern agriculture and its benefits: Trends, implications and outlook. Global Harvesting Initiative. Available at: http://www.globalharvestinitiative.org/Documents/Motes%20-%20Modern%20Agriculture%20and%20Its%20Benefits.pdf (accessed: March 27, 2013).

Nair, P. K. R. 1997. Directions in tropical agroforestry research: Past, present, and future. *Agroforest Syst* 38:223–246.

Nepstad, D. C. and C. M. Stickler. 2008. Managing the tropical agriculture revolution. *J Sustain Forest* 27:43–56.

Newman, Y. C., T. R. Sinclair, A. S. Blount, M. L. Lugo, and E. Valencia. 2007. Forage production of tropical grasses under extended daylength at subtropical and tropical latitudes. *Environ Exp Bot* 61:18–24.

Nicholas, I. D. 1988. Plantings in tropical and subtropical areas. *Agr Ecosyst Environ* 22/23:465–482.

Ogborn, J. 1969. The potential use of herbicides in tropical peasant agriculture. *Int J Pest Manag Part A* 15:9–11.

Oliver, D. P., R. S. Kookana, and B. Quintana. 2005. Sorption of pesticides in tropical and temperate soils from Australia and the Philippines. *J Agric Food Chem* 53:6420–6425.

Palladino, P. 1996. *Entomology, Ecology and Agriculture: The Making of Scientific Careers in North America, 1885–1985*. Amsterdam: Harwood Academic.

Perkins, J. H. 1982. *Insects, Experts, and the Insecticide Crisis*. New York: Plenum.

Pimental, D., U. Stachow, D. A. Takacs et al. 1992. Conserving biological diversity in agricultural/forestry systems. *Bioscience* 42:354–362.

Pimental, D., C. Harvey, P. Resosudarmo et al. 1995. Environmental and economic costs of soil erosion and conservation benefits. *Science* 267:1117–1122.

Pilatti, M. A., D. J. A. Orellana, and O. M. Felli. 2003. The ideal soil: III. Fitness of edaphic variables to achieve sustenance in agroecosystems. *J Sustain Agr* 22:109–132.

Pirarda, R. and A. Karsenty. 2009. Climate change mitigation: Should "avoided deforestation" be rewarded? *J Sustain Forest* 28:434–455.

Pretty, J. N., J. I. L. Morison, and R. E. Hine. 2003. Reducing food poverty by increasing agricultural sustainability in developing countries. *Agr Ecosyst Environ* 95:217–234.

Ramankutty, N., C. Delire, and P. Snyder. 2006. Feedbacks between agriculture and climate: An illustration of the potential unintended consequences of human land use activities. *Global Planet Change* 54:79–93.

Reddy, P. K. and R. Ankaiah. 2005. A framework of information technology-based agriculture information dissemination system to improve crop productivity. *Curr Sci* 88:1905–1913.

Richards, P. 1985. *Indigenous Agricultural Revolution*. Boulder, CO: Westview Press.

Robinson, R. A. 1996. *Return to Resistance: Breeding Crops to Reduce Pesticide Resistance*. Davis, CA: AgAccess Information Service.

Robock, A. and H. F. Graf. 1994. Effects of pre-industrial human activities on climate. *Chemosphere* 29:1087–1097.

Sachs, J. D. 2001. Tropical underdevelopment. National Bureau of Economic Research (NBER), working paper 8119. Available at: http://www.nber.org/papers/w8119 (accessed: December 12, 2012).

Sheaffer, C. C. and K. M. Moncada. 2009. Agroecosystems. In C. C. Schaffer and K. Moncada (eds), *Introduction to Agronomy, Food, Crops, and Environment*, pp. 217–246. Stamford: Cengage Learning.

Sinclair, T. R., Y. C. Newman, M. L. Lugo, E. Valencia, and A. R. Blount. 2005. Potential for year-round forage production in Puerto Rico and St. Croix. *J Agric Univ PR* 89:133–148.

Small, E. 2011. Increasing the compatibility of agriculture and biodiversity. *Biodiversity* 9:2–3.

Spriegel, G. 1992. Integrated pest management and modern agriculture. *J Biol Educ* 26:178–182.

Subbarao, G. V., O. Ito, K. L. Sahrawat et al. 2006. Scope and strategies for regulation of nitrification in agricultural systems—Challenges and opportunities. *Crit Rev Plant Sci* 25:303–335.

Tilman, D., K. G. Cassman, P. A. Matson, R. Naylor, and S. Polasky. 2002. Agricultural sustainability and intensive production practices. *Nature* 418:671–677.

Tinker, P. B., J. S. I. Ingram, and S. Struwe. 1996. Effects of slash-and-burn agriculture and deforestation on climate change. *Agr Ecosyst Environ* 58:13–22.

Tisdale, S. L., W. L. Nelson, and J. D. Beaton. 1985. *Soil Fertility and Fertilizers*. New York: MacMillan.

Thrupp, L. A. 2004. The importance of biodiversity in agroecosystems. *J Crop Improv* 12:315–337.

Valenzuela, H. R. and J. Defrank. 1995. Agroecology of tropical underground crops for small-scale agriculture. *Crit Rev Plant Sci* 14:213–238.

Vanloqueren, G. and P. V. Baret. 2009. How agricultural research systems shape a technological regime that develops genetic engineering but locks out agroecological innovations. *Res Policy* 38:971–983.

van Vliet, A., O. Mertz, A. Heinimann et al. 2012. Trends, drivers and impacts of changes in swidden cultivation in tropical forest-agriculture frontiers: A global assessment. *Global Environ Chang* 22:418–429.

Walters, B. B. 1997. Human ecological questions for tropical restoration: Experiences from planting native upland trees and mangroves in the Philippines. *Forest Ecol Manag* 99:275–290.

Welch, R. M. and R. D. Graham. 1999. A new paradigm for world agriculture: Meeting human needs: Productive, sustainable, nutritious. *Field Crop Res* 60:1–10.

Wood, S., K. Sebastian, and S. J. Scherr. 2000. Pilot analysis of global ecosystems. Agroecosystems. Washington, D.C.: International Food Policy Research Institute/World Resources Institute. Available at: http://www.ifpri.org/dataset/pilot-analysis-global-ecosystems-page (accessed: March 12, 2013).

CHAPTER 11

Agroforestry Adaptation and Mitigation Options for Smallholder Farmers Vulnerable to Climate Change

Brenda B. Lin

CONTENTS

11.1 Introduction 221
    11.1.1 Agriculture Sensitivities to Climate Variation 221
    11.1.2 Utility of Agroforestry under Future Climate Change Scenarios 223
11.2 Adaptation in Agroforestry: Examples from Coffee Agroforestry 224
    11.2.1 Coffee Agroforestry Systems 224
    11.2.2 Mitigating Temperature Variation 225
    11.2.3 Preventing Soil Water Loss and Increasing Water Capture 225
    11.2.4 Protection from Storms and Extreme Events 226
    11.2.5 Contribution to Climate Change Adaptation 228
11.3 Mitigation Potential of Implementing Agroforestry Systems 228
    11.3.1 Carbon Sequestration 228
    11.3.2 Payments for Carbon Sequestration 230
    11.3.3 Differences in Land Conversions to Agroforestry versus Row Crop Agriculture ... 231
    11.3.4 Curbing GHGs through Management in Agroforestry Systems 232
11.4 Benefits for Small Farmers 233
11.5 Conclusions 234
References 234

11.1 INTRODUCTION

11.1.1 Agriculture Sensitivities to Climate Variation

Well-documented, ongoing changes in the mean and variability of climate have led to great concern over agricultural sustainability and production under future climate scenarios (IPCC 2013). Many crops are sensitive to changes in temperature and precipitation and frequently have a narrow threshold for production success, such that threshold events that occur during key developmental stages of the crop can lead to production failure (Gregory and Ingram 2000; Mendoza and Villanueva 1997; Oram 1989).

Such sensitivities may be crucial in the tropics where most agriculture is based on rain-fed systems and where climate change has a potentially large influence on productivity (Slingo et al. 2005). Increasingly, research has focused on the importance of crop sensitivity to drought as well as periods of heat stress at particular stages of development (Challinor et al. 2005; Porter and Semenov 2005). Temperature is an important climate threshold for food crops because high temperatures that coincide with critical phases of the crop cycle can dramatically lower yield. In some crops that are well characterized, the reproductive limits are narrow, with temperatures greater than 30°C causing pollen sterility and lowering fruit-set (Porter and Semenov 2005). Observations of weather phenomena in many regions of the world have shown steady increases in daytime temperatures that challenge this threshold, pointing to an increasing vulnerability in crop production (IPCC 2013).

However, it is not just the long-term, slow-changing climate effects, such as gradually rising temperatures or gradually changing precipitation rates that are of concern. Extreme climate events that occur rapidly and over a very short time span are perhaps more threatening to agriculture. Examples of such extreme events that are often related to crop damage are the maximum temperatures on a single day or maximum one-day precipitation totals. If these extreme events coincide with an important process in the crop developmental cycle, production can be dramatically reduced. Agricultural vulnerabilities have been observed in a number of important crop species. Observations of rice production in the Philippines during an El Niño drought season showed decreases in seed weight and overall production (Lansigan et al. 2000). Studies of wheat have shown that heat pulses applied to wheat during anthesis reduced both grain number and weight, highlighting the effect of temperature spikes on grain fill (Wollenweber et al. 2003). In maize, reduced pollen viability was observed at temperatures above 36°C, which was similar to the threshold temperatures observed in a number of other crops (Porter and Semenov 2005).

Such observed agricultural vulnerabilities to changes in temperature and precipitation point to the need to develop resilient systems that can buffer crops from climate variability and extreme climate events, especially during highly important development periods such as anthesis. The development of agricultural systems that are resilient to the effects of changing temperature and precipitation is an essential topic of study as many communities greatly depend on the provisioning ecosystem services of such systems (food, fodder, fuel) for their livelihoods (Altieri 1999). Many agriculturally based economies have few other livelihood strategies (Tilman et al. 2002), and small family farms have little capital to invest in expensive adaptation strategies, thereby increasing the vulnerability of rural, agricultural communities to a changing environment. The challenge for the research community is to develop resilient agricultural systems using rational, affordable strategies such that ecosystem functions and services can be maintained and livelihoods can be protected.

Understanding the need to adapt comes with understanding the consequences of not adapting to climate change. Methods of agricultural adaptation to climate change are many and wide ranging depending on the level of stress or the scale of the operation. It should be assumed that as climate variations increase, the level of adaptation will also have to increase in order to protect and maintain production levels. Examples of progressive adaptation in agriculture include, on the first level, the ability to change varietals or planting times, and basic simple changes to management that can yield more water to plants. As climate risks increase, the adaptation options will lead to greater changes in the agricultural system as well, using not only technological techniques such as precision agriculture, but also agricultural diversification and risk management of the agricultural business system. At the more extreme level, farmers will have to adapt with more extreme measures, considering the options of changing the range of land uses across the landscape—or even selling ecosystem services as a stream of income. With greater fluctuations and more extreme variations in expected climate, adaptation options decrease and the cost and complexity of implementing adaptation increase. Yet, the need to implement adaptation and the benefits of such adaptation increase as well (Howden et al. 2010).

11.1.2 Utility of Agroforestry Under Future Climate Change Scenarios

Diversified agricultural systems provide a variety of examples in which more structurally complex systems are able to mitigate the effects of climate change for crop production and increasingly build resilience into the production system (Lin 2011). One example of a highly diversified system is the agroforestry system. Agroforestry is the production of livestock or food crops in combination with growing trees for timber, firewood, or other tree products (Montagnini and Nair 2004). Some of these systems, especially the traditional ones, can contain high species diversity within a small area of land (Kumar and Nair 2006; Leakey 1999). Not only do they provide a diversity of crops in time and space, but they also protect soil from erosion and provide litter for organic material and soil nutrients (Jama et al. 2000), reducing the need for synthetic fertilizer. Agroforestry systems are a key type of agriculture that allow for a high level of progressive adaptation from simply changing varietals to selling ecosystem services for increased economic diversification. The ability of agroforestry systems to adapt to changing climate parameters highlights the utility of this type of agriculture to maintain production levels through potentially difficult future scenarios.

Many types of agroforestry systems are employed in a number of regions of the world and at different levels of complexity. Silviculture systems are agricultural systems where trees are planted in a pasture field to provide shade for pasture animals as well as to provide food (e.g., fruit and nuts) and fuel for the farmer. Another type of agroforestry within agriculture is the intercropping of trees with row crops to provide wind shelter for the crops and to increase the soil stability of the region. Mixed-use forests are a type of agroforestry that allows for multiple crops to be produced in a small physical area because of the increased temporal and structural diversity of the ecosystem. Such examples show that agroforestry systems can allow for many different types of adaptation to occur under a range of conditions (Figure 11.1).

Agroforestry systems may also be beneficial to farmers by mitigating the agricultural production of greenhouse gases (GHG) from the crop production system. There are a number of pathways by which this occurs including increasing above- and below-ground carbon capture and preventing the unnecessary emissions of GHGs from specific management techniques that are used within the agroforestry system. Farmers are, thus, able to diversify their income by selling carbon credits while contributing to GHG reductions that increase climate variability and threaten production.

| Silviculture with trees planted in a pasture field with livestock | Intercropping of trees with row crops | Mixed-use forests with multiple crop types integrated |

Figure 11.1 (See color insert) Examples of different types of agroforestry systems with different levels of complexity. The different types of systems are implemented in different regions of the world based on the selection of crops and the climate regime.

This chapter will examine more closely the various ways in which agroforestry systems have the potential to act as an effective adaptive and mitigative strategy for agricultural systems in the face of climate change.

11.2 ADAPTATION IN AGROFORESTRY: EXAMPLES FROM COFFEE AGROFORESTRY

11.2.1 Coffee Agroforestry Systems

The coffee agroforestry system is a traditional form of cultivation in many tropical regions of the world, where the coffee plant is grown under a canopy of trees. A variety of other crops, the most well known being tea and cacao, are also commonly grown in this type of system. The shade cover over the crop is an overstory traditionally represented by a structurally complex forest canopy consisting of native trees and some cultivated trees (fruits, nuts, medicinal). However, recent economic pressure on the coffee market has led to a gradual removal of the shade cover from many systems in search of higher production (Giovannucci and Koekoek 2003), leading to a large and variable range of shade cover over coffee plantations, from systems with no shade to systems with an intact, native forest canopy (Moguel and Toledo 1999) (Figure 11.2).

Because of the geographic location of coffee, the coffee agroforestry system has been identified as an important habitat for many endemic species, a reservoir of biodiversity, and a high-quality matrix for the movement of forest species. Additionally, due to the high level of biodiversity and the structural complexity of this type of agriculture, the coffee agroforestry system has been especially noted for the many ecosystem services that it provides to the farmer, including pest regulation, pollination, and soil stability (Beer et al. 1998). Agroforestry systems also have the potential to act as an adaptation strategy to protect farmers from climate change effects including increased temperature and precipitation variability and extreme climate events. Shade cover within coffee agroforestry systems has become more important for production as climate change fluctuations have increased.

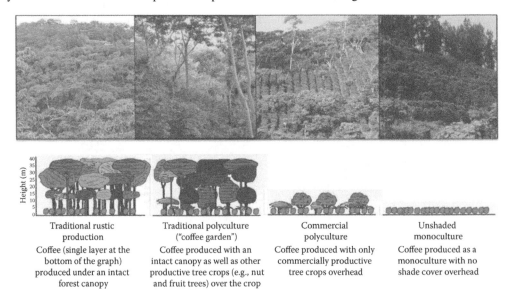

Figure 11.2 Photographic representations of the intensification gradient in the coffee agroforestry system from rustic/traditional agroforestry systems with substantial shade cover, to less-shaded systems in the commercial polyculture and shade monoculture, to the most intensively managed system in the unshaded monoculture.

Research has shown that coffee yields are sensitive to changes in temperature and precipitation, especially during the flowering and fruit development stages of production (Cannell 1976). Vital investigations, in light of current climate concerns, have shown that diverse shade canopies in coffee agroforestry systems may be able to protect coffee production through a variety of mechanisms, mitigating the effects of increased temperature and precipitation fluctuations as well as increased storm events. Such ecosystem services, which protect coffee production from future environmental change, have encouraged the readoption of highly diverse shade agroforestry.

11.2.2 Mitigating Temperature Variation

Research on coffee production and plant physiology has shown that coffee plants are quite sensitive to fluctuations in ambient temperature. The optimal temperature range for *Arabica* coffee (*Coffea arabica*) is 18°C–21°C (Alegre 1959), and experiments have shown that at temperatures above 20°C–24°C, the net photosynthesis of coffee decreases markedly, approaching 0 at 34°C (Cannell 1976; Nunes et al. 1968). At temperatures above 23°C, the development and ripening of fruit are accelerated leading to a loss of quality, and at temperatures below 18°C, growth is depressed (Camargo 1985). Additionally, under moderate levels of shade cover *Arabica* coffee plants were measured to have photosynthetic rates that were three times higher than coffee leaves under full sun (Nutman 1937). Such sensitivities to changing ambient temperature highlight the need to protect coffee production from increasing temperature variability.

Coffee agroforestry systems with high structural complexity have been shown to buffer crops from large fluctuations in temperature (Beer et al. 1998; Lin 2007; Morais et al. 2006), thereby keeping the crop closer to its optimal conditions. Greater shade cover over the crop significantly reduces the temperature fluctuations that coffee plants are exposed to under full sun by buffering solar radiation penetration of the coffee layer during the day and protecting the coffee crop from colder evening temperatures. Lin (2007) showed that the presence of increased shade cover reduced the ambient temperature at the most important point of the day when plants were most heat stressed, indicating that the advantage provided by the shade is precisely what a manager should be looking for in searching for an adaptation strategy against the effects of high temperatures on crop production. This may be useful in the future under increasing climate variability because mitigating climate fluctuations will maintain crops at their optimal temperatures and protect crops from reduced yield and associated risks (Semenov and Porter 1995).

Future changes in temperature may also present a threat to coffee production by altering pest and disease pressures in the system. At lower latitudes, the fungal disease coffee leaf rust (*Hemileia vastatrix*) may benefit from higher temperatures, increasing the spread of disease across coffee crops more readily (Lamouroux et al. 1995). Additionally, at higher temperatures, the coffee berry borer (*Hypothenemus hampei*), which is the most important coffee pest in Latin America, may be able to expand its range upward along the mountainside to infect coffee plants that were previously protected by temperature constraints (Gay et al. 2006). Agroforestry systems may be able to help prevent outbreaks of pests and diseases by controlling temperature fluctuations within the coffee layer. Soto-Pinto et al. (2002) found that the higher the number of vegetation strata, the less the incidence of the fungal disease coffee leaf rust (*H. vastatrix*), suggesting that the vegetation structure in coffee plantations affected the ecological relationships related to this important disease. Farmers may be able to control levels of shade cover within their agroforestry systems that create microclimates less suitable to pest and disease infestation.

11.2.3 Preventing Soil Water Loss and Increasing Water Capture

Coffee production is vulnerable to both the quantity of precipitation as well as the timing of precipitation events for flowering and fruit production. Coffee plants are susceptible to plant stress and damage during the dry season, yet require an extended period of drought for flower bud formation. The onset of spring rains provides a signal for flowers to blossom simultaneously over a short period of time (Magalhaes and Angelocci 1976). Water availability to the coffee plants also determines the fruit size, which generally expand and grow during the wet season (Cannell 1983). Unfortunately, farmers managing rain-fed agricultural systems do not have control over the timing or the amount of rainfall that comes at critical moments of coffee development.

With increasing climate variability, crops may be subjected to increasingly erratic precipitation events, leading to greater water stress within the system. Additionally, extreme events, such as the El Niño Southern Oscillations (ENSO), can cause an extended drought during the dry season at important cycles of development, putting a large stress on plants and reducing crop production dramatically (Salinas-Zavala et al. 2002). For example, previous ENSO years have shown a decrease in coffee crop production of 40%–80% in southern Mexico, leaving many small producers impoverished and without other means to earn an income (Castro Soto 1998).

The shade canopy of an agroforestry system can serve as an effective control against soil evaporative water loss (Jackson and Wallace 1999; Wallace et al. 1999) because canopy cover can change the microclimate near the ground surface and decrease soil evaporation rates (Teare et al. 1973). The results from Lin (2007) show that the presence of shade trees can dramatically affect the microclimate around the crop plant, thereby affecting both soil evaporation and coffee plant transpiration potential. The reduction of water loss through these two sources combined can have a large effect on the overall water loss from the system with highly shaded systems (60%–80% shade cover) preventing water loss through soil evaporation by 41% and plant transpiration by 32% when compared with less-shaded farm sites (10%–30% shade cover) (Lin 2010). Water availability in the dry season is also important for plant maintenance because when there is insufficient water in the soil, plants suffer from root and shoot dieback and drop leaves in order to prevent water loss through transpiration (Clowes 1977; Wrigley 1988), thus severely harming fruit development and fruit growth.

The ability of coffee systems with more shade cover to reduce soil water loss through microclimate mitigation is paired with their ability to capture more water from meager dry season precipitation events before the water is lost to runoff. Data from Lin and Richards (2007) show that the random roughness of the soil surface in high-shade systems is greater than in low-shade coffee systems because of the increased underground root disturbance from the shade trees. The development of roots below the surface pushes the soil upward within the soil column, creating a microtopography at the soil surface. The increased microtopography of the highly shaded coffee systems allows for the development of microrelief, pockets of water storage, providing greater infiltration of rainwater during the dry season than in sites with lower canopy cover and less soil surface complexity.

The amount of shade cover within agroforestry systems shows that high levels of shade cover within the system are able both to decrease water loss through soil evaporation and to increase water capture through greater microtopography on the soil surface. The combination of these two factors allows for greater levels of soil water to be maintained in high-shade agroforestry systems, leading to more protected flower and fruit production even during potential periods of extended drought or decreased precipitation.

11.2.4 Protection from Storms and Extreme Events

Agroforestry systems have been shown to protect crops from extreme storm events (e.g., hurricanes, tropical storms) where high rainfall intensity and hurricane winds can cause landslides, flooding, and premature fruit drop from crop plants (Stigter et al. 2002). In one example, Hurricane Stan hit the coffee-growing Soconusco region of Chiapas, Mexico, in October 2005, just at the start of the coffee harvest. The extensive rainfall (>500 mm) and wind (130 km/h) (Pasch and Roberts 2006) led to widespread fruit drop before workers could reach the fields for harvest.

Up to 50% of the Soconusco coffee harvest was lost, an estimated 170,000–280,000 ha of land were damaged, and 50%–90% of farm infrastructure was affected (Perez 2005). Coffee systems of greater farming intensity, which had little diversity or structural complexity in the shade layer, had greater percentages of farm area lost to landslides and larger quantities of coffee production lost to premature fruit drop (Philpott et al. 2008) (Figure 11.3). The results showed that farmers can actively manage their farms to reduce their environmental vulnerability to landslide damage by increasing on-farm vegetation complexity. However, regional stability would be improved dramatically if all farmers within the region increased their on-farm vegetative complexity collectively.

Increasing trees in the landscape, in general, have been shown to help create a buffer to reduce the impact on the farming system from floods and droughts. The ability of agroforestry systems to increase the infiltration of water into the soil and increase soil-holding capacity reduces the risk of flash flooding during heavy rainfall events and storms (Smith et al. 2013). In New Zealand, widely spaced poplars reduced pasture production losses due to landslides during cyclonic storms by 13.8% with, on average, each tree saving 8.4 m^2 from production loss failure (Hawley and Dymond 1988). It was also shown that if the trees had been planted in a 10 m grid with a 100% establishment rate, storm damage could have been reduced by at least 70%.

Trees in the landscape can also provide the important service of reducing damaging wind speeds associated with an extreme storm event to protect crop production (Figure 11.4). Windbreaks located perpendicular to the prevailing wind can increase farm production simply by reducing the wind and modifying the microclimate. In citrus and vegetable systems in Florida, it was found that windbreaks could reduce wind speeds up to 31 times the windbreak height on the leeside of the windbreak. However, there was great variability in the wind speed reduction depending on the height and porosity of the windbreak as well as the direction of the wind in relation to the windbreak (Tamang et al. 2010). Higher air and soil temperatures on the leeside of a shelterbelt can also extend the growing season, with earlier germination and more growth at the start of the season (Brandle et al. 2004).

Figure 11.3   Landslide in a low-shade coffee farm. The large-scale loss of coffee land from storm effects is in part due to the low level of shade cover within this farm.

Figure 11.4 Windbreaks in fruit and vegetable crops reduce wind speeds and modify the microclimate for the crops of interest. The examples shown are windbreaks used in vineyards, taro crops, and barley and malt crops.

11.2.5 Contribution to Climate Change Adaptation

From these examples, we see that coffee agroforestry is one type of mixed-forest system that may be able to protect agriculture from the effects of increasing temperature and precipitation variability as well as extreme storm events. By mitigating the temperature variation in the system, stabilizing the soil water availability for crop plants during extended dry season events, and protecting the crop from high winds and rain damage, the shade cover within the agroforestry system serves as an important buffer for the crop from changes in climate. The increased complexity of the agroforestry system both in its diversity and structure, allows for greater resilience toward changing climate effects on production and should be considered as an adaptive measure for the future.

11.3 MITIGATION POTENTIAL OF IMPLEMENTING AGROFORESTRY SYSTEMS

According to the Intergovernmental Panel on Climate Change, agriculture is responsible for a significant portion of the continuing increase in GHG emissions. However, not all agriculture has the same impact on global warming as certain agricultural practices contribute disproportionately to the overall GHG contribution of agriculture. Selecting systems that reduce the outward flux of carbon dioxide (CO_2) and nitrous oxide (N_2O) through better management will lead to increasing the mitigation potential of agricultural systems. Studies of agroforestry systems have shown huge carbon sequestration potential (Verchot et al. 2007) within these systems—with carbon capture sustained and maintained within the system over long periods of time.

11.3.1 Carbon Sequestration

Evidence is emerging that agroforestry systems have great potential for increasing both aboveground and belowground soil C stocks, reducing soil erosion and degradation, and mitigating GHG emissions (Mutuo et al. 2005), especially with an estimated 1023 million hectares under agroforestry worldwide, and substantial areas of unproductive crop, grass, forest, and degraded land that could be potentially brought under agroforestry (Ramachandran Nair et al. 2009). The extent of C sequestered in any type of agroforestry system will largely depend on a number of site-specific factors, including biological, climatic, soil, and management factors. Additionally, the profitability of C sequestration and take up for agroforestry managers will depend on the price of C in the international market, the additional income from the sale of products such as timber for farmers, and the costs related to C monitoring (Ramachandran Nair et al. 2009). The C sequestration potential of agroforestry systems is estimated between 12 and 228 Mg C/ha with a median value of 95 Mg C/ha (Albrecht and Kandji 2003). In agroforestry systems, the standing stock of the carbon aboveground is usually higher than the equivalent land use without trees (Smith et al. 2008). Established agroforestry systems maintain 22 times more carbon stored in aboveground living biomass when compared with traditional maize over a period of 7 years (Soto-Pinto et al. 2010). For smallholder agroforestry in the tropics, potential C sequestration rates range from 1.5 to 3.5 Mg C/ha/year (Montagnini and Nair 2004).

Belowground soil C sequestration occurs through inorganic chemical reactions that convert CO_2 into soil inorganic C compounds such as calcium and magnesium carbonates and through plant biomass indirectly sequestered as soil organic carbon (SOC) (Figure 11.5) during the decomposition processes (Ramachandran Nair et al. 2009). The amount of C sequestered at a site reflects the long-term balance between C uptake and release mechanisms, and large changes such as shifts in land-cover or land use practices that affect pools and fluxes of SOC have large implications for the amount of C that can be stored in agroforestry systems. In alley cropping studies from Costa Rica, 10-year-old systems with *Erythrina poeppigiana* sequestered C at a rate of 0.4 Mg C/ha/year in coarse roots and 0.3 Mg C/ha/year in tree trunks, but tree branches and leaves added to the soil as mulch contributed an additional 1.4 Mg C/ha/year in addition to 3.0 Mg/ha/year from crop residues (Oelbermann et al. 2004). In degraded soils in the subhumid tropics, improved fallow agroforestry practices have been found to increase topsoil C stocks up to 1.6 Mg C/ha/year above continuous maize cropping (Montagnini and Nair 2004).

The potential of coffee agroforestry systems for carbon sequestration has also been recognized (De Jong et al. 2000; Nelson and de Jong 2003) because they accumulate a high amount of aboveground biomass. Agroforestry systems with perennial crops, such as coffee, may be more important carbon sinks than those that combine trees with annual crops (Montagnini and Nair 2004). The results from research in Chiapas (Peeters 2001) showed 42 and 138 t/ha of aboveground biomass contained in shade vegetation in polyculture coffee and rustic coffee systems, respectively (coffee shrubs, soils, litter, and roots not included). Callo-Concha et al. (2004) estimated 195.6 t/ha for coffee with shade, similar to the amount found in home gardens. These data reveal that rustic coffee systems accumulate a significant amount of carbon, equivalent to one-third of the amount sequestered by primary forests (307 t/ha, Rice and Greenberg 2000; 465.8 t/ha, Callo-Concha et al. 2004). Additionally, soils contain 6.8% of the organic matter found in these agroecosystems

Figure 11.5   **(See color insert)** Belowground carbon storage is highly affected by site-specific conditions, but root mass and mulch from plant material contribute heavily to soil organic matter and long-term belowground carbon storage.

(Romero-Alvarado et al. 2002; Soto-Pinto et al. 2001). Agroforestry coffee systems may increase carbon sequestration by incorporating more trees, increasing their density from 229 to 393 trees/ha, which is the density in rustic coffee plantations. It has been reported that traditional densities do not affect yields, as long as shade cover is managed around 45% (Soto-Pinto et al. 2000).

11.3.2 Payments for Carbon Sequestration

The international carbon markets' offset scheme represents an opportunity to increase biodiversity through the revegetation of agricultural land. Land use change and forestry policies have taken on an increasingly important role as a result of the negotiations of the United Nation's (UN) Reducing Emissions from Deforestation and Forest Degradation (REDD+) program, part of the clean development mechanism (CDM) included in the Kyoto Protocol. This mechanism has been gaining momentum as a way to combat global warming, fund forest conservation, and deliver economic benefits to rural populations. In this program, a carbon market was created in which industries that need to meet their carbon budget can buy carbon from farmers and land managers who sequester and store carbon above and below ground. The carbon-emitting industries make payments to reward land managers for taking action to protect and plant trees or otherwise increase the carbon content of soils and vegetation on the land that they manage (Nelson and de Jong 2003). It has been estimated that cutting global deforestation rates by 10% could generate up to $13.5 billion in carbon credits under the REDD+ initiative that was approved at the UN climate talks in Bali in 2009 (Ebeling and Yasue 2008). Increased agroforestry cover in sustainable perennial systems may be one way that farmers can take advantage of the carbon markets as a means to increase the economic benefits of maintaining systems with high levels of structurally complex shade cover.

As farmers require economic benefits to be willing and able to adopt new practices, economic models that are able to predict threshold prices at which farmers begin to adopt environmental land use practices or payments for ecosystem services can also be highly effective in encouraging farmers to adopt diversified agricultural systems as adaptation options to climate change. In one model on the potential of farmers to participate in carbon sequestration contracts and increase sequestration potential through agroforestry and the terracing of fields, the analysis showed that at prices above $50 per megagram of C, adoption would increase substantially, and at prices of $100 per megagram of C, terrace and agroforestry adoption for C sequestration would have the potential to raise per capita incomes by up to 15% (Antle et al. 2007).

There has been increasing concern regarding the local mechanisms of carbon accounting and the benefactors of the CDM REDD+ payments. The increase in voluntary carbon markets internationally has been extensively criticized for their lack of contribution to sustainable development, in particular in terms of providing cobenefits to local communities (Robinson et al. 2011), thereby creating many potential unintended consequences. There are cases in which indigenous groups have lost their traditional lands because of government encroachment of lands for carbon sequestration. This has displaced native peoples from the lands that they once used to hunt, collect firewood, and graze their animals. Issues of leakage may also occur due to the displacement of GHG emissions from one carbon sequestration project area to another, leading to increased deforestation in areas that were previously conserved. Studies attempting to measure the scale of leakage related to forest-based conservation projects have recently started to emerge, showing that leakage from deforestation and degradation can be greater than 40% (Atmadja and Verchot 2012). For example, Meyfroidt and Lambin (2009) found that although there was wood regrowth of 39.1% in Vietnam as a result of logging restrictions in the country, much of this lost timber extraction in-country was displaced by forest extraction in other countries and imported to supply Vietnam's timber demands. Within the REDD+ scheme the issue of leakage is beginning to be addressed; however, cross-boundary and international leakages have thus far not really been considered, yet they are imperative to assessing the global significance of efforts to reduce GHGs (Atmadja and Verchot 2012).

11.3.3 Differences in Land Conversions to Agroforestry versus Row Crop Agriculture

The world's forests and savannas have long helped to maintain the global carbon cycle in balance, but large-scale conversion to annual crops has dismantled this particular ecosystem service. The IPCC report (2013) estimates that land use changes, mainly deforestation, contribute 20% of the CO_2 emissions globally. Latin America, Africa, and South and Southeast Asia have experienced exponential increases in cropland expansion through deforestation since 1950 (IPCC 2013), and CO_2 emissions have continued to increase over the last few decades. Emissions associated with land use changes, averaged over the 1990s, are estimated at 0.5–2.7 Gt C/year (IPCC 2013). Thus, if tropical deforestation continues unabated, not only will a significant reservoir of the earth's carbon be released into the atmosphere, but also a critical sink for near-term emissions will be destroyed.

Converting tropical rainforests into intensive farming systems, which is happening in many tropical regions of the world, also has a threefold impact on the CH_4 budget with the elimination of the CH_4 sink from actual deforestation, the emission of CH_4 from biomass burning, and the emission of CH_4 from fertilizer-based intensive agriculture significantly increasing the CH_4 emissions from the landscape (IPCC 2013). The conversion of primary tropical forests to agriculture or grassland has resulted in a loss of 370 Mg C/ha.

Human activity has heavily impacted tropical savannas, with large extensions of land converted from tree–grass mixtures to open pastures and agriculture (Solbrig et al. 1996). In moist tropical savannas, the dominant land use includes cattle production and large-scale intensive agriculture, although shifting and permanent cultivation is also practiced in Africa (Hoffmann and Jackson 2000).

If well managed, planted pastures may accumulate carbon, but many pastures are degraded and fail to serve as atmospheric C sinks (da Silva et al. 2004). Introducing grass species with higher productivity or C allocation to deeper roots has been shown to increase soil C in savannas (Fisher et al. 1994). Introducing legumes into grazing lands and mixed crop systems can also promote soil C storage (Soussana et al. 2004), potentially reducing N_2O emissions. Slowing degradation and impeding desertification with such alternative grassland management techniques could conserve up to 0.5–1.5 Pg of C annually in savannas (Dixon et al. 1994).

Reforestation through agroforestry systems may be one way to bring carbon stocks back into the system. Agroforestry systems contain 50–75 Mg C/ha compared with crops that contain <10 Mg C/ha, thereby maintaining significantly greater carbon stocks within the system (Verchot et al. 2007). Agroforestry systems such as silviculture allow for pasture and trees to be grown at the same time in the landscape (Montagnini and Nair 2004). Some of these systems, especially the traditional ones, can contain high species diversity within a small area of land (Kumar and Nair 2006; Leakey 1999), not only providing a diversity of crops in time and space, but also protecting the soil from erosion and providing litter for organic material and soil nutrients (Jama et al. 2000), thereby reducing the need for synthetic fertilizer.

Afforestation, which is the establishment of forests on land that is not forested, could also decrease atmospheric GHGs. However, afforestation typically takes the form of large-scale monocultural tree plantations, and carbon sequestration rates vary widely. Afforestation studies of agricultural land in the United States show that soil C can change from −0.07 to 0.55 Mg C/ha/year on deciduous sites and from −0.85 to 0.58 Mg C/ha/year under conifers. Under afforestation, soil N changes range from −0.1 to 0.025 Mg N/ha/year. However, even after 20–50 years of afforestation in many sites, the C/N ratios have remained more similar to agricultural systems than native forests, showing that the chemical replenishment of soils is a slow process (Paul et al. 2003). This may suggest that maintaining highly diverse agroforestry sites is more beneficial to the maintenance of nutrient-rich soils than afforestation of previously intensively farmed land.

11.3.4 Curbing GHGs through Management in Agroforestry Systems

The potential of agroforestry to curb GHG emissions is not limited to carbon sequestration. A review of agroforestry practices in the humid tropics shows that these systems were able to mitigate N_2O and CO_2 emissions from the soils and increase the CH_4 sink strength compared with annual cropping systems (Table 3; Mutuo et al. 2005). In a study of the Peruvian Amazon, a tree-based agroforestry system emitted less than one-third of the N_2O of a high (fertilizer) input annual cropping system and half that of a low-input cropping system (Palm et al. 2002). Data from several countries strongly suggest that agroforestry systems can partially offset CH_4 emissions, while conventional high-input systems exacerbate CH_4 emissions (Table 11.1).

One aspect of the mitigation potential of agroecological systems is the evidence that plant diversity within agroecosystems affects soil ecosystem processes. For example, the rate of loss of limiting nutrients from terrestrial ecosystems has been found to be lower under high plant diversity and is impacted by the composition of plant species and rotational practices (Naeem et al. 1996; Tilman 1999). Likewise, the rebuilding of soil C and N stocks in highly degraded soils can be accelerated if fields are planted with a high-diversity mixture of appropriate plant species (Knops et al. 1999). Given this, it is not surprising that diverse agroecosystems have been shown to sequester more C in soil than those with reduced biodiversity (Lal 2004). Therefore, the transformation of diverse landscape mosaics and diverse agricultural systems to large-scale monocultures not only reduces the GHG sequestration potential of the soil, but it also increases the need for fertilizer application. Because more diverse farming systems promote higher total productivity and stability, they may also reduce the necessity for synthetic fertilizer application, thereby reducing GHG emissions. For example, studies of GHG fluxes in the US Corn Belt showed that continuous corn crop rotations contributed significantly to N_2O emissions (upward of 3–8 kg/ha/year), driven by pulse emissions after N fertilization in concurrence with major rainfall events. More complex systems such as

Table 11.1   High-Input Systems Exacerbate CH_4 Emissions

Land Use System		N_2O Emissions ($\mu g\ Nm^{-2}\ h^{-1}$)	CH_4 Flux ($\mu g\ Cm^{-2}\ h^{-1}$)	CO_2 Emissions ($\mu g\ Cm^{-2}\ h^{-1}$)	Sources
Cropping system	High-input cropping	31.2	15.2	84	Palm et al. (2002)
	Low-input cropping	15.6	−17.5	66.6	Palm et al. (2002)
	Cassava/ *Imperata*	7.1	−14.8		Tsuruta et al. (2000)
Agroforestry system	Shifting cultivation	8.6	−23.5	67.5	Palm et al. (2002)
	Multistrate agroforestry	5.8	−23.3	32.6	Palm et al. (2002)
	Peach palm	9.8	−17	66.4	Palm et al. (2002)
	Jungle rubber	1	−12		Tsurata et al. (2000)
	Rubber agroforests	12.5	−27.5		Tsurata et al. (2000)
Forests	Forest	9.2	−28.8	73.3	Palm et al. (2002)
	Forest	5	−31		Tsurata et al. (2000)
	Logged forest	7.2	−38.2		Tsurata et al. (2000)

Source: Mutuo, P.K., Cadisch, G., Albrecht, A., Palm, C.A., and Verchot, L., *Nutr Cycl Agroecosys*, 71, 43–54, 2005.

corn-soybean rotations and restored prairies showed diminished N_2O emissions and contributed to global warming mitigation (Hernandez-Ramirez et al. 2009; Niggli et al. 2009).

11.4 BENEFITS FOR SMALL FARMERS

Although there are many substantial benefits to farmers when high structural and temporal diversity is maintained, farmers are still primarily concerned with production and yields. The overall goal for increased sustainability in coffee systems should include an increase in production stability as well as the maintenance of environmental services on-farm. Farmers are vulnerable to price fluctuations, but they are also vulnerable to a variety of other biotic and abiotic effects, such as pest infestations and climate change fluctuations. Agroforestry can provide a variety of management options that both protect farm economic stability and provide climate change adaptation and mitigation options.

In Section 11.2, the adaptation to climate change provided by agroforestry systems can be very effective on many fronts. The ability of agroforestry systems to mitigate the effects of increasing temperature variability, changes in rainfall amount and intensity, and wind and storm events has been shown to protect agricultural production under more extreme scenarios. The increased potential of agroforestry systems to sequester carbon within plant matter and soil stocks as well as their ability to reduce the amount of emissions that escape from the system through different management and resource cycling is also beneficial toward reducing the contributions of agriculture to GHGs.

Most importantly, agroforestry systems have the potential to stabilize and increase farm profitability for farmers. First, agricultural production is better protected from climate variation by mitigating the effects of climate on crop development and growth. Secondly, increasing the amount of structural and temporal diversity within the system leads to a *de facto* diversification of the system, where other food, fiber, and fodder products are grown to be used by the farmer or sold at market. Third, carbon sequestration, if achievable, can be included as an additional type of income diversification for farmers, further supplementing their income with a natural function of the system.

Additionally, there are other ecosystem services that can benefit farmers. Increases in ecosystem services, such as pollination and pest regulation, directly increase production and prevent crop loss. Such services, when integrated into the shaded system, are provided at low or no additional costs to the farmer. The increased diversity and shade cover within coffee agroforestry systems can also protect production from increased climate variation and from extreme climate events, directly affecting farmers' economic bottom line. Thus, some scientific and economic evidence indicates that biodiverse farms may be financially advantageous as they increase fruit-set and protect fruit production from a range of crop damage.

Agroforestry systems offer additional income to farmers from timber and nontimber products that are made available from shade trees (Somarriba et al. 2004). In Peru, shade tree products may account for ~30% of revenues—especially fruits and firewood rather than timber (R. Rice 2002 unpublished data). Escalante et al. (1987) found that fruit from the shade canopy accounted for 55%–60% of income, and timber accounted for 3% of income. In Costa Rica, fruits sales accounted for 5%–11% of income from coffee-growing areas (Lagemann and Heuveldop 1983). Reviewing a 10 year research study in Central America, Mendez and colleagues (2010) concluded that coffee households manage four different types of plant agrodiversity, shade trees, agricultural crops, medicinal plants, and epiphytes, and use the plants for food (fruits), firewood, medicine, shade, timber, and ornamentals. Farmers reported that agrobiodiversity contributed to household livelihoods by generating products for consumption and sale. Generating additional products for sale from the shade tree canopy reduces vulnerability to market fluctuations and household dependence on outside products while increasing local commerce. In Nicaragua and El Salvador, the agrobiodiversity managed by coffee households accounts for at least 50% of household income, with about half of this coming from coffee (Méndez et al. 2010). This alternative income can protect farmers from financial ruin when coffee price fluctuations are extreme.

11.5 CONCLUSIONS

In agriculture, the necessity to find appropriate win-win management strategies for farmers depends on the trade-offs between the production and ecosystem functioning of the system. Intensifying a system in such a way that destroys the beneficial ecosystem services that a farmer could use to protect his or her production would be a backward strategy. A better understanding of the potential benefits that can be gained from diverse agroforestry systems will become increasingly important as smallholders face greater environmental changes and market fluctuations.

REFERENCES

Albrecht, A. and Kandji, S. T. 2003. Carbon sequestration in tropical agroforestry systems. *Agr Ecosyst Environ* 99:15–27.

Alegre, C. 1959. Climates et cafeiers d'Arabie. *Agron Trop* 14:25–48.

Altieri, M. A. 1999. The ecological role of biodiversity in agroecosystems. *Agr Ecosyst Environ* 74:19–31.

Antle, J. M., Stoorvogel, J. J., and Valdivia, R. O. 2007. Assessing the economic impacts of agricultural carbon sequestration: Terraces and agroforestry in the Peruvian Andes. *Agr Ecosyst Environ* 122:435–445.

Atmadja, S. and Verchot, L. 2012. A review of the state of research, policies and strategies in addressing leakage from reducing emissions from deforestation and forest degradation (REDD+). *Mitig Adapt Strat Global Chang* 17:311–336.

Beer, J., Muschler, R., Kass, D., and Somarriba, E. 1998. Shade management in coffee and cacao plantations. *Agroforest Sys* 38:134–164.

Brandle, J. R., Hodges, L., and Zhou, X. H. 2004. Windbreaks in North American agricultural systems. *Agroforest Sys* 61–62:65–78.

Callo-Concha, D., Krishnamurthy, K., and Alegre, J. 2004. Carbon sequestration by Amazonian agroforestry systems. Revista Chapingo. *Serie Ciencias Forestales y del Ambiente* 8:101–106.

Camargo, A. P. 1985. O clima e a cafeicultura no Brasil. *Inf Agropec* 11:13–26.

Cannell, M. G. R. 1976. Crop physiological aspects of coffee bean yield—A review. *Kenya Coffee* 41:245–253.

Cannell, M. G. R. 1983. Coffee. *Biologist* 30:257–263.

Castro Soto, G. 1998. *The Impact of the Chiapas Crises on the Economy*. San Cristobal de las Casas: CIEPAC.

Challinor, A. J., Wheeler, T. R., Slingo, J. M., and Hemming, D. 2005. Quantification of physical and biological uncertainty in the simulation of the yield of a tropical crop using present-day and doubled CO_2 climates. *Philos Trans R Soc B* 360:2085–2094.

Clowes, M. S. J. 1977. New form of physiological die-back in coffee. *Rhod J Agr Res* 15:231.

da Silva, J. E., Resck, D. V. S., Corazza, E. J., and Vivaldi, L. 2004. Carbon storage in clayey oxisol cultivated pastures in the "Cerrado" region, Brazil. *Agr Ecosyst Environ* 103:357–363.

De Jong, B. H. J., Tipper, R., and Montoya-Gómez, G. 2000. An economic analysis of the potential for carbon sequestration by forests: Evidence from southern Mexico. *Ecol Econ* 33:313–327.

Dixon, R. K., Winjum, J. K., Andrasko, K. J., Lee, J. J., and Schroeder, P. E. 1994. Integrated land-use systems: Assessment of promising agroforest and alternative land-use practices to enhance carbon conservation and sequestration. *Clim Change* 27:71–92.

Ebeling, J. and Yasue, M. 2008. Generating carbon finance through avoided deforestation and its potential to create climatic, conservation and human development benefits. *Philos Trans R Soc B* 363:1917–1924.

Escalante, E., Aguilar, A., and Lugo, R. 1987. Identificacion, evaluacion y distribucion espacial de especies utilizados como sombra en sistemas tradicionales de cafe (*Coffea arabica*) en dos zonas del estado de Trujillo, Venezuela. *Venezuela Forestal* 3:50–62.

Fisher, M. J., Rao, I. M., Ayarza, M. A., Lascano, C. E., Sanz, J. I., Thomas, R. J., and Vera, R. R. 1994. Carbon storage by introduced deep-rooted grasses in the South-American savannas. *Nature* 371:236–238.

Gay, C., Estrada, F., Conde, C., Eakins, H., and Villers, L. 2006. Potential impacts of climate change on agriculture: A case of study of coffee production in Veracruz, Mexico. *Clim Change* 79:259–288.

Giovannucci, D. and Koekoek, F. J. 2003. The state of sustainable coffee: A study of twelve major markets. Cali, Colombia: World Bank. www.iisd.org/pdf/2003/trade_state_sustainable_coffee.pdf. (Accessed 7 August 2008.)

Gregory, P. J. and Ingram, J. S. I. 2000. Global change and food and forest production: Future scientific challenges. *Agr Ecosyst Environ* 82:3–14.

Hawley, J. G. and Dymond, J. R. 1988. How much do trees reduce landsliding? *J Soil Water Conserv* 43:495–498.

Hernandez-Ramirez, G., Brouder, S. M., Smith, D. R., and Van Scoyoc, G. E. 2009. Greenhouse gas fluxes in an eastern corn belt soil: Weather, nitrogen source, and rotation. *J Environ Qual* 38:841–854.

Hoffmann, W. A. and Jackson, R. B. 2000. Vegetation-climate feedbacks in the conversion of tropical savanna to grassland. *J Clim* 13:1593–1602.

Howden, M., Crimp, S., and Nelson, R. 2010. Australian agriculture in a climate of change. In Jubb, I., Holper, P., and Cai, W. (eds), *Managing Climate Change*, pp. 101–111. Melbourne: CSIRO Publishing.

IPCC (Intergovernmental Panel on Climate Change). 2013. Summary for policymakers. In: Climate change 2013: The physical science basis. Contribution of Working Group I to the Fifth Assessment Report of the IPCC. Cambridge, UK: Cambridge University Press.

Jackson, N. A. and Wallace, J. S. 1999. Soil evaporation measurements in an agroforestry system in Kenya. *Agr For Meteorol* 94:203–215.

Jama, B., Palm, C. A., Buresh, R. J., Niang, A., Gachengo, C., Nziguheba, G., and Amadalo, B. 2000. Tithonia diversifolia as a green manure for soil fertility improvement in western Kenya: A review. *Agroforest Sys* 49:201–221.

Knops, J. M. H., Tilman, D., Haddad, N. M., Naeem, S., Mitchell, C. E., Haarstad, J., Ritchie, M. E. et al. 1999. Effects of plant species richness on invasion dynamics, disease outbreaks, insect abundances and diversity. *Ecol Lett* 2:286–293.

Kumar, B. M. and Nair, P. K. R. 2006. *Tropical Homegardens: A Time-Tested Example of Sustainable Agroforestry*. Dordrecht: Springer.

Lagemann, J. and Heuveldop, J. 1983. Characterization and evaluation of agroforestry systems, the case of Acosta-Puriscal, Costa Rica. *Agroforest Sys* 1:101–115.

Lal, R. 2004. Soil carbon sequestration to mitigate climate change. *Geoderma* 123:1–22.

Lamouroux, N., Pellegrin, F., Nandris, D., and Kohler, F. 1995. The *Coffea arabica* fungal pathosystem in New Caledonia: Interactions at two different spatial scales. *J Phytopathol* 143:403–413.

Lansigan, F. P., de los Santos, W. L., and Coladilla, J. O. 2000. Agronomic impacts of climate variability on rice production in the Philippines. *Agr Ecosyst Environ* 82:129–137.

Leakey, R. R. B. 1999. Potential for novel food products from agroforestry trees: A review. *Food Chem* 66:1–14.

Lin, B. B. 2007. Agroforestry management as an adaptive strategy against potential microclimate extremes in coffee agriculture. *Agr For Meteorol* 144:85–94.

Lin, B. B. 2010. The role of agroforestry in reducing water loss through soil evaporation and crop transpiration in coffee agroecosystems. *Agr For Meteorol* 150:510–518.

Lin, B. B. 2011. Resilience in agriculture through crop diversification: Adaptive management for environmental change. *BioScience* 61:183–193.

Lin, B. B. and Richards, P. L. 2007. Soil random roughness and depression storage on coffee farms of varying shade levels. *Agr Water Manage* 92:194–204.

Magalhaes, A. C. and Angelocci, L. R. 1976. Sudden alterations in water-balance associated with flower bud opening in coffee plants. *J Hortic Sci* 51:419–421.

Méndez, V. E., Bacon, C. M., Olson, M., Morris, K. S., and Shattuck, A. K. 2010. Agrobiodiversity and shade coffee smallholder livelihoods: A review and synthesis of ten years of research in Central America. Special Focus Section on Geographic Contributions to Agrobiodiversity Research. *Prof Geogr* 62:357–376.

Mendoza, V. M. and Villanueva, A. 1997. Vulnerability of basins and watersheds in Mexico to global climate change. *Clim Res* 2:139–145.

Meyfroidt, P. and Lambin, E. F. 2009. Forest transition in Vietnam and displacement of deforestation abroad. *Proc Natl Acad Sci USA* 106:16139–16144.

Moguel, P. and Toledo, V. M. 1999. Review: Biodiversity conservation in traditional coffee systems of Mexico. *Conserv Biol* 13:11–21.

Montagnini, F. and Nair, P. K. R. 2004. Carbon sequestration: An underexploited environmental benefit of agroforestry systems. *Agroforest Sys* 61–2:281–295.

Morais, H., Caramori, P. H., Ribeiro, A. M. A., Gomes, J. C., and Koguishi, M. S. 2006. Microclimatic characterization and productivity of coffee plants grown under shade of pigeon pea in southern Brazil. *Pesqui Agropecu Bras* 41:763–770.

Mutuo, P. K., Cadisch, G., Albrecht, A., Palm, C. A., and Verchot, L. 2005. Potential of agroforestry for carbon sequestration and mitigation of greenhouse gas emissions from soils in the tropics. *Nutr Cycl Agroecosys* 71:43–54.

Naeem, S., Hakansson, K., Lawton, J. H., Crawley, M. J., and Thompson, L. J. 1996. Biodiversity and plant productivity in a model assemblage of plant species. *Oikos* 76:259–264.

Nelson, K. C. and de Jong, B. H. J. 2003. Making global initiatives local realities: Carbon mitigation projects in Chiapas, Mexico. *Global Environ Chang* 13:19–30.

Niggli, U., Fliessbach, A., Hepperly, P., and Scialabba, N. 2009. *Low Greenhouse Gas Agriculture: Mitigation and Adaptation Potential of Sustainable Farming Systems.* Rome: Food and Agriculture Organization of the United Nations (FAO).

Nunes, M. A., Bierhuizen, J. F., and Plorgmanm, C. 1968. Studies on the productivity of coffee. I. Effect of light, temperature, and CO_2 concentration on photosynthesis of *Coffea arabica. Acta Bot Neerl* 17:93–102.

Nutman, F. J. 1937. Studies on the physiology of *Coffea arabica.* I. Photosynthesis of coffee leaves under natural conditions. *Ann Bot* 1:353–367.

Oelbermann, M., Voroney, R. P., and Gordon, A. M. 2004. Carbon sequestration in tropical and temperate agroforestry systems: A review with examples from Costa Rica and southern Canada. *Agr Ecosyst Environ* 104:359–377.

Oram, P. A. 1989. Sensitivity of agricultural production to climatic change, an update. In *Climate and Food Security: Papers Presented at the International Symposium on Climate Variability and Food Security*, pp. 25–44. Manila: International Symposium on Climate Variability and Food Security in Developing Countries (1987).

Palm, C. A., Alegre, J. C., Arevalo, L., Mutuo, P. K., Mosier, A. R., and Coe, R. 2002. Nitrous oxide and methane fluxes in six different land use systems in the Peruvian Amazon. *Global Biogeochem Cy* 16:1073.

Pasch, R. and Roberts, D. 2006. *Tropical Cyclone Report Hurricane Stan.* Washington DC: NOAA.

Paul, E. A., Morris, S. J., Six, J., Paustian, K., and Gregorich, E. G. 2003. Interpretation of soil carbon and nitrogen dynamics in agricultural and afforested soils. *Soil Sci Soc Am J* 67:1620–1628.

Peeters, L. Y., Soto-Pinto, L., Perales, H., Montoya, G., and Ishiki, M. 2003. Coffee production, timber, and firewood in traditional and Inga-shaded plantations in southern Mexico. *Agric Ecosyst Environ* 95(2):481–493.

Perez, U.M. 2005. Temen por 4 mil jornaleros y por la cosecha de cafe. *La Jornada.* October 12. http://www.jornada.unam.mx/2005/10/12/index.php?section=estados&article=042n2est.

Philpott, S. M., Lin, B. B., Jha, S., and Brines, S. J. 2008. A multi-scale assessment of hurricane impacts on agricultural landscapes based on land use and topographic features. *Agr Ecosyst Environ* 128:12–20.

Porter, J. R. and Semenov, M. A. 2005. Crop responses to climatic variation. *Philos Trans R Soc B* 360:2021–2035.

Ramachandran Nair, P. K., Mohan Kumar, B., and Nair, V. D. 2009. Agroforestry as a strategy for carbon sequestration. *J Plant Nutr Soil Sci* 172:10–23.

Rice, R. A. and Greenberg, R. 2000. Cacao cultivation and the conservation of biological diversity. *Ambio* 29(3):167–173.

Robinson, C. J., Wallington, T., Gerrard, E., Griggs, D., Walker, D., and May, T. 2011. Draft indigenous co-benefit criteria and requirements to inform the development of Australia's Carbon Farming Initiative. A report for the Australia's Rural Industry Research Development Corporation and Australian Government Department of Department of Sustainability, Environment, Water, People and Communities, Canberra.

Romero-Alvarado, Y., Soto-Pinto, L., Garcia-Barrios, L., and Barrera-Gaytan, J. F. 2002. Coffee yields and soil nutrients under the shades of *Inga* sp. vs. multiple species in Chiapas, Mexico. *Agroforest Sys* 54:215–224.

Salinas-Zavala, C. A., Douglas, A. V., and Diaz, H. F. 2002. Interannual variability of NDVI in northwest Mexico. Associated climatic mechanisms and ecological implications. *Remote Sens Environ* 82:417–430.

Semenov, M. S. and Porter, J. R. 1995. Climatic variability and the modelling of crop yields. *Agr Forest Meteorol* 73:265–283.

Slingo, J. M., Challinor, A. J., Hoskins, B. J., and Wheeler, T. R. 2005. Introduction: Food crops in a changing climate. *Philos Trans R Soc B* 360:1983–1989.

Smith, J., Pearce, B. D., and Wolfe, M. S. 2013. Reconciling productivity with protection of the environment: Is temperate agroforestry the answer. *Renew Agr Food Syst* 28:80–92.

Smith, P., Martino, D., Cai, Z., Gwary, D., Janzen, H., Kumar, P., McCarl, B. et al. 2008. Greenhouse gas mitigation in agriculture. *Philos Trans R Soc B* 363:789–813.

Solbrig, O. T., Medina, E., and Silva, J. F. 1996. *Biodiversity and Savanna Ecosystem Processes*. Berlin: Springer.

Somarriba, E., Harvey, C., Samper, M., Anthony, F., González, J., Staver, C., and Rice, R. 2004. Biodiversity conservation in neotropical coffee (*Coffea arabica*) plantations. In Schroth, G., da Fonseca, G., Harvey, C., Gascon, C., Vasoncelos, H., and Izac, A. (eds), *Agroforestry and Biodiversity Conservation in Tropical Landscapes*, pp. 1–14. Washington, DC: Island Press.

Soto-Pinto, L., Anzueto, M., Mendoza, J., Ferrer, G., and de Jong, B. 2010. Carbon sequestration through agroforestry in indigenous communities of Chiapas, Mexico. *Agroforest Sys* 78:39–51.

Soto-Pinto, L., Perfecto, I., and Caballero-Nieto, J. 2002. Shade over coffee: Its effects on berry borer, leaf rust and spontaneous herbs in Chiapas, Mexico. *Agroforest Sys* 55:37–45.

Soto-Pinto, L., Perfecto, I., Castillo-Hernandez, J., and Caballero-Nieto, J. 2000. Shade effect on coffee production at the northern Tzeltal zone of the state of Chiapas, Mexico. *Agric Ecosyst Environ* 80(1):61–69.

Soto-Pinto, L., Romero-Alvarado, Y., Caballero-Nieto, J., and Warnholtz, G. S. 2001. Woody plant diversity and structure of shade-grown-coffee plantations in Northern Chiapas, Mexico. *Rev Biol Trop* 49:977–987.

Soussana, J. F., Loiseau, P., Vuichard, N., Ceschia, E., Balesdent, J., Chevallier, T., and Arrouays, D. 2004. Carbon cycling and sequestration opportunities in temperate grasslands. *Soil Use Manage* 20:219–230.

Stigter, C. J., Mohammed, A. E., Nasr Al-amin, N. K., Onyewotu, L. O. Z., Oteng'i, S. B. B., and Kainkwa, R. M. R. 2002. Agroforestry solutions to some African wind problems. *J Wind Eng Ind Aerod* 90:1101–1104.

Tamang, B., Andreu, M., and Rockwood, D. 2010. Microclimate patterns on the leeside of single-row tree windbreaks during different weather conditions in Florida farms: Implications for improved crop production. *Agroforest Sys* 79:111–122.

Teare, I. D., Kanemasu, E. T., Powers, W. L., and Jacobs, H. S. 1973. Water-use efficiency and its relation to crop canopy area, stomatal regulation and root distribution. *Agron J* 65:207–211.

Tilman, D. 1999. The ecological consequences of changes in biodiversity: A search for general principles. *Ecology* 80:1455–1474.

Tilman, D., Cassman, K. G., Matson, P. A., Naylor, R., and Polasky, S. 2002. Agricultural sustainability and intensive production practices. *Nature* 418:671–677.

Tsuruta, H., Ishizuka, S., Ueda, S., and Murdiyarso, D. 2000. Seasonal and spatial variations of CO_2, CH_4, and N_2O fluxes from the surface soils in different forms of land-use/cover in Jambi, Sumatra. In: Murdiyarso, D. and Tsuruta, H. (eds) *The Impacts of Land-Use/Cover Change on Greenhouse Gas Emissions in Tropical Asia*. IC-SEA, Bogor, Indonesia and NIAES, Tsukuba, Japan, pp. 7–30.

Verchot, L., Noordwijk, M., Kandji, S., Tomich, T., Ong, C., Albrecht, A., Mackensen, J., Bantilan, C., Anupama, K. V., and Palm, C. 2007. Climate change: Linking adaptation and mitigation through agroforestry. *Mitig Adapt Strat Global Chang* 12:901–918.

Wallace, J. S., Jackson, N. A., and Ong, C. K. 1999. Modelling soil evaporation in an agroforestry system in Kenya. *Agr Forest Meteorol* 94:189–202.

Wollenweber, B., Porter, J. R., and Schellberg, J. 2003. Lack of interaction between extreme high-temperature events at vegetative and reproductive growth stages in wheat. *J Agron Crop Sci* 189:142–150.

Wrigley, G. 1988. *Coffee*. Harlow: Longman Scientific and Technical.

CHAPTER 12

Agroecology for Sustainable Coastal Ecosystems
A Case for Mangrove Forest Restoration

Mona Webber, Dale Webber, and Camilo Trench

CONTENTS

12.1 Introduction and Definitions 239
    12.1.1 Mangrove Forests 240
        12.1.1.1 Unique Features of Mangrove Trees 240
        12.1.1.2 Mangrove Forest Functional Types 241
        12.1.1.3 Mangrove Tree Types 242
        12.1.1.4 Mangrove Forest Zonation 247
12.2 The Value of Mangroves 250
12.3 The Problem 251
12.4 The Solution 252
    12.4.1 Nursery Seed Banks and Mangrove Restoration 252
12.5 Nursery Propagation 254
    12.5.1 Germination 254
        12.5.1.1 Field Germination Experiment 255
        12.5.1.2 Nursery Germination 255
        12.5.1.3 Hardening of Plants and Acclimation 257
12.6 Outplanting and Field Trials 258
    12.6.1 Transplanting Considerations 259
12.7 Case Studies of Replanting Process 259
    12.7.1 Portland Bight 259
    12.7.2 Montego Bay (Bogue Lagoon) 260
12.8 Conclusions 261
References 261

12.1 INTRODUCTION AND DEFINITIONS

Agroecology is the application of ecology to the design and management of sustainable agroecosystems. It identifies the link between ecology, culture, economics, and sound agricultural practices to create and/or sustain healthy environments. Agroecology may also be viewed as the application of ecological principles to the production of food, fuel, fiber, and pharmaceuticals and the management of agroecosystems (http://www.agroecology.org/index.html). Regardless of the definition used, agroecology proposes a context for or a site-specific way of studying agroecosystems, and as such, it recognizes that there is no universal formula or recipe for ensuring the success and maximum well-being of an agroecosystem. Rather than viewing agroecology as a subset of agriculture, Wojtkowski (2002, 2006) takes a more encompassing perspective. In this perspective, natural ecology and agroecology are the major headings under ecology. Natural ecology is the study of organisms as they interact with and within natural environments. Correspondingly, agroecology is the basis for the land use sciences. Here, humans are the primary governing force for organisms within planned and managed, mostly terrestrial, environments.

> Agroecology is widely accepted as a scientific discipline that uses ecological theory to study, design, manage and evaluate agricultural systems that are productive but also employed in resource conservation. To put agroecological technologies into practice requires technological innovations, agriculture policy changes, socio-economic changes, but mostly a deeper understanding of the complex long-term interactions among resources, people and their environment. To attain this understanding agriculture must be conceived of as an ecological system as well as a human dominated socio-economic system. A new interdisciplinary framework to integrate the biophysical sciences, ecology and other social sciences is indispensable. Agroecology provides a framework by applying ecological theory to the management of agroecosystems according to specific resource and socio-economic realities, and by providing a methodology to make the required interdisciplinary connections. (Altieri 2008)

12.1.1 Mangrove Forests

Mangrove forests are assemblages of salt-tolerant coastal trees that are found in tropical and subtropical climates (Cintron-Molero 1990; Nybakken and Bertness 2005; Tomlinson 1986). The habitat associated with the mangrove forest, for example, mangrove swamps and mangrove lagoons, is often termed *the mangal* (MacNae 1968). Mangrove forests are among the most productive coastal ecosystems in the world (Hemminga et al. 1994) and thus support high biodiversity and complex food webs (Riley and Kent 1999). The mangrove is an important nursery for juvenile finfish and shellfish as the system plays a major role in the provision of oxygen, food, and shelter for these vulnerable stages (Hoilett and Webber 2002). The importance of mangrove forests to developed and developing countries (Wells et al. 2006) is made more relevant as we realize their role in coastal sediment stabilization and shoreline protection from storm waves, storm surges, and tsunamis (Gilman et al. 2008). Unfortunately, an increase in our knowledge and awareness of the importance of this natural resource has not resulted in direct or proportional benefits to the environment. The increase in coastal populations has resulted in increasing use and pressure on all the associated natural resources, including mangroves. The implications of mangrove forest loss are being more greatly emphasized as the roles of the mangal become more significant to global issues such as climate change (Ramsar Secretariat 2001).

It has been estimated that mangrove forests occupied up to 75% of the tropical coast worldwide, but human presence and pressures have reduced the global range of these forests to less than 50% of their original total cover (FAO 1994). The world's mangrove forest cover has declined by 25% over the last two decades (Valiela et al. 2001). There has been heightened urgency to replant and restore mangrove areas as we realize their value to human safety and the maintenance of coastal water quality and biodiversity (Mumby et al. 2004).

12.1.1.1 Unique Features of Mangrove Trees

The mangrove ecosystem comprises a unique assemblage of plants, animals, and microorganisms adapted to live in the dynamic environment of tropical and subtropical coastlines. The word *mangrove* is used to refer to the community as well as its constituent trees. Mangrove trees are unique plants that do not have to grow in saline soils (i.e., they are not obligate halophytes) but gain a competitive advantage over other plants by their adaptations to exclude or remove salt from their tissues. They are also adapted to overcome the effects of waterlogged, poorly oxygenated soils (Duke et al. 1998; Hogarth 2007).

Mangrove trees and forests are thus found in low-lying areas subjected to saline intrusion (Tomlinson 1986). However, mangroves can also be found along riverbanks, on river deltas, on carbonate islands (cays), and in isolated areas subjected to saline intrusion during extreme high tides. This variability in their structural and functional characteristics (Lugo and Snedaker 1974; Pool et al. 1977; Woodroffe 1992) was used to classify mangrove areas into four main functional types: fringe forest, riverine forest, overwash forest, and basin forest (Figure 12.1).

12.1.1.2 *Mangrove Forest Functional Types*

Fringe forests (Figure 12.1b) are formed along the gentle sloping shores of continents and islands and on the leeward side of sand barriers or spits (Alleng 1990; Hogarth 2007). The fringe forest does best where the wave action is moderate and the tidal activity is low. This is the dominant functional forest type found in the coastal areas of Jamaica and the Caribbean. The fringe forest is characterized at its edge by the *Rhizophora mangle* (red mangrove) tree, which is the dominant tree type in such a forest. However, it is often possible to see a range of mangrove tree types in a fringe forest with the succession of zones dominated by each type. Hence, the "idealized" fringe forest in Jamaica and the Caribbean would have the red mangrove tree (*R. mangle*) followed by the black mangrove (*Avicennia germinans*), then the white mangrove (*Laguncularia racemosa*), and finally the button mangrove (*Conocarpus erectus*), which forms the most landward zone of the forest. In most cases, however, human interference has modified the ideal zonation pattern and the only zone and tree type found is the dominant red mangrove. An example of a fringe forest is the Port Royal mangroves on the sheltered side of the Palisadoes.

Riverine forests develop on the banks of rivers (Figure 12.1c) and on the floodplains where rivers run into the sea (Hogarth 2007). They are therefore strongly affected by the unidirectional flow of

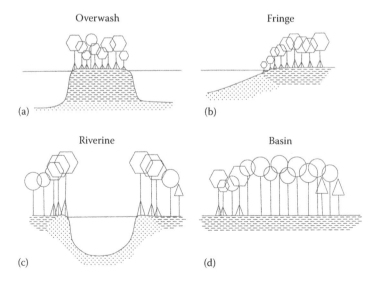

Figure 12.1 **(a–d)** Main functional types of mangrove forests. (Reproduced by permission of Robertson, A.I. and Alongi, D.M. (eds), *Coastal and Estuarine Studies: Tropical Mangrove Ecosystems*, American Geophysical Union, Washington DC. Copyright (1992) American Geophysical Union.)

freshwater and here their competitive advantage is gained by their ability to establish in waterlogged soils and to a lesser extent maintain the salt balance in their tissues. The lower "reaches" of rivers are subjected to saline intrusion with salt wedges advancing up the river below the outflowing freshwater layer. The riverine forest is highly productive (Cintron-Molero 1990) and is dominated by the red mangrove; however, black and white mangrove trees may also be found (Alleng 1990).

Overwash forests establish on small islands (Hogarth 2007) that have a flat area of sand that is exposed at low tide but is covered during periods of extreme high tide and where the current flow is not strong enough to remove the vegetation. The high tide levels must be such that they do not completely cover or drown the seedlings. The community is again dominated by the red mangrove but all three types may be found in this type of forest (Alleng 1990). As the overwash forest ages, a depression may develop in its center with a semipermanent body of water or mud. Such a "basin" would be colonized by the black mangrove. An example of an overwash forest is that found on Drunkenman's Cay in the Port Royal Cays area near Kingston, Jamaica.

Basin forests occur in inland depressions where water accumulates due to infrequent events such as storms, extreme high tides, or floods (Alleng 1990). The conditions experienced include still water, low oxygen, high salinity, and anoxic sediments. The black mangrove tends to dominate such areas, followed by the white mangrove and, where the basin is close to the sea, the red mangrove may also occur. Under extreme conditions or in particular geographic locations, variations of the basin forest may be recognized. These are called scrub mangrove forests (conditions of extreme hypersalinity) and hammock mangrove forests (specific to the Florida Everglades) where the basin is slightly elevated relative to the surrounding area and isolated from the surrounding water (Hogarth 2007). An example of a basin forest can be found in the Falmouth swamps in Trelawney, Jamaica.

Whatever the forest type, however, mangrove areas are generally characterized as a stand of woody plants growing in waterlogged and in most cases saline soils in association with a complex of other organisms (Duke et al. 1998).

12.1.1.3 Mangrove Tree Types

Four species of naturally occurring mangrove trees exist in Jamaica and the Caribbean. The red mangrove (*R. mangle*) is the most common and best-known mangrove tree found in Jamaica. The red mangrove tree is so named because of the red color under its bark, which is caused by accumulated waste products in the form of tannins. When this material leaches out of the bark, it stains the water and mud of the mangal dark red. Red mangrove trees have large, leathery elliptical leaves (Adams 1972). Its flower is a yellow-cream color with four pointed petals (Figure 12.2). However, the most familiar feature of the red mangrove tree is its root system often referred to as prop or stilt roots (Figure 12.3). The roots are aerial roots that grow from the branches (drop roots) or more commonly from the trunk (prop roots) of the tree and are supplied with pores (lenticels) filled with corky material, which allow oxygen to be taken up as the rest of the rooting system is in anaerobic conditions (Hogarth 2007).

Red mangroves maintain "normal" salt concentrations in their tissues by excluding salt at the roots (Nybakken and Bertness 2005) as well as shunting salt into the older leaves, which are then shed. An old yellow mangrove leaf about to be shed contains much more salt than a young green leaf. Another peculiar feature of the red mangrove tree is the fact that it shows viviparity (bears live young). Mangrove seeds germinate while still attached to the tree. The resultant seedling (called a propagule) has a characteristic torpedo shape (Figure 12.4) and is weighted so that it falls vertically into the mud and starts to grow, or floats upright until it touches land and becomes established. The presence of the prop root system and the production of the red mangrove propagule make this species of mangrove well suited to live on the seaward fringe or wettest part of the forest. The prop roots/stilt roots facilitate anchorage of the tree in the unconsolidated sediment (Hogarth 2007) at the lagoon–forest interface.

Figure 12.2 Leaves and flowers of the red mangrove; scale bar = 1 cm. (Courtesy of M. K. Webber.)

Figure 12.3 Red mangrove root system at the forest–lagoon interface. (Courtesy of M. K. Webber.)

The black mangrove (*A. germinans*) is so named for the dark color of its trunk due to the presence of blue-green algae, which live in the bark of the tree (Sheridan 1992). This mangrove also has a peculiar rooting system, but these roots, while also being a type of aerial root, are called pneumatophores (breathing roots) as their primary function is not anchorage but to facilitate the gaseous exchange between the plant and the air. The pneumatophores are erect branches of the horizontal

Figure 12.4   Red mangrove viviparous seedling attached to the tree. Seedling length 8 cm. Scale bar = 1 cm. (Courtesy of D. F. Webber.)

"cable" roots, which originate at the trunk of the tree and extend for long distances below ground. The pencil-like pneumatophores extend about 30 cm above the soil (Figure 12.5), are covered in lenticels or breathing pores and are comprised of aerenchyma (Hogarth 2007). Adventitious roots develop on the underside of the horizontal cable root and facilitate nutrient exchange and anchorage. The submerged portions of the pneumatophores themselves produce root hairs at their base, which are nutritive (Figure 12.6).

The leaves of the black mangrove are narrow (oblong to oblong-lanceolate according to Adams 1972) and often have salt crystals on their surface (Figure 12.7). The plant excretes excess salt through the stomatal pores of its leaves. The leaves are arranged opposite on the stem. Its flower has four rounded petals and is white (Figure 12.8). It produces an oval fruit ~2.5 cm long with a peak at one end (Figure 12.9).

The white mangrove (*L. racemosa*), like the black mangrove, is named for the color of its trunk. Its adaptations have less to do with anchorage in soft substrate and more to do with salt removal, as these species tend to grow in dryer areas of the forest where hypersaline conditions may prevail (Alleng 1990). However, where waterlogged conditions prevail, white mangroves will produce pneumatophores, which are erect but blunt or sometimes "knobby" (Figure 12.10). They rarely exceed 15 cm and function in the same manner as the black mangrove pneumatophores (McDonald et al. 2003).

White mangrove leaves are rounded (oblong or oblong-elliptical according to Adams 1972) and sometimes with pinkish stems. They are arranged opposite on the stem. A pair of large salt glands occurs on either side of the leaf petiole. They produce an inflorescence of small white flowers and the fruits that occur in clusters are green and ribbed (Adams 1972) (see Figure 12.11).

Figure 12.5 **(See color insert)** Black mangrove pneumatophores. (Courtesy of M. K. Webber.)

Figure 12.6 Black mangrove showing the structure of the rooting system. (Courtesy of H. Small.)

Figure 12.7   (See color insert) Leaves of the black mangrove with salt crystals on the upper surface; scale bar = 1 cm. (Courtesy of M. K. Webber.)

Figure 12.8   Flowers of the black mangrove; scale bar = 1 cm. (Courtesy of D. F. Webber.)

The button mangrove (*C. erectus*) is considered by some to be a "semimangrove," however it has extreme adaptations to remove salt from its tissues and is commonly found at the landward edge of the mangrove forest (Alleng 1990). The leaf of the button mangrove can be variable being linear or rounded with an alternate arrangement on the stem. Large salt glands occur on either side of the petiole as well as on the underside of the leaf along the leaf vane (Adams 1972). It has no prop roots or breathing roots and the tree bears small clusters of flowers that become clusters of fruits with rounded heads like buttons (Figure 12.12).

Figure 12.9 Black mangrove fruits; scale bar = 1 cm. (Courtesy of M. K. Webber.)

Figure 12.10 **(See color insert)** White mangrove pneumatophores, which can be wider at the tip than at the base; scale bar = 1 cm. (Courtesy of M. K. Webber.)

A second variety of button mangrove called *C. erectus* var *sericeus* has been introduced to Jamaica (Adams 1972). It has small hairs on the upper surface of its leaves, which give the leaf a silvery appearance. The plant grows in a wide range of conditions; it can be seen near the dunes along the Palisadoes road and lining the chapel lawns of the University of the West Indies Mona Campus. The characteristic button-like fruits and salt glands are very easily seen on this variety (Figure 12.13).

12.1.1.4 Mangrove Forest Zonation

Mangrove forest zonation is a well-described feature with the trees best adapted to establishing in unconsolidated waterlogged soils at the seaward edge and those preferring dryer soils at the landward side of the forest (Ellison et al. 2000). However, it is difficult to present a consistent description

Figure 12.11 White mangrove leaves and inflorescence of tiny flowers; scale bar = 1 cm. (Courtesy of M. K. Webber.)

Figure 12.12 Button mangrove leaves, fruits, and flowers; scale bar = 1 cm. (Courtesy of M. K. Webber.)

of the position of each species as the environment is unstable and several factors, including soil conditions and biotic factors, affect zonation (Lopez-Portillo and Ezcurra 1989). However, the expected arrangement of the species is for the red mangrove to be closest to the lagoon, sea, or river, followed by the black mangrove, then the white mangrove, and finally the button mangrove. Perhaps the most predictable pattern is for *R. mangle* (red mangrove) to be found closest to the water's edge, providing that the conditions are calm. The prop roots of the red mangrove further slow the water movement and encourage the deposition of more sediment. As the sediment accumulates, the level of the mud rises and is consolidated by the rooting system of the plant. The rooting system of *A. germinans*,

Figure 12.13 *Conocarpus erectus* var *sericeusa* variety of button mangrove found in Jamaica. Enlarged picture showing button-like fruits and salt glands at the base of the leaf; scale bar = 1 cm. (Courtesy of M. K. Webber.)

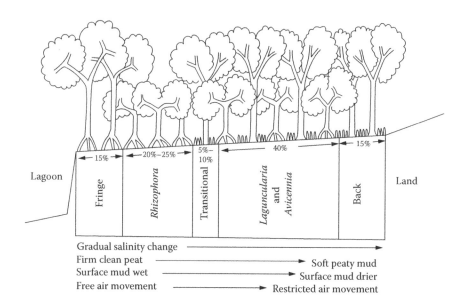

Figure 12.14 Mangrove forest zonation found along the Palisadoes and Port Royal area. (Used with permission from Warner, G.F., *J Anim Ecol*, 38, 379–389, 1969.)

the black mangrove, allows the plant to respire in the waterlogged soils. After some time, the soil becomes too compact for the growth of the black mangrove and *L. racemosa* (white mangrove) can be found. Eventually, the button mangrove (*C. erectus*) colonizes the area and the zonation/succession (Davis 1940) is considered complete. The button mangrove is also viewed as a transitional species between true mangroves and terrestrial vegetation (Thompson and Webber 2003).

Most mangrove forests in Jamaica are dominated by the red mangrove trees (M. Webber, personal observation) and it is also typical for there to be no clear zone of white mangrove or black mangrove. The white mangrove trees may simply be interspersed in the area between the red zone and the drier parts of the forest, with the trees showing root development in accordance with the nature of the sediment. Similarly, a black mangrove zone may be absent, save for small clusters of pheumatophores and a few trees growing in the area between the red zone and the drier parts of the forest. Warner (1969) was the first to describe this commonly found condition in Jamaican fringe forests (Figure 12.14) where the *R. mangle* is near the lagoon at the wetter side of the forest and the "back" or dry end is colonized by button mangroves and other xerophytes (Thompson and Webber 2003).

12.2 THE VALUE OF MANGROVES

Mangrove forests and their associated habitats support high biodiversity due to their provision of food and shelter for marine and terrestrial organisms. The "nursery" function performed by mangrove lagoons has been well documented. The prop roots of the *R. mangle* that hang into the lagoon support sessile flora and fauna, and the water column of the lagoon and the muds associated with the lagoon and forest floor are known to provide habitat and substrate for a wide variety of organisms (Kathiresan and Bingham 2001). Mangroves effect coastal stabilization due to their extensive rooting systems, which form natural "breakwaters" where the roots and trunks reduce wave energy and so mitigate coastal erosion and flooding. Red mangroves are reported to have among the strongest ability to bind sediments when tested against algae and other plants (Riley and Kent 1999). Mangroves remove especially fine terrigenous material so seagrasses and coral reefs are not deleteriously affected by the excess turbidity of freshwater. Some of the freshwater stored in the forest is lost by evaporation; the rest dilutes excess salts that accumulate in the inner fringes during periods of drought. The volume of undiluted freshwater reaching the coast is therefore reduced or it is discharged over a longer period where healthy mangrove forests exist. The environmental functions and the related ecological processes of mangroves are summarized in Table 12.1.

Mangroves also have a variety of economic uses in the following industries and sectors: construction, textile, food and drug, agriculture, fishing, and paper production. The functions or value of mangroves have been classified into regulatory, productive, carrier, and information by Macintosh and Ashton (2002). Table 12.2 uses these categories to summarize the many uses and value of mangroves.

Table 12.1   Mangrove Environmental Functions and Related Ecological Processes

Environmental Function	Ecological Processes
Production of water	Hydrological cycle
Production of food	Solar energy fixation for biomass production + storage and recycling of organic matter and nutrients
Production of other biotic resources	
Production of fuel and energy	
Maintenance of biodiversity	Maintenance of nursery and migration habitats
Production of juveniles for cultivation	
Regulation of environmental quality	
Flood mitigation and prevention of soil erosion	Sediment control Buffering from storms

Table 12.2 Mangrove Functions

Regulatory functions	• Protection against extreme weather events • Chemical composition of the coastal waters • Runoff and flood protection • Water catchment and groundwater recharge • Prevention of soil erosion and sediment control • Topsoil formation, maintenance of fertility • Solar energy fixation and biomass production • Storage/recycling of nutrients • Storage/recycling of wastes • Biological control mechanisms • Migration and nursery habitats • Biological (and genetic) diversity
Production functions	• Detritus, particulate organic matter (POM) • Food for marine organisms and other ecosystems • Oxygen • Medicinal resources • Raw materials for construction • Fuel (charcoal) • Fertilizer
Carrier functions (providing space and suitable substrate)	• Cultivation (mariculture) • Energy conversion • Recreation and tourism • Land stabilization nature protection
Information functions	• Aesthetic information • Historic information (heritage value) • Cultural • Scientific and educational information

Source: Used with permission from Macintosh, D.J. and Ashton, E.C., A review of mangrove biodiversity conservation and management, Centre for Tropical Ecosystems Research, University of Aarhus, Denmark, 2002. www.researchgate.net/publication/.../3deec529f8b3 25bf49.pdf (accessed July 15, 2014).

12.3 THE PROBLEM

There is a global challenge to restore the world's forests, including mangrove forest coverage. Present environmental and developmental pressures, coastal instability, increasing harvesting requirements with growing populations, and natural disasters impede the natural regeneration of these coastal wetlands (Teas 1977). Coastal road construction, which interferes with natural water exchange, leads to soils becoming hypersaline and forests drying out. This results in stunting and low productivity, or, for some species, the death of the parent trees. It is difficult for seedlings to establish in these conditions and therefore there is little possibility of regenerating forests by natural means. Several studies list the causes of mangrove degradation as including: agriculture, aquaculture, pollution, and coastal development, which prevent tidal and freshwater flows. In most cases, specifically designed mitigation is required for successful restoration. However, most efforts require the replacement of trees with either wild or nursery-generated seedlings.

The restoration of mangrove forests by replanting is therefore a necessary strategy to restore lost/degraded forests. Artificial replanting of mangroves has shown much success in Southeast Asia, the southern United States (Trench and Webber 2011), and some African nations and is showing a trend of increasing popularity and usefulness worldwide. This is a relatively new phenomenon to the Caribbean basin primarily due to the absence of mangrove nurseries. However, Caribbean mangrove forests require focused attention by the region as the diversity and balance of our natural ecosystems are strongly correlated to our tremendous successes in tourism. In addition, in numerous Caribbean territories, it was common practice in earlier decades to undertake wetland reclamation to accommodate numerous commercial developments. This has shown negative effects on coastal ecosystems and water quality as the ecological roles of mangrove forests have been reduced or eliminated, for example, Kingston Harbor on the south coast of Jamaica. Nevertheless, awareness of the direct and indirect importance of mangrove forests is growing. In fact, many Caribbean and American territories such as Belize are investing more in sustainable tourism and ecotourism, using wetlands and other natural ecosystems as attractions (Cooper et al. 2009).

12.4 THE SOLUTION

12.4.1 Nursery Seed Banks and Mangrove Restoration

A seed bank can be defined as a store of viable plant material usually held in an ecosystem in parental stock or in the soil. The application of this concept to mangrove restoration involves the acquisition of seed/propagule material from parental stock or the soil or both and the maintenance of its viability until required for nursery propagation. The first stage of seedling and seed selection for the mangrove species is the identification of donor trees. These are trees that are accessible and healthy, and have an abundance of flowers, fruits, or propagules. Figure 12.15 displays their attributes including the fruit used in creating the seed bank for the four mangrove forest species

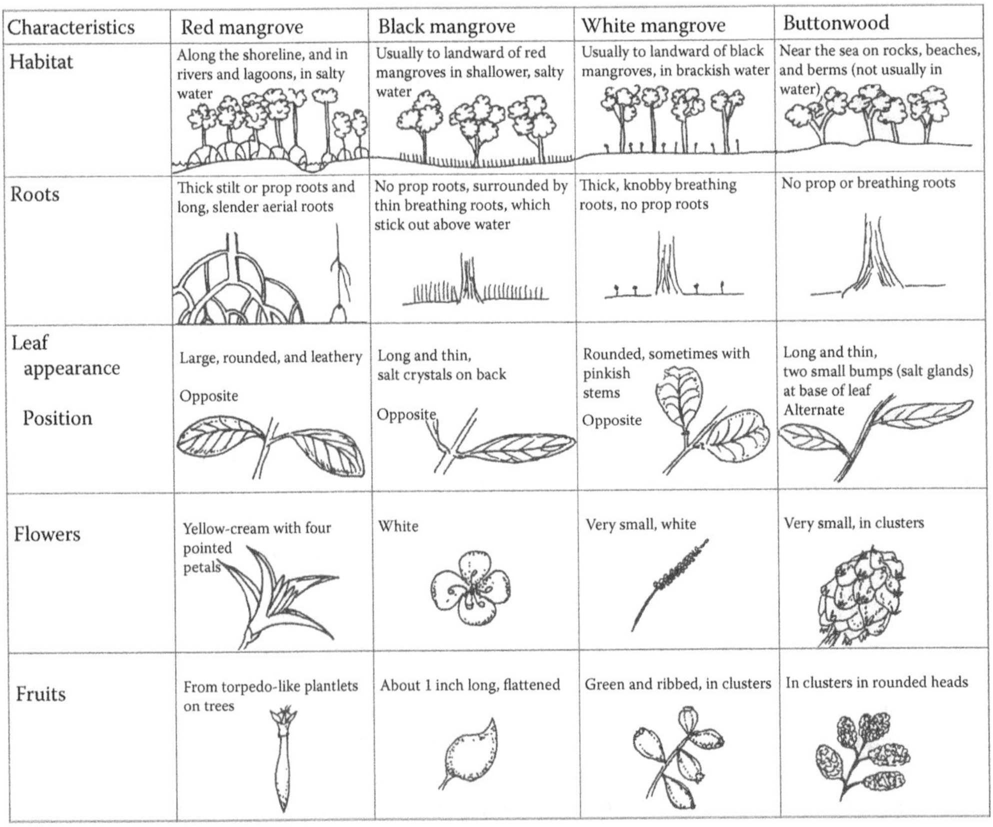

Characteristics	Red mangrove	Black mangrove	White mangrove	Buttonwood
Habitat	Along the shoreline, and in rivers and lagoons, in salty water	Usually to landward of red mangroves in shallower, salty water	Usually to landward of black mangroves, in brackish water	Near the sea on rocks, beaches, and berms (not usually in water)
Roots	Thick stilt or prop roots and long, slender aerial roots	No prop roots, surrounded by thin breathing roots, which stick out above water	Thick, knobby breathing roots, no prop roots	No prop or breathing roots
Leaf appearance Position	Large, rounded, and leathery Opposite	Long and thin, salt crystals on back Opposite	Rounded, sometimes with pinkish stems Opposite	Long and thin, two small bumps (salt glands) at base of leaf Alternate
Flowers	Yellow-cream with four pointed petals	White	Very small, white	Very small, in clusters
Fruits	From torpedo-like plantlets on trees	About 1 inch long, flattened	Green and ribbed, in clusters	In clusters in rounded heads

Figure 12.15 Identifying features and habitat of the four common mangrove species. (Used with permission from Sutton, A.H., Sorenson, L.G., and Keeley, M.A., *Wondrous West Indian Wetlands: Teachers' Resource Book*, Society for the Conservation and Study of Caribbean Birds, 2004.)

naturally occurring in Jamaica (*R. mangle* [red mangrove], *L. racemosa* [white mangrove], *A. germinans* [black mangrove], and *C. erectus* [button mangrove]).

Three of these species are known to have peak production of seeds or propagules between the months of June and November. Donor trees should be accessible and be at least 3 years old (Ravishankar and Ramasubramanian 2004). Selected donor areas should be checked weekly and mature seeds or propagules collected by hand picking. Mature seeds or propagules can also be collected from beneath the trees, but they must be checked for physical damage from herbivory. Another method of propagule collection is to suspend nets below donor trees that are observed to possess many near-ripe propagules (Duke et al. 2002). This method will keep stranded propagules cool and prevent desiccation (as they are suspended in the shade of the trees), reduce herbivory on the ground, and prevent early germination (especially in black mangroves, which have very rapid germination).

Red mangrove propagules are usually available 12 months of the year and can be collected when they begin to show brown coloration. However, this may not be the case with all mature trees in all areas. Propagules show darker coloration and swelling of the radical/root portion (Figure 12.16) when they are more mature. Propagules that are red or orange in color have been found to show poor germination success and therefore should not be collected (Trench 2007). Propagules should be checked for evidence of predation by crabs and boring by beetles. Damaged propagules can be discarded as, in a previous study, they were shown to have poor growth rates and eventual seedling failure even when they successfully established in the nursery (Trench 2007).

The seeds of the white mangrove (triangular in shape) are variable in size (1 cm on average) and mature seeds are identified by a change from their original light green to a golden brown or dark brown coloration. White mangroves show seed production between July and September, with peak production and dispersal in August (Odum et al. 1982; Rabinowitz 1978). White mangrove seeds are less subjected to herbivory (approximately <1% damage by predators observed on seeds collected) and thus are more readily and easily collected from the ground beneath the parent trees when fruiting.

The buttonwood grows at the drier upland fringe of the mangrove forest and flowers year-round, peaking in July–September. A cluster of flowers is formed and when fertilized, it forms a button-like seed case. One fruit head/button may contain 35–56 seeds (Francis and Rodrigues 1993). Seeds of the buttonwood possess extensions of the testa and are dispersed primarily by the wind. These seeds

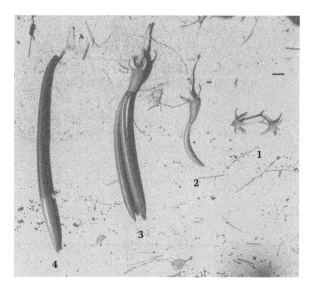

Figure 12.16 Red mangrove propagules in various stages (1: postpollination fruit; 4: mature and ready for reaping) Scale bar = 1 cm. (Courtesy of C. A. Trench.)

show the lowest germination viability of all species. Buttonwood seeds are best collected from the parent tree or immediately after falling as they will disperse quickly and become inconspicuous in the substrate.

Propagules should be inspected for holes and scars before introducing them into the nursery. Red mangrove propagules are a prime food source for the *Coccotrypes rhizophorae* (mangrove boring beetle), which is a major source of seedling failure. Black mangrove seedlings have been observed to have holes and boring scars. It is inconclusive if they are affected by the *Coccotrypes* specifically. However, propagules with holes should not be introduced into the nursery. Black mangrove plants are also subject to growth and survival limitations by a few fungal species (Nieves-Rivera et al. 2002). However, the literature does not suggest that this significantly affects germination. Black mangrove leaves are eaten by an unidentified caterpillar, but this can be prevented by disallowing butterflies and moths from entering the nursery (with shade covering) or by using any available agricultural pesticides (Diazinon, Match, Pegasus, etc.). However, the use of pesticides close to any marine environment is not recommended.

Following an inspection of the propagules, planting can proceed immediately into prepared seedling bags (with moist soil in the nursery). However, larger propagules (red, black, and white) may be subject to an initial soaking in fresh or slightly saline water in a "soaking bed" (concrete, plastic, etc.) for a period of 3–5 days. Clarke (2002) indicated that trials in Australia in which propagules were immersed in freshwater for 3–7 days enhanced germination. Soaking gray mangrove (*A. marina*) seeds in freshwater caused the outer sheath to split and appeared to enhance the germination rate by about 50% (Clarke 2002).

Extending this period was found to lead to stagnation of the water and seedling failure due to the development of anoxic conditions (Trench 2007). Soaking propagules in slowly flowing water is recommended to prevent stagnation and reduce the risk of losing the propagules that have been collected. Soaking the propagules facilitates the use of a relatively small space to germinate large quantities of seedlings and facilitates the preparation of seedling bags/planting medium over this period. In addition, seedlings germinated in a soaking bed may be left for months if space for planting is not available. However, the water must be changed weekly to prevent stagnation. Propagules may be held in the soaking bed for varying periods. However, the initial soaking of propagules is not mandatory; direct sowing is always the primary option if potting bags are ready and ample. A delay in germination with or without soaking has been previously shown to reduce germination success (Clarke 2002; Ravishankar and Ramasubramanian 2004).

12.5 NURSERY PROPAGATION

The methods associated with nursery propagation presented in this section will be separated into field germination experiment/observations and nursery germination experiment/observations. The various applications and conditions are outlined and the success of various combinations is identified. However, in keeping with the overarching concept of the context- or site-specific manner of studying agroecosystems, these are proposed only as guidelines and examples.

12.5.1 Germination

Five main objectives are usually identified to direct the germination component:

1. To observe the viability of mangrove seedlings in field and nursery conditions
2. To germinate mangrove seeds (red, black, white, and button) in a nursery under controlled conditions to determine which species germinates and grows best in a nursery and therefore provides the greatest potential for coastal restoration by replanting

3. To determine the optimum light and salinity conditions to facilitate the germination and growth of the four mangrove species
4. To determine if light or salinity is a stronger stimulus for germination
5. To grow seedlings to transplantable stage from germination

12.5.1.1 Field Germination Experiment

Areas were selected within the Port Royal mangroves where each species was observed to have mature and flowering/fruiting trees from which seeds had fallen to the substrate below, with special emphasis on obtaining samples from areas with varying moisture and light exposure conditions. Areas were selected based on high and low light climates, high and low soil moisture, degree of coastal water and freshwater introductions (storm drainage or rainwater settlement), and range of substrates (sand, gravel, mud, grass), and 0.5×0.5 m permanent quadrats were established. The identification and enumeration of seeds found within each quadrat with a proportion germinated were supplemented with light intensity (Davis light meter $\pm$ 0.5 units, salinity (AO refractometer $\pm$ 1‰), soil temperature (Taylor soil thermometer $\pm$ 0.5°C), and soil moisture (Aquaterr moisture probe $\pm$ 5%). The quadrats were checked weekly over a 6-week period to record the physicochemical data and the germination progress of the seedlings.

Naturally occurring seedlings showed lower survival than nursery-produced seedlings. *R. mangle* (red mangrove) field-"germinated" propagules showed 10.1% survival (Table 12.3), *A. germinans* (black mangrove) field-germinated seeds showed 26.4% survival (Table 12.3), and *C. erectus* (button mangrove) field-germinated seeds showed 3.1% survival (Table 12.3). Observations showed that there was a dramatic reduction in the total propagules present or a complete absence of established propagules under the trees. Areas with the greatest propagule loss had relatively high tree density and vegetative cover and showed the presence of mangrove crabs (*Aratus pisonii* and *Uca* spp.). It was further observed that the red mangrove propagules that fell into the substrate in an upright (vertical) position showed no signs of predation and remained in position for subsequent weeks. These upright propagules survived the duration of the experiment and became established.

12.5.1.2 Nursery Germination

Within the Port Royal mangroves, the seeds of three mangrove species were collected directly from trees by hand or they were picked up from below the donor trees. The seeds were inspected

Table 12.3 Percentage Survival of Mangrove Seedlings and Seeds in Forest Plots

Quadrats	% Survival		
	Red	Black	Button
1	7.1	6.7	0
2	0	0	2
3	30	0	0
4	0	0	0
5	7.1	73.3	0
6	0	100	0
7	33.3	57.7	6.25
8	0	0	0
9	13.3	0	20
Average	10.08	26.41	3.14

Note: Each plot is 1 m².

Table 12.4 Physical Treatment Conditions for Four Different Nursery Enclosures

Treatment	Salinity Treatment (‰)	Light Treatment
1. High light with high salinity	Seawater 20–40	1020–1280 (average 1184.2 lux)
2. Low light with high salinity	Seawater 20–40	450–573 (average 509.8 lux)
3. High light with low salinity	Tap water 1–3	1045–1270 (average 1178.5 lux)
4. Low light with low salinity	Tap water 1–3	284–617 (average 502.7 lux)

to ensure no external damage (no heat or predator damage) and no signs of germination (root emergence or split cotyledon). This was to ensure that the seeds placed in the nursery were viable and had not started the germination process prior to the experiment. The seeds of the *R. mangle*, *A. germinans*, and *C. erectus* were collected. The seeds of the *L. racemosa* (white mangrove) could not be collected as these seeds were not fully mature at the time of the experiment. The collected seeds were placed in four concrete nursery enclosures with four different treatment conditions as summarized in Table 12.4.

The substrate for use in the nursery was collected from an area of fine sand devoid of mangrove trees and leaf litter; with a salinity of 0‰ (the area frequently collects rainwater) in the Port Royal mangroves. Using a shovel, the sand was collected from the upper 0.5 m of the area and placed into individual planting bags. The bags were secured in an enclosed area to facilitate the watering mechanism and to protect the seeds/plants from crabs and other predators. Thirty replicates of each species were placed in individually numbered planting bags. One seed of the black mangrove was placed on the soil surface in the center of the planting bag. Red mangrove propagules were placed with the root end pushed 4–6 cm into the soil in the center of the planting bag. Multiple seeds of the button mangrove were placed in each planting bag, however. This was done because of their small seed size and to imitate the dispersal mechanism employed by this species in the field, where seeds fall from plants as one button/ball of seeds and are dispersed in close proximity to each other (Francis and Rodriguez 1993).

Plants subject to high light intensities were fully exposed to the sun throughout the day, while plants undergoing the low light intensity treatment were protected by a 60% shade cloth, which provided constant cover from the full intensity of the sun. The two salinity treatments were effected by slowly adding previously mixed seawater and freshwater to the concrete enclosures up to the soil/ sand level. These additions were made at 3 day intervals. The salinity was monitored between each water addition and was maintained by compensation additions for evaporation as necessary. An overflow mechanism was attached to the enclosures to prevent complete flooding of the seeds and nursery. The nursery tanks were located close to each other to receive similar rainfall and air conditions. The seeds of the mangroves were observed weekly for 6 weeks after the planting date for evidence of germination and changes were recorded. The parameters recorded were root or shoot emergence (germination), height of the plant in centimeters (black mangrove: height from soil level to terminal bud; red mangrove: height from shoot ring on the top of the propagule to terminal bud; button: height from soil level to terminal bud), number of leaves (cotyledon leaves ignored), and the girth of the plant at standard heights above sand level.

The propagules of the *R. mangle* (red mangrove) showed the highest survival percentage of 93% and this was achieved in low salinity and low light conditions. Treating seedlings to high light and high salinity resulted in 50% survival in the nursery (Figure 12.17). *A. germinans* (black mangrove) seeds had the shortest germination time as the roots were seen after 1 week in all cases. A 97% seed survival rate (successful germination) was achieved in conditions of low salinity with high light followed by 88% in low salinity with low light, with survival falling to <20% in high light and high salinity. *C. erectus* (button mangrove) showed the lowest survival/germination percentage of all the species tested, with values of 10%–20% recorded in low salinity with high or low light. No germination of button mangroves was achieved in high salinity conditions.

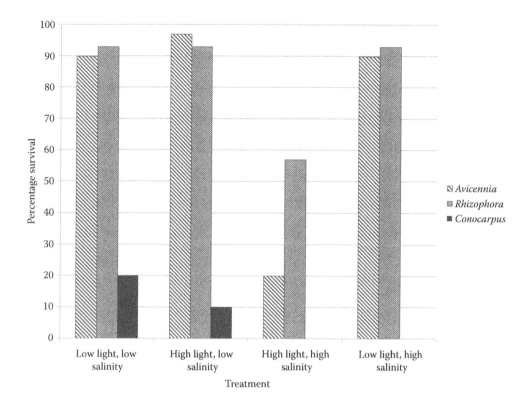

Figure 12.17 Percentage survival of three mangrove species in the four different nursery treatments.

The paired *t*-tests that were used to compare the germination results of the four field quadrats of each species with nursery treatments revealed that nursery survival was significantly greater than field survival only for *R. mangle* (*p*-value = .002). For *A. germinans* (black mangrove) the *p*-value was .1088, while *C. erectus* showed equally poor survival in both field and nursery (*p*-value = .739).

12.5.1.3 Hardening of Plants and Acclimation

Nursery design should also accommodate the removal of shade cloth over specific troughs or have fully exposed troughs to facilitate sun hardening. This is the process of exposing shaded saplings to full sun exposure at least a few months prior to transplanting. This may be done at age 12 months (Clarke and Johns 2002) onward or between 3 and 6 months before the target age.

Salinity acclimation is another key hardening procedure for plants. Saplings must be acclimated to the salinity of the area where they are to be introduced. Exposure to a different salinity will immediately send the majority of saplings into salt shock, resulting in die-offs and low survivals. Acclimation should be a slow process, changing the salinity levels in seedling troughs to a maximum of 5 parts per day. This is achieved by continuously flowing/trickling saltwater (to raise salinity) or freshwater (to lower salinity) into the concerned trough. The flow rate of this introduction will be determined by the size of the trough and the amount of water originally in the trough. A larger standing volume of water will require the introduction of a large volume of water to change the salinity. Therefore, it is a good strategy to reduce the standing volume of water (quickly dumping trough water) prior to changing the salinity. The operator may leave a minimum of 3 cm of water around the plant roots. The water should not be completely discarded as this may result in a rapid salinity change.

Acclimation may be even more simplified if the nursery is operated with troughs of different salinities. Most mangrove forests in Jamaica are usually completely saline or riverine. Therefore, having saplings in both extremes of salinity after the initial 6–12-month optimal growth period is recommended.

12.6 OUTPLANTING AND FIELD TRIALS

The agroecological approach embraced in the restoration of mangroves is further enforced in the application of the field trial component, which will be examined as outplanting exercises in different locations, each with site-specific conditions. Coastal restoration is necessary to correct natural as well as human impacts with hurricanes and storm surges, climate change issues (elevated sea levels and higher wave heights), solid waste destruction, malicious removal/destruction of mangroves, clearing for traditional uses (tourism, fishing, and housing), and pollution/chemical damage, being prime examples. It is important not to confuse wetland restoration, which means to return a degraded wetland to preexisting conditions (Schaller and Sutton 1978), with wetland creation. Wetland creation is the conversion of a nonwetland habitat type into wetlands where wetlands never existed (at least within the recent past, 100–200 years). Wetland creation has occurred worldwide to facilitate dam building and to accommodate dredge spoil, for wastewater treatment or storm water retention (Kusler and Kentula 1990).

The literature suggests that the majority of the world's coastal forest restorations have been unsuccessful, though there is a lack of statistical analyses of these events/experiments due to the nondocumentation of the majority of projects. In the 1980s especially, there were many attempts to restore wetlands, mostly arising from mitigation steps for environmental breaches. It is stated by Lewis (1990) that "hundreds of experiments in the form of restoration and creation projects in the ground have not been subject to even routine success analyses (i.e., did the plants live?)." In fact, the majority of failed mangrove restorations seemed to be hinged on the early concepts of mangrove replanting where broadscale plant introductions were carried out at sites, without the proper baseline studies or the necessary hydrological modifications. Billions of dollars have been wasted trying to grow wetlands because the physical factors were not optimum or they were antagonistic. Other reasons include ignorance of plant/forest succession, misunderstanding of salinity and the importance of plant acclimation (Lewis 1990), no drainage modification or improper modification of drainage patterns, incorrect slope calculations or use of slope values (Beever 1986; Savage 1972, 1979), overdependence on mangroves as erosion buffers, and mangroves subjected to unsuitable wave energies (Teas 1977). Despite this, the fact remains that hydrological engineering alone cannot solve some of the issues in wetland habitat restoration. This is especially true where an area is devoid of parent stock due to historical removal and conflicting environmental factors. There is much to be gained from the agroecology concept and its application in mangrove restoration.

Worldwide evidence and literature demonstrate that nursery-grown seedlings do have a growth advantage over field germination (Ellison 2000; Ravishankar and Ramasubramanian 2004). Plants may be grown in lower salinities to encourage faster initial growth and later acclimated to more saline conditions when being transplanted. High salinity treatment and acclimation, if required at any stage, is best in low light intensities (Clarke 2002; Siddique et al. 1993) of natural shade or shade cloth. In the case of the red mangrove, root and shoot establishment through upright planting of its propagules appears to be the single most important factor for achieving greater seed survival. *Rhizophora* propagules are very heavily predated (Clarke 2002; Smith et al. 1989; Worrall 2003) and are given an advantage when planted vertically in the nursery away from predators and can thereafter be transplanted.

Avicennia seedlings have a physical advantage when transplanted to the field from the nursery. Nursery-grown propagules can be planted closer to intertidal zones. At this stage, their roots will quickly spread into a sufficiently moist substrate as compared with wild propagules, which are easily carried away by tidal action and are affected by agitation, which reduces seed viability, as shown by McMillan (1971). Larger seedlings with a higher number of leaves are hardier and show better survival than younger seedlings (Lewis 1982; Ravishankar and Ramasubramanian 2004). Nursery growth may continue for 3–24 months, depending on the specific size requirements of the restoration area.

12.6.1 Transplanting Considerations

Many physical factors are of higher importance than seedling dynamics (age, position, and stage). Sapling survival in a restoration site is strongly hinged on physical and chemical factors including:

1. Slope
2. Drainage patterns and hydrology
3. Salinity
4. Tides and tidal influences (solid waste, debris)
5. Soil chemistry
6. Local environmental conditions and anomalies (e.g., goats).

Once the desired levels and the standards of these physical and chemical factors have been achieved, there are several steps that should be undertaken to ensure maximum sapling survival and ecological "mimicry." The species of saplings transplanted into an area should be determined by the salinity, tidal influences (black mangroves are more likely to recover following debris destruction and herbivory by animals), hydrological conditions (e.g., black mangroves can withstand higher salinities than red or white), historical vegetation (which species dominated the area previously), and restoration goals (e.g., restoration for animal habitat as compared with erosion control) of the area. It is recommended that simple steps be taken to maximize both sapling survival and sapling growth. Regardless of the restoration goals, optimum sapling growth and maximum survival are common indices of restoration or enhancement success. Steps to ensure good sapling survival include:

1. Using sufficiently aged saplings
2. Using sufficiently hardened saplings
3. Ensuring saplings are acclimated to the correct salinity prior to transplantation
4. Suitable positioning of species in tidal range
5. Suitable species selection
6. Satisfying optimum physicochemical conditions
7. Designing and executing an appropriate management plan (including follow-up, long-term monitoring and posttransplant mitigation)

12.7 CASE STUDIES OF REPLANTING PROCESS

12.7.1 Portland Bight

In the first of two case studies, 100 seedlings were donated to an active nongovernmental organization, Caribbean Coastal Area Management (C-CAM), who conducted the monitoring and shared their results with the Port Royal Marine Laboratory as the donor. Findings suggested incomplete survival data due primarily to the lack of an adequate continuous monitoring program accompanying mangrove replanting. C-CAM provided a 5-month assessment of saplings planting with the results as summarized in Table 12.5, indicating the significantly greater success of the red mangroves (71% survival) over the black mangroves (21% survival).

These results highlighted a number of problems with the replanting and monitoring process. These included lack of proper planning, consultation, and hydrological studies prior to planting—some saplings were planted in swash zones at low-water marks without suitable wave protection. Consequently, seedlings were transported away from the planted location.

There was also confusion between transplanted and naturally occurring recruits. Thus, the survival rate was inaccurate. This was subsequently avoided by flagging transplants at the time of planting. Finally, inadequate monitoring protocol was applied. The transplants should have been monitored at 1-, 3-, 6-, and 12-month intervals. This would give better tracking of the causes of seedling failures. Monitoring should also include basic plant measurements—height and number of leaves/nodes among other parameters.

12.7.2 Montego Bay (Bogue Lagoon)

Four hundred saplings were transplanted in a sheltered estuary in Montego Bay. The 21 month monitoring interval revealed an approximately 40%–68% survival (Tables 12.6 and 12.7). This survival was limited primarily due to the mangrove saplings being outcompeted and overgrown by freshwater plants perhaps as a result of the hydrographic conditions being more ideal for reeds and grasses.

Table 12.5   Summary of Findings of Seedling Assessment Five Months after Planting in February 2011

Species	Status	Location	Numbers	Maximum Survival Rate	Notes
Red mangrove	Live	Windward bay	62	71%	These plants were positively identified.
		Leeward lagoon	9		These plants were positively identified.
	Dead	Windward bay	25		Some may have been from last year.
		Leeward lagoon	15		Some of these may be black mangroves.
Black mangrove	Live	Windward bay	0	21%	No black mangroves in the windward bay.
		Leeward lagoon	21		Some of these may be natural seedlings.
	Dead	Windward bay	0		No black mangroves in the windward bay.
		Leeward lagoon	—		Dead red and black could be confused.
Total			132		

Table 12.6   Transect 1 Data from Montego Bay Replanting

Transect 1 (Closer to River Outflow)	January 2010	May 2010	August 2010	November 2010	November 2011
Height (cm)	41.8	47.3	51.78333	56.45	68.6
Average leaf number	11.1	9.57	9.44	8.13	13.5
Plant count/survival	25	20	18	16	10
% Survival	100	80	72	64	40

Table 12.7   Transect 2 Data from Montego Bay Replanting

Transect 2 (Closer to Tidal Influence)	January 2010	May 2010	August 2010	November 2010	November 2011
Height (cm)	38.3	37.7	41.1	40.0	60.8
Average leaf number	10.1	9.9	11.6	18.8	73
Plant count/survival	32	28	24	24	21
% Survival	100	86	75	75	68

12.8 CONCLUSIONS

The agroecological combination of seed banks, nursery propagation, and outplanting technology has and will result in successful and significant mangrove restoration and eventually sustainable coastal ecosystems. The creation of seed banks increases the regenerative capacity immeasurably as all seeds secured by the careful collection process would have otherwise been lost to the environment. With over 6000 seedlings secured by these methods and later germinated at the Port Royal Marine Laboratory Coastal Plant Nursery, the agroecological approach offers significant potential to attain a sustainable coastal ecosystem. Furthermore, due to the seed bank's ability to hold seeds for some weeks in various salinities, it offers additional potential to plant seedlings when, as well as where necessary, as a valuable ecological tool.

The nursery experiment and field observations have further proven that nursery-grown seedlings do have a growth and survival advantage over field germination as supported by Ellison (2000) and Ravishankar and Ramasubramanian (2004). *A. germinans* seedlings though showing the best field germination are given a physical advantage when transplanted to the field from the nursery.

Finally, the outplanting exercises demonstrate that survival and coastal establishment is possible and feasible with red mangroves exhibiting 71%–40% survival and black mangroves exhibiting 21% survival. While these figures appear low, they must be viewed against the backdrop of few of the other required parameters (e.g., drainage, slope, tides, salinity, and chemistry) being managed. Furthermore, these figures represent the survival of new mangroves that would not have been established without agroecological intervention.

The authors wish to add a note of caution that successful mangrove forest restoration requires careful presite preparation to ensure that the conditions are optimal for the survival of the plants. There are many documented cases of restoration attempts using mangrove transplanting without proper site investigation or preparation. For example, in 1989, over 9000 ha of saplings were planted in West Bengal, India, but only 1.52% survived (Lewis 1990). Transplanting to areas that do not have optimal conditions for mangrove forest growth can be very wasteful or will require rigorous monitoring and handling of outplants to ensure survival.

Detailed information on the history (type of mangrove forest previously existing), hydrology (water supply and drainage), the potential for natural regeneration, and the nature of the stresses likely to impact plants is essential to success.

REFERENCES

Adams, C. D. 1972. *Flowering Plants of Jamaica*. Glasgow: The University Press.

Alleng, G. P. 1990. Historical development, present status and management guidelines for the Port Royal mangal, Jamaica. MSc thesis, The University of the West Indies.

Altieri, M. A. 2008. Agroecology: Principles and strategies for designing sustainable farming systems. Agroecology in Action Project, University of California, Berkeley. http://nature.berkeley.edu/~miguel-alt/what_is_agroecology.html (accessed October 24, 2012).

Beever, J. W. 1986. Mitigative creation and restoration of wetland systems: A technical manual for Florida draft report. Tallahassee, FL: Florida Department of Environmental Regulation.

Cintron-Molero, G. 1990. *Restoration of Mangrove Systems*. Puerto Rico: Department of Natural Resources.

Clarke, A. 2002. *Mangrove Nurseries: Construction, Propagation and Planting*. Queensland: Queensland Fisheries Service.

Clarke, A. and L. Johns. 2002. Mangrove nurseries: Construction, propagation and planting: Fisheries Guidelines, Department of Primary Industries, Queensland, Fish Habitat Guideline FHG 004.

Clarke, P. J. and R. A. Kerrigan. 2002. The effects of seed predators on the recruitment of mangroves. *J Ecol* 90:728–736.

Cooper, E., L. Burke, and N. Bood. 2009. Coastal capital: Belize. The economic contribution of Belize's coral reefs and mangroves. WRI Working Paper. Washington DC: World Resource Institute. http://pdf.wri.org/working_papers/coastal_capital_belize_wp.pdf (accessed November 12, 2012).

Davis, J. H. 1940. *The Ecology and Geologic Role of Mangroves in Florida*, pp. 303–412. Washington DC: Carnegie Institution Washington, Tortugas Laboratory, Publication Number 517.

Duke, N. C., M. C. Ball, and J. C. Ellison. 1998. Factors influencing biodiversity and distributional gradients in mangroves. *Global Ecol Biogeogr* 8:95–115.

Duke, N. C., E. Y. Y. Lo, and M. Sun. 2002. Global distribution and genetic discontinuities of mangroves—Emerging patterns in the evolution of *Rhizophora*. *Trees* 16:65–79.

Ellison, A. M. 2000. Mangrove restoration: Do we know enough? *Restor Ecol* 8(3):219–229.

Ellison, A. M., B. B. Mukherjee, and A. Karim. 2000. Testing patterns of zonation in mangroves: Scale dependence and environmental correlates in the Sundarbans of Bangladesh. *J Ecol* 88:813–824.

FAO (Food and Agricultural Organization). 1994. Mangrove forest management guidelines. Forestry Paper 117, Roma: FAO.

Francis, J. K. and A. Rodriguez. 1993. Seeds of Puerto Rican trees and shrubs: Second installment. Research Note SO-374. US Southern Forest Experiment Station. Department of Agriculture, Forest Service, New Orleans, LA.

Gilman, E. L., J. Ellison, N. C. Duke, and C. Field. 2008. Threats to mangroves from climate change and adaptation options: A review. *Aquat Bot* 89:237–250.

Hemminga, M. A., F. J. Slim, J. Kazungu, G. M. Ganssen, J. Nieuwenhuize, and N. M. Kruyt. 1994. Carbon outwelling from a mangrove forest with adjacent seagrass beds and coral reefs (Gazi Bay, Kenya). *Marine Ecol Prog Ser* 106:291–301.

Hogarth, P. J. 2007. *The Biology of Mangroves and Seagrasses*. New York: Oxford University Press.

Hoilett, K. and M. K. Webber. 2002. Can mangrove root communities indicate variations in water quality? *Jamaican J Sci Technol* 12–13:16–34.

Kathiresan, K. and B. L. Bingham. 2001. The biology of mangroves and mangrove ecosystems. *Adv Mar Bio* 40:81–251.

Kusler, J. A. and M. E. Kentula. 1990. *Wetland Creation and Restoration: The Status of the Science*. Washington, DC: Island Press.

Lewis, R. R (ed.). 1982. Mangrove forests. In *Creation and Restoration of Coastal Plant Communities*, pp. 153–171. Boca Raton: CRC Press.

Lewis, R. R. 1990. Creation and restoration of coastal plant wetlands in Florida. In J. A. Kusler and M. E. Kentula (eds.), *Wetland Creation and Restoration: The Status of the Science*, pp. 73–101. Washington, DC: Island Press.

Lopez-Portillo, J. and E. Ezcurra. 1989. Zonation in mangrove and salt marsh vegetation at Laguna de Mecoacan, Mexico. *Biotropica* 21(2):107–114.

Lugo, A. E. and S. C. Snedaker. 1974. The ecology of mangroves. *Ann Rev Ecol Syst* 5:39–64.

Macintosh, D. J. and E. C. Ashton. 2002. A review of mangrove biodiversity conservation and management. Aarhus: Centre for Tropical Ecosystems Research, University of Aarhus. www.researchgate.net/publication/.../3deec529f8b325bf49.pdf (accessed July 15, 2014).

MacNae, W. 1968. A general account of the fauna and flora of swamps and forests in the Indo-West-Pacific Region. *Adv Mar Biol* 6:73–270.

McDonald, K. O., D. F. Webber, and M. K. Webber. 2003. Mangrove forest structure under varying environmental condition. *Bull Mar Sci* 73:496–501.

McMillan, C. 1971. Environmental factors affecting seedling establishment of the black mangrove on the central Texas coast. *Ecology* 52:927–930.

Mumby, P. J., A. J. Edwards, J. E. Arlas-Gonzalez et al. 2004. Mangroves enhance the biomass of coral reef fish communities in the Caribbean. *Nature* 427:533–536.

Nieves-Rivera, A. M., T. A. Tattar, and E. H. Williams Jr. 2002. Sooty mould-planthopper association on leaves of the black mangrove *Avicennia germinans* (L.) Stearn in southwestern Puerto Rico. *Arboric J* 26:141–155.

Nybakken, J. W. and M. D. Bertness. 2005. *Marine Biology: An Ecological Approach*, 6th edn. San Francisco: Pearson Education.

Odum, W. E., C. C. McIvor, and T. J. Smith III. 1982. *The Ecology of the Mangroves of South Florida: A Community Profile*. Washington DC: United States Fish and Wildlife Service, Office of Biological Services, FWS/OBS-81/24.

Pool, D. J., S. C. Snedaker, and A. E. Lugo. 1977. Structure of mangrove forests in Florida, Puerto Rico, Mexico, and Costa Rica. *Biotropica* 9:195–212.

Rabinowitz, D. 1978. Early growth of mangrove seedlings in Panama and a hypothesis concerning the relationship of dispersal and zonation. *J Biogeogr* 5:113–133.

Ramsar Secretariat. 2001. *Wetland Values and Functions: Climate Change Mitigation*. Gland, Switzerland. http://www.ramsar.org/cda/ramsar/display/main/main.jsp?zn=ramsar&cp=1-26-253%5E22199_4000_0__.

Ravishankar, T. and R. Ramasubramanian. 2004. Manual on mangrove nursery raising techniques. Chennai: M.S. Swaminathan Research Foundation. http://www.drcsc.org/VET/library/Nursery/Mangrove_Nursery_manual_HR.pdf (accessed March 24, 2012).

Riley, W. R. Jr. and C. P. S. Kent. 1999. Riley encased methodology: Principles and process of mangrove habitat creation and restoration. *Mangroves Salt Marshes* 3:207–213.

Robertson, A. I. and D. M. Alongi. 1992. *Coastal and Estuarine Studies: Tropical Mangrove Ecosystems*. Washington, DC: American Geophysical Union.

Savage, T. 1972. Florida mangroves as shoreline stabilizers. Florida Department of Natural Resources Professional Papers Series No. 19. In R. R. Lewis (ed.), *Creation and Restoration of Coastal Plain Wetlands in Florida*. Boca Raton: CRC Press.

Savage, T. 1979. The 1972 experimental mangrove planting—An update with comments on continued research needs. In R. R. Lewis (ed.), *Creation and Restoration of Coastal Plain Wetlands in Florida*, pp. 43–71. Boca Raton: CRC Press.

Schaller, F. W. and P. Sutton (eds). 1978. *Reclamation of Drastically Disturbed Lands*. Madison: American Society of Agronomy.

Sheridan, P. F. 1992. Comparative habitat utilization by estuarine macrofauna within the mangrove ecosystem of Rookery Bay, Florida. *Bull Mar Sci* 50:21–39.

Siddique, N. A., M. R. Islam, M. A. M. Khan, and M. Shahidulla. 1993. Mangrove nurseries in Bangladesh. Mangrove Ecosystem Occasional, Papers 1. Okinawa: ISME.

Smith, T. J. III, H. T. Chan, C. C. McIvor, and M. B. Robblee. 1989. Comparisons of seed predation in tropical, tidal forests from three continents. *Ecology* 70:146–151.

Sutton, A. H., L. G. Sorenson, and M. A. Keeley. 2004. *Wondrous West Indian Wetlands: Teachers' Resource Book*. Boston: West Indian Whistling-Duck Working Group of the Society of Caribbean Ornithology.

Teas, H. J. 1977. Ecology and restoration of mangrove shorelines in Florida. *Environ Conserv* 4:51–58.

Thompson, H. P. and D. F. Webber. 2003. The sand dune ecology of the Palisadoes, Kingston Harbour, Jamaica. *Bull Mar Sci* 73:507–520.

Tomlinson, P. B. 1986. *The Botany of Mangroves*. Cambridge: Cambridge University Press.

Trench, C. A. 2007. Enhancing mangrove forest restoration by nursery propagation. MSc thesis, The University of the West Indies.

Trench, C. A. and M. K. Webber. 2011. Nursery propagation of Jamaican coastal forest species. *Acta Hort* 894:185–190.

Valiela, I., J. L. Bowen, and J. K. York. 2001. Mangrove forests: One of the world's threatened major tropical environments. *BioScience* 51:807–815.

Warner, G. F. 1969. The occurrence and distribution of crabs in Jamaican mangrove swamp. *J Anim Ecol* 38:379–389.

Wells, S., C. Ravilous, and E. Corcoran. 2006. *In the Front Line: Shoreline Protection and Other Ecosystem Services from Mangroves and Coral Reefs*. Cambridge: United Nations Environment Programme World Conservation Monitoring Centre.

Wojtkowski, P. A. 2002. *Agroecological Perspectives in Agronomy, Forestry and Agroforestry*. Enfield: Science Publishers.

Wojtkowski, P. A. 2006. *Introduction to Agroecology: Principles and Practices*. New York: Food Products Press.

Woodroffe, C. D. 1992. Mangrove sediments and geomorphology. In D. Alongi and A. Robertson (eds), *Coastal and Estuarine Studies: Tropical Mangrove Ecosystems*, pp. 7–41. Washington, DC: American Geophysical Union.

Worrall, S. 2003. The effects of grapsid crabs on mangrove forest restoration. *Restor Reclam Rev* 8:1–6. http://conservancy.umn.edu/bitstream/60234/1/8.4.Worrall.pdf (accessed March 20, 2012).

CHAPTER 13

Suggesting an Interdisciplinary Framework for the Management of Integrated Production and Conservation Landscapes in Transfrontier Conservation Areas of Southern Africa

Munyaradzi Chitakira, Emmanuel Torquebiau, Willem Ferguson, and Kevin Mearns

CONTENTS

13.1 Introduction 266
13.2 Conceptual Underpinnings 266
    13.2.1 Biodiversity–Agriculture Integration Challenge 266
    13.2.2 Transfrontier Conservation Areas Model 267
    13.2.3 The Ecoagriculture Concept 267
    13.2.4 Ecoagriculture in Practice 268
13.3 Study Objectives 268
13.4 Materials and Methods 268
    13.4.1 Study Area 268
    13.4.2 Methods 270
13.5 Results and Discussion 270
    13.5.1 Potential Benefits of Systematic Ecoagriculture Strategies to TFCA Communities ...270
    13.5.2 Anticipated Challenges to Systematic Ecoagriculture 271
        13.5.2.1 Landscape-Scale Challenge..271
        13.5.2.2 The Nature of the Local Communities 271
        13.5.2.3 Location and Infrastructure 272
        13.5.2.4 Policy and Governance Issues 272
        13.5.2.5 The Interdisciplinary and Cross-Sectoral Nature of Ecoagriculture.........273
    13.5.3 How Ecoagriculture Relates to Agroecology and Related Sustainable Farming Approaches 274
13.6 Conclusion 274
Acknowledgments..274
References..275

13.1 INTRODUCTION

It is recognized that the interlinkages between society and the environment are more clearly defined than before and that the world needs approaches that cut across multiple fields of expertise to overcome future challenges (Davenport 2008; Marcu 2007). We suggest an interdisciplinary framework called ecoagriculture as a sustainable approach to managing landscapes designated for simultaneous biodiversity conservation, agricultural production, and livelihood improvement in local communities. Research has revealed positive impacts of integrated production and conservation systems on peasant communities in western Africa (Asaah et al. 2011). Examples include the feeling of empowerment from increased knowledge and success, recognition of a way out of poverty, improved nutrition, better health, and increased income.

The current study seeks to assess the benefits of systematic (or planned) ecoagriculture strategies to small-scale farmers within a transfrontier conservation area (TFCA) context. Based on a case study, we discuss how ecoagriculture provides opportunities for addressing conservation and livelihood challenges in the TFCAs of southern Africa. We acknowledge the potential challenges of ecoagriculture planning under the prevailing socioeconomic conditions and suggest how these can be addressed. In the latter part of the chapter, we discuss the position of the ecoagriculture framework relative to other disciplines of study and landscape management approaches. Particular attention is drawn to the relationship between ecoagriculture and agroecology, in recognition of some tension that surfaced between the proponents of these two approaches.

13.2 CONCEPTUAL UNDERPINNINGS

13.2.1 Biodiversity–Agriculture Integration Challenge

Protected areas (PAs) have been the major strategy for safeguarding the world's biodiversity and are still an essential part of conservation programs, particularly for sensitive habitats (Millenium Ecosystem Assessment Board 2005). However, since 40% of the earth's land area is under agriculture and only 12% is protected, the strategy of PAs alone is not sufficient to guarantee the conservation of the full range of biodiversity, particularly the numerous species in areas inhabited and utilized by human beings (Millennium Ecosystem Assessment Board 2005; Perfecto and Vandermeer 2010; Shames and Scherr 2009; Tscharntke et al. 2012). A matrix of natural habitats (such as PAs or other forms of nature reserves) connected to seminatural or managed habitats (particularly human-inhabited areas) is required (Linnell et al. 2005; Persha et al. 2010). It is increasingly acknowledged that effective biodiversity conservation is achievable through combining the strategies of "land sparing" (in which some land is set aside for conservation while other land is used intensively for agriculture) and "land sharing" (in which less land is set aside specifically for biodiversity conservation while less intensive production techniques are implemented to allow some biodiversity on agricultural areas) (Fischer et al. 2014).

Research has shown that biodiversity and agriculture are strongly interdependent (Méndez et al. 2007; Pretty 2008; Toledo and Moguel 2012). However, agriculture has tended to focus on maximizing yields from a few selected species, and this negatively impacts on biodiversity. In southern Africa, purely production-oriented agricultural practices have contributed to degradation of land, soil, and local ecosystems, thereby threatening the livelihoods of the farmers, who largely depend on natural resources (Buck and Scherr 2011; Chitakira 2011). Agriculture has thus become a major driver of biodiversity loss. To reverse this situation, there is a need to restore sustainable land management practices.

Attempts to achieve a sustainable balance between conservation and agricultural objectives have seen the emergence of land use or production strategies such as agroecology, agroforestry, organic farming, and, more recently, ecoagriculture (Gliessman 2004; Müller-Lindenlauf 2009; Nair 2007; Perfecto et al. 2009; Scherr and McNeely 2008). Efforts at the regional level have also emerged, in which national governments sharing common borders have collaborated by creating transboundary parks or TFCAs.

13.2.2 Transfrontier Conservation Areas Model

TFCAs are land areas that cross political boundaries between two or more countries and include one or more PAs and multiple resource use areas (SADC 1999). TFCAs are a model of multifunctional landscapes intended to achieve conservation, development, and sociocultural goals simultaneously through international cooperation. This model, however, faces a series of challenges, mainly because there are intrinsic contradictions. Historically, the conservation of biodiversity resources has been at odds with development, and it is only recently that some initiatives (e.g., communal areas natural resources management programs) have shown that these two objectives are not necessarily antagonistic (Murphree 2009; Taylor 2009). It is a challenge to have common conservation development objectives across national borders because socioeconomic conditions between countries often differ extensively. As such, many TFCAs exist mainly on paper with very minimal implementation.

13.2.3 The Ecoagriculture Concept

Ecoagriculture refers to "integrated conservation-agriculture landscapes where biodiversity conservation is an explicit objective of agriculture and rural development, and the latter are explicitly considered in shaping conservation strategies" (Scherr and McNeely 2007). A key term in this definition is *landscape*, defined as a spatially heterogeneous area often comprising mosaics of patches with varying size, shape, composition, and history (Wu and Qi 2000). An ideal ecoagriculture landscape is heterogeneous and multifunctional. Such landscapes are meant to achieve three interlinked goals in the same space and time: (1) to develop more sustainable and productive agricultural systems, (2) to conserve and enhance biodiversity and ecosystem services, and (3) to support viable local livelihoods. Ecoagriculture can improve the natural assets of poor people by protecting wild species important to their health and livelihoods, by ensuring the provision of environmental services critical to the people's livelihoods, and by supplementing incomes with biodiversity payments (McNeely and Scherr 2003). Countries that require a national climate change response strategy to integrate the programs of various government departments and to support key government objectives such as poverty alleviation may find an answer in the ecoagriculture approach. South Africa, for instance, requires such a response strategy (DEAT 2004). Ecoagriculture needs to be recognized as a sustainable climate change adaptation and mitigation strategy that meets various national goals (e.g., food security). Countries may consider promoting ecoagriculture implementation as a solution to the climate change challenge (Ecoagriculture Partners 2008b). Supportive legislative and governance regimes are essential for successful ecoagriculture implementation in a TFCA.

Ecoagriculture creates landscape mosaics that are balanced in terms of food production and environmental protection and that support viable local livelihoods. From a social perspective, ecoagriculture promotes diverse crop, livestock, tree, and wild species that enhance livelihood security and income generation by creating commercial opportunities in agriculture and other sectors. As noted by the Ecoagriculture Partners (2008a), ecoagriculture involves a variety of land management approaches that include:

1. Protecting and promoting local crop and livestock diversity
2. Mixing trees and crops
3. Maintaining connectivity between natural habitats within agricultural landscapes
4. Planting hedgerows around cultivated fields
5. Protecting watersheds with perennial natural and planted vegetation
6. Maintaining all-year-round soil cover to enhance rainfall infiltration

7. Managing inputs and wastes to minimize agricultural pollution of natural habitats
8. Designing farming systems to resemble natural ecosystems in terms of structure and function

13.2.4 Ecoagriculture in Practice

Several rural communities worldwide practice ecoagriculture, although not with the same name. Many farmers have adopted innovative strategies for conserving the natural resources important to their livelihoods. Example are the projects for enhancing agricultural productivity implemented on the margins of Kakamega Forest in Kenya (Ecoagriculture Partners 2009) and a holistic landscape management strategy with livestock as an integral part of the landscape in the Hwange communal lands in Zimbabwe (Neely and Butterfield 2004).

The ecoagriculture concept relates closely to UNESCO's Man and the Biosphere (MAB) Programme, intended to promote solutions to reconcile biodiversity conservation with its sustainable use (UNESCO 2014). Within the framework of the MAB Programme, areas of terrestrial and coastal marine ecosystems known as biosphere reserves are nominated by national governments for inclusion in the World Network and remain under the sovereign jurisdiction of the states where they occur. The principles of biosphere reserves deal with multifunctional landscapes in ways that support ecoagriculture practices, and, in turn, ecoagriculture can promote the MAB Programme through integrating biodiversity conservation, farming, and resource utilization in a particular landscape.

13.3 STUDY OBJECTIVES

The study aims to assess how ecoagriculture can be a sustainable interdisciplinary or cross-sectoral approach to managing multifunctional landscapes designated for simultaneous biodiversity conservation and agricultural production leading to improved livelihoods in local communities. The study attempts to address the following interrelated questions:

1. What benefits could systematically managed ecoagriculture bring to a TFCA and other communal areas under similar conditions?
2. In what ways can ecoagriculture practices meet the local livelihood and ecological goals?
3. Is the "commitment and capacity" required to build ecoagriculture landscapes available in the TFCAs?
4. What are the challenges posed by the interdisciplinary/cross-sectoral nature of ecoagriculture, and how can these be dealt with (under given circumstances)?

13.4 MATERIALS AND METHODS

13.4.1 Study Area

The ensuing discussion is based on a case study of a small-scale farming area in southern Africa known as the Mathenjwa Tribal Authority (MTA). The area is part of the Lubombo TFCA, which lies across the borders of South Africa, Mozambique, and Swaziland, stretching from 26°48′ to 26°57′ south and from 32°00′ to 32°10′ east (Figure 13.1). The area falls within the savanna biome characterized by endemic flora and is recognized at the global level as a biodiversity hotspot (Van Wyk and Smith 2001). The inhabitants are generally poor and rely significantly on harvesting natural resources, apart from government social assistance grants. A possible response to this situation is to implement landscape management strategies that simultaneously achieve conservation and livelihood goals. The MTA comprises a mosaic of unplanned ecoagriculture involving spontaneous practices such as traditional tree–crop combinations, grass strip contours, and hedgerows (see Figure 13.2).

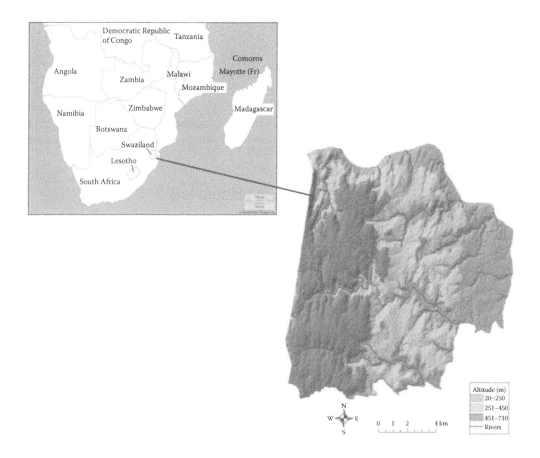

Figure 13.1 Study area location.

Figure 13.2 Nature of ecoagriculture in the MTA landscape.

13.4.2 Methods

The chapter is based on information obtained from key informants and stakeholders regarding agriculture, biodiversity, and livelihoods in the Lubombo TFCA (including local small-scale farmers, administrative personnel, and extension officers in public and private organizations). Data were collected through semistructured interviews of key informants and stakeholders ($n = 19$), questionnaire survey of farmers' households ($n = 170$), qualitative assessment of the environment (direct field observation), quantitative land-cover assessment (from satellite imagery), and policy analysis.

13.5 RESULTS AND DISCUSSION

13.5.1 Potential Benefits of Systematic Ecoagriculture Strategies to TFCA Communities

About 75% of the MTA landscape is under natural vegetative cover (Fleury 2011). The landscape is characterized by "used" land (e.g., built-up, crop fields, or roads) and interconnected patches of near-pristine natural forest, woodland, and grassland. In this condition the landscape can support significant wild and agricultural biodiversity.

However, as the human population density and demand for food and ecosystem services increase, and with climate change and variability threats, the sustainability of these unplanned agriculture–biodiversity integration practices is under threat. More systematic and better-managed approaches are, therefore, recommended. Ecoagriculture planning under the near-pristine land-cover conditions of the MTA could enhance the landscape's capacity to support biodiversity. The area's potential to produce food, conserve biodiversity, and sustain ecosystem services can be enhanced by transforming the unsystematic land management practices into planned and monitored systems. Unlike the existing unsystematic practices, the planned systems could be evaluated occasionally by stakeholders (including local farmers) and the necessary regulative measures taken.

This study identified some potential benefits of implementing ecoagriculture systematically in the MTA. Ecoagriculture could enhance resilience and self-sufficiency and create several socioeconomic and environmental benefits for the communities in the following ways:

1. Addressing production problems: improving yields through careful selection of crops or livestock combinations and adopting strategies to reduce production costs. The communities become resilient and living standards are improved when the farmers can produce more food for themselves.
2. Enhanced quality of life through conserving ecosystem services that are important for local livelihoods.
3. Employment creation and income generation leading to poverty reduction, such as through operating rural tourism projects or sale of sustainably managed forest products.
4. Farmers earning extra income from the sale of ecosystem services, for example, carbon credits and payments for environmental services from consumers or carbon-emitting organizations in the towns and cities nearby. Payments for environmental services compensate landowners for management strategies that provide ecosystem benefits to other parties but that in some way limit their own opportunities for generating revenue (Ghazoul et al. 2009). Such income can be used to purchase agricultural inputs, clothes, and supplementary food, or to pay for educational and medical services, thereby improving living standards.
5. Organic production of poultry, livestock, fruits, vegetables, cereals, and other produce. This can lead to product certification and possibly landscape eco-labeling. Certification and eco-labeling processes are associated with better product quality and higher market value. They enable better returns for the peasant farmers. Research on the feasibility of labeling the MTA as an ecoagriculture landscape showed that the area scored 0.66 on a scale ranging from 0 to 1, which implies a

high performance by the landscape, particularly in its ecosystem services provision aspect (Cholet 2010). Product certification and eco-labeling thus stand a good chance in the area. Other research confirmed a high feasibility for landscape labeling in the Lubombo TFCA (Torquebiau et al. 2012). Landscape labeling recognizes that the ecoagriculture attributes of an area add value to a range of commodities (e.g., crops or animals) and services (e.g., ecotourism or biodiversity management) and is more holistic than product labeling. Leading international retailers operating in the area, such as Woolworths, promote the production and marketing of organic products by small-scale farmers (Liquidlingo Communications 2009; SAinfo Reporter 2008). One must be careful not to overemphasize the effectiveness of agricultural certification, given the limited evidence for the benefits of certification to the environment or producers (Blackman and Rivera 2011). Cohn and O'Rourke (2011) argue that certification is a poor substitute for strong government policies and that consumers and other supply chain actors scarcely care about conservation outcomes.

An analysis of the landscape shows that systematic ecoagriculture can provide opportunities to address conservation problems in the Lubombo TFCA through:

1. Encouraging compliance with environmental regulations, particularly as local farmers realize the benefits of conserving biodiversity.
2. Protecting the rights to collect products from neighboring PAs, in particular the Ndumo Game Reserve and Usuthu Gorge Community Conservation Areas (in South Africa), the Usuthu area (in Swaziland), and the Maputo Special Reserve and the Futi Corridor (in Mozambique). Poaching problems are expected to drop significantly when local community members have legal access to resources in the PAs.
3. Reduced conflicts between local communities and conservation authorities may be possible through the preceding conditions and perceived benefits, creating a win–win situation (see Lele et al. 2010).
4. Protecting areas valued for their cultural or spiritual importance, particularly to the local communities.

13.5.2 Anticipated Challenges to Systematic Ecoagriculture

There are several challenges confronting ecoagriculture planning and implementation in the MTA. These challenges are related to the landscape scale, nature of peasant communities, location and level of infrastructure development, and policy and governance regimes.

13.5.2.1 Landscape Scale Challenge

Planning at the landscape level attempts to harmonize the plans and goals of various stakeholders while balancing economic development and conservation effort in a given geographic area (LMRC 2014). Coordinated planning by different key stakeholders (farmers, government agencies, civic groups, development planners, etc.) requires considerable effort and facilitation skills, which may not be available. Stakeholders need to agree on common goals. Having to harmonize the disparate goals and often conflicting interests of different stakeholders is a challenge.

13.5.2.2 The Nature of the Local Communities

Active involvement of local community members in ecoagriculture planning is crucial for successful implementation (Chitakira et al. 2012). However, the predominantly high illiteracy levels present a challenge. As many as 42% of surveyed household heads in MTA had never been to school, and another 35% had not gone beyond primary education. The limited scientific understanding of ecosystem functioning among these farmers may limit their capability to engage in integrated land use planning. Efforts to promote integrated production and conservation landscape management need to recognize the diversity of landscape challenges and apply locally adapted solutions.

The capacity to create locally adapted solutions in this regard in the smallholder farming areas of southern Africa may be limited because of high illiteracy rates and limited technical expertise. Provision of extension services and guidance from environmental practitioners is expected to help deal with this challenge. Providing training and technical assistance could develop and strengthen local organizational capacities.

13.5.2.3 Location and Infrastructure

A challenge revealed during interview and questionnaire surveys was the lack of access to markets for local produce. The area under focus is along international borders and is peripheral to centers of major economic development, which may explain why its infrastructure is relatively poorly developed. A lack of good transport and marketing infrastructure makes it difficult for the farmers to take their surplus produce to urban or export markets. This study identifies rural tourism management as a possible enterprise into which the local farmers could venture to diversify livelihoods and improve resilience in the community. However, the Lubombo TFCA communities, like many communal areas in southern Africa, do not have the capacity to run business enterprises such as tourism. To improve this situation, capacity building is recommended. Yet the necessary financial, technical, and other resources are not locally available. One possible way out of this situation is the extra money to be made out of landscape labeling or intervention by the central government (if this is identified as a priority) and external agencies. If the local farmers are assisted in acquiring the necessary skills and awareness, rural tourism could be feasible in the area. Small entrepreneurial projects that build on local resources (e.g., culture and handcrafts) and do not need substantial sums of capital to be operationalized can be promoted (Briedenhann and Wickens 2004).

13.5.2.4 Policy and Governance Issues

Natural resources in communal areas, including the Lubombo TFCA, are not freely accessible to anybody (open access resources) but are owned by defined member groups that have access rights. The "free rider" problem associated with shared resources (when some members consume more than their fair share of a publicly owned natural resource) can lead to excessive use and degradation of the resources in the community (Adhikari 2001; McKean and Ostrom 1995). A lack of coordination between the traditional leadership and the local government authorities makes it almost impossible to monitor and control unsustainable harvesting of natural resources or land uses such as encroachment and overgrazing. The governments of the three nations sharing the area (South Africa, Mozambique, and Swaziland) might need to agree on a plan to coordinate, monitor, and control land uses and access to natural resources within the TFCA. Gaps in policies and legal and institutional frameworks exist between these three nations. This situation derails landscape-level planning in the Lubombo TFCA, and policy harmonization is needed. Harmonizing the legislation and policies of more than two countries is a complex process, since the sovereignty of each country must be maintained and respected (Ron 2007).

Building ecoagriculture landscapes demands significant commitment in developing new skills, capacities, tools, and policies (see Scherr 2011). International cooperation on TFCA management requires consistent commitment in policy implementation. Research preceding this study produced predominantly positive results for the feasibility of ecoagriculture planning and implementation in the Lubombo TFCA (Chitakira 2013). However, a major challenge is how to ensure the commitment of the various stakeholders in supporting ecoagriculture initiatives. In the tripartite protocols of the Lubombo TFCA there are no clauses to enforce commitment to appropriate action by the parties involved, and this is a cause for concern.

13.5.2.5 *The Interdisciplinary and Cross-Sectoral Nature of Ecoagriculture*

The management of integrated production and conservation landscapes requires diverse skills and thus needs to be achieved through interdisciplinary approaches. The "science" or "art" of eco-agriculture is a meeting point of various disciplines, concepts, approaches, and goals (Figure 13.3).

Due to its multidisciplinary nature, ecoagriculture implementation promotes stakeholder collaboration. Different stakeholders from various backgrounds and across the scientific and sociocultural spectra need to bring together their experiences and expertise to develop coordinated approaches and tools to achieve related agricultural, biodiversity, and livelihood goals at the landscape scale. Ecoagriculture planning in a TFCA requires diverse expertise, and so brings together local farmers (indigenous knowledge systems), researchers (scientific perspective), practitioners, and local and international stakeholders. To get different stakeholders to work together is, however, a challenge. Collaborative research, for instance, is threatened by the mainstream thinking that tends to separate culture and nature, rather than considering human–environmental interaction as a dynamic and interpenetrative engagement (Strang 2009). The good thing, however, is that the concept of integration is gaining popularity. Given sufficient investments of time and energy, supportive policies, committed leadership, effective communication, and good facilitation skills, successful stakeholder collaboration in ecoagriculture landscape management is feasible.

Apart from bridging disciplines, ecoagriculture emphasizes sectoral integration and encourages the harmonization of policy agendas such as agriculture, food security, poverty alleviation, climate variability, ecosystems, indigenous knowledge systems, marketing, and governance approaches.

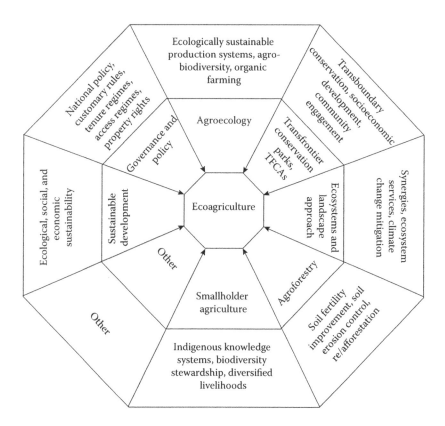

Figure 13.3 Ecoagriculture links disciplines, approaches, and goals. TFC, transfrontier conservation; TFCAs, transfrontier conservation areas.

There may be a need to review the existing applicable policies and make adjustments to permit viable sectoral integration. Biodiversity–agriculture integration policies must recognize local farmers as key stewards of ecosystems and empower them to conserve biodiversity, raise production, utilize natural resources, increase incomes, and manage climate risks.

13.5.3 How Ecoagriculture Relates to Agroecology and Related Sustainable Farming Approaches

The ecoagriculture concept has received intense criticism, particularly from leading proponents of agroecology (see Altieri 2004, 2012). The criticism is mainly attributed to a misconception of the vision of the ecoagriculture framework and the philosophy surrounding it (see Ecoagriculture Partners 2004). From an interdisciplinary perspective, the current study recognizes the significant overlaps existing between ecoagriculture, agroecology, organic agriculture, agroforestry, permaculture, and many other innovative approaches to sustainable farming.

The values and principles of all these approaches have much in common and they tend to promote, *at varying scales*, the sustainability of agricultural production systems, ecosystem services, and wild or agricultural biodiversity as well as enhancing local livelihoods. While agroecology is applicable at the small scale (e.g., plot or individual farm level) and large scale (e.g., bioregion), ecoagriculture operates at the *landscape* scale (encompassing several farms and the surrounding ecosystems). Ecoagriculture is distinguished by the following characteristics: (i) large scale—the detection and planning for interactions among different land uses are effectively carried out at the landscape scale; (ii) emphasis on the need to foster synergies among conservation, agricultural production, and rural livelihoods; (iii) emphasis on stakeholder collaboration, since ecoagriculture cannot be achieved by the efforts of individual land managers; and (iv) recognition of the importance of healthy natural ecosystems in supporting sustainable agricultural production (Ecoagriculture Partners 2014).

Although the various approaches to sustainable agriculture identified above may differ, it can be noted that the difference is mainly in scale and focus rather than in objective. As such, they need to be seen as playing complementary rather than antagonistic roles.

13.6 CONCLUSION

The present case study has shown that ecoagriculture planning and implementation can stimulate a wide range of opportunities that promote agricultural production and livelihoods while conserving biodiversity in the TFCA communities. It has shown that there are, however, factors that pose challenges to the building of ecoagriculture landscapes and their sustainable management. In this chapter we demonstrated how the "science" or "art" of ecoagriculture becomes a meeting point of various disciplines, concepts, approaches, and goals. The interdisciplinary and cross-sectoral nature of ecoagriculture has both strengths and limitations. Perhaps land managers need to take advantage of these strengths and find ways to deal with the limitations in order to achieve the goals of integrated landscape management.

ACKNOWLEDGMENTS

We appreciate the funding by the International Foundation for Science (IFS), the National Research Foundation (South Africa), Centre de coopération internationale en recherche agronomique pour le développement (CIRAD) (France), and the University of Pretoria. We also thank Ezemvelo KwaZulu-Natal Wildlife and Wildlands Conservation Trust for logistical support during data collection. A special thanks to the people of Mathenjwa community for cooperation and to all who offered their support in this study.

REFERENCES

Adhikari, B. 2001. Literature review on the economics of common property resources: Review of common pool resource management in Tanzania. NRSP Project R#7857 Report, University of York. http://www.nrsp.org.uk/database/documents/804.pdf (accessed: 27 February 2014).

Altieri, M. A. 2004. Agroecology versus ecoagriculture: Balancing food production and biodiversity conservation in the midst of social inequity. CEESP occasional papers 3. IUCN. http://www.wildfarmalliance.org/resources/ECOAG.pdf (accessed: 2 March 2014).

Altieri, M. A. 2012. Convergence or divide in the movement for sustainable and just agriculture. In E. Lichtfouse (ed.), *Organic Fertilisation, Soil Quality and Human Health. Sustainable Agriculture Reviews*, 9th edn., pp. 1–9. Houten, The Netherlands: Springer.

Asaah, E. K., Z. Tchoundjeu, R. R. B. Leakey, B. Takousting, J. Njong, and I. Edang. 2011. Trees, agroforestry and multifunctional agriculture in Cameroon. *Int J Agr Sustain* 9:110–119.

Blackman, A. and J. Rivera. 2011. Producer-level benefits of sustainability certification. *Conserv Biol* 25:1176–1185.

Briedenhann, J. and E. Wickens. 2004. Rural tourism—Meeting the challenges of the new South Africa. *Int J Tour Res* 6:189–203.

Buck, L. E. and S. J. Scherr. 2011. Moving ecoagriculture into the mainstream. In *State of the World 2011: Innovations that Nourish the Planet*, pp. 15–24. Washington, DC: Worldwatch Institute.

Chitakira, M. 2011. *Agroforestry Adoption in African Smallholder Communities: Opportunities, Challenges and Coping Mechanisms*. Saarbrucken: VDM Verlag Dr. Muller GmbH & Co. KG.

Chitakira, M. 2013. Factors affecting ecoagriculture for integrated farming and biodiversity conservation in a transfrontier conservation area in Southern Africa. PhD Thesis, University of Pretoria.

Chitakira, M., E. Torquebiau, and W. Ferguson. 2012. Unique combinations of stakeholders in a transfrontier conservation area promote biodiversity-agriculture integration. *J Sustain Agr* 36:275–295.

Cholet, N. 2010. Ecoagriculture landscape labelling: A case study from South Africa. MSc Thesis, Montpellier: SupAgro-: Institute for Tropical Agronomic Studies Montpellier.

Cohn, A. S. and D. O'Rourke. 2011. Agricultural certification as a conservation tool in Latin America. *J Sustain Forest* 30:158–186.

Davenport, T. 2008. *"Biogeochemistry" and the Need for an Interdisciplinary Approach to Business*. Boston: Harvard Business Publishing. http://blogs.hbr.org/davenport/2008/07/biogeochemistry_and_the_need_f.html (accessed: 30 May 2012).

DEAT (Department of Environmental Affairs and Tourism). 2004. *A National Climate Change Response Strategy for South Africa*. Pretoria: Government of South Africa.

Ecoagriculture Partners. 2004. Response to Altieri and Farvar articles. http://cmsdata.iucn.org/downloads/ep_response_to_ceesp_12_17.pdf (accessed: 2 April 2013).

Ecoagriculture Partners. 2008a. What is ecoagriculture? http://www.ecoagriculture.org/page.php?id=47 (accessed: 23 August 2010).

Ecoagriculture Partners. 2008b. Applying the ecosystem approach to biodiversity conservation in agricultural landscapes. *Ecoagric Policy Focus* 1:1–4.

Ecoagriculture Partners. 2009. *Ecoagriculture Snapshot: Enhancing Agricultural Productivity on the Margins of Kakamega Forest, Kenya*. Washington, DC: Ecoagriculture Partners. http://www.ecoagriculture.org/case_study.php?id=37 (accessed: 7 December 2011).

Ecoagriculture Partners. 2014. *Frequently Asked Questions*. Washington, DC: Ecoagriculture Partners. http://www.ecoagriculture.org/page.php?id=24&name=FAQs#Q1.3 (accessed: 10 January 2014).

Fischer, J., D. J. Abson, V. Butsic, M. J. Chappell, J. Ekroos, J. Hanspach, T. Kuemmerle, H. G. Smith, and H. von Wehrden. 2014. Land sparing versus land sharing: Moving forward. *Conserv Lett* 7:149–157.

Fleury, J. 2011. Agriculture et dynamiques paysageres a l'echelle du territoire Mathenjwa: Place de l'agroforesterie dans une perspective d'ecoagriculture? Cas d'étude au KwaZulu-Natal—Afrique du Sud. MSc Thesis, Montpellier: Institut des Regions Chaudes, SupAgro.

Ghazoul, J., C. A. Garcia, and C. G. Kushalappa. 2009. Landscape labelling: A concept for next-generation payment for ecosystem service schemes. *For Ecol Manag* 258:1889–1895.

Gliessman, S. R. 2004. Integrating agroecological processes into cropping systems research. *J Crop Improv* 11:61–80.

LMRC (Landscape Measures Resource Center). 2014. Landscape level planning. Cornell University and Ecoagriculture Partners. http://landscapemeasures.info/?p=98 (accessed: 14 January 2014).

Lele, S., P. Wilshusen, D. Brockingtonet, R. Seidler, and K. Bawa. 2010. Beyond exclusion: Alternative approaches to biodiversity conservation in the developing tropics. *Curr Opin Environ Sustain* 2:94–100.

Linnell, J. D., C. Promberger, L. Boitani, J. E. Swenson, U. Breitenmoser, and R. Andersen. 2005. The linkage between conservation strategies for large carnivores and biodiversity: The view from the "half-full" forests of Europe. In J. C. Ray, K. H. Redford, R. S. Steneck, and J. Berger (eds), *Large Carnivores and the Conservation of Biodiversity*, pp. 381–399. Washington. DC: Island Press.

Liquidlingo Communications. 2009. Woolworths introduces the first 100% South African organic cotton garments for summer 2009. Cape Town: The Live Eco Team. http://www.liveeco.co.za/2009/09/22/100-south-african-organic-range-for-summer-09/ (accessed: 15 March 2014).

Marcu, L. 2007. Science education: The need for an interdisciplinary approach. *Analele Universităţii din Oradea, Fasc Biol* xiv:53–56.

McKean, M. and E. Ostrom. 1995. Common property regimes in the forest: Just a relic from the past? *Unasylva* 46:3–15.

McNeely, J. A. and S. J. Scherr. 2003. *Ecoagriculture: Strategies to Feed the World and Save Wild Biodiversity*. Washington, DC: Island Press.

Méndez, V. E., S. R. Gliessman, and G. S. Gilbert. 2007. Tree biodiversity in farmer cooperatives of a shade coffee landscape in western El Salvador. *Agr Ecosyst Environ* 119:145–159.

Millennium Ecosystem Assessment Board. 2005. *Ecosystems and Human Well-Being: Current State and Trends*, vol. 1. Washington, DC: Island Press.

Müller-Lindenlauf, M. 2009. *Organic Agriculture and Carbon Sequestration: Possibilities and Constraints for the Consideration of Organic Agriculture within Carbon Accounting Systems*. Rome: FAO.

Murphree, M. 2009. The strategic pillars of communal natural resource management: Benefit, empowerment and conservation. *Biodivers Conserv* 18:2551–2562.

Nair, P. K. R. 2007. Agroforestry for sustainability of lower-input land-use systems. *J Crop Improv* 19:25–47.

Neely, C. L. and J. Butterfield. 2004. Holistic management of African rangelands. *Leisa Mag* 20:26–28.

Perfecto, I. and J. Vandermeer. 2010. The agroecological matrix as alternative to the land sparing/agriculture intensification model. *Proc Natl Acad Sci USA* 107:5786–5791.

Perfecto, I., J. Vandermeer, and A. Wright. 2009. *Nature's Matrix: Linking Agriculture, Conservation and Food Sovereignty*. London: Earthscan Publications Limited.

Persha, L., H. Fischer, A. Chhatre, A. Agrawal, and C. Benson. 2010. Biodiversity conservation and livelihoods in human-dominated landscapes: Forest commons in South Asia. *Biol Conserv* 143:2918–2925.

Pretty, J. 2008. Agricultural sustainability: Concepts, principles and evidence. *Philos Trans Royal Soc Biol Sci* 363:447–465.

Ron, T. 2007. Southern Africa Development Community (SADC) proposed framework for transfrontier conservation areas (TFCAs): Issues and options report. A report prepared for the SADC Secretariat. Gaborone: SADC.

SADC (Southern Africa Development Community). 1999. *Protocol on Wildlife Conservation and Law Enforcement*. Gaborone: SADC.

SAinfo Reporter. 2008. Woolies boost for organic farming. http://www.southafrica.info/business/trends/new-business/woolies-090108.htm (accessed: 14 May 2012).

Scherr, S. 2011. Ecoagriculture: Landscape approaches to integrate agriculture and environment. Presentation given at the USAID Summer Seminar, 13 July 2011. Washington, DC: Ecoagriculture Partners. http://rmportal.net/library/content/ecoagriculture-landscape-approaches-to-integrate-agriculture-environment/view (accessed: 10 March 2014).

Scherr, S. J. and J. A. McNeely (eds). 2007. The challenge for ecoagriculture. In *Farming with Nature: The Science and Practice of Ecoagriculture*, pp. 1–16. Washington, DC: Island Press.

Scherr, S. J. and J. A. McNeely. 2008. Biodiversity conservation and agricultural sustainability: Towards a new paradigm of "ecoagriculture" landscapes. *Philos Trans Royal Soc B* 363:477–494.

Shames, S. and S. J. Scherr. 2009. Agriculture and the convention on biological diversity: Guidelines for applying the ecosystem approach. Ecoagriculture Discussion Paper No. 4. Washington, DC: Ecoagriculture Partners.

Strang, V. 2009. Integrating the social and natural sciences in environmental research: A discussion paper. *Environ Dev Sustain* 11:1–18.

Taylor, R. 2009. Community based natural resource management in Zimbabwe: The experience of CAMPFIRE. *Biodivers Conserv* 18(10):2563–2583.

Toledo, V. M. and P. Moguel. 2012. Coffee and sustainability: The multiple values of traditional shaded coffee. *J Sustain Agr* 36:353–377.

Torquebiau, E., C. Garcia, and N. Cholet. 2012. Labelling rural landscapes. Perspective: Environmental policy No. 16. Paris: CIRAD.

Tscharntke, T., Y. Clough, T. C. Wanger, L. Jackson, I. Motzke, I. Perfecto, J. Vandermeer, and A. Whitbread. 2012. Global food security, biodiversity conservation and the future of agricultural intensification. *Biol Conserv* 151:53–59.

UNESCO (United Nations Educational, Scientific and Cultural Organization). 2014. About the Man and Biosphere Programme (MAB). http://www.unesco.org/new/en/natural-sciences/environment/ecological-sciences/man-and-biosphere-programme/about-mab/ (accessed: 15 July 2014).

Van Wyk, A. E. and G. F. Smith. 2001. *Regions of Floristic Endemism in Southern Africa: A Review with Emphasis on Succulents*. Pretoria: Umdaus Press.

Wu, J. and Y. Qi. 2000. Dealing with scale in landscape analysis: An overview. *Geogr Inf Sci* 6:1–5.

CHAPTER 14

Agroecology in Central Appalachia
Framing Problems and Facilitating Solutions

Sean Clark

CONTENTS

14.1 Introduction 279
14.2 Defining Agroecology 280
14.3 Appalachian Region..283
    14.3.1 Tobacco Dependence 284
    14.3.2 Context and Challenges in the Region..287
14.4 Changes in Farming..290
    14.4.1 Farming to Food Systems 295
    14.4.2 Facilitating Food-System Improvements..298
14.5 A Role for Agroecology...302
References...304

14.1 INTRODUCTION

Agroecology is understood by some as a field of science—a specialized subdiscipline of ecology—and by others more broadly as an approach to agriculture and food systems guided by ecological understanding and aimed at confronting environmental, economic, and social problems resulting from the detrimental side effects of agro-food industrialization. Gaining greater attention and acceptance in the United States in the 1980s, agroecology has increasingly become recognized at universities and government agencies as a field addressing concerns about the long-term sustainability of conventional agriculture. Although there has always been a clear emphasis on environmental impacts, some early works shaping current conceptions of agroecology also addressed socioeconomic aspects of agriculture and the particular challenges to and importance of small, resource-limited farmers in developing countries who often used traditional practices (Altieri 1987, 2002; Carroll et al. 1990; Gliessman 1990).

Over the past two decades, the accepted scope of agroecology in the United States has broadened to include the entire food system (Francis et al. 2003; Gliessman 2007). The influence of the accumulating and evolving body of work is widely recognizable in the establishment of the national organic program; the dramatic increase in organic food and farming driven by consumer demand;

federal and state grant programs supporting the adoption of farming practices to address environmental pollution and the loss of small farms; the emergence of a strong local foods movement across the nation; an increase in the number of farmers markets, community-supported agriculture (CSA) programs and farm-to-school programs; the emergence of academic programs in agroecology at universities and colleges; and the proliferation of student farms on college campuses. In this chapter, I examine the situation in central Appalachia, an area of the United States characterized in part by small, resource-limited farms and a history of economic hardship and underdevelopment. In particular, I attempt to identify constraints and opportunities for improving the sustainability of the region's farming and food systems and share notable examples that demonstrate the inherent trade-offs and challenges in applying agroecology to address the region's problems and build its agricultural economy.

14.2 DEFINING AGROECOLOGY

Agroecology holds a broad range of meanings today—from the application of ecological methods for studying and understanding agricultural production systems at one end, to a grassroots movement devoted to addressing societal food-system issues at the other (Wezel et al. 2009). Consequently, tensions sometimes exist between the limited body of knowledge generated by scientific methodology and efforts to promote change that are often motivated by social and political values. It is sometimes difficult to clearly delineate ideas, concepts, and principles along a continuum of scientific objectivity to popular activism. Scientists can be reluctant to extrapolate the findings of studies with narrowly defined objectives, even as practitioners and advocates of "ecological farming" crave relevant applied information to help guide and justify their decisions and actions. Despite a lack of common understanding of and agreement on precisely what agroecology is, it continues to have motivating influences while raising new questions about farming and food.

The evolution of agroecology as a multidisciplinary field is embedded within and shares parallels with that of its primary parent discipline of ecology, which can be defined simply as the study of organisms and their interactions with each other and their environment. Though the origins of ecology date back to the 1860s with Ernst Haeckel, it was about a century before it became part of the general public's consciousness in the United States, with publications like Aldo Leopold's *Sand County Almanac* in 1949 and Rachel Carson's *Silent Spring* in 1962. Both of these works were significant to the popular environmental movement beginning in the late 1960s, drawing attention to the detrimental effects of human activities and calling for a more enlightened perspective that considered more than just short-term economic gain. Carson in particular focused greater public scrutiny on the conflicts between farming and nature by arguing that indiscriminate pesticide use was having devastating effects on nontarget organisms. And her writing undoubtedly shaped the perspectives of many scholars, researchers, writers, farmers, and advocates of agroecology who followed in seeking more environmentally friendly means of food and fiber production.

Even as the ecology movement gained momentum in the 1970s, calling attention to the detrimental environmental impacts of agriculture like pesticide and fertilizer pollution and soil erosion, professional ecologists still seemed slow to direct attention toward agriculture as a subject for study. It was 1977 before agroecology was first a topic in the *Annual Review of Ecology and Systematics* (Loucks 1977) and even later in the discipline's flagship journals of the Ecological Society of America. Some peer-reviewed papers on agroecology were being published elsewhere, however, and beginning to form a body of literature on the processes of agricultural production systems viewed through the lens of ecologists. Of course, much of the emphasis was on documenting the damaging effects of common production practices and assessing the feasibility of alternatives to mitigate those negative environmental impacts.

Not all ecologists, however, were convinced that the societal and environmental costs of agricultural industrialization outweighed the benefits. In 1984, R.L. Loomis, a crop ecologist, countered the growing criticisms of industrialized farming with an article published in the *Annual Review of Ecology and Systematics* entitled "Traditional Agriculture in America." It seemed almost a direct rebuttal to those calling for a radical transformation in U.S. agriculture. Loomis pointed out the inconsistencies in romanticizing traditional agriculture without also acknowledging the extremely hard work and difficult life that accompanied it, as well as the manifold benefits to society that agricultural industrialization brought, like greater production and cheaper food. He challenged generalizations sometimes made by agricultural ecologists about diversity, stability, monocultures, mechanization, energy use, farm size, and feeding grains to ruminants. His thought-provoking analysis also dealt with topics generally considered outside the realm of the science of ecology—like child labor and the efficiency of centralized cooking—while still using ecological concepts to defend conventional agriculture.

Another decade passed before the ecology of agriculture was again a subject in the *Annual Review of Ecology and Systematics*. Vandermeer (1995) offered ecological justification for alternative agricultural practices, specifically in the areas of pest and soil management. (It is worth noting that Loomis avoided the topic of pest management in his review.) But even as he cautiously defended alternative sets of practices like "organic," "permaculture," and "holistic," he referenced relatively few concrete examples of ecological principles being used to solve farming problems. And, like much of the agroecology literature published before and after, the paper restated some rather vague and ambiguous ideas like enhancing beneficial processes and interactions, integration, ecosystem balance and health, and mimicking natural ecosystems. To its credit, the paper acknowledged that there was little ecological justification for some traditional/alternative practices (e.g., intercropping), that the problems facing farmers are more likely to be economic and political rather than ecological, and that the principles guiding alternative agriculture are "more pragmatic than ecological": like reducing costs, protecting environmental quality, and safeguarding the health of humans and livestock. Finally, Vandermeer pointed out the frequent failure that accompanied the transition from conventional to more "ecologically based" methods, suggesting an "inadequacy of our systematic ecological knowledge of agroecosystems." Indeed, the discipline has lacked the capacity to provide farmers with the kinds of information or tools that can rapidly and satisfactorily remedy problems. Instead, it has mostly offered general concepts for design that are not always well supported by economics.

A recent review of the field of agroecology takes a broader scope, referencing a greater breadth of literature, but still describes a discipline focused mostly on production practices rather than larger food-system issues (Tomich et al. 2011). Although field-plot-level experiments on soil biota and landscape-level experiments on insects demonstrate the capacity of humans to predictively manipulate organismal populations and processes, these studies are rarely connected in any direct way to the economic realm of risk and reward within which farmers operate and live. In fact, some key challenges for agroecology acknowledged by Tomich and colleagues, in addition to "mitigating environmental impacts of agriculture while dramatically increasing food production," are complex problems falling largely outside of the direct purview of most agroecologists, like "improving livelihoods" and "reducing chronic hunger and malnutrition"—issues that generally are not remedied simply with increased food production. Other areas identified in the review as important for future research are typically beyond the comfort zones of agroecologists trained in the traditional agricultural sciences—like dealing with different values and ethical perspectives, assessing how production systems affect human health and animal welfare, and attempting to measure and predict the resilience of agriculture and food systems.

One group of prominent university academics has attempted to coax agroecologists into these newer frontiers by formally proposing the widening of the definition of agroecology to be "the ecology of food systems" (Francis et al. 2003). This seemingly simple proclamation of an idea and understanding already widely held in the alternative-agriculture community was important because it recognized that the problems of farming often do not involve production—something critics and writers outside the field of agroecology have noted for decades. It was also a more open acknowledgement, in both the statements made and the language used, of the implicit shared values of agroecology as a science and a movement (Wezel et al. 2009). Some ecologists will undoubtedly question the level of objectivity and justification for some of the generalizations made by Francis et al., but without an attempt to analyze how pre- and postproduction sectors of the food system (input manufacturing, food processing, transportation, aggregation, marketing, preparation, consumption, regulation, and waste management) affect farming—and vice versa—agroecology is unlikely to offer much in addressing the complicated agro-food problems that communities and regions face. At the same time, this broader definition opens the realm of agroecological study to almost anything, creating methodological challenges for agroecologists in identifying system boundaries for analysis—since almost anything can be justified as having some influence on the food system.

Here, I examine the conditions in the central Appalachian region and use this broad definition of agroecology proposed by Francis et al. (2003), while acknowledging some fundamental value-based assumptions inherent in my examination (Table 14.1). Further, in keeping with the multidisciplinary approach and acknowledging the importance of information and knowledge generated outside of the traditional peer-reviewed, academic literature, I rely on a wide variety of sources—including case studies, articles in the popular press, and reports from governmental and nongovernmental organizations—to document and describe indicators and trends, identify the sources of constraints and challenges, and call attention to innovations and creative solutions as possible models to be borrowed from or replicated. Agroecologists frequently point out the relevance of scale—temporal and

Table 14.1 Agroecological Indicators Used in This Analysis Generally Assumed to Be Positive (Desired) or Negative (Undesired) Trends in Describing and Analyzing the State of Farming and Food in the Appalachian Region

Indicator	Positive	Negative
Number of farms and farmers	Stable or increasing number	Declining number
Farm income	Increasing	Declining
Farm energy and material inputs	Reduced consumption	Increased consumption
Agro-food employment	Increasing number of jobs	Declining number of jobs
Farming economy and infrastructure	Diversifying	Dependence on a few agricultural commodities
	Expanding into value-added products	Dependence on low-value commodities
	Alternative, localized marketing and distribution	Food retail sales dominated by globalized chains
Agro-food entrepreneurship	Examples of successful agro-food entrepreneurial endeavors	Failure of recently established agro-food enterprises
Environmental quality	Adoption of practices on and beyond farms aimed at protecting environmental quality and community quality of life	Continued use of practices demonstrated to compromise environmental quality and community quality of life
Cooperation	Examples of productive partnership benefits throughout the food supply chain	Examples in which benefits accrue to a few and the costs are borne by others in the food supply chain
Public health	Decreases in obesity and diabetes	Increases in obesity and diabetes
	Decrease in pesticide use	Stable or increasing pesticide use
Animal welfare	Consumer- and farmer-driven adoption of practices generally considered to improve animal welfare	Lack of public concern or consumer preference for improved animal welfare

spatial—in studying and understanding ecological processes and phenomena. For this chapter, the spatial scale is the central Appalachian region of the United States (described below) and the temporal scale is the past few generations (100 years), which have shaped the present, and possibilities for the next generation (over the coming 20–25 years).

A premise of this analysis is that practices, enterprises, and efforts that promote the economic viability of the region's farms and improve rural livelihoods in the region are generally viewed as or assumed to be positive or desirable. I envision an ideal situation or trajectory to include stronger and more diversified rural economies in the coming decades with (1) a greater degree of regional food self-sufficiency; (2) agricultural production practices that have sufficient economic efficiency while not compromising environmental quality, farmers and farmworkers, community members and consumers; (3) a reduction in poverty; and (4) improvements in public health (Table 14.1). Few would argue with any one of these objectives independently, but trying to prioritize them will undoubtedly be problematic. And when we dissect each one further to understand how any or all of them could be facilitated, we reach a level of complexity that can seem unresolvable—perhaps explaining why the field of agroecology has been slow to move beyond field-plot and landscape studies.

In field-based agricultural studies, it is safe to assume that if researchers discover a particular new agronomic practice that increases soil organic matter levels or reduces the need for pesticides, they would conclude that the practice is an improvement. Likewise, research indicating that particular low-input animal husbandry methods reduce livestock production costs and improve economic returns for farmers would likely be taken as positive. But if local food prices decline at a regional produce auction due to higher crop productivity and farmers begin to struggle financially as a result, the conclusion would be mixed—less expensive food but at the cost of farm financial viability. Likewise, if mechanization results in improved financial performance for a farm but also generates more greenhouse gases (GHG) per unit of product, it is debatable whether the net change is positive or negative. Indeed, opinions would depend on one's perspective and scope of analysis. Francis et al. (2003) and Tomich et al. (2011) acknowledge the importance of factoring in "quality of life" and "livelihoods" into the field of agroecology, but avoid a deeper exploration of what this means practically for agroecologists. No one would argue against "mitigating adverse impacts of agriculture on the environment," "ensuring food security," or "sustaining livelihoods." But we should forthrightly acknowledge that there are trade-offs to be made in pursuing these broad objectives and that human values—shaped by politics, history, religion, education, and other institutions—play an important role.

14.3 APPALACHIAN REGION

The Appalachian region encompasses the Appalachian mountain range of the eastern United States, an area occupied mostly by species-rich, deciduous forests when and where human disturbance is absent or minimal. It is often referred to as "Appalachia," though that name is sometimes reserved for the cultural region with its traditions and shared history as opposed to the physical land and its natural features. Here, I will use the terms mostly interchangeably since the focus is on agriculture and food systems—land-dependent, human enterprises shaped by culture and history. There are also differences of opinion on where the boundaries of the region lie. The Appalachian Regional Commission (ARC), established in 1965 by the federal government to promote and support economic development, defines it as "a 205,000-square-mile region that follows the spine of the Appalachian Mountains from southern New York to northern Mississippi," which includes all or parts of 13 states and 420 counties, amounting to nearly 6% of the U.S. land area and about 8% of the population (Pollard and Jacobson 2013).

Williams (2002) offered a more restrictive 164-county area, corresponding largely to the "central" subregions of Appalachia identified by the ARC (2009), which he referred to as the "core region." In this analysis, I concentrate on this core or central region, including eastern Kentucky,

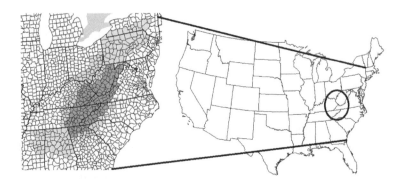

Figure 14.1. **(See color insert)** The Appalachian region in the eastern United States according to the Appalachian Regional Commission (*yellow*) and Williams (2002) (*orange*). The focus of this analysis is on the central Appalachian area surrounded by the oval. (Source of national and county maps: Wiki Commons, Creative Commons.) (Data from Williams, J.A., *Appalachia: A History*, University of North Carolina Press, Chapel Hill, 2002; ARC (Appalachian Regional Commission). Subregions of Appalachia. ARC. http://www.arc.gov/research/MapsofAppalachia. asp?MAP_ID=31. 2009.)

eastern Tennessee, western North Carolina, southwestern Virginia, southeastern Ohio, and most of West Virginia (Figure 14.1). In addition to mountainous terrain, which limits the range of possible farming options, this region is also characterized by a predominance of small farms, many of which raise cattle, and a historic but declining economic dependence on tobacco as a major cash crop. In fact, according to the U.S. Department of Agriculture (USDA)'s Farm Resource Regions classification, the "Eastern Uplands," which closely corresponds to this central Appalachian Region, has the greatest number of small farms of any of the eight regions in the contiguous 48 states (Heimlich 2000).

14.3.1 Tobacco Dependence

It is hard to overstate the importance of tobacco to the agricultural economy of Appalachia during the twentieth century and the impact that its decline is having in the twenty-first century (Figure 14.2). By the beginning of the twentieth century, there were clear signs that the region had reached its human carrying capacity. According to Gragson et al. (2008), "the overall viability of small farms in Appalachia was in jeopardy, and many full-time farmers had to seek part-time wage employment in mining and timbering to ensure their families' survival." The Agricultural Adjustment Act of 1938 established price supports and quotas for a range of commodities, including tobacco, with the goal of preventing wide fluctuations in supply and price, thus stabilizing income for farmers. Tobacco production was actively promoted by USDA personnel in a region seen as having "too many people and too little land" (Williams 2002). With government supports in place and demand for tobacco rising with the increasing popularity of cigarettes, producing just a few acres (1 acre = 0.4 ha) of tobacco offered Appalachian farmers and their families an opportunity to stay on their land and reliably generate enough cash to buy what they could not produce on their own. Tobacco production certainly did not eliminate the impoverished conditions that many suffered or stop the steady outward migration to cities in the Midwest in search of work, but it allowed many more small farms to persist than would have been possible otherwise (Figure 14.3).

The number of tobacco farms and associated acreage began to decline slowly in the 1950s, but tobacco production—particularly burley, a light, air-cured tobacco used in cigarettes—remained an important part of the agricultural economy and farmers' livelihoods in Appalachia for decades due in large part to government regulation. As cigarette consumption per capita in the United States peaked and began to decline in the 1960s, tobacco acreage also gradually declined (Capehart 2004).

Figure 14.2 Tobacco barns like this one in West Virginia are ubiquitous in central Appalachia. Although many are still in use, quite often—like this one—they have fallen into disrepair, a visible sign of the decline of tobacco in the region.

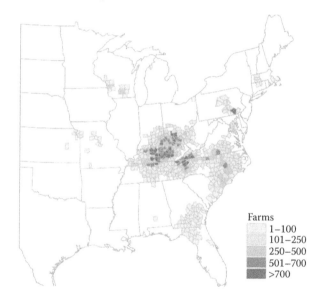

Farms
1–100
101–250
250–500
501–700
>700

Figure 14.3 **(See color insert)** The number of tobacco farms per county. The concentration of tobacco production, particularly burley tobacco, in the Appalachian region has helped to maintain the abundance of small farms. There is uncertainty about what, if anything, will replace tobacco in the region. (From Capehart, T., Trends in US tobacco farming, Economic Research Service, US Department of Agriculture, 2004, using data from the 2002 Census of Agriculture.)

Throughout the 1970s and until the mid-1990s, tobacco acreage in the United States remained relatively stable (Figure 14.4) and concentrated in Kentucky, Tennessee, North Carolina, and Virginia, even as the number of farms gradually declined and the average size of tobacco farms increased (Capehart 2004).

The anticipation and final passage of the Tobacco National Settlement Agreement of 1998 and the Fair and Equitable Tobacco Reform Act of 2004—referred to as the "tobacco buyout"— dramatically changed the economic situation for tobacco farmers. The intent of the buyout was to help tobacco farmers transition to a free-market tobacco economy by providing them with direct annual payments—funded by tobacco manufacturers—over a 10-year transitional period (2005– 2014). Presumably, tobacco farmers could use the funds to invest in new skills, equipment, facilities,

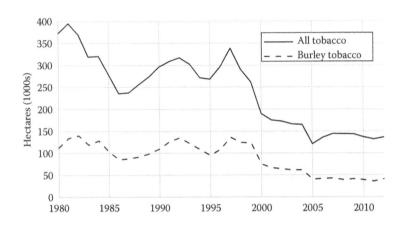

Figure 14.4 Decline in land area producing tobacco (total and burley) in the United States since 1980. The central Appalachian region accounts for most burley tobacco production. (From USDA, Census of Agriculture. National Agriculture Statistics Service, 2007. http://www.agcensus.usda.gov/index.php (accessed: February 18th, 2014).)

land, and business and marketing relationships to diversify their farms' income streams and reduce or eliminate their dependence on tobacco.

Indeed, the amount of U.S. farmland planted to tobacco has fallen sharply since the late 1990s (Figure 14.4), but few legal alternative crop enterprises have emerged to reliably generate the same level of revenue as that of tobacco with government price supports. A survey by Mendieta et al. (2012) of burley tobacco producers, mostly from Kentucky, North Carolina, Tennessee, and Virginia, indicated an inclination toward substituting greater hay and livestock production for tobacco—enterprises to which many were already accustomed. However, neither of these options can generate similar returns per hectare to tobacco. There was much less interest among survey respondents in fruits, vegetables, or grain crops, except in North Carolina, where some considered fruits and vegetables to have high potential. Similarly, Craig (2008) studied income diversification among Kentucky tobacco farmers in 2002 and found some of them already expanding their hay, cattle, and grain production as well as using their skills in off-farm work. But Craig concluded that "[p]roactive efforts by burley growers to shape their farming futures were woefully lacking, partly because economic necessity had not yet provided the stimulus for many to act."

It appears that many former tobacco producers in Appalachia may simply be choosing to exit farming. Agricultural Census data revealed that despite a rise in the total number of farms in the United States by over 75,000 between 2002 and 2007—the first increase in that statistic since the 1930s—the counties of central Appalachia suffered among highest concentrations of farm losses in the nation during that period, particularly in Tennessee and to a lesser extent in Kentucky and North Carolina (Figure 14.5). This trend corresponds closely with the decline in tobacco acreage in the region. Some production shifted within the United States to areas outside of Appalachia (e.g., eastern North Carolina and western Kentucky), but mostly it was replaced with production outside of the United States where costs, particularly for labor, are less.

Even as tobacco production has declined, it continues to make an important contribution to the overall agricultural economies of these central Appalachian states, especially in North Carolina and Kentucky, generating over $1 billion or 3.6% of farm receipts (Table 14.2). A wide range of alternatives has been suggested by land-grant universities and cooperative extension offices in the region—including vegetables, vegetable transplants, small fruits, tree fruits, culinary and medicinal herbs, hops, ornamental plants, nontimber forest products, sheep, goats, aquaculture, and agroforestry—but most simply cannot match the income generated by tobacco with government price supports (Gale et al. 2000). Those products can have other issues like risk due to perishability, pest

1 Dot = 10 Farms increase
1 Dot = 10 Farms decrease

Figure 14.5 **(See color insert)** Changes in the number of farms between the agriculture censuses of 2002 and 2007. The decrease in farm numbers has been particularly acute in the central Appalachian region. (From USDA, National Agricultural Statistics Service, http://www.agcensus.usda.gov/Publications/2007/Online_Highlights/Ag_Atlas_Maps/Farms/Size/07-M004-RGBDot2-largetext.pdf.)

Table 14.2 **Relative Importance of Tobacco to the Agricultural Economies of the Six States with Area in the Central Appalachian Region**

State	Farm Total Sales ($1,000)	Percentage of Farm Total Sales						
		Tobacco	Grains	Cattle	Hogs	Poultry and Eggs	Vegetables	Fruits
North Carolina	10,313,628	5.3	6.8	2.8	30.1	39.6	3.2	0.8
Ohio	7,070,212	0.1	47.5	8.0	8.1	12.5	1.9	0.6
Kentucky	4,824,561	6.5	18.0	19.4	1.9	20.3	0.4	0.1
Virginia	2,906,188	2.3	9.3	19.8	2.0	33.4	3.2	2.3
Tennessee	2,617,394	2.7	19.0	24.2	1.3	21.9	2.7	0.1
West Virginia	591,665	0.1	2.0	27.9	0.4	51.0	1.0	2.4
Six-state total	28,323,648	3.6	20.2	11.2	13.6	27.5	2.3	0.8

Source: USDA (United States Department of Agriculture), Census of Agriculture. National Agriculture Statistics Service, 2007, http://www.agcensus.usda.gov/index.php (accessed: February 18th, 2014).

pressures, lack of consistent and predictable markets; specialized equipment and mechanization requirements; lack of storage, processing and distribution infrastructure; and the required learning curve for producing something new. Any of these obstacles might be addressed with technological assistance and economic investment, but broader societal issues may present even more significant challenges.

14.3.2 Context and Challenges in the Region

Beyond the rugged topography that limits agricultural options and a history of dependence on tobacco as the primary cash crop, there are socioeconomic and cultural factors that may hinder or preclude the development of more sustainable farming and food systems in Appalachia. It is important to point out, however, that the agricultural economies and food systems of the six states making up the central Appalachian region differ vastly from each other (Table 14.2) as they include substantial areas outside of Appalachia. The Appalachian counties are therefore affected by the policies, programs, economic resources, and trends of their states—which means that the opportunities and challenges differ depending on the state of which they are a part. Nevertheless, there are some common characteristics and conditions that warrant consideration.

The Appalachian region has struggled with high unemployment more often than not. Although the region's unemployment rate has recently been comparable to that of the national average, it still has a lower average household income and a higher percentage of people living in poverty than most of the rest of the nation (Pollard and Jacobson 2013). In the central Appalachian states, the percentage of individuals living below the poverty level is consistently higher in Appalachian counties compared to non-Appalachian counties in those same states as well as to the national average (note: the entire state of West Virginia falls inside of Appalachia) (Table 14.3.) The situation in eastern Kentucky is particularly dramatic, with nearly one in four individuals living below the poverty level. Income support through federal benefits is high throughout the region, but particularly so in eastern Kentucky, northeastern Tennessee, southern Ohio, southwestern Virginia, and West Virginia (White et al. 2012). The counties of the central Appalachian region, which include only 3.4% of the U.S. population, accounted for nearly a quarter of the 100 counties with the highest dependency rates on the Supplemental Nutritional Assistance Program ("food stamps") in 2009. In many of these counties, more than a third of adults and well over half of children receive this assistance (Bloch et al. 2009).

The education and public health conditions are equally troubling. The Appalachian counties of these states lag behind their non-Appalachian counterparts and the nation in the percentage of adults with a bachelor degree (Table 14.3). Likewise, these counties have a higher percentage of adults who lack even a high school diploma. These conditions—low-paying jobs, high poverty rates, and a lack of education among a significant fraction of the population—create fundamental obstacles to building sustainable agro-food systems in central Appalachia.

Public health conditions present serious issues as well, putting heavy demands on limited financial resources. According to the Centers for Disease Control and Prevention, the Appalachian counties of Kentucky, Tennessee, and West Virginia make up one of two areas in the nation with distinctly higher incidences of obesity and diabetes (Gregg et al. 2009). There are also higher overall rates of cancer in central Appalachia compared the U.S. average, the result of higher incidences of lung cancer, other tobacco-related cancer, and colorectal cancer (Wingo et al. 2008). Heart disease, a leading cause of premature death throughout the United States, is especially prevalent in central Appalachia (Halverson and Bischak 2008). Given this dismal public health situation, it is perhaps not surprising that the central Appalachian region was identified as one of three areas of the United States with the highest concentration of fast-food businesses per capita (Stimers et al. 2011). And Morton and Blanchard (2007) identified the Appalachian regions of Kentucky and West Virginia as among the nation's rural "food deserts." Tobacco use is also high in this tobacco-producing region, with Kentucky and West Virginia ranking first and second in adult cigarette use in the United States (CDC 2012).

Despite these discouraging statistics, there are encouraging examples throughout the region to educate, inspire new ideas, prompt conversation, and serve as models of how plans and actions can be put into place to bring about desirable changes. These examples range from creative and visionary farmers and entrepreneurs to nonprofit organizations and educational institutions working at different scales, with different segments of society, and at different points in the food supply chain—in effect, creating an informal network that can assist others in finding needed resources—including information, ideas, skill-building opportunities, markets, equipment, land, and even financial capital—to support new and ongoing efforts. These varied efforts also provide important grounding and reality checks so that the nostalgic and romantic pursuits of agrarian idealism are balanced with the pragmatism that comes from on-the-ground experience.

Table 14.3 Percentages Living below the Poverty Level and with a Bachelor's Degree, 2007–2011

State	Living below Poverty Level (%)[a]		Bachelor's Degree or More (%)[b]		Lacking a High School Diploma (%)	
	Appalachian	Non-Appalachian	Appalachian	Non-Appalachian	Appalachian	Non-Appalachian
North Carolina	16.9	15.9	23.0	27.3	17.2	15.6
Ohio	16.7	14.3	15.7	26.4	15.3	11.5
Kentucky	24.8	15.6	12.7	23.7	27.4	14.8
Virginia	18.1	9.9	17.5	36.2	21.8	12.5
Tennessee	17.5	16.4	20.0	25.5	18.7	15.3
West Virginia	17.5		17.6		17.4	
United States	14.3		28.2		14.6	

Source: Pollard, K. and L. A. Jacobson. 2013. The Appalachian Region: A data overview from the 2007–2011 American Community Survey Chartbook. Appalachian Region Commission. http://www.arc.gov/research/researchreportdetails.asp?REPORT_ID=103 (accessed: February 18th, 2014).

Note: All of West Virginia is considered part of Appalachia.

[a] Poverty level: $22,811 for a family of two adults and two children in 2011.

[b] Adults 25 years and older.

14.4 CHANGES IN FARMING

When viewed with a sufficiently broad and long-term perspective, the loss of tobacco farming has to be considered a mostly positive change for the region—a reduction in the acreage and production of a pesticide-intensive crop (GAO 2003) that has debilitating health consequences for its users and high public health costs for society. The loss of small tobacco farms seems to be an unfortunate but possibly necessary trade-off for creating the conditions for substantive changes to occur in the region's agriculture while reducing support for one of the more deleterious public health burdens to society. Now, providing future generations of potential farmers and food-system innovators with relevant experiential education in agroecology seems fundamental for a transition away from tobacco. As older tobacco farmers who are either uninterested, unwilling, or unable to make changes decide to exit farming, this creates opportunities for younger people to get into farming. Certainly it takes much more than ecologically minded and socially conscious farmers and activists to implement this kind of substantive societal change. There also has to be widespread general education, enhancement in job prospects and wages, some protection from risk, and improvements to public health and the general quality of life. And all of this requires thoughtful investment and planning at multiple points in the food system.

Educational institutions addressing agroecology in some capacity include land-grant universities, the institutions traditionally charged with "generating knowledge" and extending it to rural communities, as well as other colleges and universities that are increasingly emphasizing agroecology in their curricula and offering students opportunities to "learn by doing" on campus farms and gardens (Sayre and Clark 2011). Two universities in central Appalachia offering relatively new undergraduate degrees explicitly in agroecology are West Virginia University and Appalachian State University (Table 14.4). Other institutions have major or minor programs that emphasize "sustainability" and commonly include a distribution of courses across multiple disciplines as an alternative to a highly specialized and technical program concentrated in the natural sciences. Agroecologists tend to favor this kind of broad course composition and experiential learning because it is assumed to promote "integration," "systems understanding," and creative problem solving—emphasizing holism over reductionism.

One criticism of the hands-on, experiential components of some of these programs is the predominance of labor-intensive, market-gardening practices and small-scale operations, often guided by particular philosophies or brands, like permaculture and biointensive production. The high initial costs of appropriate mechanization are a significant obstacle for small academic programs and concerns about student safety with farm equipment undoubtedly exist, but the educational relevance of teaching farming practices that are uneconomical under real market conditions due to insufficient scale or undercapitalization is questionable. Agrarian idealism has to be balanced with practical realism. Another related concern is that some of these programs are located in or around larger metropolitan areas that can and do support small-scale agriculture to a greater extent than rural central Appalachia. What can be appropriate in these urbanized areas where disposable incomes are higher may not work in poor rural areas. To their credit, these programs tend to address broader societal concerns—beyond field production—that have an important bearing on the future of agriculture and food in the region; for example, fair trade, animal welfare, food security, and nutritional wellness. This kind of breadth helps to bridge agriculture out and demonstrate its broad impact on a society disconnected from the land. But the educational experiences offered by these programs may not be sufficient to adequately support younger farmers trying to become successfully established and stable.

This is where the Cooperative Extension Service has historically filled an important role by interacting directly with farmers. It has had a profound effect in shaping agriculture and rural society for a century—though some observers note that it has sometimes been unsupportive of

Table 14.4   Key Institutions and Organizations Addressing Agroecological Concerns and Needs in
Central Appalachia through Research, Education, Assistance, Promotion, and Providing
Access to Product-Development Facilities

Research Universities and Experiment Stations	Colleges Offering Academic Programs	Nongovernmental Organizations	Shared-Use Commercial Food-Processing Facilities
North Carolina State University Mountain Horticultural Crops Research and Education Center, Mills River, NC	Appalachian State University, BS in Sustainable Development with Agroecology and Sustainable Agriculture concentration, Boone, NC	Appalachian Sustainable Agriculture Project, Asheville, NC	Appalachian Center for Economic Networks (ACEnet), Athens, OH
West Virginia University Program in Agroecology, Morgantown, WV	Warren Wilson College, BS in Environmental Studies with Sustainable Agriculture conc., Swannanoa, NC	Organic Growers School, Old Fort, NC	Blue Ridge Food Ventures, Candler, NC
University of Tennessee Organic and Sustainable Crop Production Program, Knoxville, TN	Berea College, BS in Agriculture and Natural Resources, Berea, KY	Appalachian Sustainable Development, Abingdon, VA	Jackson County Regional Food Center, Annville, KY
University of Kentucky Sustainable Agriculture program	Ferrum College, BS in Agriculture with Agroecology concentration, Ferrum, VA	Community Farm Alliance, KY	
Virginia Tech Civic Agriculture and Food Systems minor program		Grow Appalachia, Berea, KY	
North Carolina State University Agroecology minor program		Carolina Farm Stewardship Association, NC	

farmers attempting to transition to more ecologically based practices, like organic farming (Lotter 2003). The Extension Service had a fundamental role in getting small farmers in central Appalachia "hooked" on growing tobacco and now, with the expertise and support of land-grant universities and agricultural experiment stations, has been encouraging tobacco farmers to find alternatives. There has been a strong emphasis in promoting new "specialty" crops, particularly vegetables and fruits, in anticipation of the decline in tobacco. Defined as "fruits and vegetables, tree nuts, dried fruits, horticulture, and nursery crops (including floriculture)" according to the Specialty Crop Competitiveness Act of 2004 and the Food, Conservation, and Energy Act of 2008, so-called specialty crops have a couple of potential advantages among the various possible alternatives to tobacco: greater value per hectare than livestock and hay and many can be produced using practices and equipment similar to those used for tobacco. Interestingly, many specialty crops are considered essential foods for a healthful diet and rank among the most commonly consumed fruits and vegetables in the United States; for example, lettuce, spinach, tomatoes, peppers, potatoes, and onions. Their production in these central Appalachian states, however, still lags behind that of tobacco (Table 14.2).

Promotion of horticultural crops has not been limited to cooperative extension efforts—some nonprofit organizations have also been actively assisting farmers with transitioning as cooperative extension's role declines due to shrinking budgets. One successful example is the Appalachian Harvest program established in 2000 by Appalachian Sustainable Development, a nonprofit organization based in southwestern Virginia. Flaccavento (2010) described the program as "an integrated field-to-table network in southwest Virginia and northeast Tennessee," which included mostly former tobacco farmers who transitioned to organic vegetables. Appalachian Sustainable Development offered training, assisted farmers with organic certification, established wholesale markets, and built a facility for aggregating, washing, packing, and distributing to large regional grocery chains, often greater obstacles to farmers than producing new crops. Indeed the risk of growing highly perishable crops with uncertain markets is a clear contrast to producing tobacco with guaranteed price supports.

An examination of extension publications and field-trial research results from Kentucky sheds light on the potential risks and rewards of producing specialty crops, the high degree of uncertainty accompanied by a lack of control over external factors, and the essential importance of market and timing for securing sales. Crop profiles from the University of Kentucky's Center for Crop Diversification (University of Kentucky 2013a) offer growers a range of income scenarios, including optimistic, conservative, and pessimistic. For example, optimistic returns for bell peppers (*Capsicum annuum*) grown on black plastic mulch can generate net returns exceeding $7000 per hectare when sold through wholesale markets—a value comparable or even superior to tobacco. But while experiment-station results lend support to these expectations (e.g., Law et al. 2006), on-farm grower trials demonstrate that serious financial losses can also occur, even with lower-than-predicted production costs (Spalding and Coolong 2012). Certified organic production adds the potential for even greater returns and losses since market premiums may exist but production costs are also greater. Crop profiles show a range of $(1,173)–$4,989/ha for conventional tomatoes and $(5,459)–$13,462/ha for certified organic tomatoes. It is important to note that there is generally not a strong market demand for organic products in the central Appalachia states. These states account for 17% of the nation's farms and 7% of its farmland but only 7% of its organic farms and 2% of certified organic farmland (Table 14.5). Given the difficult economic situation in many central Appalachian counties, the weak market demand for certified organic foods is not surprising. Metropolitan areas within these states have stronger organic markets but not sufficiently so to substantially influence production in the region.

If strong and reliable wholesale or retail markets for vegetable crops are located, the next most important concerns for growers are probably labor and soil protection. Like tobacco, most vegetable production leaves soil exposed for long periods and vulnerable to crusting, erosion, and the loss of organic matter. On the one hand, farmers need to protect their crops from weed competition. But leaving the soil surface devoid of cover via cultivation and/or herbicides also has a cost—one that is harder to calculate and less immediately visible. Plastic mulches offer a cost-effective means of suppressing weeds but require an investment in machinery, generate plastic waste at the end of use, and leave the exposed soil around the plastic mulch vulnerable to degradation. Organic mulches

Table 14.5 Demographic, Farm, and Organic Agriculture Statistics for the Six States with Area in the Central Appalachian Region

State	Population[a]	Number of Farms[b]	Farmland Acreage (Million Acres)[b]	Average Farm Size (Acres)[b]	Number of Certified Organic Farms[c]	% of Farms Certified Organic	Certified Organic Acreage[c]
Kentucky	4,339,367	86,000	14	163	60	0.07	6,256
North Carolina	9,535,483	52,000	9	183	174	0.33	20,593
Tennessee	6,346,105	78,000	11	139	28	0.04	1,009
Virginia	8,001,024	47,000	8	182	93	0.20	23,361
West Virginia	1,852,413	23,000	4	170	5	0.02	337
Ohio	11,536,504	75,000	14	187	505	0.67	57,117
United States	313,914,040	2,075,510	920	446	12,880	0.62	5,383,119
Six-state total as a percentage of United States	13%	17%	7%		7%		2%

[a] U.S. Census, 2010.
[b] U.S. Department of Agriculture, National Agricultural Statistics Service, 2010.
[c] U.S. Department of Agriculture, Economic Research Service, 2011.

offer an alternative but can be cost prohibitive (e.g., Law et al. 2006; Clark and Meyer 2013) except in unusual circumstances, such as when nearby straw bales spoil or yard wastes from suburban areas are available free of charge. But farmers should be particularly cautious with such materials since persistent herbicides used in pastures and yards can carry over and cause damage to vegetable crops (Davis et al. 2010). Another option is to seed areas between plastic beds into cover crops to protect against erosion and build soil organic matter (Figure 14.6). Planting cover crop seed is less expensive, in terms of materials and labor, but effectively managing the cover crop so that it does not become a weed later is necessary and sometimes risky.

For some vegetable crops, the production period is so short and the need to have an extremely clean product so great that organic mulches are not an option and plastic mulches may not be necessary or cost effective. Leafy greens, like lettuce, spinach, kale, and mustards, for example, may require only weeks rather than months from transplanting to harvest and between successive crops. For these crops, a weed-free microenvironment around the plants is important to prevent unwanted plant material from contaminating the final products. Season-extension technologies like high tunnels (hoop houses) and low tunnels offer opportunities for nearly year-round production and these structures can protect soil from heavy rains and wind. With leafy green production, most risks can be minimized as long as reliable markets exist, but labor costs for transplanting (if used instead of direct seeding) and weed suppression, as well as harvesting, can be high. Investing in a packing shed with sufficient cooler space is a significant investment but can provide a grower with some flexibility and protection in dealing with a high-value, perishable crop. Production of leafy greens for local markets, like grocery stores, restaurants, farmers markets, and CSA programs offers growers competitive advantages over products shipped in from distant states, even when grown on a relatively small scale.

The fact that vegetable and fruit production still does not match tobacco in farm receipts in the six-state region suggests that there are serious obstacles to successful horticultural production (Table 14.2). The required investments in new equipment and infrastructure as well as the market risks associated with producing these crops explains, in large part, the tendency of former tobacco farmers to instead expand cattle and/or hay production. From several perspectives, cattle (and other ruminant) production seems highly compatible with the Appalachian region. Scaglia et al. (2008) consider the Appalachian region to be "ideally suited for grassland-based beef production" since tall fescue thrives in the hilly terrain throughout much of the year, providing reliable forage and

Figure 14.6 Annual crops, in this case strawberries, grown on beds with black plastic mulch and rye (*Secale cereale*) seeded as a cover crop in the strips between the beds to protect soil and build soil organic matter.

preventing erosion on slopes. Indeed, according to the 2007 Agricultural Census, cattle densities are relatively high throughout much of the central Appalachian region, particularly in Virginia, Tennessee, and part of Kentucky, and cattle sales account for a sizable fraction of total farm income in these three states as well as in West Virginia (Table 14.2). But that does not mean that cattle production is necessarily economically viable or environmentally sound. Young (2006) examined the economics of small beef cattle operations in Appalachian Virginia and found the enterprise unlikely to be profitable at all. Farms using management-intensive grazing with forage stockpiling performed better financially than traditional practices relying on hay production because they could sustain higher stocking rates and had lower machinery costs, but none of the scenarios examined generated a profit.

Environmentally, grass-based, beef cattle production in Appalachia would seem to make sense as long as overgrazing is avoided. Grass-based production, as opposed to finishing on grain, has been promoted as a more ecologically sound means of production with lower fuel consumption and GHG emissions as well as improved protection of soil and water quality (Clancy 2006; Gurian-Sherman 2011). But some recent research calls these claims into question. Pelletier et al. (2010) used life cycle assessment (LCA) methods to compare grass-based beef cattle to the more conventional grain-finishing in feedlots in the northern Midwest of the United States. They concluded that on a live-weight basis, grass-finished beef actually consumed more energy, had a larger ecological foot-print, and generated more GHG emissions than the grain-based feedlot system due in large part to the methane emissions over the longer production period on grass. An important conclusion from the research was that none of the beef production systems were considered "ecologically efficient" in comparison to other food production systems when the food returns on investment were calculated. More recent research suggests that methane production by ruminant livestock and manure may be even greater than previously known (Miller et al. 2013), raising more doubts about the ecological rationality of cattle production, even on grass.

If beef cattle production on small Appalachian farms is only marginally viable economically, and its environmental performance is inferior to other food production systems, are there any social benefits from supporting its prevalence in the region? To many small farmers in Appalachia, cattle production may be more about a chosen lifestyle than a business endeavor to support livelihoods (McBride and Mathews 2011). "Residential/lifestyle" farms are dependent largely on off-farm income, generate an average of only a few thousand dollars in net farm income annually, and are nearly as likely to generate negative net returns as a profit (Hoppe and Banker 2010). They are therefore maintained as a lifestyle choice or hobby, subsidized with off-farm income and the farm owners' unpaid labor. In a sense, these farmers subsidize the entire beef industry and its consumers by willingly or unknowingly accepting financial losses while supplying beef to the market. This situation would apply to most sheep and goat farmers in Appalachia as well. Enterprise budgets from land-grant universities in the region show very marginal returns at best for these species, and the risks of predators as well as the costs for parasite and disease management create little prospect for profitability.

From a public health perspective, there is probably not the same conscious association between ruminant livestock production and the health consequences of red meat consumption as there is between tobacco production and smoking. But diet-related illnesses are prevalent in Appalachia, as they are across the entire United States. Red meat consumption is associated with higher mortality rates due to cardiovascular disease, cancer, and diabetes (Pan et al. 2012; Walker et al. 2005). Research, however, indicates that grass-finished beef has measurable nutritional benefits over grain-finished beef, including more desirable fatty acid profiles and higher cancer-fighting antioxidant content (Daley et al. 2010). The risks of cardiovascular disease and cancer may therefore be diminished, at least somewhat, by choosing grass-finished over grain-finished beef. This raises an intriguing question about the relative health benefits of simply reducing red meat consumption versus choosing grass-finished red meat over the conventional, grain-finished equivalent. Both could contribute to improvements in public health and reductions in health-care costs.

Taken in total, there are some compelling agroecological arguments in favor of and against producing beef cattle and other ruminants in central Appalachia. There are trade-offs that have different costs and benefits depending on the perspective assumed. The risks to producers are relatively low in comparison to other options, like vegetables, and there are well-established channels, even for small producers, for getting animals to market. This is often not the case for other livestock species—like hogs, chickens, and fish—or the grains and legumes for feeding these livestock. The hog and poultry industries are highly vertically integrated and concentrated, and therefore leave few opportunities for raising these species on a small scale, except for specialty, niche markets where consumers seek out humanely raised meats, such a pastured pork (Figure 14.7). So cattle is a safe choice that is scalable to available resources, like land and labor, and therefore an attractive option if expectations for profit are minimal.

14.4.1 Farming to Food Systems

Some of the most notable and influential efforts to improve food systems in central Appalachia come from organizations and individuals working in sectors of the food supply chain other than crop and livestock production. These include local and regional branding and marketing campaigns (Reul 2012), value-added processing using shared-use facilities, and innovative businesses addressing the "triple bottom line" of "people, planet, and profits." Not only do these endeavors have the capacity to create greater market demand for agricultural products from central Appalachia, they also serve as an important link between farmers and consumers, facilitating communication and education. Although the Appalachian region has a slightly larger percentage of its population employed in farming and forestry (2.5%) compared to the entire United States (1.9%), it is still a small fraction of the total population (ARC 2011). Mechanisms are needed to facilitate interactions between farmers and their actual or potential consumers.

Arguably the most successful branding and marketing campaign in the region is the "Appalachian Grown" program of the Appalachian Sustainable Agriculture Project (ASAP), a nonprofit organization with the stated goal of supporting family farms and rural communities in western North Carolina

Figure 14.7 Hogs raised on pasture offers consumers an alternative to products derived from industrial confinement production and offer small farmers a means of production without the significant capital investment. Advantages to outdoor production include more humane conditions and absence of subtherapeutic antibiotics in the feed. In some ways, such production methods are similar to traditional practices in the region with the following two major exceptions: the addition of portable electric fencing as a management tool and the disappearance of American chestnut (*Castanea dentata*) as a food source.

(Kirby et al. 2007). Since 2006, it has offered certification to farms and businesses in western North Carolina and neighboring counties of adjacent states to verify product location of origin while strongly promoting certified products as a means of strengthening the regional economy for rural communities. It markets to consumers, restaurants, grocery stores, and institutions with a readily recognizable logo, local food buying guide, farm tour program, automobile bumper stickers, and other promotional items and events that facilitate linkages, direct and indirect, between farmers and consumers (Figure 14.8). In addition, the organization provides staff support to farmers markets in Asheville, North Carolina, the city where it is based, and sponsors educational workshops and conferences for small farmers, particularly on the business of farming. It has no requirements for and makes no claims about production practices—the focus is solely on supporting the community by buying local products.

Probably in large part the result of ASAP's work, the level of public interest in and support for local agriculture and food systems in Asheville and the surrounding region of western North Carolina is unusually strong, especially for Appalachia (Figure 14.9). There are several farmers' markets scheduled for different locations and days throughout the week as well as dozens of CSAs in and around the city, offering a wide range of fresh produce and meats throughout most or all of the year. But the small city also boasts an impressive array of less common examples of food-system entrepreneurship. For example, the Carolina Ground Flour Mill was established as a project in 2008, with support from the Carolina Farm Stewardship Association, to address issues affecting an often overlooked part of local food systems: staple grains. The purpose of establishing the milling business was "closing the gap between the farmer and baker" while creating markets for organic grain growers and reducing food miles (Lapidus 2013). Similarly, the Riverbend Malt House,

Figure 14.8 Branding and marketing campaigns can be effective mechanisms to educate consumers and increase demand for local and regional products. The "Appalachian Grown" program was developed by ASAP, a nonprofit organization with the stated goal of supporting family farms and rural communities in western North Carolina.

Figure 14.9 A painted mural in an alley in downtown Asheville calls attention to and raises awareness about the importance of agriculture to the area's history and culture and celebrates not only the cherished final products, like honey, but also the less pleasant stages of getting food to consumers, like livestock slaughter. To both support and take full advantage of the thriving local foods scene that has emerged, the local government has dubbed the city "Foodtopia" and the first "Foodtopian Society."

established a couple of years later, created a link between grain farmers and craft beer brewers as an artisanal malting business using regionally produced, and often certified organic, barley, wheat, and rye (Ackley 2012). Though these small businesses occupy a different position in the food supply chain from farmers, they struggle with many of the same challenges. They require high-quality inputs (grains), need sufficient storage capacity, must protect the grains from pests to ensure an uncontaminated product for consumers, need to balance costs of labor and mechanization, and are often subjected to price pressures between their grain suppliers (farmers) and customers (bakers and brewers), who may expect prices comparable to commodities. Because of this situation, a commitment to the shared values of a local food economy are essential for such alternative food value chains to function successfully within the larger globalized food system.

That kind of common vision is apparent throughout Asheville's local food economy and receives strong support from a broad assemblage of cooperating stakeholders: nonprofit organizations, schools and colleges, businesses, and the city government. One example of this is Blue Ridge Food Ventures, an incubator kitchen—a commercially equipped facility housed in a community college to assist entrepreneurs in launching food and natural-product businesses (Lefler 2008). Established in 2005, the idea was inspired by a similar organization in Athens, Ohio, called the Appalachian Center for Economic Networks (ACEnet), which was the first food-focused, small business incubator in central Appalachia (Morris and Nogrady 1999). These organizations provide necessary facilities for food manufacturing, training, and consulting services to support new businesses. Like ACEnet, Blue Ridge Food Ventures has played a vital role in establishing dozens of food-product businesses, many of which use locally produced agricultural ingredients in their products. These business thereby generate economic benefits for farmers, directly create new food manufacturing jobs, and lead to greater retail revenue in the community.

Examined from an agroecological perspective, the exceptional developments that have occurred in Asheville over the past decade appear to yield the kinds of socioeconomic benefits that advocates of local food systems typically ascribe to them: farm product and income diversification; new jobs; stimulation for rural economies; cooperation and partnerships among different sectors of the food economy for public and private benefits; entrepreneurial opportunities; and greater availability of fresh, healthful foods (Martinez et al. 2010; O'Hara 2011). But questions remain about the environmental benefits of these trends. It is still unclear what the net effects are of localizing food systems on energy consumption, GHG emissions, environmental contamination, and degradation of ecosystem services and qualities. Weber and Matthews (2008), for example, reported that transportation accounts for a relatively small percentage of GHG emissions in the U.S. food system, and that a shift away from red meat and dairy products would be a more effective means of reducing environmental impact than buying locally. And there is not necessarily a reduction in pesticide use in crops or antibiotic use in livestock, or even improvements to animal welfare, with food products simply because they are locally produced. The assumption may be that the shortening of food supply chains, and the improved and more open communication it allows between producers and consumers, would bring about the same kinds of environmental and livestock-welfare improvements associated with programs like USDA Certified Organic, Certified Naturally Grown, and Animal Welfare Approved.

While questions remain about net environmental effects and inherent trade-offs that come with food-system localization or regionalization, the socioeconomic questions are likely of more immediate concern for central Appalachia. There is little opportunity to address the environmental effects of farming and food systems when confronting daunting economic and social obstacles. Ideally, agroecologists would attempt to understand and explain how economic, social, and environmental problems relate to each other and identify opportunities for implementing solutions that address all of them systematically. This, of course, is an enormous challenge that has seemed to elude the substantial and coordinated efforts of many organizations and individuals in central Appalachia for decades. So, it may be useful to consider the major constraints or obstacles, as well as the opportunities or strengths that may hinder or aid the development of stronger local food economies in the region. Understanding these factors seems a fundamental first step for identifying potential roles for agroecology to contribute to strategies and plans for more sustainable food systems.

14.4.2 Facilitating Food-System Improvements

Clearly there are signs that agroecology, as a science and movement, is being embraced in places within central Appalachia: at educational institutions, at farmers' markets, in city tourism campaigns, and in the multifaceted efforts of numerous nonprofits working to support farmers and communities. But there are also pervasive obstacles to change, and the most substantial is economic (Table 14.6). The relatively low incomes and high dependence on government programs for food and medical care described previously not only limit the potential for generating greater market demand for local products, they also hinder investment in new farming and food endeavors—and this situation is unlikely to improve in the near future. In fact, the coal-mining industry, known for being one of the few sources of high-paying jobs in central Appalachia, even for individuals with limited education, is in decline due to the depletion of the most accessible coal, competition with other coal and fossil energy sources, and environmental regulations aimed at addressing the detrimental effects of mountaintop removal practices, air pollution, and climate change (McIlmoil and Hansen 2010). As with tobacco, attitudes toward the coal industry among central Appalachian residents likely depend on how they perceive it to affect their livelihood as well as those of their families and communities. But despite the high-paying jobs and associated tax revenue generated, the levels of poverty and unemployment in coal-producing counties are among the highest in the region. Higher mortality rates associated with environmental pollutants, smoking, poverty, and lack of education are also concentrated in these coal-mining counties and have considerable societal costs that are often overlooked when tallying the wages and taxes (Hendryx and Ahern 2009). Nevertheless, the loss of these jobs in the coming decades is expected to have "dramatic effects on local and state economies," particularly in West Virginia and Kentucky (McIlmoil and Hansen 2010), and therefore must be factored into any kind of plan for developing farming and food economies.

Table 14.6   Obstacles, Strengths, and Needs Pertaining to the Development of More Sustainable Food Systems in Central Appalachia

Obstacles	Strengths	Needs
Poverty and a predominance of low-wage jobs	Small farms are still abundant	Development of agritourism
Poor public health situation	Cultural heritage in farming and gardening with a preservation of historical practices	Support for the development of unique regional products for export (out of the region)
Historical dependence on tobacco production with government support	Agrobiodiversity: heirloom and heritage crops and livestock	Better understanding of the pros and cons of policies and programs on farming and food systems
Cultural resistance to changes that could yield economic benefits	Scenic landscapes and natural areas	Analysis of the costs and benefits of current investments by governments and nonprofits
Economic dependence on coal industry and its decline	Emerging models of successful agro-food development in a few areas of the region	Development of appropriate technologies, infrastructure, and management systems for small Appalachian farms and rural communities that mitigate risk

Given the current and expected economic conditions in central Appalachia, it seems reasonable to look outside of the region for both investment and markets. Indeed, even the thriving local food economy of Asheville depends heavily on tourists, investors, and entrepreneurs from outside the region. They are drawn there by the creativity and collective enthusiasm for sustainability and local foods. And the city has embraced this grassroots movement in its "Foodtopia" tourism campaign, emphasizing its regional agricultural and food heritage, commitment to small farms and locavorism, and creative and adventurous cuisine. There is also a strong emphasis on environmental awareness and livable wages throughout the food supply chains, from producers to consumers. But creating this kind of situation in other Appalachian cities and towns may be limited as much by politics and culture as it is by economics.

For example, despite the repeal of alcohol prohibition in the United States over 80 years ago, dozens of counties, largely in politically and religiously conservative areas of the country, still prohibit the sale of alcoholic beverages. There is a particularly high concentration of these so-called "dry" counties in central Appalachia, though the number is gradually declining as more voters recognize the economic costs of alcohol prohibition to communities (Herships 2013). Without legal alcohol sales, restaurants and other businesses are reluctant to open, and those that do often struggle without the income generated by beer, wine, and spirits. This creates a serious obstacle to cultivating tourism, especially that centered on agriculture and food, and facilitating the benefits it can bring, such as increases in employment and wages, reductions in poverty, and improved education and health (Reeder and Brown 2005). It is certainly true that the societal costs of alcohol abuse are substantial. But they are much less than those attributed to tobacco, poor diet, and physical inactivity (Mokdad et al. 2004) and are not closely associated with either alcohol production or sales the way that tobacco is. Those who want to purchase alcoholic beverages simply go to counties where it is legal or buy from bootleggers who transport and sell those products illegally. Thus, dry counties bear the costs of alcohol consumption without sharing in the direct and indirect economic benefits.

Similarly, the evangelical religiosity and conservative political views that are prevalent throughout central Appalachia tend to be associated with negative attitudes toward science and the environment (Guth et al. 1995; Pew Research Center 2007). This is a potential barrier for the adoption of more environmentally sound agricultural production practices based on science as well as for building consumer demand for ecolabeled food products. Producers and consumers of certified organic products may be motivated by a range of factors, including environmental concern, personal and family health and nutrition, and support for family farms, as well as the quality and characteristics of products (Hughner et al. 2007). But marketing research indicates that higher income and educational levels are associated with purchasing organic products. Both of these factors—income and education—are lacking in central Appalachia. Thus, the educational, economic, and cultural situation in the region creates conditions that are not expected to strongly support ecologically based agriculture or the food businesses that might depend on it.

Assuming these socioeconomic and cultural barriers can be gradually addressed, what are the region's strengths and opportunities with respect to agriculture and food, and how can they be sufficiently supported and harnessed for sustainable economic development? In a report for the ARC, Jean Haskell, coeditor of the *Encyclopedia of Appalachia* (Abramson and Haskell 2006), wrote that Appalachia "boasts many assets for community and economic development—beautiful landscapes, sparkling waterways, abundant flora and fauna, internationally famous music and craft, a good work force, natural energy resources, and a rich and complicated history" (Haskell 2012). While it is reasonable to question the specific meanings of some of these vague descriptions, they reflect commonly held sentiments about Appalachia and should be given serious consideration. Haskell specifically identified "an ongoing tradition of small farming and home gardening; the Region's vast food diversity, knowledge of seed saving and growing of heirloom varieties of local foods" as particularly important assets that deserve attention.

Indeed, Veteto et al. (2011) concluded that "southern and central Appalachia has the highest documented levels of agrobiodiversity" in North America after identifying over 1400 distinctly named heritage and heirloom vegetable, fruit, and grain cultivars. A number of small, specialty seed companies throughout the region contract with small growers to produce many of these cultivars for sale to gardeners through catalogs. Other rare cultivars are available only through collectors/propagators and seed-saving organizations. The Appalachian Staple Foods Collaborative in Athens, Ohio, was established by a group of individuals concerned in part about the preservation and continued use of older cultivars of staple crops, like beans and corn, as well as maintaining the necessary knowledge, networks and infrastructure for producing, harvesting, processing, distributing, and using crops that have historically provided a significant fraction of human dietary energy and protein. Their efforts, in collaboration with ACEnet, resulted in the establishment of Shagbark Seed and Mill, a company serving as a link between area farmers—certified organic or farming without synthetic fertilizers and pesticides—and consumers within and beyond the Appalachian region (Pressman and Sharp 2010). The company cleans, packages, and sells unprocessed grains (seeds) and produces value-added products like corn chips, crackers, and pasta from them. In doing so it is reestablishing commerce linkages between farms and communities in the region that have diminished in recent decades and supporting the *in situ* preservation of the region's agrobiodiversity.

Interest in preserving and celebrating Appalachian foodways is not limited entirely to heirloom crops. Along with corn and beans, hogs were a common and important component of small Appalachian farms prior to the rapid consolidation of the hog industry several decades ago. They were often raised on a mixture of grain, by-products and food waste—then processed, preserved, and used for home consumption or sold to generate needed cash (Colyer 2001). Today, likely in response to public concerns over industrial hog production—including excessive antibiotic use (Kennedy 2013), animal welfare (Heleski et al. 2004), environmental pollution, and corporate control of the food supply (Food and Water Watch 2012)—there is growing curiosity about traditional hog production and processing methods and market demand for those products. The owners and operators of Black Oak Holler Farm in West Virginia are attempting to address these concerns and preserve the desirable characteristics of the heritage breeds that have been supplanted by those selected for high-density, confinement production. They raise heritage breeds, like Ossabaw and Large Black, on pasture and woodlots to supply Woodlands Pork, a company specializing in the production of traditional dry-cured pork products for sale in large urban markets on the east coast of the United States (Black 2007; Hiersteiner 2012). Market demand for such specialty products may be limited in rural Appalachia but is strong in large, diverse metropolitan areas. In support of endeavors like this one, the Livestock Conservancy, a nonprofit organization in North Carolina dedicated to conserving and promoting

Figure 14.10 A heritage-hog butchering demonstration and pork tasting in Kentucky as part of a traveling event called Cochon 555, which promotes heritage breeds, family farms, and local agriculture.

endangered livestock breeds, coordinated a project to evaluate growth and meat qualities of rare and endangered hog breeds in order to assist small- to mid-scale producers interested in pursuing this kind of enterprise. Expanding public interest in heritage hogs is evident in the proliferation of meat-processing demonstrations, charcuterie courses, and heritage pork tasting events (Figure 14.10).

Taking advantage of the region's cultural and biological assets to brand agricultural and food products sold outside of Appalachia, as well as to bring people into the region for agritourism, seems a fundamental first step for any kind of agricultural-based economic development. In areas where coal mining has not devastated the landscape and scenic natural areas have been protected from degradation, garbage, and litter, the region holds broad appeal to visitors and offers substantial opportunities for the kinds of agro-food tourism seen in Asheville and now emerging elsewhere (Figure 14.11). But scenic beauty and heritage are likely insufficient to spur the kind of transformation needed to address the deep socioeconomic problems that plague the region. Next, I suggest three interrelated areas where agroecologists—working with community activists, entrepreneurs,

Figure 14.11 The Appalachian region's scenic mountain landscapes along with its cultural and biological heritage are valuable assets for spurring economic development from agritourism, such as this mountain farm tour in North Carolina.

investors, politicians, regulators, educators, marketers, and farmers—could make useful contributions toward improving farming and food systems in central Appalachia (Table 14.6).

14.5 A ROLE FOR AGROECOLOGY

First, there is a need for developing and evaluating appropriate technology, infrastructure, and management systems for small farmers, food processors, distributors, and others throughout the food system to ensure sufficient economies of scale to be efficient and competitive while still taking into consideration the obstacle of limited capital for investment and the need for environmental and community protection. Small-scale production relying largely on hand tools and manual labor (e.g., permaculture practices) may have minimal negative environmental impacts and the maximum food return per unit of fossil energy expended, but will likely be uneconomical under current conditions except when it supplies high-end, urban markets with specialty products— the average expendable incomes in central Appalachia are simply too low. On the other hand, substantial investment in machinery and infrastructure is risky when markets, competition, and weather conditions are uncertain, which is almost always. Agroecologists could make contributions by quantifying the economic, social, and environmental trade-offs of distinct production and business management models. These distinctions could be based on different production philosophies (organic, biodynamic, permaculture, local, etc.), levels of mechanization and human labor, certification and branding efforts, and organizational and distribution structures (CSAs, food hubs, cooperatives, regional auctions, etc.). Useful analyses could include assessing financial costs and energy use, environmental pollution and protection, wages and employment, profitability, farmworker health and safety, food quality, and even farmer quality of life and community well-being.

A second area where agroecologists could contribute is in assessing the likelihood and magnitude of risks and rewards, and identifying ways of mitigating the risks and fairly sharing the rewards of entrepreneurial and collaborative endeavors, particularly when public resources are used. University experiment stations and research farms are usually well endowed with equipment and personnel, as well as protected from market risk. Thus, published economic analyses and enterprise budgets generated under these favorable conditions may not adequately portray the level of risk and potential for loss to individual producers. In the past, tobacco farmers were protected with guaranteed markets and price supports—a situation that helped to maintain so many small farms. There may be a general reluctance now to make farming investments with so much uncertainty. It is not that the people of central Appalachia are less able to handle risk than others. To the contrary, there are some heavy risk takers in the region—for example, marijuana growers, moonshine producers, ginseng and mushroom poachers, and dairy farmers who produce and sell raw milk. It is interesting to note that marijuana is actually the largest cash crop in the United States based on economic value and Kentucky, Tennessee, and West Virginia rank among the leading producers (Clines 2001; Gettman 2006). But with limited capital and a need for legal employment with fair wages throughout the region, there must be mechanisms to fairly share costs, risks, and rewards in developing the regional agro-food economy. The commercial incubator kitchens in Athens, Ohio, and Asheville, North Carolina, are relevant examples. Both depend on public and private funds to operate and would not be viable if they relied solely on user fees. A similar facility, the Jackson County Regional Food Center, opened in eastern Kentucky in 2010, and though it has struggled to attract local users in one of the poorest counties in the United States, it is serving some small farmers and entrepreneurs from several states as they attempt to create marketable value-added food products. If this operation were a for-profit business relying on private investment, it would certainly not be viable. Therefore measuring and documenting the long-term impacts of these services and facilities will be important to determine if they truly "incubate" creative ideas and visions that later build the economy. This same kind of analysis could be applied to farmers markets, CSAs, and food hubs as well to figure out the best ways to minimize risk, share costs, and promote success and sustainability.

A third potential area where agroecologists may be able to contribute is in carefully identifying and measuring how different programs, regulations, and policies currently or potentially benefit and/or cost different stakeholders in the agro-food economy. For example, tobacco production, and the federal programs that supported it, protected small farms but also contributed to a serious public health burden. Today it is reasonable to claim that tobacco is the primary reason that the region is still well endowed with so many small farms. But arguments over whether tobacco's net effect has been positive or negative might never be resolved. A more tractable and relevant problem might be in assessing the impacts of different types of support or assistance by governmental and non-governmental organizations, like cost-sharing, grants, and technical assistance. A relatively new organization called Grow Appalachia is attempting to address poor diets, food insecurity, and the declining interest in farming and gardening with its multistate outreach program. This ambitious undertaking, with its multiple, interrelated objectives, provides religious and community organizations with funding for wages, seeds, equipment, and advice to establish community gardens, the food from which can be distributed to members or sold for income. How does Grow Appalachia's efforts affect the eating habits and financial conditions of those families involved, the incomes of other farmers and gardeners in the area who may have previously sold vegetables to those families, and area businesses like farm-supply and grocery stores? And do the program's local impacts persist after the funds have discontinued? How does this form of assistance compare to alternative means of investment, like the establishment of a new business that provide jobs? These are the kinds of relevant and challenging questions that should and could be addressed to assist in future policy creation and decision making.

Some might argue that in so strongly revering its cultural heritage and traditions, the Appalachian region is resistant to change—even to that which could yield tangible economic, social, and environmental benefits. The sometimes dogmatic following and impassioned defense of heirloom seeds and crops may be a reflection of this. Haskell (2012) mentioned the term "heirloom" more than two dozen times in her assessment of the food systems of Appalachia. Similarly, Reul (2012) pointed out the significant potential branding value that "heirloom" might have for Appalachian products. And Grow Appalachia requires that its participants grow heirlooms in the community gardens "to preserve and strengthen Appalachia's horticultural heritage" (Grow Appalachia 2013). But despite their allure, heirlooms typically offer both advantages and disadvantages relative to their nonheirloom and hybrid counterparts (Tortorello 2011). A willingness to honestly acknowledge this fact allows for more informed, rational decision-making in seed selection and does not have to comprise cultural values.

Likewise, farming and food policy options should be approached in the same systematic way we might weigh the pros and cons of vegetable cultivars to plant on a farm. Laws governing alcoholic beverage sales or the ability of a farmer to make and sell a product, for example, can be weighed in terms of public safety, economic impact, job creation, quality of life, risks to stakeholders, and other variables. Changes can then be justified when the benefits reasonably appear to outweigh the costs. In 2003, for example, Kentucky House Bill 391 was enacted to allow limited sales at farmers markets of value-added products made in home kitchens (University of Kentucky 2013b). A decade later the impact can be seen across the state at farmers' markets where vendors sell baked goods, sauces, and jams—products that were once illegal at such markets. A seemingly small modification of a state policy created opportunities for thousands of producers and has given more choice to consumers. And yet, no one has quantitatively measured the full impact of this law on the state's food system. Such an exercise could yield valuable information for assessing the costs and benefits to stakeholders of similar policy changes in the future.

A closer examination of other food laws and regulations intended to protect consumers might show that they also create unnecessary bureaucratic impediments that present serious obstacles to small-scale producers—in effect favoring industrialization and consolidation. For example, small, independent poultry producers have limited options for getting their chickens, turkeys, and ducks processed. Because simple, on-farm processing is prohibited in Kentucky, producers often must travel long distances at considerable cost to get small batches of birds processed to legally sell. Similarly, small-scale aquaculture has been promoted for decades by land-grant universities and cooperative extension offices as a possible alternative to tobacco. Yet farmers have no reasonable markets for small batches of live fish. Processing those fish into value-added products like fillets could be an option but requires an approved Hazard Analysis and Critical Control Points (HACCP) plan, courses for certifications offered only sporadically in distant cities, costly new facilities or upgrades, multiple inspections by health department officials, and associated inspection fees. These requirements are imposed even to simply de-head freshwater prawns for sale directly to consumers. Such examples demonstrate that regulatory demands can be serious disincentives, raising the costs and risks beyond what most reasonable small farmers and artisanal food processors would or could accept.

A willingness to look at new possibilities and creative, unconventional solutions while accepting that change is inevitable and necessary seems to be a fundamental prerequisite for improving the food system of central Appalachia—from supporting opportunities for small farms to curbing obesity rates. There are linkages to be understood between dietary habits and the regional macro-level environment, including food production and distribution, food and agriculture policies, food assistance programs, land use, and societal norms and values (Story et al. 2008). Change efforts that recognize these interactions may provide the most benefit with limited available resources. The unsettling regional socioeconomic indicators and trends as well as the inevitable but largely unknown impacts of climate change portend a daunting challenge for the communities of Appalachia over the next generation. The changes within human capacity that occur can either be guided by our best understanding or simply be allowed to happen at the discretion of others who may have little appreciation for or understanding of farming, food, or science. Agroecologists therefore need to engage with citizens, organizations, communities, regulators, and policy makers—bringing with them truly practical understandings of farming and broad skills in analyzing complex systems—if they want and reasonably expect the ecology of food systems to be given serious consideration in central Appalachia.

REFERENCES

Abramson, R. and J. Haskell. 2006. *Encyclopedia of Appalachia*. Knoxville, TN: University of Tennessee Press.

Ackley, D. 2012. Supporting sustainable beer in Asheville, NC: Riverbend Malt House. Triple Pundit: People, Planet, Profit. June 28. http://www.triplepundit.com/2012/06/supporting-sustainable-beer-asheville-nc-riverbend-malt-house/ (accessed: February 18th, 2014).

Altieri, M. A. 1987. *Agroecology: The Scientific Basis for Alternative Agriculture*. Boulder, CO: Westview Press.

Altieri, M. A. 2002. Agroecology: The science of natural resource management for poor farmers in marginal environments. *Agr Ecosyst Environ* 93:1–24.

ARC (Appalachian Regional Commission). 2009. Subregions of Appalachia. ARC. http://www.arc.gov/research/MapsofAppalachia.asp?MAP_ID=31 (accessed: February 18th, 2014).

ARC (Appalachian Regional Commission). 2011. Economic overview of Appalachia. ARC. http://www.arc.gov/images/appregion/Sept2011/EconomicOverviewSept2011.pdf (accessed: February 18th, 2014).

Black, J. 2007. Better cured meat begins on the hoof at home. *Washington Post*. November 28. http://www.washingtonpost.com/wp-dyn/content/article/2007/11/27/AR2007112700620.html.

Bloch, M., J. DeParle, M. Ericson, and R. Gebeloff. 2009. Food stamp usage across the country. Interactive map using data from the USDA Economic Research Service. *The New York Times*. November 28. http://www.nytimes.com/interactive/2009/11/28/us/20091128-foodstamps.html (accessed: February 18th, 2014).

Capehart, T. 2004. Trends in US tobacco farming. Economic Research Service, US Department of Agriculture.

Carroll, C. R., J. H. Vandermeer, and P. M. Rosset. 1990. *Agroecology*. New York: McGraw-Hill.

CDC (Centers for Disease Control and Prevention). 2012. Tobacco control state highlights. Atlanta, GA: CDC. http://www.cdc.gov/tobacco/data_statistics/state_data/state_highlights/2012/ (accessed: February 18th, 2014).

Clancy, K. 2006. Greener pastures: How grass-fed beef and milk contribute to healthy eating. Cambridge, MA: Union of Concerned Scientists. http://www.ucsusa.org/assets/documents/food_and_agriculture/greener-pastures.pdf (accessed: February 18th, 2014).

Clark, S. and J. Meyer. 2013. Financial comparison of organic potato production using different integrated pest management systems. In S. K. Saha, J. Snyder, and C. Smigel (eds), *2013 Fruit and Vegetable Crops Research Report*, pp. 33–36. Lexington, KY: University of Kentucky College of Agriculture. http://www2.ca.uky.edu/agc/pubs/PR/PR673/PR673.pdf (accessed: February 18th, 2014).

Clines, F. X. 2001. Kentucky Journal; Fighting Appalachia's top cash crop, marijuana. *The New York Times*, February 28.

Colyer, D. 2001. Agriculture in the Appalachian Region: 1965–2000. In West Virginia University, Department of Agricultural Resource Economics Conference Papers, No. 19101. http://ageconsearch.umn.edu/bitstream/19101/1/cp01co02.pdf (accessed: February 18th, 2014).

Craig, V. A. 2008. Tobacco grower livelihoods during agricultural restructuring. *J Rural Community Develop* 3:23–40.

Daley, C. A., A. Abbott, P. S. Doyle, G. A. Nader, and S. Larson. 2010. A review of fatty acid profiles and antioxidant content in grass-fed and grain-fed beef. *Nutr J* 9:1–12. http://www.nutritionj.com/content/pdf/1475-2891-9-10.pdf (accessed: February 18th, 2014).

Davis, J., S. E. Johnson, and K. Jennings. 2010. Herbicide carryover in hay, manure, compost and grass clippings. North Carolina State University and North Carolina A & T Cooperative Extension special publication. http://www.ces.ncsu.edu/fletcher/programs/ncorganic/special-pubs/herbicide_carryover.pdf (accessed: February 18th, 2014).

Flaccavento, A. 2010. The transition of Appalachia. *Solutions* 1:34–44. http://www.thesolutionsjournal.com/node/718 (accessed: February 18th, 2014).

Food and Water Watch. 2012. The economic cost of food monopolies. Washington, DC: Food and Water Watch. http://www.foodandwaterwatch.org/doc/CostofFoodMonopolies.pdf (accessed: February 18th, 2014).

Francis, C., G. Lieblein, S. Gliessman et al. 2003. Agroecology: The ecology of food systems. *J Sustain Agr* 22:99–118.

Gale Jr, H. F., L. F. Foreman, and T. C. Capehart Jr. 2000. Tobacco and the economy: Farms, jobs, and communities. No. 34007. United States Department of Agriculture, Economic Research Service.

GAO (United States Government Accountability Office). 2003. Pesticides on tobacco: Federal activities to assess risks and monitor residues. Report GAO-03-485. March. U.S. Government Accountability Office. http://www.gao.gov/products/GAO-03-485 (accessed: February 18th, 2014).

Gettman, J. 2006. Marijuana production in the United States (2006). *Bulletin of Cannabis Reform* issue 2. http://www.drugscience.org/Archive/bcr2/MJCropReport_2006.pdf (accessed: February 18th, 2014).

Gliessman, S. R. 1990. *Agroecology: Researching the Ecological Basis for Sustainable Agriculture*. New York: Springer.

Gliessman, S. R. 2007. *Agroecology: The Ecology of Sustainable Food Systems*. Boca Raton, FL: CRC Press.

Gragson, T. L., P. V. Bolstad, and M. Welch-Devine. 2008. Agricultural transformation of southern Appalachia. In C. L. Redman and D. R. Foster (eds), *Agrarian Landscapes in Transition: Comparisons of Long-Term Ecological and Cultural Change*, pp. 89–121. New York: Oxford University Press.

Gregg, E. W., K. A. Kirtland, B. L. Caldwell et al. 2009. Estimated county-level prevalence of diabetes and obesity: United States, 2007. *MMWR Morb Mortal Wkly Rep* 58:1259–1263.

Grow Appalachia. 2013. What we do. Grow Appalachia. http://www.berea.edu/grow-appalachia/what-we-do/ (accessed: February 18th, 2014).

Gurian-Sherman, D. 2011. Raising the steaks: Global warming and pasture-raised beef production in the United States. Cambridge, MA: Union of Concerned Scientists. http://www.ucsusa.org/assets/documents/food_and_agriculture/global-warming-and-beef-production-report.pdf (accessed: February 18th, 2014).

Guth, J. L., J. C. Green, L. A. Kellstedt, and C. E. Smidt. 1995. Faith and the environment: Religious beliefs and attitudes on environmental policy. *Am J Polit Sci* 39:364–382.

Halverson, J. A. and G. Bischak. 2008. Underlying socioeconomic factors influencing health disparities in the Appalachian Region. Report to the Appalachian Regional Commission. http://www.arc.gov/research/researchreportdetails.asp?REPORT_ID=9 (accessed: February 18th, 2014).

Haskell, J. 2012. Assessing the landscape of local food in Appalachia. A report for the Appalachian Regional Commission. http://www.arc.gov/assets/research_reports/AssessingLandscapeofLocalFoodinAppalachia.pdf (accessed: February 18th, 2014).

Heimlich, R. 2000. Farm Resource Regions. USDA ERS Agriculture Information Bulletin No. (AIB-760). USDA. http://www.ers.usda.gov/publications/aib-agricultural-information-bulletin/aib760.aspx#.UpimOdLBP90 (accessed: February 18th, 2014).

Heleski, C. R., A. G. Mertig, and A. J. Zanella. 2004. Assessing attitudes toward farm animal welfare: A national survey of animal science faculty members. *J Anim Sci* 82:2806–2814.

Hendryx, M. and M. M. Ahern. 2009. Mortality in Appalachian coal mining regions: The value of statistical life lost. *Public Health Rep* 124:541–550.

Herships, S. 2013. There are communities where prohibition still exists, and it's costing them. *Marketplace Morning Report*. December 5. http://www.marketplace.org/topics/economy/there-are-communities-where-prohibition-still-exists-and-its-costing-them (accessed: February 18th, 2014).

Hiersteiner, S. 2012. Some pig. *Flavor*. March 23. http://flavormagazinevirginia.com/somepig/ (accessed: February 18th 2014).

Hoppe, R. A. and D. E. Banker. 2010. Structure and finances of US farms. US Department of Agriculture, Economic Research Service, EIB #6 6.

Hughner, R. S., P. McDonagh, A. Prothero, C. J. Shultz, and J. Stanton. 2007. Who are organic food consumers? A compilation and review of why people purchase organic food. *J Consum Behav* 6:94–110.

Kennedy, D. 2013. Time to deal with antibiotics. *Science* 342:777.

Kirby, L. D., C. Jackson, and A. Perrett. 2007. Growing local: Expanding the western North Carolina food and farm economy. Asheville, NC: Appalachian Sustainable Agriculture Project. http://asapconnections.org/downloads/growing-local-expanding-the-western-north-carolina-food-and-farm-economy-full-report.pdf (accessed: February 18th, 2014).

Lapidus, J. 2013. Farmer + miller + baker. *Food Arts*. November 11. http://www.foodarts.com/menu/the-pros/29472/farmer-miller-baker (accessed: February 18th, 2014).

Law, D. M., A. B. Rowell, J. C. Snyder, and M. A. Williams. 2006. Weed control efficacy of organic mulches in two organically managed bell pepper production systems. *HortTechnology* 16:225–232.

Lefler, S. 2008. Blue Ridge food ventures. *Smoky Mountain Living*. September 1. http://www.smliv.com/features/1408-blue-ridge-food-ventures.html (accessed: February 18th, 2014).

Loomis, R. L. 1984. Traditional agriculture in America. *Annu Rev Ecol Syst* 15:449–478.

Lotter, D. W. 2003. Organic agriculture. *J Sustain Agr* 21:59–128.

Loucks, O. L. 1977. Emergence of research on agro-ecosystems. *Annu Rev Ecol Syst* 8:173–192.

Martinez, S., M. Hand, M. Da Pra et al. 2010. Local food systems: Concepts, impacts, and issues, ERR 97, U.S. Department of Agriculture, Economic Research Service Report Number 97. http://www.ers.usda.gov/publications/err-economic-research-report/err97.aspx#.UrJQtdJDvOs.

McBride, W. and K. Mathews. 2011. The diverse structure and organization of U.S. beef cow-calf farms. USDA Economic Information Bulletin # 73. http://www.ers.usda.gov/publications/eib-economic-information-bulletin/eib73.aspx#.UqSxl9JDvOs.

McIlmoil, R. and E. Hansen. 2010. The decline of central Appalachian coal and the need for economic diversification. Morgantown, WV: Downstream Strategies. http://www.downstreamstrategies.com/documents/reports_publication/DownstreamStrategies-DeclineOfCentralAppalachianCoal-FINAL-1-19-10.pdf (accessed: February 18th, 2014).

Mendieta, M. P., M. Velandia, D. M. Lambert, and K. Tiller. 2012. The potential of other crop and livestock enterprises to replace tobacco: Perceptions of U.S. burley producers. *J Extension* 50(5). http://www.joe.org/joe/2012october/rb8.php (accessed: February 18th, 2014).

Miller, S. M., S. C. Wofsy, A. M. Michalak et al. 2013. Anthropogenic emissions of methane in the United States. *Proc Natl Acad Sci USA* 110:20018–20022.

Mokdad, A. H., J. S. Marks, D. F. Stroup, and J. L. Gerberding. 2004. Actual causes of death in the United States, 2000. *J Am Med Assoc* 291:1238–1245.

Morris, L. and S. Nogrady. 1999. ACEnet replication manual. Athens, OH: Appalachian Center for Economic Networks. http://www.whyhunger.org/uploads/fileAssets/d527aa_4fcc1c.pdf (accessed: February 18th, 2014).

Morton, L. W. and T. C. Blanchard. 2007. Starved for access: Life in rural America's food deserts. *Rural Realities* 4:1–10. http://www.ruralsociology.org/wp-content/uploads/2012/03/Rural-Realities-1-4.pdf (accessed: February 18th, 2014).

O'Hara, J. K. 2011. Market forces: Creating jobs through public investment in local and regional food systems. Washington, DC: Union of Concerned Scientists. http://sustainableagriculture.net/wp-content/uploads/2011/08/market-forces-report.pdf (accessed: February 18th, 2014).

Pan, A., Q. Sun, A. M. Bernstein et al. 2012. Red meat consumption and mortality: Results from 2 prospective cohort studies. *Arch Intern Med* 172:555–563.

Pelletier, N., R. Pirog, and R. Rasmussen. 2010. Comparative life cycle environmental impacts of three beef production strategies in the upper midwestern United States. *Agr Syst* 103:380–389. http://www.leopold.iastate.edu/sites/default/files/pubs-and-papers/2010-04-comparative-life-cycle-environmental-impacts-three-beef-production-strategies-upper-midwestern-unite.pdf (accessed: February 18th, 2014).

Pew Research Center. 2007. Science in America: Religious belief and public attitudes. September 18. Washington, DC: Pew Research Center. http://www.pewforum.org/2007/12/18/science-in-america-religious-belief-and-public-attitudes/ (accessed: February 18th, 2014).

Pollard, K. and L. A. Jacobson. 2013. The Appalachian region: A data overview from the 2007–2011 American Community Survey Chartbook. Appalachian Region Commission. http://www.arc.gov/research/researchreportdetails.asp?REPORT_ID=103 (accessed: February 18th, 2014).

Pressman, A. and H. Sharp. 2010. Appalachian Staple Foods Collaborative share information about staple beans, grains, and oilseeds. NCR-SARE's Field Blog. January 21. http://ncrsare.blogspot.com/2010/01/appalachian-staple-foods-collaborative.html (accessed: February 18th, 2014).

Reeder, R. J. and D. M. Brown. 2005. Recreation, tourism, and rural well-being. USDA Economic Research Service Report Number 7, Washington, DC.

Reul, L. 2012. Branding study for Appalachian local food economies. Central Appalachian Network. http://www.ams.usda.gov/AMSv1.0/getfile?dDocName=STELPRDC5105334 (accessed: February 18th, 2014).

Sayre, L. and S. Clark. 2011. *Fields of Learning: The Student Farm Movement in North America*. Lexington, KY: University Press of Kentucky.

Scaglia, G., W. S. Swecker, J. P. Fontenot et al. 2008. Forage systems for cow-calf production in the Appalachian region. *J Anim Sci* 86:2032–2042.

Spalding, D. and T. Coolong. 2012. On-farm commercial vegetable demonstrations in central Kentucky. In T. Coolong, J. Snyder, and C. Smigel (eds), *2012 Fruit and Vegetable Crops Research Report*, pp. 6–7. Lexington, KY: University of Kentucky College of Agriculture. http://www2.ca.uky.edu/agc/pubs/pr/pr656/pr656.pdf (accessed: February 18th, 2014).

Stimers, M., R. Bergstrom, T. Vought, and M. Dulin. 2011. Capital vice in the Midwest: The spatial distribution of the seven deadly sins. *J Maps* 7:9–17. http://dx.doi.org/10.4113/jom.2011.1133.

Story, M., K. M. Kaphingst, R. Robinson-O'Brien, and K. Glanz. 2008. Creating healthy food and eating environments: Policy and environmental approaches. *Annu Rev Publ Health* 29:253–272.

Tomich, T. P., S. Brodt, H. Ferris et al. 2011. Agroecology: A review from a global-change perspective. *Annu Rev Env Resour* 36:193–222.

Tortorello, M. 2011. Heirloom seeds or flinty hybrids? *The New York Times*. March 23.

University of Kentucky. 2013a. Center for Crop Diversification. http://www.uky.edu/Ag/CCD/.

University of Kentucky. 2013b. Homebased processing and microprocessing. http://www2.ca.uky.edu/agcomm/micro/ (accessed: February 18th 2014).

USDA (United States Department of Agriculture). 2007. Census of Agriculture. National Agriculture Statistics Service. http://www.agcensus.usda.gov/index.php (accessed: February 18th, 2014).

Vandermeer, J. 1995. The ecological basis of alternative agriculture. *Annu Rev Ecol Syst* 26:201–224.

Veteto, J. B., G. P. Nabhan, R. Fitzsimmons, K. Routson, and D. Walker. 2011. Place-based foods of Appalachia: From rarity to community restoration and market recovery. College Park, MD: Sustainable Agriculture Research and Education (SARE). http://www.sare.org/Learning-Center/Project-Products/Southern-SARE-Project-Products/Place-Based-Foods-of-Appalachia (accessed: February 18th, 2014).

Walker, P., P. Rhubart-Berg, S. McKenzie, K. Kelling, and R. S. Lawrence. 2005. Invited paper: Public health implications of meat production and consumption. *Public Health Nutr* 8:348–356.

Weber, C. L. and H. S. Matthews. 2008. Food-miles and the relative climate impacts of food choices in the United States. *Environ Sci Technol* 42:3508–3513.

Wezel, A., S. Bellon, T. Doré, C. Francis, D. Vallod, and C. David. 2009. Agroecology as a science, a movement and a practice. *Agron Sustain Dev* 29:503–515.

White, J., R. Gebeloff, F. Fessenden, A. Tse, and A. McLean. 2012. The geography of government benefits. Interactive map using data from the USDA Economic Research Service. *The New York Times*, February 11. http://www.nytimes.com/interactive/2012/02/12/us/entitlement-map.html?_r=1& (accessed: February 18th, 2014).

Williams, J. A. 2002. *Appalachia: A History*. Chapel Hill, NC: University of North Carolina Press.

Wingo, P. A., T. C. Tucker, P. M. Jamison et al. 2008. Cancer in Appalachia, 2001–2003. *Cancer* 112:181–192.

Young, D. C. 2006. Profitability analysis of forage based beef systems in Appalachia. Doctoral dissertation, The Virginia Polytechnic Institute and State University.

CHAPTER 15

Can Agroecological Practices Feed the World?
The Bio- and Ecoeconomic Paradigm in Agri-Food Production

Lummina G. Horlings

CONTENTS

15.1 Introduction 309
15.2 Bioeconomy in Agri-Food Production 311
15.3 Ecoeconomy in Agri-Food 314
    15.3.1 Agroecology 315
    15.3.2 Multifunctionality and the Deepening, Broadening, and Regrounding of Agriculture... 316
15.4 Framings of Agri-Food Production in the Bio- and Ecoeconomic Paradigm..................... 316
15.5 Ecoeconomic Contribution to Food Security 317
    15.5.1 Organic Agriculture.. 318
    15.5.2 Agroecological Projects.. 318
15.6 Discussion and Conclusions.. 319
References.. 320

15.1 INTRODUCTION

This chapter starts with the challenge to critically consider agroecological perspectives for food production in a period of what might now be regarded as a new era of agri-food productionism. How can we "feed the world" in a truly sustainable way? Can sustainable agricultural farm systems and practices contribute to an efficient, productive, and profitable agriculture, meeting the demands of a growing world population and adapting to climate change?

The challenge to increase food production has become more urgent than ever. The expectation is that the World Food Summit (WFS 1996) target to reduce the number of undernourished between 1990/1992 and 2015 by half will not be met (FAO 2006). More than one billion people are now hungry and undernourished worldwide (FAO 2009).

Agricultural production has shown a spectacular conventional growth since the start of the first green revolution in the 1960s. In per capita terms, it has outpaced population growth and resulted in an extension of irrigated area. Compared to food consumption in 1961, each person today has (pro rata) 25% more food. However, these aggregate figures hide important differences between regions.

The growth has differed across continents: in Africa it rose by 140%, in Latin America by almost 200%, and in Asia by 280%. The greatest increases have been in China, where a fivefold increase of production occurred, mostly during the 1980s and 1990s (FAO 2009).

World population will increase up to at least the mid-twenty-first century, and absolute demand for food will rise. Estimates of population increases over the coming decades vary, for example, depending on the expected average number of children per woman (IAASTD 2009a); but the emerging consensus is that the world will have approximately 9 billion people by about 2050 (UN 2008). Predictions of future food demand also differ, but even the most optimistic scenarios require increases in food production of at least 50% (The Royal Society 2009). Food demands will both grow and shift in the coming decades not only as a result of population growth but also because uneven economic growth increases consumer purchasing power, especially for meats; growing urbanization encourages people to adopt new diets; and climate change variations and events threaten both land and water resources (Pretty et al. 2006).

The successes and limitations of the first green revolution in the 1960s have led to many calls for renewed investment and collaboration. There have been calls for a "greener revolution," a "double-green revolution" (Conway 1997), an "evergreen revolution" (Swaminathan 2000), a "blue revolution" (Annan 2000), and an "African green revolution" (Sanchez et al. 2009). A variety of influential international reports on agriculture and food have been published (see for example World Bank 2008; IAASTD 2009a). These reports express some optimism that the necessary increases in food production can be achieved. Opinions vary, however, about the best way to address these challenges. Others (Ambler-Edwards et al. 2009; Evans 2009) indicate there is a stronger need to invest in what Parrot and Marsden (2002), somewhat earlier, called a "real green revolution." Evans (2009) argues, for instance, for a knowledge-intensive approach based on ecologically integrated approaches that put more power in the hands of farmers rather than seed companies.

In recent decades, many scholars have discussed the side effects of the dominant food paradigm and its myths of efficiency (Morgan et al. 2006). Policies that promote industrial agriculture are justified by their proponents by the claim that large-scale, high-chemical-input, mechanized agriculture is the most efficient form of farming. These premises have attained new power and status as a result of the 2007/2008 food crisis, which saw significant price rises and volatilities in basic food goods, in combination with rises in oil and agricultural inputs; and the emergence of biofuels as a palliative to peak oil concerns. There is a new orthodoxy developing in the global food system and in national policy circles that countries need to regain the postwar priority of intensive productivism as a way of "feeding the 9 billion by 2050" (Evans 2009; The Royal Society 2009).

We argue in this chapter that two dynamic and contesting paradigms for agri-food production are evolving: the ecoeconomic paradigm and the bioeconomic paradigm, each underpinning alternative economic models for sustainable development and food production.

The ecoeconomic paradigm (EEP) may be seen as an essentially sociospatial understanding of both production and consumption spheres consisting of complex networks or "webs" of new viable businesses and economic activities. These activities utilize the varied and differentiated forms of environmental resources in more sustainable ways, which, rather than resulting in a net depletion of resources, provide cumulative net benefits that add value to the environment. Thus, EEP potentially realigns production–consumption chains and captures local and regional value between rural and urban spaces.

The bioeconomic paradigm (BEP) focuses on the (largely corporately controlled) production of biomass and biofuels, together with other strands including biotechnology, genomics, chemical engineering, and enzyme technology. Rather than a local, value-adding phenomenon, the BEP operates at more global, corporate economic levels.

The bio- and ecoeconomy paradigms both make substantive sustainability claims, but identify different pathways for development and unfold different notions of space and place. This raises questions concerning which pathways, if any, will become dominant, how, and why. Further questions surround who will gain control of these pathways, and on what theoretical and conceptual basis will this control rely. This is pertinent given the tendency for sustainability to be co-opted within neoliberal modes of governance (Drummond and Marsden 1995; Hajer 1995).

We would argue here that the dominant food regime has responded to the challenge of food security with a growing emphasis on policy, research, and practice in the bioeconomic paradigm. This strategy may decrease environmental effects to a certain extent, but also exposes some important missing links, as we will show in this chapter. The aim therefore is to outline if and how the EEP may offer an alternative pathway for sustainable agri-food systems, more specifically expressed in the form of agroecological and multifunctional rural practices. The central questions are how do the bio- and ecoeconomic paradigms frame sustainable agri-food production, and what empirical evidence is there that ecoeconomic expressions such as agroecological practices can contribute to food security?

These questions are theoretical as well as empirical. Theoretically, we frame these arguments by conceptualizing the bio- and ecoeconomy in the Sections 15.2 and 15.3. We will show the differences in framing by both paradigms (Section 15.4). A key question, which Terry Marsden and I have addressed in an article in *Global Environmental Change* (Horlings and Marsden 2011a), is the search for empirical examples that can function as alternatives for the dominant food system contributing to food security. Based on this work, we will describe some expressions of the ecoeconomy in agri-food in Section 15.5. Some conclusions will be drawn in Section 15.6. Our analysis is based on an extensive critical review of the available literature (Horlings and Marsden 2010, 2011a; Horlings et al. 2010).

15.2 BIOECONOMY IN AGRI-FOOD PRODUCTION

The bioeconomy can be defined as those economic activities that capture the latent value in biological processes and renewable bioresources to produce improved health and sustainable growth and development. The term "biobased economy" is also used at times, which deals more narrowly with industrial applications: it is an economy that uses renewable bioresources, efficient bioprocesses, and ecoindustrial clusters to produce sustainable bioproducts, jobs, and income (OECD 2006). It covers and indeed actively merges areas such as medicine, nutrition, agriculture, industrial biotechnology, the environment, and security and expresses itself in the (largely corporate-controlled) production of biomass and biofuels, together with other strands including biotechnology, genomics, chemical engineering, and enzyme technology (Juma and Konde 2001; EC 2005).

Different driving forces are enhancing the bioeconomy. Bioeconomic processes have the potential for, and in some cases have already accomplished, the transformation of nature at a fundamental and genetic level. Genetic modification changes the reproductive and biological processes within nature itself, such that it obeys different rules and time–space parameters and reduces some former environmental externalities. In changing the "nature of nature itself," building on Smith (1984, 2007) and Boyd et al. (2001), the bioeconomic paradigm begins to establish a "third nature" (Marsden 2012). This can be seen as a next step after the use of nature by humans ("first nature") and the commoditization and economic exploitation of nature through the capitalist process ("second nature").

The effects of this transformation are further enhanced by technological innovation and the application of industrial ecology principles. Industrial ecology offers an avenue toward energy independence and a more "green economy" (Jordan et al. 2007). Two main characteristics of industrial ecology can be identified. First, there is the ongoing decoupling of production systems from the natural environment. An example is the utility in synthetic biological systems of standardized, composable parts, parts with defined, and therefore predictable, functionalities that enable design and assembly based on specifications alone (Carlson 2007). A second characteristic is ecoefficiency. The trend toward more ecoefficiency has stimulated the clustering of highly efficient firms with closed loops of energy, water, and waste. Bioeconomic efficiency can speed up product cycles. In other words, the bioeconomy frames notions of time and speeding of life cycles (Horlings and Marsden 2011b).

Agri-food systems have been influenced by the combination of bioeconomic developments, technology development, state regulation, and extension services in western European countries in the last decades. Bioeconomic applications are rapidly increasing in agri-food in the form of biotechnological engineering of crops, molecular techniques, and the production of biobased products such as plastics and biofuels. The Organization for Economic Cooperation and Development (OECD) has estimated that the bioeconomy is likely to make a significant contribution to economic activity. Indeed, it estimates that by 2030, the use of biotechnologies will contribute up to 35% of the output of chemicals and other industrial products that can be manufactured using biotechnology, up to 80% of pharmaceuticals and diagnostic production, and some 50% of agricultural output.

With regard to biotechnology, it has been estimated that by 2015, approximately half of global production of the major food, feed, and industrial feedstock crops could come from plant varieties developed using one or more types of biotechnology (OECD 2006). Even without new policies or major breakthroughs, biotechnology could contribute up to approximately 2.7% of GDP in the OECD by 2030. For developing countries, this share could be higher, thanks to the greater importance of primary and industrial production in overall output.

Moreover, as these figures assume business as usual, they probably underestimate the potential effects on energy, health, and farming where a wide range of R&D activities are maturing at a remarkably rapid pace. These include improved healthcare technologies drawing on genetics, genomics, and proteomics; more sustainable and higher-value-added food and fiber production systems; cleaner, more ecoefficient biofuels; enzymatic processing that cuts energy and water consumption as well as the generation of toxic wastes during manufacturing; and stronger bio-(nano)-materials (OECD 2006).

We can identify several different arenas for the unfolding of the bioeconomy in agri-food:

1. The *improvement of efficiency* by traditional breeding and farming techniques.
2. The *manipulation of environmental conditions* such as soil fertility (with fertilizers), land (drainage, irrigation, changing land systems, parceling), and ecological relations (chemical management of pests and weeds).
3. The *decoupling* of agricultural products from the environment (glass houses and intensive animal husbandry), the decoupling of production and processing in agribusiness, and the decoupling of land management and ownership of resources (such as seeds and patents).
4. *Biological engineering* of the specific intrinsic characteristics of nature itself and the industrialization of food products, based on the reduction of raw products in standardized, composable parts. Biotechnology has led to new species, new ecological effects, and the industrial production of "food products," such as artificial meat.

Industrial ecology has created conditions for many new biobased products. The use of raw biobased products for all kinds of applications has been given strong government support in some countries. Biological science has an increasing influence on agricultural production in the form of genetic analysis and genomics, marker technology, genetic modification, and phenotype analysis. The innovation in industrial processing of agricultural production has given way to the large-scale production of biomass. Furthermore, biofuel production is a still-developing bioeconomic sector. It is now dominated by ethanol from sugar and starch-rich plants and fatty acid methyl esthers or biodiesel from vegetable oil. New bioeconomic industrial opportunities lie in the development of technologies for what are often called second-generation biofuels (Nilsson et al. 2009; Schubert 2006).

The application of genetically modified (GM) techniques in crop plants has, however, also raised controversial opinions. In the United States, Argentina, Brazil, India, and Canada, GM crops are grown widely (125 million ha in 2008), whereas in Europe and Africa (except South Africa), they are largely absent (ISAAA 2008). Critiques expect that new biotechnologies will not meet their expectations and will affect farming communities (see, e.g., Altieri and Rosset 1999). One of the major critiques of biotechnology is that it can create new environmental problems and risks we cannot yet foresee:

> By unlocking the secrets of the genome, biotechnology may allow us to sidestep some environmental and resource problems, but it may create wholly new ones. (Anex 2004, p. 2)

The trend of industrial production using plant-derived raw materials may have complex social and environmental impacts. Increased reliance on agriculture as a source of industrial foodstuffs creates conflicts with the need for food production as well as increasing agricultural impacts on water and air quality (Anex 2004). It is resulting in new international "land-grabbing" appropriations, whereby key industrializing countries, such as China, Japan, Korea, and Saudi Arabia, are seeking out land and agricultural resources in parts of Africa (IFPRI 2013; World Bank 2010). We see the competition for land especially increasing in the area of biofuel production.

A relatively recent development is the production of synthetic fuel from microbes, produced by companies like Amyris Biotechnologies. The economic considerations of scaling up microbial production of biofuels are fundamentally and radically different from those of traditional petroleum production and refining, and the costs are probably lower. Costs will fall even further as production eventually moves from alcohols to hydrocarbon biofuels that are completely immersible in water. Eventually it will be possible to treat biomass or waste material as feedstocks for microbes producing more than just fuels. As a result of these processes, more and more of the total production of economically important compounds will soon be miniaturized within biological systems, internalized within single-celled (and eventually multicelled) organisms. There is substantial funding behind these efforts (Carlson 2007).

The bioeconomy model is increasingly seen as a multidimensional panacea to the global onset of climate change, population growth, and energy, water, and food shortages (Horlings et al. 2010). It holds increasingly strong and wide political and economic power and the potential for a more significant spatial development in both urban and rural contexts. We would argue here that bioeconomic developments in agri-food have led to a decrease in environmental problems to some extent; however, there have also been negative side effects. These effects illustrate the exclusion or at best underestimation of the wider and much more diverse social, cultural, political, and spatial dimensions of agriculture. We can outline at least four missing dimensions (Horlings and Marsden 2011a):

Socially, we have seen a large decrease of agricultural employment and a loss of farmers' freedom, with more dependency on privately regulated global markets, retailers, research, and policy measures (Van der Ploeg 1991; Horlings 1996). Primary producers at the end of the chain bear the responsibility for the quality of products but are excluded from the often more lucrative, value-added, retailer-led food markets. For those that do gain entry, the degree of informal control over their operations severely constrains their room to maneuver (Marsden 2004). Weak ecological modernization in this sense can lead to loss of autonomy at the local scale.

Culturally, "the environment" is reduced to a series of fragmented or "closed-box" concerns about resource inputs, waste, and pollution emissions. As cultural needs and nonanthropocentric values cannot be reduced to monetary terms, they tend to be marginalized or excluded from consideration (Christoff 1996). The culture of agri-"culture" itself, expressed in craftsmanship and a large variety of locally embedded farming styles, has become marginalized as the influence of external institutes such as extension services and scientific research has increased (Horlings 1996).

Politically, in agriculture a "hygienic mode of regulation" has become dominant in agri-food in the form of the proliferation of arms-length bureaucratic forms of environmental safeguards and instruments. Private and public forms of regulation have led to a schematization that creates new regulatory barriers to market entry for many smaller producers and processors. Farms (and farmers) have taken the brunt of this scientific risk management strategy (see Marsden et al. 2010).

Spatially, intensive agricultural production has been decoupled and fragmented from space and place, visible in the form of footloose production, international food transport, and the deconstruction of food into different value-added food components. This gives the industrial producer and processor the power to exchange resources worldwide, making farmers more vulnerable to global markets (Van der Ploeg 1992).

In the Section 15.3, we will show some of the contours of an alternative paradigm to the prevailing bioeconomic agri-food model.

15.3 ECOECONOMY IN AGRI-FOOD

We can conceive of the ecoeconomy as an alternative and more diverse and fragmented arena for the development of new production and consumption chains and networks. It places an emphasis on the recalibration of microeconomic behavior and practices that, added together, can potentially realign production–consumption chains and capture local and regional value between rural and urban spaces.

The ecoeconomy model involves the rise of complex networks or webs of viable businesses (many of them small- and medium-sized new businesses) and economic activities that utilize ecological resources in more sustainable and ecologically efficient ways (e.g., new renewable energy firms, agritourism, food processing and catering, social enterprises). Importantly, these do not result in a net depletion of resources, but instead provide cumulative net benefits that add value to rural and regional spaces in both ecological and economic terms. Kitchen and Marsden (2009, p. 289) put forward a definition that captures these characteristics:

> The effective social management of environmental resources (as combinations of natural, social, economic and territorial capital) in ways designed to mesh with and enhance the local and regional ecosystem rather than disrupting and destroying it. The eco-economy thus consists of cumulative and nested 'webs' of viable businesses and economic activities that utilize the varied and differentiated forms of environmental resources of rural areas in sustainable ways. They do not result in a net depletion of resources but rather provide net benefits and add value to the environment and to the community.

The ecoeconomy is strongly rooted in agroecological principles and notions of multifunctionality, as we will show in the Sections 15.3.1 and 15.3.2. It can be seen as a reaction to key driving forces in agriculture and in a sense the bioeconomic dominance. Driving forces involved in affecting the growth of the ecoeconomy are:

1. The cost-prices squeeze in agriculture (as well as other land-based activities, such as forestry), which requires new answers embedded in rural ecoeconomic development.
2. The crises in agriculture such as foot and mouth disease, swine fever, bovine spongiform encephalopathy, and food scandals, have created the momentum for transition.
3. The changing societal and more urban demands in the rural arena with the entry of new actors. Agriculture responds to these demands by producing new rural products and services.

In agri-food, we can see an increasing number of empirical examples of the EEP in the form of agroecological practices and the spatial deepening, broadening, and regrounding of multifunctional rural practices. Faced with the continuing cost squeeze on agriculture and globalization processes associated with the bioeconomy, many farmers have responded to these developments and are being encouraged toward more "value-adding" and multifunctionality and new production–consumption networks. The ecoeconomy expresses itself in agri-food networks in the form of locally embedded sustainable agriculture, based on variable agroecological principles, as well as new production–consumption networks and multifunctional practices, as will be described next (Horlings and Marsden 2012).

15.3.1 Agroecology

Ecoeconomical critique in the sphere of agri-food challenges the emerging neoproductivist orthodoxy that has, somewhat paradoxically, arisen out of the current global food crisis and its wake-up calls about food security and ecological risks. It does this by attempting to further develop the sustainable rural development paradigm as part of a revised ecological modernization process. This is a critical debate to begin at this juncture given the onset of rapid forms of bioindustrial nanotechnologies that are currently being developed and applied to solve the forthcoming "peak food" problem.

The arguments about the efficacy of the agroindustrial model are based on criteria that are considered too restrictive, being two-dimensional in taking account of yields per unit of surface area. They do not consider the effects on soil, the third dimension, or the agroecosystem's capacity for future production, with time being the fourth dimension (Fernandez et al. 2002). Standard agroeconomic criteria are also monofunctional, considering only crop yield prices, while neglecting the effects of industrial farming on social well-being and culture and valuable crop and animal genetic diversity.

Ecoeconomic approaches, on the other hand, broaden the agri-food debate by including principles on human ecology and agroecology. Agroecology, defined as a science that provides ecological principles for the design and management of sustainable and resource-conserving agricultural systems, offers several advantages over the conventional agronomic or agroindustrial approach, according to Altieri et al. (2001): First, agroecology relies on indigenous farming knowledge and selected modern technologies to manage diversity, incorporate biological principles and resources into farming systems, and intensify agricultural production. Second, it offers the only practical way to restore agricultural lands that have been degraded by conventional agronomic practices. Third, it provides for an environmentally sound and affordable way for smallholders to intensify production in marginal areas. Finally, it has the potential to reverse the anti-peasant bias of strategies that emphasize purchased inputs as opposed to the assets that small farmers already possess, such as their low opportunity costs of labor.

Agroecologists also argue that the agroindustrial model does not consider its effects on soils and water sources or the agroecosystem's capacity for future production (Fernandez et al. 2002). These scholars have discussed the principles and practices of a wide variety of more sustainable and organic forms of agriculture as an alternative to the agroindustrial model. The agroecological literature describes a wide variety of agroecological practices, labeled by different terms such as low-input farming, agroforestry systems, multi- and intercropping farming, polycultures, natural systems agriculture, organic production, and so on (see, e.g., Altieri 2002; Altieri et al. 2001; Jackson 2002; Mäder et al. 2002).

Agroecology has contributed to agri-food systems by responding to environmental challenges, including a broader variety of ecological aspects, including soil fertility and biodiversity. Agroecological approaches also have relevance for rural economies. Sonnino et al. (2008) give an overview of its key principles and features such as the coevolution of society and natural factors, the use of indigenous knowledge and endogenous potentialities, systematic strategies that reflect an integral farming approach including biophysical factors and social actions, collective forms of actions, ecological and cultural diversity, and agricultural multifunctionality, while encompassing the social ecology component.

15.3.2 Multifunctionality and the Deepening, Broadening, and Regrounding of Agriculture

Macroeconomically, rural economies seem to be caught in the process of a continuous squeeze between the prices and costs associated with land-based production and the growing market and consumer expectation of high-quality or natural rural resource-based goods and services. Faced with these far-reaching concerns and issues, farmers are being encouraged toward more *value adding* and *multifunctionality* (see, e.g., Marsden and Sonnino 2008). As Kitchen and Marsden (2009, p. 275) state, "agricultural and wider land-based perspectives need to be reintegrated with broader questions of rural eco-economic development."

As a reaction to this cost squeeze, as Van der Ploeg et al. (2002) explains, agricultural activities are deepened, transformed, and expanded by the linkages and associations with new actors and agencies, as farm enterprises seek to deliver new products that entail more added value because they better meet the demands of the society at large. This is referred to as the *deepening* process, which is exemplified by organic farming, high-quality food production, and the creation of new, short food supply chains. Second, the interactions with the rural environment are broadened through the inclusion of new nonagricultural activities that are located on the interface amongst society, community, landscape, and biodiversity. This constitutes the process of *broadening*, which occurs through activities such as agritourism, nature and landscape management, new on-farm activities, and diversification. Finally, through the process of *regrounding*, farm enterprises are grounded in new or different sets of resources and become involved in new patterns of resource use. Van der Ploeg et al. (2002) indentify pluriactivity and farming economically as two major examples of this process. The challenge here, we argue, is to give more attention and potential value to the evolution of the ecoeconomy, which can be considered as more "place-based" and to critically explore new ecoeconomic perspectives.

15.4 FRAMINGS OF AGRI-FOOD PRODUCTION IN THE BIO- AND ECOECONOMIC PARADIGM

An important conclusion relevant for future spatial policies is that both the bio- and ecoeconomy lead to differential framings (Table 15.1).

Both paradigms identified here are assemblages of different modes of science, politics, technologies, and governance systems, and they have profound implications for sustainable agri-food systems. We have to realize that both of the models have some missing links. On the regional level, we see a misfit of the bioeconomy with the locationally specific characteristics of ecology and landscape. This is especially visible in the emergence of large agricultural megafarms, planted like islands in the landscape disconnected from local, environmental conditions. The bioeconomy also blocks off direct producer–consumer contexts and is poorly embedded in social local and regional networks. The ecoeconomy faces problems of scaling up and lack of expertise, skills, and professionalism on the business level (Marsden 2012).

The question remains: can the EEP expressed in agroecological and multifunctional practices contribute to an efficient, productive, and profitable agriculture, meeting the demands of a growing world population and adapting to climate change? We will address this question and illustrate it with empirical case studies based on an extensive literature review (Horlings and Marsden 2010, 2011a) that expresses a rich empirical variety of locally embedded and innovative farm practices.

Table 15.1 Framing by the Bio- and Ecoeconomy

Dimensions	BEP	EEP
Economic	Corporatization Productivity (yield) oriented Cost-prize squeeze on agriculture	Agri-food networks Integral approach of food production Value adding at farm level
Technological	Technology development economically driven Technological environmental solutions Closed loops of energy, waste, and minerals	Technological generation as a demand-driven process and spatially sensitive
Ecological	Ecological and genetic engineering (industrial ecology)	Based on agroecological principles, flexible and adaptive to ecologies and places
Social–cultural	Dependency, scientification, rational man–nature relation, loss of farmers' freedom/ agricultural employment	Autonomy Synergy between man and nature Demand-driven research Labor-intensive
Spatial	Globalized Export-oriented Use of external resources	Locally embedded in the community Use of local resources
Political	Top-down steering One-direction communication by extension services Power concentrated at multinationals and large retailers Privatized research and development	Enabling policy Participatory approaches Influence of communities in agri-food networks Local and regional institutional actors

Source: Horlings, I. and Marsden, T.K., *Global Environmental Change*, 21, 441–452, 2011.

15.5 ECOECONOMIC CONTRIBUTION TO FOOD SECURITY

The sustainability of any agricultural system is dependent on the specific context of space and place. Within agroecological systems, it is important to recognize that they utilize space and place in ways that augment sustainability (see Morgan et al. 2006). Indeed, they explicitly rely on their local conditions as a means as well as a condition of production; and, not least, to maintain and enhance diversity. The emphasis on diversity means that it is rarely possible to generalize or to genericize sustainable production technologies, however strong the scientific urge. Nevertheless, we can identify some elements that are shared among agroecological systems. Furthermore, it is important to question the role of current and historic institutional frameworks in nurturing, or otherwise, place-based agricultural initiatives.

Research shows that there is a multitude of such "real" sustainable agricultural systems worldwide. They range from small subsistence farms to small-scale and large commercial operations across a variety of ecosystems and encompassing very diverse production patterns. In Africa alone, there are at least 20 major farming systems combining a variety of agroecological approaches, small or large scale, irrigated or nonirrigated, crop or tuber based, hoe or plough based, in highland or lowland situations (Spencer et al. 2003).

Sundkvist et al. (2005) speculate that local-scale food systems are more sustainable because they have "tight feedback loops" linking consumers, producers, and ecological effects. In such systems, positive adaptive responses are more possible because of earlier and stronger signaling of negative effects requiring a change in behavior in the system. In their view, intensification, specialization, distancing, concentration, and homogenization are trends that can be identified as major constraints for tightened feedback loops (Sundkvist et al. 2005). This suggests locally embedded food systems are more resilient. But they tend not to be measured in this way.

From a critical review of the literature relating to the rural South, more specifically, we can broadly distinguish the following sustainable farming systems that, to varying degrees, espouse the "strong ecological modernization (EM)" and agroecological principles outlined in the Section 15.3.1. These are (1) organic agriculture, (2) urban and periurban agriculture, (3) conservation agriculture or zero tillage, (4) low-input agriculture (we use this as an overarching term for all kinds of farming techniques that use less external inputs and reduce negative environmental effects), (5) agroforestry and multifunctional agriculture, and (6) aquaculture. In the next sections, we will give some examples of the productivity of these sustainable farming systems (FAO 2002; Horlings and Marsden 2010, 2011a)

15.5.1 Organic Agriculture

During an international conference on organic agriculture and food security in 2007 in Italy, it was stated that organic agriculture could produce enough food on a global per capita basis (Scialabba 2007), based on models of Badgley et al. (2007) and Halberg et al. (2006). High-yield ratios in the developing world are obtained when farmers incorporate intensive agroecological techniques such as crop rotation, cover cropping, agroforestry, addition of organic fertilizers, and more efficient water management (Badgley et al. 2007). A Food and Agriculture Organization (FAO) analysis based on more than 50 cases in the United States and Europe, and just over a dozen studies in developing countries, showed that organic farms are more economically profitable, despite frequent yield decrease (Nemes 2009). Higher outcomes are due to premium prices and predominantly lower production costs. These conclusions can also be drawn from studies in developing countries, but there, *higher* yields combined with high premiums are the underlying cause for higher relative profitability.

15.5.2 Agroecological Projects

Pretty and Hine (2001) (see also Pretty 2003; Pretty et al. 2006) undertook an extensive study of 208 agroecological projects in 52 countries, which showed how farmers have improved crop productivity and at the same time increased water-use efficiency and carbon sequestration and reduced pesticide use. Their dataset contains reliable data on yield changes in 89 projects. The relative yield increases are greater at lower yields, indicating greater benefits for poor farmers and for those missed by the recent decades of modern agricultural development. The team found improvements are occurring through four different mechanisms:

- Intensification of a single component of a farm system
- Addition of a new productive element to a farm system
- Better use of nature to increase total farm production
- Improvements in per hectare yield of staples through the introduction of new regenerative elements into farm systems.

While this needs far more in-depth treatment than space in this chapter allows, it is clear that on different continents we can see examples of farming systems that are (more) locally embedded in communities, more resilient toward external threats and global processes, more environmentally friendly, contributing to biodiversity and, not the least important, productive in terms of yields. Some of the most interesting examples of sustainable agriculture come from developing countries in Africa, Asia, and Latin America, as described below, focusing on the ecological dimension and productivity (see also FAO 2002).

In Ethiopia, The Ensete agroforestry system is a 5000-year-old farm system, practiced by the Gedeo people in the highlands of southern Ethiopia (Kippie 2002). The system is able to produce a large variety of products such as Ensete, a high-quality food, coffee, honey, timber, highland sheep, and a variety of crops. The perennial cropping system has good resilience against droughts, thanks to the Ensete plant, which captures water with its fan-shaped leaves and has a fibrous root system that also prevents erosion. Since only a small proportion of the farm area is harvested and replanted, damage to the site by rainwater erosion or by sunburn is minimized. There is a biological differentiation in cropping between the three zones in the highlands. High productivity of Ensete and judicious use of accompanying crops result in a very high carrying capacity. Ensete proves to be the highest-yielding Ethiopian food crop, yielding over 5.6 t/ha/year in agroforests.

In Brazil, there are some 15 million hectares under *plantio direto*, also called zero tillage, even though there is some disturbance of the soil. In Argentina, there are more that 11 million hectares under zero tillage, up from <100,000 ha in 1990. Paraguay has another 1 million hectares of zero tillage. Many of the Clubes Amigos da Terra, literally "friends of the land clubs," which are essentially farmer groups, have been closely involved in this transformation. The principles of zero tillage include no mechanical soil disturbance, permanent soil cover, and judicious choice of crop rotations (Benites and Aschburner 2001, cited in Vaneph and Benites 2001; see also Pieri et al. 2002). Although zero tillage is a monocropping system including the use of pesticides, it can be regarded as positive for both biodiversity and sustainable agriculture. The approach has led to higher yields in crop production, declines in labor costs, and diversification into livestock as well as agroprocessing, resulting in improved food security for small farmers (UNCED 2002).

In China, despite the onset of agricultural intensification and urbanization, it is also recognized that many regions display a fertile basis for agroecological and organic agricultures. The interest of local state entrepreneurs in value-added agricultural development, in combination with a growing export market for organic food, has led to the rapid expansion of self-identified organic agricultural products in rural China (Thiers 2002). More than 20 provinces have formulated "Measures for Green Food Management" and "Measures for Administration of Green Food Labelling." The export value (90% under the AA grade) increased from US$71 million in 1997 to US$2140 million in 2007 (Lin et al. 2009). This has benefited from the establishment of the Green Food Development Center. In 2004, there were 2836 certified enterprises, producing 3142 different products on 8.94 million hectares of land (Bin et al. 2006). In addition to organic "green production," more emphasis is being placed on Chinese Ecological Agriculture (CEA), following agroecological principles (Prändl-Zika 2008; Ye et al. 2002; Zhen et al. 2005).

The common factor among the above examples is that they all address ecological problems by developing agricultural practices that are embedded in the characteristics of the local physical, economical, and social context. They also use natural resources in a (more) sustainable way, are based on farmers' expertise, and are more resilient by broadening and interlinking different spheres of production and recycling. They follow the principle of Sundkvist et al. (2005) of developing more embedded and tighter feedback loops.

15.6 DISCUSSION AND CONCLUSIONS

We have argued here that both the bioeconomy and the ecoeconomy may be seen as examples of sustainable pathways in agri-food. The bioeconomy has a firm position in agri-food corporate-controlled business. Biological innovation has created new generic biobased products, a wide range of genetic modification applications (termed as "third nature"), and the production of biofuels. The use of microbes for energy production is an expression of the trend that more and more of the total production of economically important compounds will soon be miniaturized within biological systems. These new developments have led to international land grabbing, unforeseen risks of biotechnically engineered crops, and social protests against new large-scale, intensive agricultural production. The increase of production scale and miniaturizing of biological processes are both an expression of the decoupling of agri-food from the environmental context and natural processes.

The bioeconomy is increasingly seen as a multidimensional panacea to the global onset of climate change, population growth, and carbon and food shortages. Nevertheless, we have argued that this agri-food model may mitigate environmental effects to a certain extent but may also cause new negative side effects and expose some important social, cultural, political, and spatial missing links.

Examples are the standardization and hygienic bureaucratization of agriculture, the distancing of food production and consumption geographically and institutionally, and the marginalization and fragmentation of the role of agriculture in local communities.

The central question posed has been whether there is evidence in practice that an ecoeconomic model, based on agroecological practices, can contribute to meeting future food demands, especially in developing countries. The sustainable practices we described show a rich variety of these often more multifunctional approaches, mostly implemented by small-scale farmers. On different continents, we see examples of farming systems that are locally and ecologically embedded in communities, more resilient toward external threats and globalization, environmentally friendly, contributing to biodiversity, and, not the least important, enhance productivity.

This should not simply be interpreted as a plea for supporting smallholder agriculture. We have also argued that agroecological initiatives can contribute to food security—which is an urgent challenge for the next decades—and a value place-based ecoeconomy. This requires, however, not only analyzing and favoring these initiatives as alternatives to weak ecological modernization, but also a political reorganization of current markets and institutions in ways that reduce the barriers to their wider dissemination and agglomeration outlined above. It is conceptually and methodologically difficult to scale up diversity without giving way to generic systems approaches. A starting point, therefore, should be the development of a more sophisticated comparative understanding and dissemination of the agroecological and socioeconomic conditions under which alternatives are adopted and implemented at the local level.

Agroecological practices can potentially be up-scaled, but this depends on three major conditions. First, we have to turn the "problem" of diversity and context dependency of agricultural practices into a real ecological and social virtue. The second condition is an enabling policy. The IAASTD (2009a,b) report has especially given useful recommendations on this point. We can also witness a new set of counter-logics for the current food regime in, for example, nongovernmental organization initiatives, some FAO reports, and new institutions like Global Gap and Fair Trade networks. The third challenge is the redirection of agricultural research, development, and knowledge transfer.

By reviewing key strands of the literature, we have begun to trace the contested processes of the bio- and ecoeconomies. One of the key questions that remains is how these contested contingencies between the bioeconomy and the ecoeconomy will play themselves out in rural spaces. We would argue that these current discussions need to be informed by the concepts and debates contained in this paper, especially how a more vibrant rural-based ecoeconomy can be stimulated by forthcoming sustainable agri-food policies.

REFERENCES

Altieri, M. A. 2002. Agro-ecology: The science of natural resource management for poor farmers in marginal environments. *Agriculture, Ecosystems and Environment* 93: 1–24.

Altieri, M. A. and P. Rosset. 1999. Ten reasons why biotechnology will not ensure food security, protest the environment and reduce poverty in the developing world. *AgBioForum* 3–4: 155–162.

Altieri, M. A., P. Rosset, and L. A. Thrupp. 2001. The potential of agro-ecology to combat hunger in the developing world. In P. Pinstrup-Andersen and R. Pandya-Lorch (eds), *The Unfinished Agenda: Perspectives on Overcoming Hunger, Poverty, and Environmental Degradation*, pp. 123–127. Washington DC: IFPRI.

Ambler-Edwards, S., K. Bailey, A. Kiff, T. Lang, R. Lee, T. Marsden, D. Simons, and H. Tibbs. 2009. Food futures: Rethinking UK strategy. Chatham House Report. London: The Royal Institute of International Affairs.

Annan, K. 2000. We the peoples: The role of the United Nations in the 21st century. New York: United Nations. Available at: https://www.un.org/en/events/pastevents/pdfs/We_The_Peoples.pdf (access date: 22 August 2013).

Anex, R. 2004. Something new under the sun? The industrial ecology of biobased products. *Journal of Industrial Ecology* 3–4: 1–4.

Badgley, C., J. Moghtader, E. Quintero et al. 2007. Organic agriculture and the global food supply. *Renewable Agriculture Food Systems* 22: 86–108.

Benites, J. R. and J. E. Aschburner. 2001. FAO's role in promoting conservation agriculture. Draft paper for the First World Congress on Conservation Agriculture: A world-wide challenge. October 1–5, 2001, Madrid.

Bin, Y., X. Ni, T. Wentao, H. Linquing, D. Peng, L. Liaoyuan, and Y. Jihong. 2006. Chinese food industry and market report. LIFS, Lund University, Sweden and Zhongnan University of Economics & Law, China.

Boyd, W., W. S. Prudham, and R. A. Schurman. 2001. Industrial dynamics and the problem of nature. *Society and Natural Resources* 14: 555–570.

Carlson, R. 2007. Laying the foundations for a bio-economy. *Systems and Synthetic Biology* 3: 109–117.

Christoff, P. 1996. Ecological modernisation, ecological modernities. *Environmental Politics* 5(3): 476–500.

Conway, G. 1997. *The Double Green Revolution*. London: Penguin Books.

Drummond, I. and T. K. Marsden. 1995. Regulating sustainable development. *Global Environmental Change* 5: 51–64.

EC (European Commission). 2005. New perspectives on the knowledge-based bio-economy. EU: Science and Research Conference paper.

Evans, A. 2009. Feeding the nine billion. Chatham House Report. London: Royal Institute of International Affairs.

FAO (Food and Agriculture Organization). 2002. Land and agriculture: From UNCED, Rio de Janeiro 1992 to WSSD, challenges and perspectives for the World Summit on Sustainable Development Johannesburg 2002: A compendium of recent sustainable development initiatives. Rome: FAO Corporate Document Repository.

FAO (Food and Agriculture Organization). 2006. Mid-term review of achieving the world food summit target. CFS: 2006/3. Rome: FAO.

FAO (Food and Agriculture Organization). 2009. The state of food insecurity in the world, economic crises-impacts and lessons learned. Rome: FAO.

Fernandez, E., A. Pell, and N. Uphoff. 2002. Rethinking agriculture for new opportunities. In N. Uphoff (ed.), *Agro-Ecological Innovations: Increasing Food Production with Participatory Development*. London: Earthscan.

Hajer, M. A. 1995. *The Politics of Environmental Discourse: Ecological Modernisation and the Policy Process*. Oxford: Oxford University Press.

Halberg, N., H. F. Alroe, M. T. Knudsen, and E. S. Kristensen. 2006. *Global Development of Organic Agriculture: Challenges and Prospects*. Wallingford: CABI Publishing.

Horlings, I. 1996. Duurzaam boeren met beleid; innovatiegroepen in de Nederlandse landbouw, Dissertation, Nijmegen: Katholieke Universiteit Nijmegen. Also published in Studies van Landbouw en Platteland no. 20, Wageningen: Landbouw Universiteit.

Horlings, I., L. Kitchen, G. Bristow, and T. K. Marsden. 2010. Exploring the potential contributions of the bio-economy and eco-economy to agri-food and rural regional development. BRASS working paper, no 59. Cardiff: Cardiff University.

Horlings, I. and T. K. Marsden. 2011a. Towards the real green revolution? Exploring the conceptual dimensions of a new ecological modernisation of agriculture that could feed the world. *Global Environmental Change* 21: 441–452.

Horlings, I. and T. K. Marsden. 2011b. Rumo ao desenvolvimento espacial sustentável? Explorando as implicações da nova bioeconomia no setor de agroalimentos e na inovação regional. *Sociologias* 27: 142–178.

Horlings, L. G. and T. K. Marsden. 2010. Towards the real green revolution? Exploring the conceptual dimensions of a new ecological modernisation of agriculture that could "feed the world". BRASS Working Paper. Cardiff: Cardiff University.

Horlings, L. G. and T. K. Marsden. 2012. Exploring the "New Rural Paradigm" in Europe: Eco-economic strategies as a counterforce to the global competitiveness agenda. European Urban and Regional Studies, published online May 30, 2012.

IAASTD (International Assessment of Agricultural Knowledge, Science and Technology for Development). 2009a. Global report. Washington, DC: Island Press.

IAASTD (International Assessment of Agricultural Knowledge, Science and Technology for Development). 2009b. Executive summary of the synthesis report. Washington, DC: Island Press.

IFPRI (International Food Policy Research Institute). 2013. "Land grabbing" by foreign investors in developing countries. Washington, DC: IFPRI. Available at: http://www.ifpri.org/publication/land-grabbing-foreign-investors-developing-countries (access date: 22 August 2013).

ISAAA (International Service for the Acquisition of Agri-biotech Applications). 2008. Global status of commercialized biotech/GM crops: 2008. ISAAA Briefs 39-2008. Available at: http://www.isaaa.org.

Jackson, W. 2002. Natural systems agriculture: A truly radical alternative. *Agriculture, Ecosystems and Environment* 88: 111–117.

Jordan, N., G. Boody, W. Broussard et al. 2007. Sustainable development of the agricultural bio-economy. *Science* 316: 1570–1571.

Juma, C. and V. Konde. 2001. *The New Bioeconomy: Industrial and Environmental Biotechnology in Developing Countries*. New York: United Nations Conference on Trade and Development.

Kippie, T. K. 2002. *Five Thousand Years of Sustainability? A Case Study on Gedeo Land Use (Southern Ethiopia)*. Heelsum: Treemail Publishers.

Kitchen, L. and T. K. Marsden. 2009. Creating sustainable rural development through stimulating the eco-economy: Beyond the eco-economic paradox? *Sociologia Ruralis* 49: 273–293.

Lin, L., D. Zhou, and C. Ma. 2009. Green food industry in China: Development, problems and policies. *Renewable Agriculture and Food Systems* 25(1): 69–80.

Mäder, P., A. Fliessbach, D. Dubois, L. Gunst, P. Fried, and U. Niggli. 2002. Soil fertility and biodiversity in organic farming. *Science* 296(5573): 1694–1697.

Marsden, T. K. 2004. The quest for ecological modernisation re-spacing rural development and agro-food studies. *Sociologia Ruralis* 44: 129–147.

Marsden, T. 2012. Third natures? Reconstituting space through placemaking strategies for sustainability. *International Journal of Sociology of Agriculture and Food* 19(2): 257–274.

Marsden T. K., R. Lee, A. Flynn et al. 2010. *The New Regulation and Governance of Food: Beyond the Food Crisis?* London: Routledge.

Marsden, T. K. and R. Sonnino. 2008. Rural development and the regional state: Denying multifunctional agriculture in the UK. *Journal of Rural Studies* 24: 422–431.

Morgan, K., T. K. Marsden, and J. Murdoch. 2006. *Worlds of Food; Place, Power and Provenance in the Food Chain*. Oxford: Oxford Geographical and Environmental Studies.

Nemes, N. 2009. Natural Resources Management and Environment Department. *Comparative Analysis of Organic and Non-Organic Farming Systems: A Critical Assessment of Farm Profitability, Natural Resources Management and Environment*. Rome: FAO.

Nilsson, M., A. Varnäs, C. Kehler Siebert, L. J. Nilsson, B. Nykvist, and K. Ericsson. 2009. A European eco-efficient economy, governing climate, energy and competitiveness. Report for the 2009 Swedish Presidency of the Council of the European Union, Stockholm: Stockholm Environment Institute.

OECD (Organisation for Economic Co-operation and Development). 2006. *The New Rural Paradigm: Policies and Governance*. Paris: OECD.

Parrot, N. and T. K. Marsden. 2002. *The Real Green Revolution: Organic and Agro-ecological Farming in the South*. London: Greenpeace Environmental Trust.

Pieri, C., G. Evers, J. Landers, P. O'Connell, and E. Terry. 2002. No-till farming for sustainable rural development. ARD Working Paper, Washington DC: World Bank.

Prändl-Zika, V. 2008. From subsistence farming towards a multifunctional agriculture: Sustainability in the Chinese rural reality. *Journal of Environmental Management* 87: 236–248.

Pretty, J. 2003. Agro-ecology in developing countries: The promise of a sustainable harvest. *Environment* 45: 8–20.

Pretty, J. N., A. D. Noble, D. Bossio, J. Dixon, R. E. Hine, F. W. T. Penning de Vries, and J. I. L. Morison. 2006. Resource-conserving agriculture increases yields in developing countries. *Environmental Science and Technology* 40: 1114–1119.

Pretty, P. and R. Hine. 2001. Reducing food poverty with sustainable agriculture: A summary of new evidence. Primal report from SAFE-World Research Project (The Potential of Sustainable Agriculture to Feed the World). Colchester: University of Essex.

Sanchez, P. A., S. Ahamed, F. Carre et al. 2009. Digital soil map of the world. *Science* 325: 680–681.

Scialabba, N. E-H. 2007. Organic Agriculture and Food Security, International Conference on Organic Agriculture and Food Security, 3–5 May 2007. Rome: FAO.

Schubert, C. 2006. Can biofuels finally take centre stage? *Nature Biotechnology* 24: 777–784.

Sonnino, R., Y. Kanemasu, and T. K. Marsden. 2008. Sustainability and rural development. In J. D. Van der Ploeg and T. K. Marsden (eds), *Unfolding Webs: The Dynamics of Regional Rural Development*, pp. 29–52. Assen: Royal Van Gorcum.

Smith, N. 1984. *Uneven Development: Nature, Capital and the Production of Space*. Oxford: Blackwell.

Smith, N. 2007. Nature as accumulation strategy. In L. Panitch and L. C. Leys (eds), *Coming to Terms with Nature*, pp. 16–36. London: The Merlin Press.

Spencer, D. S. C., P. J. Matlon, and H. Löffler. 2003. *African Agricultural Production and Productivity in Perspective*. Amsterdam: InterAcademy Council.

Sundkvist, A., R. Milestad, and A. Jansson. 2005. On the importance of tightening feedback loops for sustainable development of food systems. *Food Policy* 30: 224–239.

Swaminathan, M. S. 2000. An evergreen revolution. *Biologist* 47: 85–89.

The Royal Society. 2009. *Reaping the Benefits: Science and the Sustainable Intensification of Global Agriculture*. London: The Royal Society.

Thiers, P. 2002. From grassroots movement to state-coordinated market strategy: The transformation of organic agriculture in China. *Environment and Planning C: Government and Policy* 20 (3): 357–373.

UN (United Nations). 2008. World population prospects: The 2008 revision population database. UN. Available at: http://www.un.org/esa/population/publications/wpp2008/wpp2008_highlights.pdf.

UNCED (United Nations Conference on Environment and Development). 2002. *World Summit on Sustainable Development*. Johannesburg: UNCED.

Van der Ploeg, J. D. 1991. *Landbouw als mensenwerk; arbeid en technologie in de agrarische ontwikkeling*. Muiderberg: Coutinho.

Van der Ploeg, J. D. 1992. The reconstitution of locality: Technology and labour in modern agriculture. In T. K. Marsden et al. (eds), *Labour and Locality; Uneven Development and the Rural Labour Process. Critical Perspectives on Rural Change*. Series 4, pp. 19–43. London: David Fulton.

Van der Ploeg, J. D., A. Van der Long, and J. Banks. 2002. *Living Countrysides: Rural Development Processes in Europe. The State of the Art*. Doetinchem: Elsevier.

Vaneph, S. and J. Benites. 2001. An unexpected success; from zero tillage to conservation agriculture. *Leisa Magazine* 17(3): 22.

WFS (World Food Summit). 1996. *World Food Summit*. Rome: FAO.

World Bank. 2008. *World Development Report 2008: Agriculture for Development*. Washington, DC: World Bank.

World Bank. 2010. *Rising Global Interest in Farmland. Can It Yield Sustainable and Equitable Benefits?* Washington, DC: World Bank.

Ye, X. J., Z. Q. Wang, and Q. S. Li. 2002. The ecological agriculture movement in modern China. *Agriculture, Ecosystems and Environment* 92: 261–281.

Zhen, L., J. K. Routray, M. A. Zoebisch, G. Chen, G. Xie, and S. Cheng. 2005. Three dimensions of sustainability of farming practices in the North China Plain: A case study from Ningjin County of Shandong Province, PR China, Agriculture, *Ecosystems and Environment* 105: 507–522.

CHAPTER 16

Vermont Agricultural Resilience in a Changing Climate
A Transdisciplinary and Participatory Action Research (PAR) Process

Rachel Schattman, Ernesto Méndez, Katherine Westdijk, Martha Caswell,
David Conner, Christopher Koliba, Asim Zia, Stephanie Hurley,
Carol Adair, Linda Berlin, and Heather Darby

CONTENTS

16.1 Introduction 326
    16.1.1 Background 326
    16.1.2 Research and Initiative Objectives 327
    16.1.3 Agroecology and PAR Frameworks 327
16.2 Our Approach 329
    16.2.1 Climate Change Best Management Practices (CCBMPs) 329
    16.2.2 Economics of CCBMPs 329
    16.2.3 Mitigation Potential of CCBMPs 329
    16.2.4 Governance and Policy through Agent-Based Models 330
    16.2.5 Landscape Visualization and CCBMPs 330
    16.2.6 Stakeholders and the PAR Approach 332
16.3 Methods 332
    16.3.1 Initial Investigation, Year 1 332
    16.3.2 Agent-Based Models, Years 1 and 2 334
    16.3.3 On-Farm Research, Years 2 and 3 335
        16.3.3.1 Farm Selection 335
        16.3.3.2 Farmer Interviews 335
        16.3.3.3 Economic Analyses 336
        16.3.3.4 C Sequestration and GHG Emissions 336
        16.3.3.5 Landscape Visualization 337
16.4 Selected Preliminary Results 337
    16.4.1 Survey 337
    16.4.2 Reports from the Field 338
    16.4.3 Agent-Based Model 339

16.5 Discussion..339
        16.5.1 PAR Process: Taking Stock of Transdisciplinary Process Participation.................339
        16.5.2 Lessons Learned and Future Directions...341
Summary..342
Acknowledgments..343
References..343

16.1 INTRODUCTION

16.1.1 Background

It is widely acknowledged that global climate change will lead to increasing global temperatures, rising sea levels, and decreasing snow and ice cover on land and over bodies of water within the next 50–100 years. Current projections from the Intergovernmental Panel on Climate Change (IPCC) range from low emissions scenarios (projected atmospheric carbon concentrations of 550 parts per million, or ppm) to high emissions scenarios (projected atmospheric carbon concentrations of 880 ppm) (Walthall et al. 2012). Increasing global temperatures will have numerous effects on both natural and human systems, including those associated with food and agriculture. Higher atmospheric temperatures will have an effect on the frequency and volume of rain events in addition to influencing plant and animal geographic ranges and interactions. While the full range of climate change implications for ecosystem and human communities is yet unknown, it is widely accepted that the emissions of today will influence how our world might change in the latter half of this century (Bernstien et al. 2007; Frumhoff et al. 2007).

Future interactions between climate change and agro-food systems can be expected to be dynamic and complex (Eriksen et al. 2009) with agricultural systems both contributing to and becoming increasingly vulnerable to the effects of climate change. Agricultural land use contributes to climate change by emitting approximately 31% of greenhouse gas (GHG) emissions globally (roughly 15 billion tons of CO_2 equivalents) (Scherr and Sthapit 2009; Smith et al. 2008), which does not include the additional emissions related to the processing, transportation, marketing, and consumption of food. On the other hand, projected changes in temperatures, precipitation regimes, and natural hazard frequencies will have an impact on the production capacity and resilience of different agricultural systems (Smith and Olesen 2010). The ability of an agroecosystem to adapt and mitigate or contribute to climate change largely depends on the types of components it includes, its management regime, and external factors such as policies and markets (Smith et al. 2008; Tubiello et al. 2008).

In the northeastern United States, climate change is expected to severely affect rural populations and farming communities (Lal et al. 2011). Growers already implement farming practices that have the potential for climate change mitigation and adaptation through sustainable agriculture (Wall and Smit 2008), but whether these practices have the greatest potential to limit risk and reduce vulnerability at the farm level remains an untested question. We refer to climate change adaptation as a farmer's adjustment to the conditions and effects of climate change, which leads to a reduction of risk at the farm level (Smith et al. 2008). Some of the climate change impacts that are anticipated for this region, and to which farmers will be required to respond, include an increase in the number of heavy storms and floods, changes in the suitability for growing traditional crops (e.g., apples, blueberries, and cranberries), changes in insect and plant communities, and decreases in milk production due to hotter summers (Frumhoff et al. 2007; Wolfe et al. 2007). Although the IPCC has responded to criticism that there is insufficient evidence that recent upticks in disasters such as floods and droughts at a regional level are directly caused by climate change, this is primarily due to a lack of monitoring at local scales. There is an acknowledgement that a changing climate increases vulnerability and risk associated with extreme weather and climatic events (IPCC 2012). The recent devastation of tropical storm Irene in Vermont has exposed the need for stakeholders to develop strategies that respond to extreme climatic events.

16.1.2 Research and Initiative Objectives

The Vermont Agricultural Resilience in a Changing Climate Initiative seeks to make contributions to agroecology through (a) researching and implementing a transdisciplinary, participatory action research (PAR) framework; and (b) reporting on that process with a special focus on stakeholder participation. Specifically, this work in progress is inclusive of multiple stakeholders (researchers with a wide range of foci, a professional advisory committee that includes farmers and other collaborators, farmers who cultivate a wide range of products, and policy makers). Our research approach is to work with diverse stakeholder groups to identify the best management practices (BMPs) that will (1) best help farmers adapt to climate change now and in the future; (2) provide information on how farmers can contribute to GHG mitigation; (3) work with outreach professionals to deliver information about these practices to a broad community of farmers and other professionals; (4) assess the future needs related to climate change of stakeholders in the Vermont agro-food system; and (5) create and utilize tools to inform policy and governance that are specifically related to climate change and agriculture issues.

16.1.3 Agroecology and PAR Frameworks

As a conceptual framework, agroecology has the capacity to address problems at multiple scales (plot, farm, ecosystem, region, state, global), while simultaneously engaging stakeholders and enabling interaction with broader influences, including social, ecological, and economic factors (Francis et al. 2003; Guzmán and Woodgate 2013). Climate change and its relationship to agriculture and agro-food systems is a highly complex interaction, which presents challenges to the economic viability of businesses, the ecological balance, and social well-being. We propose that agroecology can contribute to addressing some of these issues. Many of the political and social components of agroecological theory are concerned with the rural setting, specifically attempting to reconcile conceptual and social factors at the plot, field, and farm scales (Amekawa 2011). Vermont, a small rural state with a long agricultural history (Albers 2002), is an appropriate location to apply this theory. It is worth noting that an agro-food systems approach to agroecology is a relatively recent theoretical and practical application (Francis et al. 2003; Gliessman 2007; Wezel et al. 2009), and prior to current efforts, it has been identified as the weakest contribution to agroecology thus far, and with the most opportunity for contributions to be made in the future (Tomich et al. 2011).

Recent contributions to agroecological theory and practice have argued that transdisciplinary and PAR approaches are well suited to reach a better and more balanced understanding of the social, economic, and ecological forces in agricultural and agro-food systems (Méndez et al. 2013). Transdisciplinary research integrates multiple knowledge systems, including academic disciplines and nonacademic knowledge (e.g., local or indigenous), to seek solutions to complex, real-world issues and problems (Belsky 2002; Francis et al. 2003; Godemann 2008; Stokols 2006). PAR similarly emphasizes exchange and collaboration across knowledge systems, and involves a diversity of stakeholders as active participants in an iterative process that integrates research, reflection, and action. PAR seeks to provide a voice to actors, such as farmers, who have traditionally been excluded from the scientific research process (Bacon et al. 2005; Kindon et al. 2010; Méndez et al. 2013). The integration of multiple academic disciplines and nonacademic knowledge through the participation of key stakeholders is necessary to identify and address threats to ecological and human health at all levels, and to contribute to greater long-term climate change resilience in the agro-food system. The degree and implications of how and when different stakeholders participate in a PAR process represent areas of scholarship with many unanswered questions. Varying levels of participation in problem identification and data collection and analysis, and subsequent action at the community or policy level reflect stages of empowerment and have the capacity to influence the agency of different groups. Although PAR has been criticized for not sufficiently shifting the locus of power from the researcher to other stakeholders (Kindon et al. 2010), it is still rare for papers based on a PAR process to include an analysis of stakeholder participation and subsequent power relationships (Manzo and Brightbill 2010). To date, examples of this power analysis have been published primarily in literature that addresses PAR theory (Figure 16.1), as in Kindon et al. (2010) and Pretty (1995). The assumption of the participation continuum presented in Figure 16.1 is that higher levels of participation lead to greater benefit for stakeholders, and higher levels of empowerment lead to greater interest in and execution of participation. Through this chapter, we attempt to provide an in-process review of how these dynamics have evolved in our initiative and provide reflections on how these dynamics can shift in the second half of this multiyear effort.

Additionally, it is necessary to evaluate the context in which PAR processes occur and potentially compete with other parallel processes. In the case of Vermont, a state that boasts a citizen legislature, farmers and representatives of the university and membership associations are often invited to give testimony on key legislation. Additionally, several senators and representatives themselves own and operate commercial farms. This state's deep connection to agriculture sets the stage for sympathetic legislation, such as the Farm to Plate Investment Program (F2P), which was approved by the Vermont State Senate and House in 2009. One of the results of the F2P process was the development of a network of farmers, funders, service providers, researchers, and policy makers, which allows diverse parties to work together, on an annual basis, to review key problems in the Vermont food system, set goals, and measure progress on those goals. Our team is involved with the F2P process, allowing us to share ideas with individuals and organizations working on related efforts throughout the state.

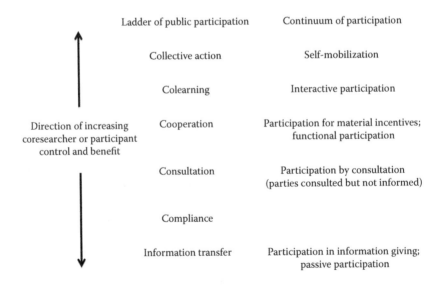

Figure 16.1 Participation continuums. (Adapted from Kindon, S., Pain, R., and Kesby, M. (eds)., *Participatory Action Research Approaches and Methods: Connecting People, Participation and Place*, Routledge, New York, 2010.)

16.2 OUR APPROACH

16.2.1 Climate Change Best Management Practices (CCBMPs)

Farmers constantly innovate in their daily practice, making decisions based on multiple factors on a daily, monthly, seasonal, and yearly basis. Much can be learned by identifying and analyzing existing agricultural management practices that have the potential to adapt to and/or mitigate climate change. *Best management practices* is a commonly used term to describe those approaches that have been tested and proven to have a positive impact on some part of an agricultural system. We identify those BMPs specifically related to addressing climate change as climate change BMPs (CCBMPs). Agroecological analysis has demonstrated how ecologically based and locally designed farming practices can increase the resilience of agroecosystems to extreme climate events. In Central America, Holt-Giménez (2006) compared the impacts of Hurricane Mitch between paired agroecologically and conventionally managed farms. Farms that had established agroecological practices (i.e., soil conservation practices and organic management) fared much better than their conventional counterparts did in terms of soil erosion, economic losses, and vegetation cover. Another study in Chiapas, Mexico, documented that increased vegetation complexity mitigated the damage in coffee farms from one hurricane (Philpott et al. 2008). In Canada, Wall and Smit (2008) documented sustainable agricultural practices that farmers had adopted as an adaptation response to climate change. These included crop and enterprise diversification, land resource management (e.g., conservation tillage and use of shelterbelts), water resource management (e.g., irrigation and use of ponds), and livestock management (e.g., intensive grazing).

One of the outcomes of this work will be to identify what qualities make a BMP a CCBMP. In other words, we are trying to answer the question: "What agricultural practices have the greatest potential to both mitigate GHGs and/or reduce the vulnerability and risk faced by farmers due to climate change?"

16.2.2 Economics of CCBMPs

No examination of CCBMPs and how farmers utilize them would be complete without an economic analysis. Understanding the costs and benefits associated with CCBMPs is particularly necessary given that the majority of farms in Vermont, and in the United States, generally earn negative net income (USDA-NASS 2007). Farmers often do not keep detailed and accurate records of their costs of production. Traditional crop or enterprise budgets are typically constructed by consulting academic experts who assign typical practices and costs to each category. However, this approach fails to represent the heterogeneity of the scales, practices, and experiences of farmers. Short-cut "rules of thumb" such as target revenues for each hour spent harvesting and packing can be dangerously misleading (Conner 2004). Cost measurement is especially difficult on diversified farms, which grow many crops at relatively small scales (Conner and Rangarajan 2009), which is of particular concern in an agroecological context where diversity and small-scale agriculture are often lauded. Knowing costs is also important in pricing decisions and in ensuring that revenues gained from the adoption of practices cover costs. The costs of the implementation of BMPs, and potentially a new suite of CCBMPs, will inform efficient resource allocation toward any future potential scenarios involving adaptation to climate change, carbon trading (i.e., climate change mitigation), or payment for ecosystem service provisions.

16.2.3 Mitigation Potential of CCBMPs

In future climate scenarios, interactions between differing agricultural management practices and projected changes in precipitation regimes and temperature will result in a diversity of potential feedback loops between climate and land use change. In this context, an important research question is whether and how current CCBMPs and conventional farming practices affect carbon and GHG balances. Depending on the diversity of the habitats and the characteristics of the plants that make up a given grower's land, such as maintaining forested areas, high-productivity or high-diversity ecosystems (e.g., Fornara and Tilman 2008), farms may store more carbon (C) to offset GHGs. However, even if a farmer's lands act as a C sink by increasing C storage in biomass and soils (taking up more carbon dioxide [CO_2] from the atmosphere than is released), their net effect on climate will be determined by trace gas emissions (methane [CH_4] and nitrous oxide [N_2O]). Both CH_4 and N_2O are more potent GHGs than CO_2, trapping 25 and 298 times more heat over 100 years than CO_2, respectively (IPCC 2007). The primary sources of N_2O are denitrification and nitrification. Losses of N_2O via denitrification are transient, driven by precipitation events that produce anoxic conditions in the topsoil, which also inhibit nitrification (Parton et al. 1996). CH_4 may also be produced in anoxic soils via microbial methanogenesis. Denitrification is considered to be the primary source of N_2O from agricultural land, but Panek et al. (2000) reported the equal contribution of both processes to total N_2O emissions. N_2O emissions from fertilized agricultural lands may range from 9 to 17 kg N_2O ha^{-1} yr^{-1} (6–11 kg N ha^{-1} yr^{-1}; Frolking et al. 1998), and emitting 1 kg of N as N_2O offsets the permanent storage of 54 kg of C. Thus, a crucial question for managing the C and GHG balances of farmlands now and in the future is how such systems affect not only C storage, but also the production of these potent GHGs.

16.2.4 Governance and Policy through Agent-Based Models

Another key component of our research focuses on better understanding the way that farmers make decisions as they relate to CCBMPs, and how this process interacts with existing or future policies related to climate change. Public policies are designed and executed using multiple sources of information, and there is a growing appreciation for the contribution of complex governance networks in these processes. We argue that computer models increase the power and capacity with which we are able to advance governance theories and frameworks. Governance is defined here as the "means by which an activity or an ensemble of activities is controlled, steered or directed" (Koliba and Zia forthcoming). Heterogeneously acting and interacting agents work within and across organizations; the description of how these actors interact is called *governance infomatics*. Understanding these complex interactions helps network managers to better understand the forces at play, and assists in solving seemingly intractable problems. This is especially relevant to climate change as a large complex problem that exists across multiple scales and involves and affects numerous networks of organizations, governing bodies, and populations.

Considering this, we have utilized an agent-based modeling (ABM) approach, which is a computer simulation experimental method for modeling the emergence of system-wide outcomes that arise from the complex interaction between landscape-level changes and institutional agent decision making (Koliba et al. 2011; Koliba and Zia forthcoming; Zia et al. 2013). In ABM, a system is modeled as a collection of autonomous decision-making entities called agents. Each agent individually assesses its situation and makes decisions on the basis of a set of rules. Agents may execute various behaviors appropriate for the system that they represent—for example, producing, consuming, or selling. The ABMs are premised on describing a system from the perspective of its constituent units (North and Macal 2007). Computer models of this nature can account for uncertainty and the adaptability of agents and eventually support scenario planning and the nonlinear analysis of farming practice dynamics. These kinds of simulated process-based models allow knowledge to emerge and be utilized throughout the interactive analytic process.

16.2.5 Landscape Visualization and CCBMPs

Our research group has begun to develop a series of landscape visualizations that will enable farmers and other stakeholders to envision the potential impacts and resiliencies associated with the adoption of CCBMPs at both the farm and landscape levels. Both eye-level and orthophoto (map-view) images of photo-realistic landscapes are presented to stakeholder groups to both demonstrate the spatial and visual effects of CCBMP implementation and gauge the utility of this form of imagery within PAR processes. Figures 16.2 and 16.3 show examples of the type of "existing versus proposed" landscape views that we are developing to share with stakeholders. This type of visualization has been increasingly employed in the environmental planning field as a means to communicate the distinctions between different policy, land use, and land management scenarios. Landscape visualizations have become an increasingly important component of environmental decision making and public participation processes, including in natural resource management studies (Pettit et al. 2011), in rural landscape settings (Appleton and Lovett 2003), and in public dialogues about visualizing the impacts associated with climate change (Sheppard and Meitner 2005; Sheppard et al. 2011). Landscape visualizations complement other forms of communication and have been found to be

Figure 16.2 **(See color insert)** Photograph of existing vegetable field in a floodplain (*top*) and the proposed condition with hoop houses, crop diversification, and wetlands (*bottom*).

Figure 16.3 **(See color insert)** Photograph of an existing cornfield on a riverbank (*top*) and the proposed condition with riparian buffer (*bottom*).

accessible to audiences from an array of backgrounds, including laypersons (Lewis and Sheppard 2005). Lewis and Sheppard (2006) describe realistic landscape visualizations as a beneficial element in decision making, with demonstrable influences on human behavior and policy structure around climate change. With increased interfacing of landscape visualization techniques using geographic information systems (GIS) and with the recent development of numerous 3-D visualization models, researchers have had increased success in their efforts to communicate current and future land use scenarios to diverse audiences (Ghadirian and Bishop 2008; Griffon et al. 2010). Accordingly, landscape visualization is a logical component of our transdisciplinary research approach.

Figures 16.2 and 16.3 are examples of the type of visualizations of BMPs that we will present to farmers and use to guide discussions in 2014–2015. Both photos were taken in an agricultural area at the Intervale, a conserved agricultural area in Burlington, Vermont, that is highly vulnerable under current climate change scenarios. The land is a floodplain, and is therefore susceptible to flooding, erosion, and contamination from upstream sources. Figure 16.2 depicts a farm field in the Intervale in a partially flooded condition, paired with a visualization of what that same parcel could look like if it was managed with a constructed wetland and alternative drainage practices. Hoop houses that would extend the growing season for farmers are also shown. Figure 16.3 depicts a riverbank, also in the Intervale floodplain, paired with a visualization of what this area could look like with a vegetated riparian buffer and stream bank erosion-prevention BMPs added to it. The riparian buffer would both help sequester carbon and protect the farmland from erosion and flood hazards in the case of extreme weather. These images will be used to facilitate conversations between researchers, outreach professionals, landowners, and managers about the implications of using and not using BMPs and the type of impact that these practices may have under specific climate change scenarios.

16.2.6 Stakeholders and the PAR Approach

The emergence of this initiative in 2011 relied on input from key stakeholders such as the Vermont Agency of Agriculture, the VT Natural Resources Conservation Service, the University of Vermont (UVM) Extension, the Vermont Farm to Plate Network (F2P), the Vermont State Climatologist, Stone Environmental, certified crop advisors, and researchers at other US universities. The input from these key groups and individuals is formalized in our project through an advisory group. Members of the advisory committee include vegetable, dairy, livestock, and diversified farmers, representing trade organizations such as the Vermont Vegetable and Berry Grower's Association, the Farmer's Watershed Alliance (dairy farmers), and the Vermont Grass Farmer's Association (non-dairy, pastured livestock farmers). We convene this group two to three times per year to ensure that our goals and potential impacts are relevant, to contribute to the interpretation of research findings, and to contribute to project assessment. We have selected our advisory group based on their interest in this project and their ability to represent farmers, agricultural service providers, researchers, and policy makers to address the impacts of climate change on agriculture in Vermont and nationwide. In addition, we conducted secondary analysis of the reports from farmers submitted through the Vermont Vegetable and Berry Growers Association's "Reports from the Field." We reviewed farmer submissions from 1998 to 2012 to determine if and when this group of farmers was talking about climate change with their peers. Their concerns and attitudes have helped inform our work.

16.3 METHODS

16.3.1 Initial Investigation, Year 1

Figure 16.4 details the progression of research and outreach in this project. What the figure does not show is the informal discussion and problem identification that took place prior to the funding

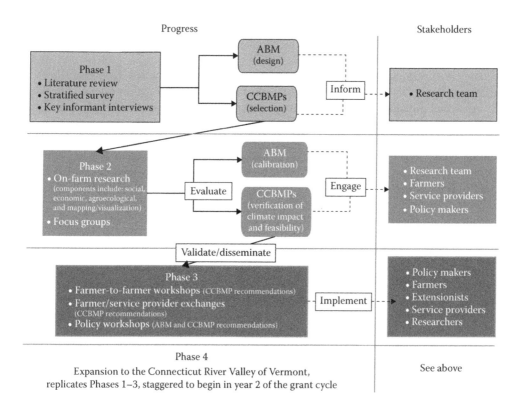

Figure 16.4 Diagrammatic representation of the proposed project, including phases, activities, and stakeholders.

of the project and the development of the research team. The discussions that led to the initiation of this work included conversations among researchers and farmers in the wake of Tropical Storm Irene and the devastation from the storm on Vermont's farms and infrastructure. Additionally, the Agroecology and Rural Livelihoods Research Group (ARLG) facilitated a daylong event in May 2012, targeted toward Vermont agricultural service providers, which attempted to capture information about how these service providers approached the topic of climate change with farmers, and how farmers differentiated between climate change and extreme weather as influential concepts. Preceding the event, we conducted a survey of Vermont agricultural service providers, which captured their initial thoughts and concerns about climate change and how farmers integrate these concerns into their decision-making processes (Schattman et al. 2012). In addition, several members of our project team are extension and outreach professionals, who work with farmers, other technical service providers, and community members on a regular basis. These team members serve as key informants, and are invaluable to our efforts because of the degree to which they represent the concerns and perspectives of these stakeholder groups.

In 2012, the research team began by identifying goals and opportunities for collaboration, identifying working norms, and conducting a literature review on BMPs related to climate change which were most applicable to farms in the northeastern United States. All principal investigators contributed to the development of a stratified survey, which was conducted in the Champlain Valley of Vermont, specifically in the Lamoille and Missisquoi watersheds. These watersheds were selected for two reasons: (1) farms located in the Lamoille watershed are representative of farms in Vermont as a state (Lovell et al. 2010a), and (2) the principal investigators who lead the ABM portion of this project are conducting additional research in the Missisquoi watershed and hoped to use this survey to enhance the depth of their investigations. The survey was tested with five farmers in the Champlain Valley of Vermont in the winter of 2013, and was revised to incorporate their feedback.

In order to ensure our survey was delivered to farmers within the specified watersheds, our group contracted the National Agriculture Statistics Service (NASS) to administer an anonymous survey. An initial postcard was sent to farms that met the location criteria. The mailing inquired about land use, ownership, and primary sources of income, and concluded by asking the participant if he or she would be willing to fill out a longer questionnaire. This screening survey was sent to 1104 farms, with a total response rate of 20%. Of those who replied, 128 responded that they would be willing to fill out the longer questionnaire. The full survey questionnaire was mailed to these respondents between April and July 2013. Of these, 48 completed surveys were returned by mail, and 31 were collected over the phone by NASS enumerators, for a total of 79 complete responses. This resulted in a confidence interval of 10.49% at a 95% confidence level.

Simultaneous to the survey, we conducted a secondary document review of Reports from the Field. These reports are submitted by growers on a semimonthly basis to the Vermont Vegetable and Berry Growers Association, who then publishes the reports through a list serve and in print through the Vermont Agency of Agriculture, Food and Market's monthly newspaper. The reports range in topics from evaluating crop varieties to commenting on the weather to markets and customers. We used a double-coding approach (Boyatzis 1998) to investigate if and when farmers were sharing their thoughts about climate change, and in what context. In addition, we coded for specific plant diseases, pests, dry and wet weather, and extreme weather events. The coding and analysis were conducted using HyperRESEARCH (Researchware Inc. 2013), a qualitative data analysis software.

In addition to the survey and document review efforts, and equally as important, Year 1 was the period in which our research team began to build relationships within our own ranks and with professional partners (including farmers), initiate outreach, and solicit additional funding to support our work. While our group is committed to a PAR process, it is exceedingly challenging to secure the funding that is needed to support the reflective and relationship-building components of this approach. We firmly emphasize that the time required to build relationships and trust within the research team and with our external partners is the foundation on which the quality of our research and outreach depends.

16.3.2 Agent-Based Models, Years 1 and 2

A multilevel ABM was developed using AnyLogic Professional Version 6.6 (AnyLogic 2013). Farmers will be modeled as farm-level agents, who will exist under the institutional jurisdictions of various town, county, regional, and state government agencies. The higher-level agents are described as institutional agents. The decision heuristics for both farm-level and institutional agents will be derived from analyzing existing datasets, focus groups, the farmer survey, interviews, and policy documentation analysis. The ABM will be built on land use datasets of the study area and will be calibrated to the observed land use and carbon emission patterns from 2000 to 2010. The calibrated models will then be used to generate and test experimental simulations for alternate policy and decision behavioral scenarios. The findings from calibrated ABMs for various scenarios will be shared with broader stakeholder groups in mediated modeling sessions. Further, emergent scenarios will be derived through stakeholder inputs and will be tested in the calibrated ABMs. The decision rules of the decision-making agents such as farmers, households, businesses, and organizations/entities will be derived/simulated based on empirical datasets. Empirical datasets are being used to calibrate these models, including the utilization of new farmer surveys and interviews, analysis of existing datasets including the US Department of Agriculture (USDA) census of agriculture, the US census, the National Land Cover Database (NLCD), and permitting data collected by the state's Agency of Agriculture.

16.3.3 On-Farm Research, Years 2 and 3

16.3.3.1 Farm Selection

Four components of this investigation require researchers to engage directly with farmers as coinvestigators. It was important to choose farmers carefully for this stage of the research, as we not only want collaborators in the research, but partners who will also support our outreach efforts. In order to select which farmers to approach, in the initial screening, we used a multistage process as follows:

1. We sourced names from key contacts, including members of our research team, as well as professional and technical services providers. The survey described in Section 16.3.1 also concluded with a question asking respondents to indicate if they were willing to partake in on-farm research, and this provided a sample of willing farmers.
2. The farmers were sorted by type of farm (vegetable, dairy, meat producers, or diversified), with a goal of 12 total participating farmers, with 3 replicate farms in each category. Maple and hay producers were included if they also produced goods in one of the four listed categories, but were excluded if not. This was because there were a limited number of BMPs that our group could address for those who exclusively produced maple and hay.
3. In order to ensure that our economic analysis would ultimately be useful to commercial producers in the Northeast, farmers were sorted by gross income in 2011 (if they were survey respondents), and those who grossed <$10,000 were excluded.
4. The BMPs employed by each farm were listed (if they were survey respondents), and cross-referenced with those BMPs of most interest to the on-farm research team. The BMPs of high priority were no-till cultivation, cover cropping, storm water runoff management, rotational grazing, and conservation buffers. These were selected based on data collected (surveys and interviews), as well as the experience, professional interest, and expertise of the principal investigators.
5. Farmers were then contacted, asked to participate, and offered compensation at an hourly rate. They were given an outline of the project that detailed the on-farm research activities proposed and an estimate of the amount of time that they would be expected to contribute to each component.

16.3.3.2 Farmer Interviews

The on-farm research portion of this project will include qualitative interviews to assess farmer and technical service provider knowledge about climate change, reasons for adopting specific BMPs, and decision-making processes. Interviews will be transcribed and analyzed using HyperRESEARCH (Researchware Inc. 2013). A double-coder, constant comparison approach to analysis will look for emergent themes, using a grounded theory approach (Charmaz 2005; Glaser 1992; Glaser and Strauss 1967; Strauss and Corbin 1990). We will also utilize structuration theory, which asserts that both the agent, or the actor, and the social or organizational structure in which that agent operates are equally important for understanding behavior and outcomes (Giddens 1984; Held and Thompson 1989). In this light, we will examine critical institutional, governmental, and organizational influences that impact the levels of risk experienced at the farm level. As an addition to the interview tool, we will use a modified version of an evaluation tool developed by Lovell et al. (2010b) to record farmer and service provider perceptions of BMPs and how these BMPs may or may not help to mitigate the risk of climate change at the farm level. This evaluation tool addresses the social/cultural, economic, and ecological aspects of BMPs. Farmers are asked to rank each aspect on a scale of −2 (extreme negative impact) to +2 (extreme positive impact), with a score of 0 being no impact or not applicable. This will help us to create a relative qualitatively analyzed ranking (Boyatzis 1998) of CCBMPs from the farmer and the technical service provider perspective, which will be critical for an intentional reflection of our investigation and will also help to inform future outreach, information sharing, and research.

16.3.3.3 Economic Analyses

The economists in our group will work directly with farmers to conduct an economic analysis of potential CCBMPs, including cost analysis and projections, with the goal of evaluating their viability and barriers for adoption by farmers in the Northeast (over a 3 year period). In addition, this activity will yield information on potential ways to improve CCBMPs and make them more attractive to farmers from an economic standpoint. We will develop cost functions for each of the identified farms and CCBMPs that show promise, using the following basic formula:

$$C_{ij} = \sum_{1}^{F} \frac{Pf * Xf}{Tf} + \sum_{1}^{V} Piv * Xiv$$

where the cost (C) of implementing mitigation or adaptation practices i on farm j in a given year is the sum of the fixed (f) costs (quantity [X] times price [P]) amortized over T years of service plus the sum of variable costs. Fixed costs include the installation of infrastructure, vegetation, and so on, with an expected service of more than 1 year. Variable costs include those with a single year of service. For each CCBMP, input and labor costs, as well as machinery and fuel use, as applicable, will be calculated. If owner-operator or family labor is used, an opportunity cost will be assigned. The measurements will constitute a series of snapshots over farms and years, with attention to the phase of adoption (new, continuing) and various farm attributes (crops, scales, tenure), in order to understand the costs of CCBMP use in a variety of settings. The data will be collected via paper or electronic forms according to the farmer's choice. Farmers will record all relevant costs each month and provide completed forms for data processing and analysis. Any revenues resulting from the adoption of CCBMPs will be recorded as well. Each year, the data will be compiled into annual cost and revenue functions for each farm with key expenses and categories highlighted (Conner and Rangarajan 2009; Conner et al. 2010).

16.3.3.4 C Sequestration and GHG Emissions

To quantify the climate change mitigation potential and begin to understand the GHG balance of specific CCBMPs, we will measure the C storage and GHG emissions of selected farms and CCBMPs. To calculate the climate change mitigation potential of CCBMPs within a given farm type, we will measure the C stored (in CO_2 equivalents [CO_2E]) in soils and aboveground biomass (AGB) and GHG emissions (CO_2, N_2O, and CH_4). The carbon storage in AGB and soils will be measured in all farm type and CCBMP combinations. Herbaceous AGB will be estimated by clipping peak standing AGB in a 0.075 m^2 area in three or four locations within each farm type. Woody AGB will be estimated using allometric equations developed for northeastern forests. Soil C will be measured in each plot to a depth of 1 m increments of 0–10, 10–20, and 20–60 cm. The bulk density will also be determined using these soil cores (drying and weighing each soil core's known soil volume prior to compositing). The soil texture in one location per farm type and CCBMP combination will be determined by the hydrometer method (0–20 cm layer only). GHG emissions will be sampled in three or four locations per farm type. On each sampling date, we will measure CH_4 and N_2O fluxes using the vented, closed chamber method (Hutchinson and Mosier 1981). CO_2 fluxes will be measured using an LI-COR 8100A soil respiration survey system with a 20 cm diameter chamber. Inorganic soil N, soil temperature, and soil moisture (gravimetric) will be measured concurrently, as GHG flux covariates. We will measure inorganic soil N (as a covariate for N_2O fluxes) by taking three or four soil cores (0–20 cm) per farm type and CCBMP combination. Cores will be composited and subsampled for 2 M KCl extraction. These extracts will be analyzed for inorganic soil N at UVM's Agricultural and Environmental Testing Lab (Lachat QuickChem FIA). Year-round measurements (as described above) will be taken across 3 years.

16.3.3.5 *Landscape Visualization*

Photo-simulated landscape visualizations using both landscape-scale (high-resolution ortho-photos) and site-scale perspectives (photos at eye level) will be developed for at least one of each farm type. Adobe Photoshop and ArcMap GIS software will be used to create the scenario visualizations. These visualizations will provide more in-depth descriptions of CCBMP potential on-farm, which will complement in-depth farmer-to-farmer outreach activities. Photo simulations will be debuted at several winter farm conferences in the region in 2014, where i-clickers (survey tools) will be used to gauge stakeholder preferences in response to (1) the acceptability and practicality of the CCBMPs shown and (2) the utility of these visualizations in the knowledge-sharing pieces of a transdisciplinary research and outreach approach. In 2015, smaller focus groups will also have an opportunity to react to a library of landscape visualizations representing the different CCBMPs implemented across the four major farm types explored in this study.

16.4 SELECTED PRELIMINARY RESULTS

16.4.1 Survey

Respondents to our survey own 15,106 acres and lease an additional 1,891 acres in the area of study. The average acreage owned by a respondent is 220 acres (SD of 219), and the average acreage leased is 67 acres (SD of 62). The median acreage owned is 150 acres, and the median acreage leased is 55 acres. Table 16.1 shows that the respondents to this survey have a wide variety of management approaches, including certified organic, organic but not certified, conventional, biodynamic, and integrated pest management. Growers who identified their management practices as conventional were represented more heavily than other categories. Of greater interest is the number of respondents who reported multiple management strategies including conventional approaches paired with certified and noncertified organic practices. Our survey did not address growers' understanding of what qualifies as a noncertified organic approach.

Table 16.1 Description of Survey Respondents

Farm Management Type, $n = 76$	
Certified organic	16
Certified organic and conventional	1
Conventional	34
Organic, not certified	18
Organic, not certified and conventional	2
Integrated pest management (IPM)	1
Organic, biodynamic, and nutrient-dense soil management	1
Sustainable	1
Number of Years Spent Farming, $n = 76$	
<3	0
3–7	4
8–10	3
11–20	13
21–30	16
31–40	26
41–50	8
50+	6

In relation to the number of years farming, responses were categorized into decades of experience with the exception of those who have farmed for fewer than 10 years. These "beginning farmers" were separated out into three categories according to the beginning farmer typology described by Scheils (2002). This typology will be used in subsequent analysis when examining farmers' perceptions of how climate change will affect their business and livelihood.

16.4.2 Reports from the Field

Frequency reports from the secondary analysis of grower reports (Reports from the Field) indicate that growers mention climate change in passing, but without great frequency. Table 16.2 demonstrates the typical comments submitted by growers, distinguished between comments that directly reference climate change and those that address extreme weather events. Of the comments that address climate change, the two selected comments demonstrate a laissez-faire attitude that thinly veils a willingness (or perhaps a need) to test new conditions, push seasonal limits, and take risks in diversified operations. A review of these comments showed that growers were much more likely to discuss extreme weather events. While heat and dryness were associated with low plant disease pressure, excess rain was commonly linked with evidence of foliar and root diseases.

Table 16.2   Farmer Voices from Reports from the Field, 1998–2012

Topic	Year	Farmer Voices
Climate Change	1998	"There's still time to squeeze out a few bucks to lose on… a late planting of radishes, arugula, cilantro, and spinach. If September temperatures are going to be in the 80s, might as well take advantage of global warming."
	1998	"We are finally conceding to global warming and putting out 75 peach trees. If they make it and fruit 2 out of 4 years they will be worth the investment. If the winter is so cold it whacks them then maybe it will be cold enough to reduce some of the overwintering insect pests we have been seeing in large numbers the last couple of years: squash bugs, striped cucumber beetles, Colorado potato beetles, and first-generation corn borer."
Extreme Weather Events	1998	"The ice storm and wet weather caused loss of 75% of my newly planted raspberries this year (2,000 plants)… They started out great in the spring and then they started dying back probably due to severe winter injury after an open winter without snow cover."
	2000	"Fourteen inches of snow plus several days of rain have not helped our seedings of field crops, and to date we have only 2 seedings in of peas, carrots, beets, turnips, and radishes."
	2000	"On Friday evening June 5, a hail storm blew through the Connecticut Valley at high velocity. The storm raged for about 20 minutes with high winds, heavy rains, and large hail. All of our spring crops were shredded or buried in mud. We lost peas, strawberries, lettuce, tomatoes, melons, etc.. However, I haven't seen a flea beetle yet, and I don't dare complain about the few cutworms I've seen. We have postponed our first CSA (Community Supported Agriculture program) distribution for a month."
	2004	"This has been a difficult spring. First we have a couple of intense heat days in April. Then excessive wind drying things out, and May gave us 10 inches of rain and lots of grey cool cloudy weather. That was followed by 2 days of 90° weather that gets blown out by a storm that deposits 1.5 inches of rain and some trees in about an hour. When that's done we have frost warnings on June 10th and 11th. The strawberries have just plain freaked out. They are ripening the earliest in recent memory, yet we can't find a beet green or radish close to harvest. Despite my best efforts (and a second mortgage to pay for fungicides and stickers) 10 inches of rain has taken its toll… Greenhouse sales were strong, thank goodness…."

Reports on weather events included mild winters and lack of snow cover damaging overwintering plants. On the other side of the coin, growers also reported too much snow in May. High winds and hail featured in several reports, while some growers wrote during the years when the range of weather events seemed to affect farms. Other reports revealed that while some growers are susceptible to flooding, and managing too much water can be a problem, others are deeply reliant on irrigation to ensure both crop availability and quality. The key question raised by this review of the Reports from the Field is whether farmers distinguish between weather and climate, and how their decision making is influenced by their understanding of these two concepts. A review of the reports indicates that weather and the effects of climate and weather combined are of immediate concern to fruit and vegetable growers in Vermont, but long-term planning based on grower knowledge of climate change is discussed less frequently. The implications of this are related to farmers' ability to cope with changing climatic conditions. This is a line of inquiry that we will follow with additional analysis of the survey and qualitative interviews.

16.4.3 Agent-Based Model

Details of the initial ABM results can be found in a recent report by members of this project (Tsai et al. 2012). The hypothesis posed by members of this team was that financial considerations in combination with factors such as climate change and public policy are the primary influences on farmers' land use decisions. Researchers constructed six scenarios to represent the varying presence of exogenous factors (climate change, public policy, etc.) on the financial conditions of farmers. By running these scenarios through the ABM, two conclusions were reached: (1) the primary factor influencing farmers' decisions regarding land use is their financial condition, and (2) exogenous factors that reduce financial stress among farmers have the greatest potential for limiting the shrinkage of agricultural lands and the growth in forested lands in Vermont.

16.5 DISCUSSION

16.5.1 PAR Process: Taking Stock of Transdisciplinary Process Participation

Critical to the PAR process is the inclusion of stakeholders in multiple phases of the project, as well as an examination of the levels of stakeholder engagement. Figure 16.4 illustrates not only the phases of this project, but also at what stages of the work the different stakeholders are involved. As discussed previously, Kindon et al. (2010) outline a continuum of participation in PAR projects as compared with a ladder of public participation. The assumption of this continuum is that greater levels of participation lead to greater benefit for stakeholders, and that greater degrees of empowerment lead to greater interest in and execution of participation (see Figure 16.1). This is of particular interest in an agroecological framework that prioritizes the empowerment of the disenfranchised (Tomich et al. 2011). Using their framework, we have evaluated the degree of participation of each stakeholder group in our project, and identified areas in which we can improve our facilitation of stakeholder involvement.

As Table 16.3 demonstrates, not all stakeholders in this PAR effort participate equally, and by extension, not all stakeholders have parity of power within the research. It is critical to note that, while much is written about increasing the empowerment of disenfranchised groups in both agroecology (as a social movement) and PAR, an analysis of the ability or willingness to participate in decisions affecting a particular group's own condition is rarely conducted at the onset of the work. Questions can be raised about both the level of power held by stakeholders prior to the project (Stillman 2013), and the degree to which a PAR process can change the level of empowerment held by a particular stakeholder group. For example, our project focuses on the resilience and risk

Table 16.3 Stakeholder Participation Analysis Summary: A Snapshot of Years 1 of 3

Stakeholder Group	Level of Participation	Roles and Responsibilities	Team Goals for Years 2 and 3
Farmers	Participation for material incentives; functional participation	Provide on-farm research setting and time, inform research outcomes and benefit from research for management decisions, participate in advisory committee, test interview and survey instruments, and give feedback	Increase participation and investment in the project (to what rung on the ladder and how?). Empower farmers through farmer-to-farmer training on specific CCBMPs, provide input into the next iteration of our work
Technical service providers	Co-learning	Contribute to defining research goals and approach, contribute to framing the issue, key team members for outreach portion of the project	Contribute to data analysis, engage other technical assistance (TA) providers in Train the Trainer workshops, maintain relationships with researchers and farmers, provide input into the next iteration of our work
University-based researchers and extension/outreach professionals	Co-learning	Contribute to defining the research goals and approach, contribute to framing the issue, the key team members for research, and the outreach portions of the project	Complete the on-farm portion of the research, conduct analysis, deliver results to outreach professionals, and collaborate with them to provide training that places farmers and TA providers in leadership roles. Apply for additional funding for the next iteration of our work
Policy makers	Information transfer	Participate in information giving/receiving	Receive information from our project to inform future policy decisions

management among farmers, but to date the farmers involved in our project participate functionally (testing interview and survey instruments, providing feedback, and serving on the advisory committee) and for incentives (in return for providing a setting for on-farm research), but not by setting research goals and objectives. Future iterations of our efforts are structurally designed to allow for and facilitate greater degrees of farmer participation in key decision-making processes, including the direction of new research objectives. Since participation serves as a proxy for the empowerment experienced by stakeholders, attention must be paid to how participation changes through iterative PAR cycles. Greater inclusion of the knowledge and opinion of stakeholders, especially those not normally included in agenda-setting processes, benefits not only these stakeholders but the work as a whole (Stillman 2013).

This framing of agroecology and PAR requires us to pay more rigorous attention to how power is distributed in our process, and how our process interacts with notions of social justice and equity (Gatenby and Humphries 2000). It is through PAR that we can address the key ethical and moral concerns of our research. Specifically, we draw from the work of Emanuel et al. (2000), which examined many international standards for ethical research and articulated the following criteria:

> To be ethical, research must have social or scientific value, demonstrate scientific validity, be conducted using fair subject/participant selection, have a favorable risk–benefit ratio, be subject to independent review, practice informed consent of research participants, and demonstrate respect for potential and enrolled participants. (Emanuel et al., 2000, p. 2703)

Khanlou and Peter (2005) review PAR approaches in light of these criteria, and encourage us to carefully examine the following key factors: (1) whether our PAR efforts truly have emancipatory potential; (2) whether our motivating foci are based on rigorously examined scientific knowledge; and (3) that we do not select participants solely because of their level of disenfranchisement or privilege. In our work, the emancipatory potential of our efforts is grounded in the assumption that stakeholders (Kania and Kramer 2011), and researchers in particular (Francis et al. 2008; Rosenfield 1992), are constrained by narrow understandings of complex problems. By bringing together researchers and stakeholders from a variety of backgrounds and disciplines, we seek to broaden our understanding of the problem (in this instance, the effects of climate change on the agro-food system) and increase the creativity with which we conceptualize and apply solutions (Rosenfield 1992). Through the integration of economic analysis, biogeochemistry, and qualitative and policy analysis, we seek to bring scientific rigor to the community level and let further inquiry be based on the needs of the community, as identified by the community (Bacon et al. 2005).

While our transdisciplinary approach is designed to maximize the effectiveness of our research and outreach, it is not without its challenges. Kessel and Rosenfield (2008) identify many potential benefits and challenges to transdisciplinary research. Among those highlighted, we have experienced an openness and appreciation of other team members' knowledge and level of expertise, as well as a shared understanding of the problem at hand. Prior to our project, many of the participating researchers and extension educators knew one another and had a positive rapport, though most had a limited depth of knowledge about their colleagues' research. One of the first challenges that we faced was getting researchers to find the time so that team members could get to know each other and their work in more depth. Fry (2001) identifies building this rapport, deep understanding, and appreciation of other's disciplines as one of the key elements to a successful transdisciplinary research process. To facilitate this, we provided time at full team meetings for individual team members to present their work, a practice that we will continue in the second half of the project and in future iterations of the PAR process. This seemed to work well for everyone to become more familiar with the components of the research that each individual or group was working on, and it also facilitated the integration of the different approaches utilized. This is something that we will also do with farmers in order to integrate the knowledge derived from experience (rather than academic), though we are still in the process of finalizing the methods that we will use.

The challenges face by our team, which were predicted by Kessel and Rosenfield (2008), include concerns about the diffusion of work because of multiple foci and the lack of a preexisting research framework. The differences in how different stakeholders are evaluated can also present a challenge, such as the difference between how extension professionals and tenure-track faculty are evaluated by the chairs of their department or their supervisors (McDowell 2001). We address these concerns by relying on strong facilitation to keep the group informed of individual and group efforts in both research and publication, and by carefully documenting and reviewing our emerging process (Alrøe and Kristensen 2002). In addition, one of the greatest challenges to a project that is both transdisciplinary and based on PAR is the friction between scientific knowledge and local knowledge, as well as the conflict between differing goals and agendas, which can potentially derail trust and collaboration between researchers and other stakeholders, and bears special sensitivity when conducting PAR.

Finally, Khanlou and Peter (2005) encourage researchers to be attentive to the possible risks that research and its resulting social action pose to all stakeholders; to seek an ethical and independent review of the research at each iterative cycle of the PAR process; to require informed consent from stakeholders involved in the research process; and to address stakeholder concerns with the research process in a transparent and open manner.

16.5.2 Lessons Learned and Future Directions

This thoughtful analysis in each PAR process and conversation between stakeholders are needed for two reasons: (1) to increase the dialogue among parties and identify those areas where power dynamics can result in intentional or unintentional oppression (Chatterdon et al. 2010) and (2) to address the concerns that PAR processes may be biased by the social agendas of the participants (including the researchers). We subscribe to the perspective that all research is biased to varying degrees (Alrøe and Kristensen 2002). The transparency of bias is one tool that we employ to address concerns about research validity, while simultaneously developing the trust and openness between collaborators that is necessary to succeed using a PAR approach (Kessel and Rosenfield 2008). In light of this, we wish to make two points that will add to how we have understood and employed PAR in an agroecological context.

First, we acknowledge that, at its inception, this was not a farmer-generated project. Rather, it was conceived of within the context of the university, and because of this, we were required to invest time and resources into proposal writing, resources garnering, and team building. To integrate multiple stakeholder views into the initial stages of this project, we shaped our initial research goals through meetings with agricultural service providers and policy makers in Vermont (such as the Vermont Natural Resources Conservation Service) and relied heavily on the extension educators on our team. We also drew on the input of team members who are otherwise embedded in the agricultural community of our region. While the farmers themselves were not well represented at this stage in the research, PAR is an iterative process and future opportunities for defining agendas will incorporate their voices more actively. We strive to be attentive to the needs of farmers in the context of climate change, since PAR as an approach has emerged as a response to top-down, academic, and policy-driven research (Fernandez et al. 2013). Ultimately, we hope that this will guide our work, making it of real value to farmers and the public at large, in accordance with the original mission and goals of public research institutions (McDowell 2001).

Secondly, temporal factors play a significant role in our conceptualization of empowerment in a PAR process. While Table 16.3 illustrates the roles and levels of participation in our project to date, it is only as representative as a snapshot. PAR processes are long-term, committed endeavors with a multiplicity of dimensions that are designed to address complex problems such as climate change (as is the case in this project). Our ultimate hope is that we can contribute to a process following PAR principles that brings us all to a place where everyone has a more equal voice in the dialogue. To do this, methods of tracking and reporting the levels of participation, empowerment, and investment in research processes should be developed. These will lead to a deeper understanding of how the power dynamics in PAR efforts shift over time, or differ depending on who is involved at what point in the process. This could also lead to a framework for evaluating when PAR is most applicable and of value to stakeholders versus when more straightforward research approaches are appropriate. Currently, we are not aware of any precedent for assessing the levels of stakeholder participation and empowerment in PAR processes over time.

In this light, and humbly accepting that we could not do it all, we are working to build relationships, generate data, and contribute as much as we can to both Vermont agriculture and relevant policy dialogue and practice. It is our hope that our experiences and reflections on them will contribute to the efforts of others seeking to use transdisciplinary approaches to find grounded, innovative solutions for complex problems.

SUMMARY

Complex problems, such as agricultural resiliency in the face of climate change, require multistakeholder and transdisciplinary approaches. This chapter presents an innovative research and outreach effort employed in Vermont to address the challenges associated with climate change that farms may face in the near and distant future. Our research team is composed of eight faculty with a diversity of specialties including agricultural economics, agroecology, climate change science, extension, sustainable agriculture, governance, and policy. In this chapter, we present both conceptual and empirical contributions to PAR and agroecological thought, which are drawn from our experiences with the Vermont Agricultural Resilience in a Changing Climate Initiative, an in-progress, multiyear effort. This chapter discusses the successes and challenges of this approach, with special attention given to the levels of stakeholder participation and empowerment, as well as our experiences working with a highly diverse team of researchers and stakeholders on a highly complex problem. We find that a framework for evaluating change in stakeholder power and parity in PAR processes over time is needed. In addition, our approach to transdisciplinary work related to agriculture and climate change can be used as a blueprint, to be adapted and improved on by other groups. The richness of this effort comes from an integration of theory and practice, shown through our reflections.

ACKNOWLEDGMENTS

This work was funded in part by UVM Extension and UVM College of Agriculture and Life Sciences as part of the UVM Food Systems Initiative that focuses on transdisciplinary campus-wide efforts to address critical issues related to supporting community-focused food systems. Additional funding was provided by the University of Vermont Extension, the Vermont Community Foundation, the Highmeadows Fund, the Gund Institute for Ecological Economics, the University of Vermont Environmental Program, the University of Vermont Department of Plant and Soil Science, EPSCOR Research on Adaptation to Climate Change (RACC), and the US Department of Agriculture (VT Agricultural Experiment Station—HATCH Program).

Thanks to the following individuals who contributed to this project: our advisory committee, Vicky Drew, Jim Wood, Lesley-Ann Dupigny Giroux, Jake Claro, Vern Grubinger, Rich Smith, Eric Noel, Ben Brown, Julie Moore, and Andy Jones. Postdoctoral researchers and graduate and undergraduate student research assistants include Rebecca Fox, Jennifer C. Miller, Tyler Goeschel, Ann Hoogenboom, Rachel DiStefano, Yu-Shiou Tsai, Stephanie Cesario, Martha Waterman, and Maija Lawrence.

REFERENCES

Albers, J. 2002. *Hands on the Land: A History of the Vermont Landscape.* Cambridge: MIT Press.

Alrøe, H. F. and E. S. Kristensen. 2002. Towards a systemic research methodology in agriculture: Rethinking the role of values in science. *Agr Hum Values* 19(1):3–23.

Amekawa, Y. 2011. Agroecology and sustainable livelihoods: Towards an integrated approach to rural development. *J Sustain Agr* 35:118–162.

AnyLogic. 2013. AnyLogic multimedia simulation software, professional version 6.6. http://www.anylogic.com/ (accessed: February 12, 2014).

Appleton, K. and A. Lovett. 2003. GIS-based visualisation of rural landscapes: Defining "sufficient" realism for environmental decision-making. *Landscape Urban Plan* 65(3):117–131.

Bacon, C., E. Méndez, and M. Brown. 2005. Participatory action research and support for community development and conservation: Examples from shade coffee landscapes in Nicaragua and El Salvador. Santa Cruz, CA. http://escholarship.org/uc/item/1qv2r5d8.pdf (accessed: September 28th, 2013).

Belsky, J. M. 2002. Beyond the natural resource and environmental sociology divide: Insights from a transdisciplinary perspective. *Soc Natur Resour* 15(3):269–280.

Bernstien, L., P. Bosch, O. Canziani et al. 2007. Climate change 2007: An assessment of the Intergovernmental Panel on Climate Change. http://www.ipcc.ch/pdf/assessment-report/ar4/syr/ar4_syr.pdf (accessed: December 15, 2013).

Boyatzis, R. E. 1998. *Transforming Qualitative Information: Thematic Analysis and Code Development.* Thousand Oaks, CA: Sage Publications.

Charmaz, K. 2005. Grounded theory in the 21st century: Applications for advancing social justice studies. In Y. S. Denzin and N. K. Lincoln (eds), *The Sage Handbook of Qualitative Research*, pp. 507–535. Thousand Oaks, CA: Sage Publications.

Chatterdon, R., D. Fuller, and P. Routledge. 2010. Relating action to activism: Theoretical and methodological reflections. In S. Kindon, R. Pain, and M. Kesby (eds), *Participatory Action Research Approaches and Methods: Connecting People, Participation and Place*, pp. 216–222. New York: Routledge.

Conner, D. 2004. Shortcuts to measuring crop profitability: Are they misleading? Marketing, May. http://marketingpwt.dyson.cornell.edu/SmartMarketing/pdfs/conner5-04.pdf (accessed: December 15, 2013).

Conner, D. and A. Rangarajan. 2009. Production costs of organic vegetable farms: Two case studies from Pennsylvania. *Horttechnology* 19:193–199.

Conner, D. S., K. B. Waldman, A. D. Montri, M. W. Hamm, and J. A. Biernbaum. 2010. Hoophouse contributions to economic viability: Nine Michigan case studies. *Horttechnology* 20:877–894.

Emanuel, E. J., D. Wendler, and C. Grady. 2000. What makes clinical research ethical? *J Am Med Assoc* 283:2701–2711.

Eriksen, P. J., J. S. I. Ingram, and D. M. Liverman. 2009. Food security and global environmental change: Emerging challenges. *Environ Sci Policy* 12:373–377.

Fernandez, M., V. E. Méndez, and C. Bacon. 2013. Seasonal hunger in coffee communities: integrated analysis of livelihoods, agroecology, and food sovereignty with smallholders of Mexico and Nicaragua. In Conference Paper #42, Food Sovereignty: A Critical Dialogue. International Conference Yale University. September 14–15.

Fornara, D. A. and D. Tilman. 2008. Plant functional composition influences rates of soil carbon and nitrogen accumulation. *J Ecol* 96:314–322.

Francis, C., G. Lieblein, T. A. Breland et al. 2008. Transdisciplinary research for a sustainable agriculture and food sector. *Agron J* 100:771–776.

Francis, C., G. Lieblein, S. Gliessman et al. 2003. Agroecology: The ecology of food systems. *J Sustain Agr* 22:99–118.

Frolking, S. E., J. L. Bubier, T. R. Moore et al. 1998. Relationship between ecosystem productivity and photosynthetically active radiation for northern peatlands. *Global Biogeochem Cy* 12:115–126.

Frumhoff, P. C., J. J. McCarthy, J. M. Melillo et al. 2007. An integrated climate change assessment for the northeast United States. *Mitig Adapt Strateg Glob Chang* 13:419–423.

Fry, G. L. A. 2001. Multifunctional landscapes—Towards transdisciplinary research. *Landscape Urban Plan* 57:159–168.

Gatenby, B. and M. Humphries. 2000. Feminist participatory action research: Methodological and ethical issues. *Womens Stud Int Forum* 23:89–105.

Ghadirian, P. and I. D. Bishop. 2008. Integration of augmented reality and GIS: A new approach to realistic landscape visualisation. *Landscape Urban Plan* 86:226–232.

Giddens, A. 1984. *The Constitution of Society: Outline of the Theory of Structuration*. Berkeley, CA: University of California Press.

Glaser, B. G. 1992. *Basics of Grounded Theory Analysis*. Mill Valley, CA: Sociology Press.

Glaser, B. G. and A. L. Strauss. 1967. *The Discovery of Grounded Theory: Strategies for Qualitative Research*. Chicago: Aldine Press.

Gliessman, S. R. 2007. *Agroecology: The Ecology of Sustainable Food Systems*, 2nd edn. Boca Raton, FL: CRC Press.

Godemann, J. 2008. Knowledge integration: A key challenge for transdisciplinary cooperation. *Environ Educ Res* 14:625–641.

Griffon, S., A. Nespoulous, J. Cheylan, P. Marty, and D. Auclair. 2010. Virtual reality for cultural landscape visualization. *Virtual Real* 15:279–294.

Guzmán, E. S. and G. Woodgate. 2013. Foundations in agrarian social thought and sociological theory agroecology. *Agroecol Sustain Food Syst* 37:32–44.

Held, D. and J. B. Thompson. 1989. *Social Theory of Modern Societies: Anthony Giddens and His Critics*. New York: Cambridge University Press.

Holt-Giménez, E. 2006. *Campesino a Campesino: Voices from Latin America's Farmer to Farmer Movement for Sustainable Agriculture*. Oakland, CA: Food First Books.

Hutchinson, G. L. and A. R. Mosier. 1981. Improved soil cover method for field measurement of nitrous oxide fluxes. *Soil Sci Soc Am J* 45:311–316.

IPCC (Intergovernmental Panel on Climate Change). 2007. Climate change 2007—The physical science basis: Working group I contribution to the fourth assessment report of the IPCC volume 4 of Assessment report. http://www.ipcc.ch/pdf/assessment-report/ar4/syr/ar4_syr.pdf (accessed: December 15, 2014).

IPCC (Intergovernmental Panel on Climate Change). 2012. Summary for policymakers: Managing the risks of extreme events and disasters to advance climate change adaptation. http://ebooks.cambridge.org/ref/id/CBO9781139177245 (accessed: December 19, 2013).

Kania, B. J. and M. Kramer. 2011. Collective impact. Stanford Social Innovation Review. http://www.cap-sonoma.org/downloads/AgendaPacketV1.pdf (accessed: January 13, 2014).

Kessel, F. and P. L. Rosenfield. 2008. Toward transdisciplinary research: Historical and contemporary perspectives. *Am J Prev Med* 35:S225–234.

Khanlou, N. and E. Peter. 2005. Participatory action research: Considerations for ethical review. *Soc Sci Med* 60:2333–2340.

Kindon, S., R. Pain, and M. Kesby (eds). 2010. Participatory action research: Origins, approaches and methods. In *Participatory Action Research Approaches and Methods: Connecting People, Participation and Place*, pp. 9–18. New York: Routledge.

Koliba, C. and A. Zia. forthcoming. Governance informatics: Using computer simulation models to deepen situational awareness and governance design considerations. In E. DeSouza and K. Johnston (eds), *Policy Informatics*. Cambridge, MA: MIT Press.

Koliba, C., A. Zia, and B. Lee. 2011. Governance informatics: Utilizing computer simulation models to manage complex governance networks. *Innov J* 16:1–26.

Lal, P., J. R. R. Alavalapati, and E. D. Mercer. 2011. Socio-economic impacts of climate change on rural United States. *Mitig Adapt Strateg Glob Chang* 16:819–844.

Lewis, J. L. and S. R. J. Sheppard. 2005. Ancient values, new challenges: Indigenous spiritual perceptions of landscapes and forest management. *Soc Natur Resour* 18:907–920.

Lewis, J. L. and S. R. J. Sheppard. 2006. Culture and communication: Can landscape visualization improve forest management consultation with indigenous communities? *Landscape Urban Plan* 77:291–313.

Lovell, S. T., V. E. Méndez, D. L. Erickson, C. Nathan, and S. DeSantis. 2010a. Extent, pattern, and multifunctionality of treed habitats on farms in Vermont, USA. *Agroforest Syst* 80:153–171.

Lovell, S. T., S. DeSantis, C. A. Nathan et al. 2010b. Integrating agroecology and landscape multifunctionality in Vermont: An evolving framework to evaluate the design of agroecosystems. *Agr Syst* 103:327–341.

Manzo, L. C. and N. Brightbill. 2010. Toward a participatory ethics. In S. Kindon, R. Pain, and M. Kesby (eds), *Participatory Action Research Approaches and Methods: Connecting People, Participation and Place*, pp. 33–40. New York: Routledge.

McDowell, G. R. 2001. *Land-Grant Universities and Extension into the 21st Century: Renegotiating or Abandoning a Social Contract*. Ames, IA: Iowa State University Press.

Méndez, V., C. Bacon, and R. Cohen. 2013. Agroecology as a transdisciplinary, participatory, and action-oriented approach. *Agroecol Sustain Food Syst* 37:37–41.

North, M. J. and C. M. Macal. 2007. *Managing Business Complexity: Discovering Strategic Solutions with Agent-Based Modeling and Simulation*. Oxford: Oxford University Press.

Panek, J. A., P. A. Matson, I. Ortiz-Monasteria, and P. Brooks. 2000. Distinguishing nitrification and denitrification sources of N_2O in a Mexican wheat system using ^{15}N. *Ecol Appl* 10:506–514.

Parton, W. J., A. R. Mosier, D. S. Ojima et al. 1996. Generalized model for N_2 and N_2O production from nitrification and denitrification. *Global Biogeochem Cy* 10:401–412.

Pettit, C. J., C. M. Raymond, B. A. Bryan, and H. Lewis. 2011. Identifying strengths and weaknesses of landscape visualisation for effective communication of future alternatives. *Landscape Urban Plan* 100:231–241.

Philpott, S. M., B. B. Lin, S. Jha, and S. J. Brines. 2008. A multi-scale assessment of hurricane impacts on agricultural landscapes based on land use and topographic features. *Agr Ecosyst Environs* 128:12–20.

Pretty, J. N. 1995. Participatory learning for sustainable agriculture. *World Dev* 23:1247–1263.

Researchware Inc. 2013. HyperRESEARCH 3.5.2. http://www.researchware.com (accessed: December 13, 2013).

Rosenfield, P. L. 1992. The potential of transdisciplinary research for sustaining and extending linkages between the health and social sciences. *Soc Sci Med* 35:1343–1357.

Schattman, R., V. E. Méndez, and K. Westdjik. 2012. Vermont farm resilience in a changing climate: Survey of Vermont agricultural service providers. http://www.uvm.edu/~agroecol/SchattmanR&VEMendez_VTServiceProviderClimateChangeSurveyReport_2012.pdf (accessed: January 3, 2014).

Scheils, P. 2002. What does the term "new farmer" mean? Growing new farmers, professional development discussion series #101. http://www.smallfarm.org/uploads/uploads/Files/GNF_PD_-What_does_the_term_new_farmer_mean.pdf (accessed: December 15, 2013).

Scherr, S. J. and S. Sthapit. 2009. *Mitigating Climate Change through Food and Land Use Change*. Washington, DC: Ecoagriculture Partners and Worldwatch Institute.

Sheppard, S. R. J. and M. Meitner. 2005. Using multi-criteria analysis and visualisation for sustainable forest management planning with stakeholder groups. *Forest Ecol Manag* 207:171–187.

Sheppard, S. R. J., A. Shaw, D. Flanders et al. 2011. Future visioning of local climate change: A framework for community engagement and planning with scenarios and visualisation. *Futures* 43:400–412.

Smith, P., D. Martino, Z. Cai et al. 2008. Greenhouse gas mitigation in agriculture. *Philos Trans Roy Soc B* 363:789–813.

Smith, P. and J. E. Olesen. 2010. Synergies between the mitigation of, and adaptation to, climate change in agriculture. *J Agr Sci* 148:543–552.

Stillman, L. 2013. Participatory action research and inclusive information and knowledge management for empowerment. In *Proceedings of the Sixth International Conference on Information and Communications Technologies and Development Notes—ICTD'13*, vol. 2, pp. 163–166. New York: ACM Press.

Stokols, D. 2006. Toward a science of transdisciplinary action research. *Am J Commun Psychol* 38:63–77.

Strauss, A. L. and J. M. Corbin. 1990. *Basics of Qualitative Research: Grounded Theory Procedures and Techniques*. Thousand Oaks, CA: Sage Publications.

Tomich, T. P., S. Brodt, H. Ferris et al. 2011. Agroecology: A review from a global-change perspective. *Annu Rev Environ Resour* 36:193–222.

Tsai, Y., A. Zia, C. Koliba et al. 2012. *Impacts of Land Managers' Decisions on Land-Use Transitions within Missisquoi Watershed Vermont: An Application of Agent-Based Modeling System*. Burlington, VT: Vermont EPSCoR.

Tubiello, F., P. G. Neofotis, and E. Fernandes. 2008. Climate change response strategies for agriculture: Challenges and opportunities for the 21st century. Paper #42, Agriculture and Rural Development Discussion. Washington, DC: The World Bank.

USDA-NASS. 2007. Census of Agriculture. http://www.agcensus.usda.gov/ (accessed November 23, 2013).

Wall, E. and B. Smit. 2008. Climate change adaptation in light of sustainable agriculture. *J Sustain Agr* 27:37–41.

Walthall, C. L., P. Hatfield, L. Backlund et al. 2012. *Climate Change and Agriculture in the United States: Effects and Adaptation*. Washington, DC: USDA. http://www.usda.gov/oce/climate_change/effects_2012/CC and Agriculture Report (02-04-2013)b.pdf (accessed: December 15, 2013).

Wezel, A., S. Bellon, T. Doré et al. 2009. Agroecology as a science, a movement and a practice. A review. *Agron Sustain Dev* 29:503–515.

Wolfe, D. W., L. Ziska, C. Petzoldt et al. 2007. Projected change in climate thresholds in the northeastern U.S.: Implications for crops, pests, livestock, and farmers. *Mitig Adapt Strateg Glob Chang* 13:555–575.

Zia, A., C. Koliba, and Y. Tian. 2013. Governance network analysis: Experimental simulations of alternate institutional designs for intergovernmental project prioritization processes. In L. Gerrits and P. K. Marks (eds), *COMPACT I: Public Administration in Complexity*, pp. 144–165. Litchfield Park, AZ: Emergent Publications.

CHAPTER 17

Experiential Learning Using the Open-Ended Case
Future Agroecology Education

Charles A. Francis, Lennart Salomonsson, Geir Lieblein,
Tor Arvid Breland, and Suzanne Morse

CONTENTS

17.1 Introduction 347
17.2 Importance of Holistic Systems and Experiential Learning to Agroecology 348
17.3 Open-Ended Cases as a Strategy for Learning and Capacity Building 349
17.4 Examples of Learning in Farming and Food Systems: Norway Model 351
17.5 Long-Term Implications of Applied Systems Education for Future Food Systems 354
17.6 Summary and Conclusions 355
Acknowledgments.. 356
References 356

17.1 INTRODUCTION

Educators in agriculture and food systems are seeking meaningful methods to actively engage students in acquiring the knowledge and skills needed for tomorrow's unpredictable climatic, resource, and economic environments. It is essential to stimulate new professionals to recognize the complex challenges that stakeholders face and the multiple and often conflicting goals of key players in the food system. Graduate agroecologists need the tools and confidence to move into real-world situations with an assurance that they have the capacity to deal with people who are much more experienced in the details of farming as well as food processing and marketing, yet to offer their competence in systems studies to help analyze contemporary challenges. Graduates need knowledge and skills, without doubt. Yet to help them better understand the motivations of their clients, we add the dimensions of exploring values and worldviews that often guide decisions, as well as a capacity for long-term visioning and planning for a more desirable and sustainable future. In this chapter, we summarize the experiences of the past decade in designing experiential learning curricula, with special emphasis on the open-ended case as a valuable tool for introducing realism in the education of agroecologists.

Important to education beyond the information, skills, tools, and systemic methods that are necessarily embedded in learning is the appreciation of context, the ecological principles of niches, and the special characteristics of place (Lieblein et al. 2005). Although many of the principles of science in agriculture and the mechanisms of biological processes can be recognized and applied in a range of situations, it is uniqueness of place that is central to the study of agroecosystems (Picasso et al. 2011). When we define agroecology as the *ecology of food systems* (Francis et al. 2003), our area of study and concern includes understanding the complexity of production, processing, marketing, and consumption of food. Further, we embrace the process of food production and its economic evaluation, plus the environmental and social consequences of decisions made by stakeholders in the food system, both on farms and in rural communities. To approach a challenge with this breadth and complexity, students need a comprehensive research and learning strategy. We have found the open-ended case method as central to fill this educational need in agroecology (Francis et al. 2009).

The concept and practice of experiential learning grew from a tradition of engaging each new generation in the hunting and gathering that were essential to early food systems and to survival, and this time-honored concept can be applied to the food production process in today's complex and organized agriculture (Moncure and Francis 2011). Noted educator and philosopher John Dewey (1966) described how moving students out of the classroom was one key to building relevance and stimulating learning, with the critical element of incorporating new information into prior experiences. Case studies were documented as important in Harvard Law School in the nineteenth century (Langdell 1871), and were soon adopted in business and medical education. Decision cases have been used in agriculture in recent decades (ASA 2006; Simmons et al. 1992). Practical cases help students deal with questions in context, build an understanding of the complexity and dynamics of systems, and cope with uncertainty. However, most such cases are built around an event or situation where a decision has already been made, with a solution known to the instructor and client, and the challenge for students is to be clever enough to find out what these mentors already know. We have modified this method to more closely resemble what graduates will encounter when they leave the university, as described in Section 17.3 on open-ended case studies (Francis et al. 2009).

17.2 IMPORTANCE OF HOLISTIC SYSTEMS AND EXPERIENTIAL LEARNING TO AGROECOLOGY

Our experiences in offering doctoral short courses in 1995–1997 (Lieblein et al. 1999) and semester-long immersion courses since 2000 as the first component of an agroecology MSc curriculum (Lieblein et al. 2008) have provided opportunities to learn about designing diverse educational opportunities for students to study systems. From the start, we have been inspired by principles developed in the Hawkesbury program in New South Wales, Australia, where students were assigned to farms and lived with farm families for an extended period while they studied the whole-farm system. Students interacted with the farmers as they made decisions, performed analyses of the entire farm operations, and then related this experience to what they were learning in the classroom on campus (Bawden 1991; Bawden et al. 2007; Packham and Sriskandarajah 2005). The primary guidelines were to see things firsthand, to learn by doing, and to combine theory in the classroom with practical applications in the field. The value of farmer experience was elevated to comparable status with the theory and information learned in the university. In many ways, this was expanding the boundaries of the university to include the realm of the farmer and his or her experiences as vital components of learning systems agriculture.

This strategy of "phenomenology" as a foundation for the study of systems assumes that learning starts in the field, and that important elements, connections, and emergent properties will be revealed through the system itself (Østergaard et al. 2010). Starting on the farm, student understanding includes the biophysical components, socioeconomic elements, and most importantly the goals and long-term ambitions of the farmer who is key to the design of farming systems to reach those goals. The boundaries of the system are explored, farms are illustrated with a rich picture of components and interactions, and driving forces both internal and external are identified. Discussion activities in plenary as well as in small student teams provide opportunities to share ideas, observations, and meaning, and also allow students to introduce their prior relevant experiences and build on these as described by Dewey (1966). A similar learning strategy can be implemented at the community, regional, or national level, while important differences are the challenge of drawing appropriate boundaries around the system of study and the increasing level of complexity when multiple stakeholders and goals are involved.

This process of experiential learning is guided by the principles published by David Kolb (1984 and prior papers), whose *Kolb Learning Cycle* is widely recognized and applied in education. The design of learning experiences is also informed by the soft systems methodology articulated by Checkland (1981) and Checkland and Scholes (2001). In our program, students are mentored in the process of making key observations in the field, organizing and reflecting on information from the fieldwork, and identifying key issues that currently impact the success and shortcomings of the farm enterprises and overall system. Beyond production and economics, students are urged to explore the environmental impacts of present operations and how these could be altered by changes in practices or system design. They especially focus on such social elements as family involvement in labor and management, interactions with neighbors and community, and the overall organization of rural infrastructure as it encourages or impedes agricultural success in the local area. The methods for achieving these goals are described in Section 17.4.

Since agroecology for us is defined as the *ecology of food systems* (Francis et al. 2003), we urge the students to observe farm-level issues that are influenced directly or otherwise by the surrounding landscape and community. A farm does not operate in isolation. The study of holistic systems must necessarily involve thoughtful consideration of boundary issues, what is impacting the farm from nearby as well as from regional and national levels such as regulations and subsidies. One of the most difficult challenges is deciding where the most relevant boundaries are situated, whether this is a farm, a community, or a region. The property lines of a farm are only one somewhat arbitrary distinction of boundary or limit to that farm, since many inputs come in from the larger community and the sale of products is highly dependent on the local, national, and global economy. In the systemic approach advocated by Howard Odum (1994, 2007), the boundary issue is described as the "window of interest" or the "window of focus" that informs the drawing of diagrams of integrated complex systems of nature and societies. For example, such diagrams can be used to illustrate a community that is made up of local farms and other institutions, as well as multiple businesses and consumers. That community is linked to the larger region and the national economy from which many factors impact local decisions. Thus, the principle of systems boundaries is an essential question to define as students deal with the farm or the community. Ultimately, everything in the global food system is related, as weather, markets, and political decisions impact commodity prices and thus the economic success at the local level. These are among the holistic systems approaches used in experiential learning to help students appreciate and understand the functioning of farming and food systems. The open-ended case approach has proven useful in education about the complexities of contemporary systems, and is described next.

17.3 OPEN-ENDED CASES AS A STRATEGY FOR LEARNING AND CAPACITY BUILDING

Decision case studies have long been used in colleges of business, law, and medicine as a way to help students to integrate the information that they have learned in university classes and to prepare them for confronting the complexity and multiple dimensions of real-world situations that they will be facing in a professional career. This type of "learning in context" provides practical application of theory and brings reality into the course of study in a relatively "safe space" where students can learn from experiences without suffering undue consequences or causing any damage. What characterizes typical decision cases is the certainty of outcomes, since almost invariably, each case is taken from an event or a decision-making circumstance that has already happened, and both the answers and the outcomes are known to the instructor. In short, the students need to be clever enough to figure out what the instructor already knows, and will be graded on "getting the correct answer." For historical reference to case studies in the Harvard Law School, see Langdell (1871).

In agriculture, educators in the University of Minnesota have developed and tested a number of decision case studies (Simmons et al. 1992). These include farmer decisions related to crop cultural practices, farming system design, and marketing of agricultural products. A number of such case studies were authored by the Minnesota group and others, and have appeared in a compendium of practical cases for use in the university agricultural classroom (American Society of Agronomy [ASA 2006]). Most such cases are focused on resolving single-issue challenges, often from the perspective of a single specialized discipline, and involving a relatively simple cause and effect set of circumstances. They come from soils and agronomy, horticulture and plant protection, agricultural economics, or natural resource management. Of the 47 cases that were published in the ASA collection, only one included a substantial literature review of the published theory and practice of decision making (Tan et al. 2001, in ASA 2006).

One case described by Simmons et al. (1992) involved a real-world scenario on a currently unresolved issue, and this was termed a *now case* to distinguish its uniqueness from the others. As described in a previous paper, "now cases are presented to clients and students who are personally concerned with the outcomes and impacts of a decision and course of action," and they involve "key stakeholders in a process of consensus building or group problem solving, while dealing with an unresolved, current issue in an effective way of engaging students in the process of decision making" (Francis et al. 2009). This case is a precedent for the approach that we currently use with open-ended cases in the Norway agroecology program, and provides a model that could be used in other practical disciplines where important outcomes of education are internalized in graduates who are prepared for responsible action in their future careers (Lieblein and Francis 2007).

The open-ended case in agroecology places students in circumstances where clients in the field have clear challenges but the answers are not yet known to any of the players (Francis et al. 2009). Student teams visit farmers to become acquainted with their farm operations, local environmental and social contexts, and goals for the future, and then derive multiple scenarios that could help the farmer reach those goals. In a parallel exercise, student teams visit communities to assess their goals, learn about local resources and motivations, and again develop several scenarios that could help communities reach their goals. In this chapter, we summarize the experiences of over a decade involving students in such an immersion learning experience in the Nordic MSc agroecology program, and present an evaluation of their educational outcomes as reported by the students themselves and by their stakeholders in the field.

We consider this educational tool especially important for graduates who move immediately into administrative and other project leadership positions. Many MSc and PhD graduates from developing countries arrive home to resume administrative duties or to be promoted into other positions of higher responsibility. Although they may be particularly well trained in the detailed investigative methods of a specific discipline—for example, breeding of transgenic crops, determination of mechanisms of resistance to insecticides, manipulation of complex econometric analyses—there may have been little opportunity to learn the basic skills needed to administer a complex department in the Ministry of Agriculture or in a university. Among these skills are the capacity to do project planning and future visioning for programs, challenges that involve the integration of information from many disciplines and the reliance on the opinions of multiple stakeholders who will be charged with the implementation of the results or will be impacted by the consequences. Rarely are students trained to facilitate a process of change in such "fuzzy" complex situations, especially when there are also goal conflicts among stakeholders (Saarikoski et al. 2012).

For higher-order decisions that involve real-world challenges, complexity, and always a degree of uncertainty, a different type of education is required. The open-ended case approach that involves intimate interactions with stakeholders including farmers and other specialists involved in the food system would appear to be valuable as part of graduate-level education. Here, we present two examples from the Norway agroecology program that illustrate the application of this teaching method. A parallel program for doctoral candidates in agroecology and capacity building is currently under development to serve mid-career professionals in developing and developed countries (Salomonsson and Francis 2013).

17.4 EXAMPLES OF LEARNING IN FARMING AND FOOD SYSTEMS: NORWAY MODEL

One goal of the agroecology MSc program in Norway is to prepare students in agriculture and other food-related fields to deal with unresolved challenges that they will meet in the future. The textbook by Wilson and Morren (1990) has provided a technical foundation for the study of whole-farm systems, and several chapters are useful to students each year in agroecology classes. These and other literature resources are assigned as readings prior to students arriving on campus. They are encouraged to integrate new information and perspectives with prior experiences, and to reflect on the importance of the integration of enterprises on the farm and diverse ways to deal with complexity.

Readings are put into practice immediately as students move quickly through preparations for fieldwork, learning about methods and tools for observation and interviews. For example, in the first week, students are sent on transect walks across the landscape to observe the natural, farmed, and built environments and to reflect on their structure and function (Francis et al. 2012a). Extracurricular activities add to the intense classroom interactions to help build a coherent and trusting learning community of students and instructors (Francis et al. 2011). Students practice visual diagramming to illustrate the interactions of people, goals, stakeholders, and physical infrastructure (Breland et al. 2012). Then students and instructors travel together to spend 4 days living on a biodynamic dairy farm and working on farms nearby. They are guided to look for integration efficiencies in several well-managed farms that become case studies for whole-farm analysis. Students work for one full day carrying out priority tasks under the direction of the farmers. The shared experience on farms provides a rich context for class discussion and further exploration of systems theory and farming systems analysis. This field learning experience lays the foundation for the later student team projects, when each group will spend a week in September and another in November in a community where they will interview farmers and food system professionals and implementers to develop potential scenarios to create a desired future situation. These scenarios are provided back to the clients for their consideration and adoption of the most useful ideas.

Using a research approach that has been called phenomenology (Østergaard et al. 2010), when students are immersed in the farm and the community environment in Norway, they begin to observe and interact with residents to establish for themselves the current context of farming and food systems. Their questions and priorities emerge from the phenomenon itself, rather than from the study of books and journal articles about the situation or challenges. Early in the course, students are introduced to major players in the farming and food system: practicing farmers, food processors and marketers, food handlers and nutritionists, politicians, and educators in the academic disciplines at the university. This is often done in the field and in the rural community, and the perspective taken is a whole systems approach. Students are urged to bring their prior experiences into the situation, to combine them with new knowledge learned about systems theory and practice, but to reserve judgment about solutions as they interview people in the field. They work in teams of four to six students, and we have found that the rich mixture of information from their former studies in different disciplines and cultures and the interaction among members of each team all add to the value of this group experience (Lieblein et al. 2008).

To better illustrate the learning landscape in agroecology: farming and food systems (PAE 302) at the Norwegian University of Life Sciences (NMBU, Ås, Norway), we describe here how students go into the field to practice observational skills in the farm, community, and natural environments. With minimal prior practice, they interview farmers and food system clients to learn about the local context and to begin to wrap their minds around the complexity in which people currently make decisions. From the farmers, they learn about current successes in farming enterprises as well as the reality of the resource base within which they operate. Farmers help them understand the primary driving forces on the farm, in the environment, and in the marketplace that have shaped the current system. Through interactions with the farmer and his or her family, they identify major challenges that are perceived in planning for the future and reaching their goals. For example, a dairy farmer may want to become certified organic and to market milk locally, and will need to secure adequate financing for on-farm improvements as well as information about accessible markets. Through these interactions with farmers and using extensive discussions among the student team members, the current picture of the farm, the desired future situation, and the potential methods of arriving there begin to come into focus. Yet, we still urge students to refrain from being prescriptive, and to wait until they learn more about the larger food system.

One Canadian student expressed her impressions of this learning environment:

> I think that the most important content I gained from the farming and food systems courses came in the areas of learning, soft systems methodology, and food systems analysis. Using SSM to analyse farming and food systems was quite an exciting discovery. For the first time in school it didn't feel like I was being asked to study a seemingly random piece of the puzzle, but rather to look at the entire "big picture" and determine which pieces were most important to zoom in on. Tools like mind mapping, using metaphors, and thinking in terms of systems all helped me re-learn the way to approach problem solving, in a manner that was at once systematic and highly creative. (Francis et al. 2013)

In rural communities, the student teams meet leaders in the procurement and preparation of food for schools, hospitals, municipal canteens, and other public facilities. They talk with managers of food processing companies such as slaughterhouses, commercial and cooperative dairies, and grain mills, as well as agricultural input suppliers and crop advisors. Students meet supermarket managers and learn about national meat and dairy monopolies that are large national cooperatives in Norway. In this process, they learn about the overall economic and social context of the community, and about people's concerns about how to improve their local food supply and food security.

Needless to say, learning about consensus or goals at the community level is far more complex than at the farm level, since there are many stakeholders and multiple, often conflicting points of view. Just as the challenges of multiple goals emerge from the teams' interactions with people in the communities, some of the potentials for a solution also become clear as students work with people to better understand the local "phenomenon" (Francis et al. 2012b). Most of the differences in opinion revolve around methods to achieve goals, since residents often agree on the need for an affordable, safe, nutritious food supply for the long term. Some of the goals that students have found in Norwegian communities include increasing the reliance on local foods from nearby farms, increasing the availability of organic foods, getting local or organic foods or both into public cafeterias, and reducing the environmental impact of chemical pesticides and fertilizers that pollute waterways and lakes. Students have been impressed not only with the range of opinions among people and groups that they have met, but also with the shared concern about the future that they have encountered in these communities. There is a spirit of innovation among those who are active in community planning, and they have used both the ideas and the enthusiasm generated by the agroecology students to help consolidate interest and energy around community initiatives. One prime example is the *Lokalmat* (local food) celebration that was first planned by the 2010 student team in Tolga that resulted in a one-day exhibit and sale of local foods as part of a community-wide celebration. Two events have now been held in 2011 and 2012, with nearly 500 people attending each year, and the community intends to continue this as an annual event. One indicator of the value of this approach to learning has been an offer by communities to provide partial support from their own funds for student teams in subsequent years, used by the project for travel and lodging.

Our strongest indicators of success are the observations that we make of student teams in the field, the reading of their final farm and community project reports, and reading their reflection documents about the learning process that are submitted at the end of the autumn semester. Initially, many students are hesitant about this apparently "unstructured" approach to learning where each student and team is essentially responsible for their own learning. The instructors provide a rich learning landscape with some direction on how to navigate through the semester, as well as some key tools such as team dynamics, interviewing skills, and guides to relevant literature (course syllabus is available from the authors), but the rest is up to the students and teams. In general, students are held accountable both to the instructors and to the clients in the field for designing their own projects and developing results. This is a new approach for many who come from a more traditional, structured learning environment with lectures, assigned readings, and examinations to test their comprehension of the material. Our program is all about applications of the theory and past experiences to new situations. The final reflection documents from students frequently describe their "aha moments" when they appreciated how the process works, what they have learned about farms and food systems in context, and how they will be able to apply this strategy in the future. They also learn about their own learning styles and ability to function well within a team. We have additional evidence of this application phase as students develop their thesis plans and actually put the methods to use in their own research.

As described by one Canadian student:

> The project work carried out in both the farming and food systems courses simulated a very realistic working environment. Group dynamics were representative of what one faces in virtually any work environment. The strong focus of working as a team to discover complex solutions for even more complex situations was excellent practice for future work with environmental and agricultural issues in Canada. In addition, the exercises that were undertaken on how individuals learn were helpful in understanding group dynamics more clearly. It made for a more respectful group interaction with a better awareness of what motivates people and where strengths and weaknesses occur. (Francis et al. 2013)

To summarize the results of this venture in experiential learning through open-ended cases, we provide a list of the characteristics of the method as compared with the more conventional decision case method for learning as applied in other disciplines. Table 17.1 compares 17 ways in which open-ended cases differ from conventional decision cases, including goals, the process of gathering and using information, who owns the educational process and how it is evaluated, and the roles of both students and instructors in this learning landscape (from Francis et al. 2009).

If we can imagine the preconceived ideas of students coming to the program from a more formally structured, linear learning, and hierarchical system of education, it is easy to understand some of the challenges that they face when confronted with this more open and interactive form of learning. Students from the biological and physical sciences find that many issues are fluid and complex, and there are not obvious linear, cause and effect relationships. While experienced in dealing with the objectivity of studying relatively simple systems and addressing the individual components of these systems, students find that the real world of clients' decision making involves much more than facts and figures. It is relatively easy to collect a soil sample, run a chemical test in the laboratory, and derive recommendations on how much nitrogen should be applied to the maize crop. There are multiple challenges in solving complex, higher-order questions on this same farm with the same crop when questions arise about alternative sources of nutrients to feed the crop, potentials and costs of integrating livestock with the maize field and processing residues, and whether maize is really the best crop to grow given the requirements for water, pesticides, nutrients, and fossil fuels needed for success. In the community it is relatively easy to quantify the amount of food needed each year

Table 17.1 Summary Comparison of Conventional Decision Case Learning and Open-Ended Learning Strategies Used in Courses in the US and Nordic Agroecology Programs

	Conventional Decison Case Method Learning	Open-Ended Cases for Learning in Agroecology Courses
Goal	Develop solutions from a predetermined situation	Envision potential solutions to real-world situations
Process	Follow a series of defined steps to uncover known solution	Follow a discovery process to envision alternatives
Information	Provided by instructors in a logical/sequential manner	Students seek out needed information from key clients in field/community
End product	Rational solution that may correspond to actual situation	Multiple possible future scenarios and their potential impacts
Type of learning	Closed learning cycle to seek what is known to instructor	Open colearning by students and instructor to explore unknown
Evaluation of learning	How closely does solution relate to the "real answer"	How creative are future scenarios and evaluations of potential impacts
Ownership of process	Instructors know the answers and determine student success	Students own the learning and set their own criteria for success
Learning culture	Conventional search by students to find fixed answers	Open-ended cosearch to develop future options and predict impacts
Institutional setting	Stimulus from teacher and response from students	Multiple sources of stimulation, continuous interaction toward goals
Role of instructor	Design the logical steps to reach the known (right) answers	Open a learning landscape for creative discovery of alternatives
Role of student	Active learner, engaged in a comfortable process	Autonomous learners find discomfort in a stressful open-ended situation
Responsibility for learning	Starts with instructor, passed to students in case study	Primarily rests with students, who are free to pursue different options
Applicable mainly to	Past and present situations that are known	Future situations that are complex and unknown
Appropriate mostly for	Lower hierarchical system levels	Higher-order hierarchical system levels
Most useful for	Simple, well-defined systems and situations	Complex, ill-defined systems and situations
Answers and solutions	Mostly fixed and predetermined	Mostly open and dependent on multiple factors and context
Major sources of inspiration	Hard facts and discrete systems that are well known	Hard facts and social methods, plus human judgment and creativity

Source: Francis, C., King, J., Lieblein, G., Breland, T.A., Salomonsson, L., Sriskandarajah, N., Porter, P., and Wiedenhoeft, M., *J Agric Educ Extension*, 15, 385–400, 2009.

depending on the number of residents, the quality of diets, and the levels of activity, but it is more complex to consider cultural preferences, ways to improve diets to reduce levels of obesity, or to find alternative food systems that are more locally based and involve changes in the availability and cost of local versus imported foods. It is these types of challenges that our students undertake to understand and then design potential scenarios for individual farmers and decision makers in communities to consider while pursuing their long-term goals.

17.5 LONG-TERM IMPLICATIONS OF APPLIED SYSTEMS EDUCATION FOR FUTURE FOOD SYSTEMS

Based on over a decade of experience using the open-ended case approach to learning, we have observed many problems that students encounter, as well as the benefits, when embarking on experiential learning, perhaps for the first time with an in-depth, client-oriented experience. As we work through these challenges each semester, it becomes ever more obvious that this type of experiential learning really does work and that students are pleased with what they have learned—both in the content and the process. We think that this type of learning agenda is essential for students to truly understand the complexity of systems and the multiple dimensions of problems that their clients will have to consider when improving farming and food systems for the future.

The classical strategy in academia for the past century has been to divide up the challenges into components that can be addressed by specialists who are well trained to use careful laboratory and field methods to study the mechanisms of biology or economics that are influencing systems. Although these are complicated systems, we have been relatively successful in understanding the parts and improving them. However, when we look at the multiplicity of driving forces and the large numbers of vested interests in society that would like our research and development agenda to come up with favorable outcomes for their specific interest or investment, the objectivity of the research and education establishment may begin to break down. From the obvious funding of university research on specific chemicals or seed varieties that will benefit specific companies, to the more subtle impacts of funding research on molecular mechanisms that can lead to profitable patents, there is a danger of skewing the overall research agenda away from whole systems analysis. When we focus only on the biology of a problem, and direct massive resources toward the study of specific mechanisms, we may incur a large opportunity cost by not attempting to understand the economics, the environmental impacts, or the social implications of the adoption of one new technology. There is little incentive for industry to support farming systems research—crop rotations, tillage alternatives, cover crop mixtures, analyses of trade-offs among technologies—when these will have limited or no potential for commercial payoff in the technology marketplace.

We believe that a focus on whole production systems on the farm and food systems in the community will help students appreciate the complexity of future challenges, including the need for more and well-funded systems research. They will also understand the need for thoughtful incorporation of systems methods and perspectives into all applied fields where experience from the field needs to be combined with historical study of research results and the application of science to solve practical problems (Odum 1994; Wilson and Morren 1990). This is the best possible orientation that we can provide for our agroecology graduates, including open minds to consider multiple possible scenarios to improve systems and how to communicate with clients as well as learn from them to build a shared future.

17.6 SUMMARY AND CONCLUSIONS

The experiences of a team of educators in agroecology have been directed toward the design of an experiential learning course in agroecology for MSc students. Building on the observations of successful activities in three doctoral short courses of 1-week duration in the 1990s, a full semester course in agroecology of farming and food systems has been offered since 2000 at the UMB (NMBU) in Norway. Based on the concepts of John Dewey from a century ago, and the advances in education using phenomenology and case-based learning, our students have used interactions with key players on the farm and in rural communities to further their understanding of complex systems and design future scenarios to help clients meet their goals in development. The open-ended case has been the primary method used as a vehicle for learning and serving their clients.

A learning landscape that employs open-ended cases in agroecology is designed by modifying the popular decision cases that are commonly used in education in business, medicine, and law to bring real-world challenges into the classroom for students to study and resolve. But unlike decision cases where the answer is known to instructor and client, the open-ended cases deal with contemporary challenges where answers are not known, and a team of students + client + instructor is faced with the task of designing multiple scenarios for resolving a problematic situation. At first, such a learning strategy is difficult for students who have been trained to understand simple cause and effect relationships and to quickly diagnose problems and provide solutions. We have found their learning to be enhanced by investigating as many dimensions as possible of a current problem, to fully understand the goals and resources of clients, and to provide alternative potential strategies for how to solve a problem along with *a priori* assessment of the consequences of implementing each strategy. This learning method has been used for over a decade with agroecology students in the NMBU program, and has been positively evaluated by students in their final reflections on the course. Stakeholders on the farm and in rural communities have responded by adopting some of the scenarios, and have even provided some monetary support for team activities in subsequent years. This is a strong indicator of success.

Another long-term indicator of successful learning has been our observation of how students have applied experiential methods in their MSc thesis research projects and in employment after graduation. A student working in Tanzania used group learning to envision and plan a long-term strategy to expand organic agricultural production in the country. Another expanded a taxonomic study of forest species in Panama to include a range of social and economic indicators of family income and welfare, and found a multiplicity of products they harvested from the tropical forest that were worth several times the value of their principal cacao product. Working with a number of local public and private agencies, a student in Victoria, BC, Canada, designed a local food council that is now promoting the production and sale of local foods to enhance the economy of the island. In the first year of operation, farmers markets in eight locations in Norway were studied by a student in the MSc program to determine the success of vendors and the satisfaction of customers. Urban gardens are the primary source of vegetables in the large cities of Cuba, and one of our students studied the potentials of the capture and use of gray water and rainwater for the irrigation of these plots. A student worked with two women's cooperatives in Senegal to see how they could improve vegetable production to service a large resort, a strategy that brought them regular income during the tourist season and reduced costs to the client for quality produce raised near the place where it was consumed. In these and many more cases, students have interacted with multiple stakeholders to deal with very real and immediate challenges that were of high priority. Each provides a small success story, and there are now over 50 students who have completed the MSc degree and moved on to employment in the private or public sector, or have continued in graduate study. They provide the most important indicators of success in a program of experiential learning.

ACKNOWLEDGMENTS

Because of the innovative organization and success of the agroecology program in the Nordic Region, the Nordic Veterinary and Agriculture University (NOVA) Educational Programme Award was presented on May 31, 2007, to the team of instructors in a ceremony in Uppsala, Sweden. In addition, the All-University Excellence in Education Award from the Norwegian University of Life Sciences (UMB) was presented to the instructor team in 2011 in Ås, Norway.

REFERENCES

ASA (American Society of Agronomy). 2006. Case studies published in *Journal of Natural Resources and Life Science Education*, 1992–2005. Madison: American Society of Agronomy (ASA).

Bawden, R. J. 1991. Systems thinking and practice in agriculture. *J Dairy Sci* 74:2362–2373.

Bawden, R., B. McKenzie, and R. Packham. 2007. Moving beyond the academy: A commentary on extra-mural initiatives in systemic development. *Syst Res Behav Sci* 24:129–141.

Breland, T. A., G. Lieblein, S. Morse, and C. Francis. 2012. Mind mapping to explore farming and food systems interactions. *NACTA J Teaching Tips* 56:90–91.

Checkland, P. B. 1981. *Systems Thinking, Systems Practice*. New York: Wiley.

Checkland, P. B. and J. Scholes. 2001. Soft systems methodology in action. In J. Rosenthal and J. Mingers (eds), *Rational Analysis for a Problematic World Revisited*, pp. 61–90. Chichester: Wiley.

Dewey, J. 1966. *Democracy and Education: An Introduction to the Philosophy of Education*. New York: Free Press.

Francis, C., T. A. Breland, E. Ostergaard, G. Lieblein, and S. Morse. 2012a. Phenomenon-based learning in agroecology: A prerequisite for transdisciplinarity and responsible action. *J Agroecol Sustain Food Sys* 37:60–75.

Francis, C., J. King, G. Lieblein, T. A. Breland, L. Salomonsson, N. Sriskandarajah, P. Porter, and M. Wiedenhoeft. 2009. Open-ended cases in agroecology: Farming and food systems in the Nordic Region and the U.S. Midwest. *J Agric Educ Ext* 15:385–400.

Francis, C., A. Lawseth, A. English, et al. 2013. Adding values through practical education in agroecology: Review of Canadian student experiences. *Int J Agric Food Res* 2(2):7–17.

Francis, C., G. Lieblein, S. Gliessman, et al. 2003. Agroecology: The ecology of food systems. *J Sustain Agric* 22:99–118.

Francis, C., S. Morse, T. A. Breland, and G. Lieblein. 2012b. Transect walks across farms and landscapes. *NACTA J Teaching Tips* 56:92–93.

Francis, C., S. Morse, G. Lieblein, and T. A. Breland. 2011. Building a social learning community. *NACTA J Teaching Tips* 55:99–110.

Kolb, D. 1984. *Experiential Learning: Experience as the Source of Learning and Development*. Englewood Cliffs: Prentice-Hall.

Langdell, C. C. 1871. *Selection of Cases on the Law of Contracts*. Cambridge: Harvard Law College.

Lieblein, G., T. A. Breland, L. Salomonsson, N. Sriskandarajah, and C. Francis. 2008. Educating tomorrow's agents of change for sustainable food systems: Nordic agroecology MSc program. *J Hunger Environ Nutr* (Special issue on Sustainable Food Systems) 3:309–327.

Lieblein, G. and C. Francis. 2007. Towards responsible action through agroecological education. *Ital J Agron/Riv Agron* 2:79–86.

Lieblein, G., C. A. Francis, L. Salomonsson, and N. Sriskandarajah. 1999. Ecological agriculture research: Increasing competence through PhD courses. *J Agric Educ Ext* 6:31–46.

Lieblein, G., E. Østergaard, and C. Francis. 2005. Becoming an agroecologist through action education. *Int J Agric Sustain* 2:147–153.

Moncure, S. and C. Francis. 2011. Foundations of experiential education for agroecology. *NACTA J Teaching Tips* 55:75–91.

Odum, H. T. 1994. *Ecological and General Systems: An Introduction to Systems Ecology*. Boulder: Niwot University Press of Colorado.

Odum, H. T. 2007. *Environment, Power, and Society for the Twenty-First Century: The Hierarchy of Energy*. New York: Columbia University Press.

Østergaard, E., G. Lieblein, T. A. Breland, and C. Francis. 2010. Students learning agroecology: Phenomenon-based education for responsible action. *J Agric Educ Ext* 16:23–37.

Packham, R. and N. Sriskandarajah. 2005. Systemic action research for postgraduate education in agriculture and rural development. *Sys Res Behav Sci* 22:119–130.

Picasso, V. D., E. Brummer, M. Liebman, P. M. Dixon, and B. J. Wilsey. 2011. Diverse perennial crop mixtures sustain higher productivity over time based on ecological complementarity. *Renew Agr Food Syst* 26:317–327.

Saarikoski, H., K. Raitio, and J. Barry. 2012. Understanding "successful" conflict resolution: Policy regime changes and new interactive arenas in the Great Bear Rainforest. *Land Use Policy* 32:271–280.

Salomonsson, L. and C. Francis. 2013. *Agroecology and Capacity Building Doctoral Programme Proposal*. Uppsala: Swedish Agricultural University.

Simmons, S. R., R. K. Crookston, and M. J. Stanford. 1992. A case for case study. *J Nat Res Life Sci Educ* 21:2–3.

Wilson, K. K. and G. E. B. Morren Jr. 1990. *Systems Approaches for Improvements in Agriculture and Resource Management*. New York: Macmillan.

Index

A

ABA, *see* Abscisic acid (ABA)
ABM approach, *see* Agent-based modeling (ABM) approach
Aboveground biomass (AGB), 336
Abscisic acid (ABA), 168–169
Acidobacterium, 17
Aconitum heterophyllum, 157
Activated carbons, 77
ADE, *see* Anthropogenic dark earths (ADE)
Adobe Photoshop software, in landscape visualizations, 337
Adsorbent (activated carbons), use of carbon, 77
Aerogels, 72
AESs, *see* Agri-Environment Schemes (AESs)
Afforestation, 231
AGB, *see* Aboveground biomass (AGB)
Agent-based modeling (ABM) approach, 330, 334
Agricultural business system, 222
Agricultural ecosystems, typical natural *vs.* high-production, 204
Agricultural heritage, 81
Agricultural intensification
    generalized process of, 140
    and weed response, 142–143
Agricultural management on SOC, effect of
    CAPs, 44
    crop residue
        and its quality for SOC stabilization, 44–45
        management, 45–46
        and tillage, 46
    fertilizer management, 47
    organic and conventional farming systems, 47–48
Agricultural production
    climate change issue, 153
    in different continents, 310
    in per capita terms, 309
    in tropics, high-tech and alternative, 210–212
Agricultural residues, 28
Agriculture, 121
    broadening process, 316
    climate sensitivity analysis of, 132
    cultural dimensions, 313
    deepening process, 316
    legumes role in, 164–167
    multifunctionality, 316
    organic agriculture, 318
    political dimensions, 314
    problems in, 158
    regrounding process, 316
    sensitivities to climate variation, 221–222
    social dimensions, 313
    spatial dimensions, 314
    vulnerabilities in, 222
Agri-Environment Schemes (AESs), 140, 143, 144
Agri-food
    bioeconomic applications, 312
    ecoeconomy
        agroecology (*see* Agroecology)
        broadening process, 316
        deepening process, 316
        definition, 314
        growth affecting factors, 314
        multifunctionality, 316
        regrounding process, 316
        value adding, 316
Agroecological approaches, to agricultural production, 81
Agroecology, 239–240
    advantages of, 315
    Appalachian region
        ARC definition of, 283
        cultural and socioeconomic factors, 287
        education, 288
        farming (*see* Farming, Appalachian region)
        poverty level, 288, 289
        public health conditions, 288
        tobacco dependence, 284–287
        unemployment rate, 288
    and applications of new technologies, 214–215
    challenges for, 281
    definition of, 281–282, 315
    ecoagriculture relates to, 274
    evolution of, 280
    experiential learning, 349
    of food system, 182, 186–193
    graduates, 347
    holistic systems, 348–349
    indicators, 282
    projects, 318–319
    scale and dimension of research in, 182
    as scientific discipline, 182
    socioeconomic aspects, 279
    in Vermont, 327
Agroecosystems, 139, 141
    approach, 182
    communication in, 88
    definitions, 203
    managed, 207–210
    of Plain of Valence, 188
    research, 48
    strategies to integrate beneficial rhizosphere microorganisms in, 112–114
    tropical, 205–207, 212–214
    types of, 203–205
Agroforestry systems
    adaptation in, 224–228
    CH_4 emissions, 232
    under climate change, utility of, 223–224
    climate change adaptation, 228
    to curb GHGs emission, 232–233
    mitigating temperature variation, 225
    mitigation potential of implementation, 228–233
    protection from storms, 226–228
    vs. row crop agriculture, 231

soil water loss prevention and water capture, 225–226
    types of, 223
Agromicrobes, 82
    diversity of, 84
    poor agroecological knowledge of, 83
    potentialities of, 83
Agromicrobial diversity, 83–84
    communication in agroecosystems, 88
    culture-based methods, 84–85
    culture-independent methods, 85–88
Agronomic traits of crop genotypes, 110
Agrostemma githago, 142
ALCA, *see* Attributional LCA (ALCA)
AM, *see* Arbuscular mycorrhiza (AM)
Amazonia, terra preta sites in, 4, 5
Amazonian soils, 60
Ambrosia artemisiifolia, 142
AMF, *see* Arbuscular mycorrhizal fungi (AMF)
Amino acid-changing variants, 153
Ammonium-based fertilizers, 89
Amplified PCR products, 14
Amplified ribosomal DNA restriction analysis
            (ARDRA), 86
Ancient DNA, 13
Anthropogenic dark earths (ADE), 60
Antinutritive proteins, 167
Appalachian Harvest program, 291–292
Appalachian region, agroecology in
    ARC definition of, 283
    cultural and socioeconomic factors, 287
    education, 288
    farming (*see* Farming, Appalachian region)
    poverty level, 288, 289
    public health conditions, 288
    role for agroecology, 302–304
    tobacco dependence, 284–287
    unemployment rate, 288
Appalachian Regional Commission (ARC), 283
Appalachian Sustainable Agriculture Project (ASAP),
            295–296
Appalachian Sustainable Development, 291–292
Aquatic *vs.* terrestrial plant material, 7–8
Arabica coffee, 225
Arabidopsis thaliana, 153–154, 156–157
Arable weeds, 139, 140
Arbuscular mycorrhiza (AM), 159
Arbuscular mycorrhizal fungi (AMF), 105–106
    and PGPR bacteria, interactions between, 108–109
    and *Trichoderma*, interactions between, 108
ARC, *see* Appalachian Regional Commission (ARC)
ArcMap GIS software, 337
ARDRA, *see* Amplified ribosomal DNA restriction
            analysis (ARDRA)
Artificial inoculation
    nitrogen acquisition of microbial origin, 92–94
    phosphorus acquisition through microbe-mediated
            processes, 94–95
Artificial replanting, of mangroves, 251
Associations pour le Maintien de l'Agriculture Paysanne
            (AMAP) system, in France, 186, 189–191

Attributional LCA (ALCA), 26
    vs. consequential LCA, 27
Avicennia germinans, 241, 243–247
Azospirillum, 89–91
    colonization of, 93
Azotobacter, 93

B

BAC system, *see* Bacterial artificial chromosome (BAC)
            system
Bacteria(al), 89, 158
    antagonists, 160
    N fixation, 107
Bacterial artificial chromosome (BAC) system, 83
Basin forests, 242
BAT, *see* Biologically active time (BAT)
BCA, *see* Biological control agent (BCA)
Belterra Plateau, soils of, 3
Beneficial agromicrobes for sustainable agriculture,
            utilization of, 88–89
    artificial inoculation, 92–95
    engineering agricultural practices, 89–92
Beneficial microbial populations, natural engineering of, 89
Beneficial rhizosphere microorganisms
    agricultural practice, impact of, 110–112
    in agroecosystems, strategies to integrate, 112–114
Beneficial species, 144
Benzenepolycarboxylic acid (BPCA), 14, 23
BEP, *see* Bioeconomic paradigm (BEP)
Best management practices (BMPs), 327; *see also* Climate
            change best management practices (CCBMPs)
Bifidobacterium, 157
Bile acids, chemical structure of, 10
BIND, *see* Biomolecular Interaction Network
            Database (BIND)
Biobased economy, 311
Biochar-rich soils, 15
Biochars
    biological interaction, 31–32
    cation exchange capacity of, 24
    chemical and physical properties of, 16
    economy *vs.* ecology, 24–25
    environmental toxicology, 32–33
    fast, reliable, and cheap analytical tools, 20–23
    greenhouse gas fluxes, effects on, 68
    inter- and transdisciplinary biochar research, 33
    lack of reactions in environment, 29–30
    LCA, 25–29
    material properties, 30–31
    mechanisms, theoretical concepts of, 29
    production, 69
    research, limitations of, 29
    soil, nutrients, and water, physicochemical interactions
            with, 31
    terra preta concept, copying, 62–64
    upscaling in space and time, problem of, 19–20
Biocontrol of plant diseases, PGPR, 108
Biodiversity, 152–154
Biodiversity–agriculture integration challenge, 266–267

Biodiversity loss, counteracting, 143–144
Bioeconomic paradigm (BEP)
    agri-food production, 316, 317
    description of, 310
Bioeconomy
    definition of, 311
    efficiency of, 312
    model, 313
Biofertilization, 92
Biofertilizers, plant probiotics as, 157–160
BioGRID, *see* Biological General Repository for
        Interaction Datasets (BioGRID)
Biolog™, 84, 85
Biological control agent (BCA), 159
Biological General Repository for Interaction Datasets
        (BioGRID), 171
Biologically active time (BAT), 31
Biological nitrogen fixation (BNF), 158
Biomolecular Interaction Network Database (BIND), 171
Biopesticides, plant probiotics as, 157–160
Biotechnology, 312, 313
Black carbon, 61
Black mangroves, 241, 243–247
    percentage survival of, 257
    rooting system of, 248–249
    seeds of, 254, 256
BNF, *see* Biological nitrogen fixation (BNF)
Bones, from mammals and fish, 13–14
"Boot polish" effect of urea, 82
Borlaug, Norman, 154
BPCA, *see* Benzenepolycarboxylic acid (BPCA)
Brassicaceae, 110–111
Brazil, soybean agroecosystems in, 113–114
Broadening process, in agriculture, 316
Buildings in carbon capture and use
    construction materials, 70–72
    economic value (value chains), 73–75
    life cycle assessment, 75
    scalability, 76
Burkholderia cepacia, 108
Bush fallow system, 81
Button mangroves, 241, 246–249
    percentage survival of, 257

C

CAM, *see* Caribbean Coastal Area Management (C-CAM)
CAPs, *see* Conservation agriculture practices (CAPs)
Carbon
    as adsorbent (activated carbons), use of, 77
    by-product, 71
    markets, with long-term sequestration potential, 77–78
    offsets, 67
    sequestration, 58–59
        agroforestry systems, 228–230
        ecological value, 67–68
        payments for, 230
        potential, 27
    to soil, economic value of, 66–67
    stabilization, aggregation model of, 49

Carbon capture and storage (CCS), 59–60
Carbon capture and use
    biochar systems, 62–64
    in buildings and civil infrastructure
        construction materials, 70–72
        economic value (value chains), 73–75
        life cycle assessment, 75
        scalability, 76
    ecological value (ecosystem services), 67–69
    economic value of carbons to soil, 66–67
    plasma carbon, 64–66
        ongoing and further research on, 70
    scalability, 69–70
    terra preta phenomenon, 60–62
Carbon dioxide emissions, 58
Carbon fiber-reinforced plastics, 72
Carbon fibers, 72
Carbon nanofibers, 72
Carbon nanotubes, 72
Carbon preference index (CPI), 7
Caribbean Coastal Area Management (C-CAM), 259–260
Caribbean mangrove forests, 251–252
Carson, Rachel, 280
Cation exchange capacity (CEC) of biochars, 24
Cavity ring down spectroscopy (CRDS) methodology, 32
CCBMPs, *see* Climate change best management practices
        (CCBMPs)
CCS, *see* Carbon capture and storage (CCS)
CDM, *see* Clean development mechanism (CDM)
CEC of biochars, *see* Cation exchange capacity (CEC) of
        biochars
Cereal crops, proteomics in, 162–164
Charred organic material, 14–16
Chemical-based agricultural production systems, 88
Chemical weed control, 139
Citric acid, 167
Civil infrastructure in carbon capture and use
    construction materials, 70–72
    economic value (value chains), 73–75
    life cycle assessment, 75
    scalability, 76
Civil Transactions Act (1984), 127
Clavicipitaceae, 107
CLCA, *see* Consequential LCA (CLCA)
Clean development mechanism (CDM), 230
Clearance of forest, 128–130
Climate change, 152–154
    environmental pollution and, 154
    second green revolution and, 155–160
    wildlife problems, 153
Climate change best management practices (CCBMPs)
    agent-based modeling approach, 330
    description of, 329
    economics of, 329
    governance infomatics, 330
    and landscape visualizations, 330–332, 337
    mitigation potential of, 329–330
Climate sensitivity analysis of agriculture, 132
CLPP, *see* Community-level physiological profiling
        (CLPP)

^{13}C nuclear magnetic resonance (NMR) spectroscopy, 61
Coastal restoration, 258
Coastal road construction, 251
Coffee agroforestry systems, 224–225
    for carbon sequestration, 229
Coffee berry borer, 225
Communication in agroecosystems, 88
Community-level physiological profiling (CLPP), 83–85
Community-supported agriculture (CSA), 186, 189–191
Complexity reduction method, 162
Complex multicrop systems, 205
Comprehensive Peace Agreement (CPA) 2005, 127
Computational approaches, for reconstruction, modelling,
            and analysis of -omes, 170–171
Concrete, 75
Conocarpus erectus, 241, 246–249
Consequential LCA (CLCA)
    vs. attributional LCA, 26
Conservation agriculture practices (CAPs), 44
Conservation of beneficial rhizosphere microorganisms,
            113–114
Construction-grade solid carbon, 75
Conventional agriculture, 211
Conventional farming systems, 47–48
Coomassie Brilliant Blue (CBB)-stained gel, 169
Cooperative Extension Service, 290–291
Coprostanol, 11
CPI, *see* Carbon preference index (CPI)
"Cradle-to-grave" approach, 25
CRDS methodology, *see* Cavity ring down spectroscopy
            (CRDS) methodology
Crop(s)
    cultivated in tropical regions, 208
    genotype, 110
    rotation, 110–111
    and varieties, 207–208
    yields, 68–69
Cropping systems, 188, 208–210
    soil, 202
    sustainability of, 210
Crop productivity, microbial inoculants for, 92
Crop residue
    management, 45–46
    and quality for SOC stabilization, 44–45
    and tillage, 46
CSA, *see* Community-supported agriculture (CSA)
Culture-based methods, 84–85
Culture-independent methods, 85–88
Culture-independent molecular techniques, 83
"Customary" land tenure systems, 126
Cuticular waxes of terrestrial plants, 7
Cyanobacterial biofertilizer, 94

D

Darfur Peace Agreement (DPA) 2006, 127
DArT, *see* Diversity arrays technology (DArT)
Database of Interacting Proteins (DIP), 171
DCFC, *see* Direct carbon fuel cell (DCFC)
Deepening process, in agriculture, 316

De facto diversification, 233
Deforestation, 130, 205
Denaturing gradient gel electrophoresis (DGGE), 86
Denitrification, 330
Desertification, in Sudan, 131–132
Deterioration of soil fertility, 130
Dewey, John, 348, 355
DFS, *see* Diversified Farming Systems (DFS)
DGGE, *see* Denaturing gradient gel electrophoresis
            (DGGE)
Diara soil, 91
Diastrophic (nitrogen fixer) microorganisms, 92
Diazotrophs, physical location of, 92
Diois agroecosystem, 187–188
DIP, *see* Database of Interacting Proteins (DIP)
Direct carbon fuel cell (DCFC), 72
Direct counting, 84
Diversified Farming Systems (DFS), 211
Diversity arrays technology (DArT), 162
Diversity of agromicrobes, 84
DNA microarrays, 87
DNA reassociation profiles, 85
Donor trees, 253
Drinking water catchments and organic agriculture, 186,
            191–193
Drought, 162
Drought-tolerant near isogenic lines (NILs), 163
Drylands
    ecology of, 123–124
    nature of, 133

E

East Peace Agreement (EPA) 2006, 127
Ecoagriculture
    characteristics, 274
    concept of, 267–268
    interdisciplinary and cross-sectoral nature of,
            273–274
    landscapes, 272
    in practice, 268
    systematic (*see* Systematic ecoagriculture)
Ecoeconomic paradigm (EEP)
    agri-food production, 316, 317
    description of, 310
Ecoefficiency, 312
Ecological value (ecosystem services), 67–69
Ecology
    definition, 280
    vs. economy, 24–25
Economic assessment of biochar production, 26
Economic value (value chains)
    in buildings and civil infrastructure, 73–75
    of carbons to soil, 66–67
Economy *vs.* ecology, 24–25
Ecoregion Kaindorf (Austria), 67
Ecosystem services, 67–69
Ectomycorrhizal fungi, 95
EEP, *see* Ecoeconomic paradigm (EEP)
El Niño Southern Oscillations (ENSO), 226

Endophyte behavior of *Neotyphodium* (Clavicipitaceae), 107
Engineering agricultural practices
  plant genotype, 89–90
  soil organic matter, 89
  soil type and land use history, 91–92
  tillage, 90–91
The Ensete agroforestry system, 318–319
ENSO, *see* El Niño Southern Oscillations (ENSO)
Entomopathogenic fungi (EPF), 106–107
Environmental change, impacts of, 152
Environmental pollution and climate change, 154
Environmental toxicology, 32–33
Environment-friendly BCA, 159
EPF, *see* Entomopathogenic fungi (EPF)
Erythrina poeppigiana, 229
European agroecosystems, 144
European Community Regulation, 33
Exotic inocula, 113
External mycelium, 105

F

FAE, *see* Forestry agroecosystems (FAE)
"False time series" concept, 20
FAME, *see* Fatty acid methyl ester (FAME)
FAO, *see* Food and Agriculture Organization (FAO)
Farmers, benefits for, 233–234, 270
Farming
  Appalachian region
    Appalachian Harvest program, 291–292
    beef cattle production, 294
    bell peppers, 292
    Cooperative Extension Service, 290–291
    cover crop seed planting, 293
    demographic, farm, and organic agriculture
      statistics, 292
    food systems, 295–302
    grass-based production, 294
    hog and poultry industries, 295
    institutions and organizations, 290, 291
    land-grant universities, 290
    loss of tobacco farms, 290
    organic mulches, 292–293
    plastic mulches, 292
    specialty crops, 291
  community, 83
  Norway model, 351–354
  practices, analysis of, 193
Farm to Plate Investment Program (F2P), 328
Fast, reliable, and cheap analytical tools, 20–23
Father of the green revolution (Norman Borlaug), 154
Fatty acid methyl ester (FAME)
  analysis, 87–88
  profiling, 83
Fecal indicators in terra preta, 11
Fertility, soil, 212–213
Fertilization, 111
Fertilization practices, 193
Fertilizer, use efficiency, 28
Fertilizer management, 47

Fiber-reinforced plastics, 72
Field-based agricultural studies, 283
Field-germinated seed, 255
Field germination experiment, 255
Fish, bones from, 13–14
Fluorescence microscopy, direct counting using, 84
Fluvial sedimentation, 4
FNC, *see* Forests National Corporation (FNC)
Food and Agriculture Organization (FAO), 121
Food demands, 310
Food security, 152
  ecoeconomic contribution, 317–319
Food system approach, 182–183
  community-supported agriculture, 189–191
  conceptual analysis of, 183
  natural sciences approach, 183–184
  organic agriculture and drinking water catchments,
    191–193
  organic grain production, 186–188
  prerequisites for, 194
  social sciences approach, 184–185
Food-system improvements, in Appalachian region, 298–302
Food systems
  applied systems education for, 354–355
  Norway model, 351–354
  research (*see* Food system approach)
Forest, clearance of, 128–130
Forestry agroecosystems (FAE), 205
Forests National Corporation (FNC), 130
Fossil fuels, 64
Fourier transformation (FT), 22
Fourier-transformation infrared spectroscopy analysis, 77
F2P process, *see* Farm to Plate Investment Program (F2P)
Francis, C., 282
Fringe forests, 241
FT, *see* Fourier transformation (FT)
Fungi, 89
  arbuscular mycorrhizal fungi, 105–106
  entomopathogenic fungi, 106–107
  mycorrhizal, 94–95
  Trichoderma, 106

G

Gas chromatography flame ionization detection
    (GC-FID), 14
GasPlas reactor, 65, 66
GC-FID, *see* Gas chromatography flame ionization
    detection (GC-FID)
GDP, *see* Gross domestic product (GDP)
Gel-based assays, 161
Gene manipulation, 156
Genetically engineered plants, 156–157
Genetically modified (GM) techniques, in crop plants, 313
Genomes, 170–171
Geogenic *vs.* anthropogenic origin, 3
Germination, 254–255
  field germination experiment, 255
  nursery germination, 255–257
  plant hardening and acclimation, 257–258

GHG, *see* Greenhouse gases (GHG)
Global climate, impacts of, 152
Global climate change
    adaptation and mitigation, 326
    agricultural land use, 326
Global warming, 155
Glomeromycota, 105–106
Glomus intraradices, 108–109
Glomus mosseae, 109
Gluconacetobacter diazotrophicus, 93
Governance infomatics, 330
Grabbing land for public and private use, 127
Grape, 168–169
Greenhouse gas (GHG) emissions, 28, 58, 223, 326, 336
Green revolution, 215, 310
    first, 154–155
    second, 155–160
Gross domestic product (GDP), 121
Growth-promoting hormones, 106

H

Haeckel, Ernst, 280
Hammock mangrove forests, 242
Hartt, Charles, 2
Haskell, Jean, 300
Hazardous compounds, 69
H+CCU concept, 71, 74
Heat shock proteins (HSPs), 163
Herbaceous aboveground biomass, 336
Herbicide (flumioxazin)-treated grape wine berry
            proteins, 168
Herbicides, use of, 142
High-altitude plant habitats, 156–157
Higher cation exchange capacity, of terra preta, 24
High-tech agroecosystems, 210–212
High-throughput approaches, 160–162
High-throughput omics technologies, 152
Himalayan glaciers, 152
Horticultural crops, proteomics in, 167–170
HSPs, *see* Heat shock proteins (HSPs)
Human activity, in tropical savannas, 231
Humid tropics, 205–206
Hydrocarbon fossil fuels, 70
Hydrochars, 69
HyperRESEARCH software, 334
Hyphal network, 112
Hypothenemus hampei, 225

I

IARCs, *see* International Agricultural Research
            Centers (IARCs)
IEF gels, *see* Isoelectric focusing (IEF) gels
Industrial agrosystems, development of, 215
Industrial ecology
    biobased products, 312
    characteristics of, 311–312
Industrialized agricultural systems, 139
Industrial Revolution, 152

Infrared spectroscopy, 21, 22
INM mode, *see* Integrated nutrient management (INM)
            mode
Inoculate legume crops, 93
Inoculation, 112–113; *see also* Artificial inoculation
In silico approach, of plant protein interactomes, 170–171
IntAct, 171
Integrated nutrient management (INM) mode, 92
Intensive farming systems, 231
Inter- and transdisciplinary biochar research, 33
Intergovernmental Panel on Climate Change (IPCC),
            155, 326
International Agricultural Research Centers (IARCs),
            207–208
International carbon markets' offset scheme, 230
International Land Coalition, 127
In vitro approach, of plant protein interactomes, 170
In vivo approach, of plant protein interactomes, 170
IPCC, *see* Intergovernmental Panel on Climate Change
            (IPCC)
Isoelectric focusing (IEF) gels, 166
Isotope labeling, 31

K

Katzer, Friedrich, 2
Köppen, Wladimir, 205

L

Labile C fractions, 48
Labile SOM pools, 48
Lactobacillus, 157
Laguncularia racemosa, 241, 244, 247, 248
Land grabbing, in Sudan, 129
Land management approaches, 267–268
Landscape(s)
    defined as, 267
    heterogeneity, 143
    scale challenge, 271
    visualizations and CCBMPs, 330–332
Landslides, 227
Land tenure situation in Sudan, fragility of, 126–128
Land use, 43–44
    history, 91–92
Large-scale mechanized farming, 131
LCA methods, *see* Life cycle assessment (LCA) methods
LF, *see* Light fraction (LF)
Life cycle assessment (LCA) methods, 25–29, 294
Light fraction (LF), 48
LMWOS, *see* Low-molecular-weight organic substances
            (LMWOS)
Local community nature, 271–272
Localized agro-food system, 185
Lolium rigidum, 142
Long-term carbon sequestration concepts, 76–78
Loomis, R. L., 281
Low-cost carbon materials, 75
Low-input animal husbandry methods, 283
Low-molecular-weight organic substances (LMWOS), 31

Low-shade coffee farm, 227
Lubombo transfrontier conservation area
    communities, 272

M

Macroaggregates, 49
Macromolecular interaction networks, 170
"Maize basket" of Bihar (India), 91
Malaysian-African Agriculture Company, 126
Mammals, bones from, 13–14
Managed agroecosystems
    cropping systems, 208–210
    crops and varieties, 207–208
Man and the Biosphere (MAB) Programme, 268
Mangal, 240
Mangrove forests, 240
    environmental functions and ecological processes, 250
    features of, 240–241
    functional types, 241–242
    functions, 250
    nursery seed banks and mangrove restoration, 252–254
    outplanting and field trials, 258–259
    problems in, 251–252
    transplanting considerations, 259
    tree types, 242–249
    values of, 250–251
    zonation, 247–250
Marker-assisted backcrossing, 161
Marker-assisted selection (MAS), 161
Marsden, T., 311
MAS, see Marker-assisted selection (MAS)
Mathenjwa Tribal Authority (MTA), 268, 269
Matrix-assisted laser desorption/ionization time-of-flight/
    time-of-flight (MALDI-TOF/TOF) mass
    spectrometry analysis, 163–164
Mature seeds, 253
MBCD, see Methylated cyclodextrins (MBCD)
MCSs, see Multiple cropping systems (MCSs)
Mean residence time (MRT), 27, 62
Mechanized agriculture in Sudan, development of,
    124–126
Mechanized farming, 90
Mechanized Farming Corporation (MFC), 123, 125
MeJA, see Methyl jasmonate (MeJA)
Meta-analysis, of biochar systems, 28
Metabolism-related proteins, 165
Methionine-rich protein (MRP), 167
Methylated cyclodextrins (MBCD), 169
Methyl jasmonate (MeJA), 169
MFC, see Mechanized Farming Corporation (MFC)
Microaggregates, 49
Microbe-mediated process, phosphorus acquisition, 94–95
Microbes, 82
Microbial biofertilizers, 157
Microbial community composition, 91
Microbial diversity, of terra preta, 18
Microbial inoculants, 92
Microbial inoculations, nontarget effects of, 109
Microbial interactions, in rhizosphere, 108–109

Microbial origin, nitrogen acquisition of, 92–94
Microbial pest control agents (MPCA), 104
Microbial plant growth promoters (MPGP), 104
Microbial processes in terra preta, 18
Microbial production, of biofuels, 313
Microbial rhizosphere consortia, 109
Microbial technologies, 83
Microbiology, of terra preta, 17–18
Microwave plasma technology, 66
Mid-infrared (MIR) spectroscopy, 21, 22
Mineralization, of labile SOM, 61
MINTs, see Molecular Interaction Database (MINT)
MIR spectroscopy, see Mid-infrared (MIR) spectroscopy
Mixed crop–livestock farms, 193
Mixed farming, 210
Modern hybrid varieties, development of, 90
Modern land tenure systems, 126
Modern terra preta research, 3–7
Molecular breeding, for crop improvement, 160–162
Molecular Interaction Database (MINT), 171
Molecular markers, 85, 161
Molecular microbial ecology, 85
Monocropping, 111
Montego Bay (Bogue Lagoon), 260, 261
MPCA, see Microbial pest control agents (MPCA)
MPGP, see Microbial plant growth promoters (MPGP)
MRP, see Methionine-rich protein (MRP)
MRT, see Mean residence time (MRT)
MTA, see Mathenjwa Tribal Authority (MTA)
Multifunctionality, in agriculture, 316
Multiple cropping systems (MCSs), 209
Multiple trait indices, 161
Multivariate analysis, 85
Multivariate statistical tools, 22
Mycoparasitism, 106
Mycorrhiza formation, 113
Mycorrhizal associations, 105
Mycorrhizal biofertilizer, 95
Mycorrhizal fungi, 94
Mycorrhizal network, 112

N

Natural ecosystems, disturbances of, 214
Natural engineering of beneficial microbial
    populations, 89
Natural sciences approach, 183–184
Natural sciences-oriented research program, evolution of, 188
Near-infrared reflectance spectroscopy (NIRS), 21
Neotyphodium (Clavicipitaceae), endophyte behavior of, 107
Nimeiri government (1969–1985), 127
NIRS, see Near-infrared reflectance spectroscopy (NIRS)
Nitrate-based fertilizers, 89
Nitrogen
    acquisition of microbial origin, 92–94
    fertilization, 187
    fertilizer, 47
    fixation, 158, 159
    fixer, 92
    nutrition, 159

Nitrous oxide (N$_2$O), primary sources of, 330
Non-CO$_2$ GHG emissions, 68
Noncrop arable plants, 139
Nursery germination, 255–257
Nursery propagation, 254–258
Nursery seed banks and mangrove restoration, 252–254
Nutrient retention of biochar systems, 24–25
Nutrient stocks in terra preta, 6–7
Nutrient supply system, 90
Nutrition, plant, 212

O

Odd-over-even predominance (OEP), 7
O$_3$-exposed wheat plants, 164
Omics technologies, 160
On-farm research
    on organic wheat, 186–188
    in Vermont
        C sequestration and GHG emissions, 336
        economic analysis, 336
        farmer interviews, 335
        farm selection, 335
        landscape visualizations, 337
Open-ended case method
    vs. conventional decision case learning, 353, 354
    for learning and capacity building, 349–350
    Norway model, 351–354
    now cases, 350
    phenomenology, 348, 351, 355
Organic agriculture, 318
    and drinking water catchments, 186, 191–193
Organic farming, 47–48, 81, 143
Organic grain production, 186–188
Organic integrated crop/animal system, 211
Organization for Economic Cooperation and Development
        (OECD), 312
Oryza sativa L., 154
Overwash forests, 242
Oxidation–reduction proteins, 163
Oxyfuel concept, 59

P

Paired *t*-tests, 257
Participatory action research (PAR) process, 327
    empowerment, 342
    ethical research, 340
    participation continuums, 328, 339
    and stakeholders, 332, 339, 340
    transdisciplinary research, 341
Particulate organic matter (POM), 44
Pathogenesis-related (PR) upregulation, 163
Pathogens, suppression of, 160
PCA, *see* Permanently cropped agroecosystems (PCA)
PCR, *see* Polymerase chain reaction (PCR)
Peanut, proteomic studies in, 167
Pear, 169–170
PEMFCs, *see* Proton-exchange membrane fuel cells
        (PEMFCs)

Peptide mass finger-printing (PMF), 169
Permanently cropped agroecosystems (PCA), 204–205
Pest management
    applications, 112
    with chemicals and alternatives, 213–214
Phenotypic methods, 83
Phosphate-solubilizing bacteria (PSB), 158
Phosphate-solubilizing microorganisms (PSMs), 94
Phospholipid fatty acid (PLFA) profiling, 83
Phosphorus (P)
    nutrition, 158
    solubilization, PGPR, 108
Phosphorus acquisition, through microbe-mediated
        processes, 94–95
Photo-simulated landscape visualizations, 337
Physicochemical stabilization, of SOC, 49
Phytophagous arthropods feed, 140
Plant-beneficial rhizosphere microorganisms, functional
        groups of
    AMF, 105–106
    EPF, 106–107
    plant growth-promoting rhizosphere bacteria, 107–108
    Trichoderma, 106
Plant biotechnology, 161
Plant breeding, 160–162
Plant breeding programs, 110
Plant genotype, 89–90
Plant growth–promoting rhizobacteria (PGPR), 91, 92,
        107–108
    AMF and, interactions between, 108–109
Plant hardening, 257–258
Plant health and nutrition, 104
Plant nutrition, 212
Plant probiotics
    application of, 158
    as biofertilizers and biopesticides, 157–160
Plant–protein interaction maps, 170
Plant rhizosphere, 94
Plasma carbon, 64–66, 70
Plasma carbon platform technology, 65
Plate counting, 84
PLFA profiling, *see* Phospholipid fatty acid (PLFA)
        profiling
PMF, *see* Peptide mass finger-printing (PMF)
Podophyllum hexandrum, 157
Pollinators, 140
Polymerase chain reaction (PCR), 14, 86, 161
Polymerase chain reaction (PCR)-based molecular marker
        techniques, 161
POM, *see* Particulate organic matter (POM)
Port Royal mangroves, 255–256, 259–260
P34 protein, 166–167
Probiotics, 157
Propagules, 253–254
Proteins, 154, 162–164
Proteomes, 170–171
Proteomics, 154
    in cereal crops, 162–164
    in horticultural crops, 167–170
    in model legume species, 164–167

in omics approaches toward next-generation crops,
170–171
and omics technologies, 160
Proton-exchange membrane fuel cells (PEMFCs), 72
PSB, *see* Phosphate-solubilizing bacteria (PSB)
Pseudomonas fluorescens, 109
PSMs, *see* Phosphate-solubilizing microorganisms
(PSMs)
Pyrolysis, 62
Pyrolysis-biochar systems, 28

Q

QS, *see* Quorum sensing (QS)
QTL, *see* Quantitative trait locus (QTL)
Quantitative label-free shotgun proteomic analysis, 163
Quantitative tandem mass tag proteomic analysis, 163
Quantitative trait locus (QTL), 163
Quorum sensing (QS), 88

R

Radiocarbon dates, of terra preta, 4
Rain-fed agriculture types, 122
REDD+ program, *see* Reducing Emissions from
Deforestation and Forest Degradation
(REDD+) program
Red mangroves, 241–244
percentage survival of, 257
propagules of, 253–254, 256
prop roots of, 250
Reduced representation shotgun (RRS), 161
Reducing Emissions from Deforestation and Forest
Degradation (REDD+) program, 230
Reforestation, 231
Regrounding process, in agriculture, 316
Replanting process, case studies of, 259–261
Rep-PCR (microsatellite region), 87
Research, in food system approach, 186–193
Restoration, of mangrove forests, 251–254
Restriction fragment length polymorphism (RFLP),
86, 161
Rhizobium inoculation, 93
Rhizobium spp., 159
Rhizodeposition, 104
Rhizophora mangle, 241–244
Rhizosphere, 104–105
communities of, 110
microbial interactions in, 108–109
Ribosomal interspace analysis (RISA), 87
Ribulose-1,5-bisphosphate carboxylase/oxygenase
(RuBisCO), 164
Rice, 154
artificial inoculation in, 94
proteomics contribution, 162–164
Rio Earth Summit, 155
RISA, *see* Ribosomal interspace analysis (RISA)
Riverine forests, 241–242
Root nodule (RN) symbiosis, 88
Row crop agriculturevs. agroforestry systems, 231

RRS, *see* Reduced representation shotgun (RRS)
RuBisCO, *see* Ribulose-1,5-bisphosphate carboxylase/
oxygenase (RuBisCO)

S

Salinity acclimation, 257–258
Saplings, survival of, 259
SCA, *see* Seasonally cropped agroecosystems (SCA)
Scalability, 69–70, 76
Scrub mangrove forests, 242
Seasonally cropped agroecosystems (SCA), 203–204
Seed(s)
bank, defined as, 252
case study in, 164–167
development, 164
storage proteins, 166
Semimechanized farming on environment and
development, impact of, 128
clearance of forest, 128–130
desertification in Sudan, 131–132
mechanization and its socioecological impact, 132–134
soil depletion and collapse of yield production, 130–131
Sequestration of solid carbon in (polluted) ocean
sediments, 76
Shade canopy, of agroforestry system, 226
Shannon index, 17
Shifting cultivation, 81
Silviculture systems, 223
Simple sequence repeats (SSRs), 161–162
Single-nucleotide polymorphisms (SNPs), 161–162
Single-stranded conformational polymorphism (SSCP), 86
"Slash and burn" agroecosystems, 213
Small-farm systems, 209
SNPs, *see* Single-nucleotide polymorphisms (SNPs)
SOC, *see* Soil organic carbon (SOC)
Social sciences approach, 184–185
Soil, 202
acidification, 82
biogeochemistry
biochar systems (*see* Biochar systems)
terra preta (*see* Terra preta)
biota, 140
density fractionation, 15
depletion, 130–131
erosion, 208
fertility, 68–69
deterioration of, 130
in tropical agroecosystems, 212–213
microorganisms, 89–90
nutrients, and water, physicochemical interactions
with, 31
quality, 69
tillage, 90–91
types of, 43, 91–92
water loss prevention, 225–226
Soil organic carbon (SOC), 23, 28, 41–42, 229
agricultural management, effect of (*see* Agricultural
management on SOC, effect of)
factors affecting, 42–44

Soil organic matter (SOM), 42, 60, 89
 decomposition of, 42
 dynamics, 213
 mineralization, 58
 stocks, 4–5
Sole-source carbon use (SSCU) patterns, 84–85
Solid carbon in (polluted) ocean sediments, sequestration of, 76
SOM, *see* Soil organic matter (SOM)
Sombroek, Wim, 3
Soybean
 agroecosystems in Brazil, 113–114
 proteomic studies in, 165–167
Specialty crops, Appalachian region, 291
Split root systems, 163
18S rDNA, 86
16S rRNA gene sequences, 17
SSCP, *see* Single-stranded conformational polymorphism (SSCP)
SSCU patterns, *see* Sole-source carbon use (SSCU) patterns
SSRs, *see* Simple sequence repeats (SSRs)
Stakeholders, social sciences approaches, 185
Stanols, chemical structure of, 10
Storms, agroforestry systems, protection from, 226–228
Strawberries, 167
Stress tolerance, in rice, 164
Stubble retention, 45
Sudan
 ecology of, 123–124
 land tenure situation in, 126–128
 mechanized agriculture development in, 124–126
 states, map of, 122
Supply chain analyses, 193
Sustainable agriculture, 82
 climate change, biodiversity and, 152–154
 first green revolution, 154–155
 high-throughput approaches, 160–162
 second green revolution, 155–160
 soil fertility, 212
 utilization of beneficial agromicrobes for, 88–89
  artificial inoculation, 92–95
  engineering agricultural practices, 89–92
Sustainable farming approaches, ecoagriculture relates to, 274
Sustainable production systems, development of, 215
Sustainable soil fertility, 19
Swidden agriculture systems, 214
Symbiont arbuscular mycorrhizal fungi, 82
Symbiosis-receptor-kinase-gene (SYMRK), 159
Systematic ecoagriculture
 challenges, 271–274
 landscape scale challenge, 271
 location and infrastructure, 272
 policy and governance issues, 272
 strategies benefits, 270–271

T

TAR, *see* Terrestrial-to-aquatic n-alkane ratio (TAR)
Taxonomic cluster analysis, 17

Temperature, for food crops, 222
Temperature gradient gel electrophoresis (TGGE), 86
Terminal restriction fragment length polymorphism (T-RFLP), 86
Terra preta
 concept, 62–64
 milestones of
  discovery, 2–3
  geogenic *vs.* anthropogenic origin, 3
  modern research, 3–7
 molecular markers
  aquatic *vs.* terrestrial plant material, 7–8
  ash and charred organic material, 14–16
  composted garbage, 8
  human and animal excrements, 8–12
  mammals and fish, bones from, 13–14
  microbiology, 17–18
 phenomenon of, 60–62
Terrestrial-to-aquatic n-alkane ratio (TAR), 7, 8
Tetrazolium salt, 85
TGGE, *see* Temperature gradient gel electrophoresis (TGGE)
Thematic modules, 30
Thermolysis, 65
Tillage, 45
 agromicrobes for sustainable agriculture, 90–91
 beneficial rhizosphere microorganisms, 111–112
 crop residue and, 46
 system, 142
Tobacco dependence, in Appalachian region, 284–287
Tomich, T. P., 281
Traditional agroecosystems, 202
Transcriptomes, 170–171
Transfrontier conservation area (TFCA)
 conceptual underpinnings, 266–268
 ecoagriculture, 274
 materials and methods, 268–270
 model, 267
 systematic ecoagriculture (*see* Systematic ecoagriculture)
T-RFLP, *see* Terminal restriction fragment length polymorphism (T-RFLP)
Trichoderma, 106
Trichoderma harzianum, 108
Tropical agroecosystems, 205–206
 biodiversity, 206–207
 human influences on, 214
 soil fertility in, 212–213
Tropical crops, 207
Tropical forests, 205
Tropical savannas, human activity in, 231

U

UNCCD, *see* United Nations Convention to Combat Desertification (UNCCD)
UNEP, *see* United Nations Environment Programme (UNEP)
UNFCCC, *see* United Nations Framework Convention on Climate Change (UNFCCC)

United Nations Convention to Combat Desertification (UNCCD), 123
United Nations Environment Programme (UNEP), 155
United Nations Framework Convention on Climate Change (UNFCCC), 155
Unplanned mechanized farming, 128
Unregistered Lands Act of 1971, 126–127
Upscaling in space and time, problem of, 19–20

V

Vandermeer, J. H., 281
Vapor-deposited carbon (VDC) materials, 71
Variable number of tandem repeats (VNTRs), 87
Vascular plants, 143
VDC materials, *see* Vapor-deposited carbon (VDC) materials
Vermont agricultural resilience
    agent-based models, 330, 334, 339
    agroecology, 327
    best management practices, 327
    CCBMPs (*see* Climate change best management practices (CCBMPs))
    F2P process, 328
    on-farm research
        C sequestration and GHG emissions, 336
        economic analysis, 336
        farmer interviews, 335
        farm selection, 335
        landscape visualizations, 337
    PAR process, 327–328
    preliminary results
        reports from the field, 338–339
        survey respondents, 337–338
    research
        diagrammatic representation of, 332, 333
        double-coding approach, 334
        and objectives, 327
        surveys, 333–334
    transdisciplinary research, 327

VNTRs, *see* Variable number of tandem repeats (VNTRs)
Volcanic sedimentation, 4

W

Waste biomass streams, 28
Waste feedstock, 66
Water-deficit stress effect, 168
Water retention of biochar systems, 25
Weed(s), 139, 140
    community, 140, 141, 142
    diversity, recent trends in, 140–142
    seed bank, 140
Wet–dry tropics, 206
Wheat, 162–164
Wheat-flour–bread chain, 188
Wheat-flour food chain, 188
Wheat production, 186–188
White mangroves, 241, 244, 247, 248, 253
Windbreaks, 227, 228
Wine grapes, 168
WMO, *see* World Meteorological Organization (WMO)
Woody aboveground biomass, 336
World Bank, 124
World Food Summit (WFS 1996), 309
World Meteorological Organization (WMO), 155

Y

Yield production and soil depletion, semimechanized farming, 130–131

Z

Zech, Wolfgang, 3
Zero tillage, 319

Printed and bound by CPI Group (UK) Ltd, Croydon, CR0 4YY

23/10/2024

01778254-0012